Preface

The first edition of this book was published (1968) in a volume which combined a treatment of both glacial and periglacial geomorphology. Two reasons prompted us to separate these topics when the book fell due for revision. First, because of the enormous expansion of research in these fields, we felt an adequate account at this level demanded a lengthier treatment; this in turn meant a division into two volumes if the work was not to become unduly bulky. Secondly, although there are close links between these two branches of geomorphology, they are not so close that separation involves any great disadvantage, nor does it involve any significant repetition. The majority of research workers in these fields have concentrated their attention on one or the other, rarely both, and some schools of geomorphology have developed particularly strong interests in one, as for instance, the long tradition of Polish research in periglacial geomorphology.

The ice age has left a direct imprint on a large part of the world's land surface. Ice is also still active in many areas today, modifying the landscape and creating distinctive landforms. This book deals with glacial landscapes in which the unifying theme is provided by ice. The emphasis throughout is on the landforms and processes of erosion and deposition. Many aspects of glaciology, such as the study of sea ice or glacier surface features, are either omitted or only touched on briefly, nor is it the authors' intention to provide any comprehensive account of Pleistocene geology, stratigraphy and chronology. The indirect effects of glaciation, such as changes in land and sea level, were important and widespread, but limitations of space prohibit more than an outline of these aspects. In 1957, two major works in the English language, both largely concerned with Pleistocene geology and stratigraphy, were published: the second edition of R. F. Flint's *Glacial and Pleistocene Geology* and the two-volume survey of *The Quaternary Era* by J. K. Charlesworth. The former was extensively revised and enlarged in 1971. Although both of these works contain sections dealing with glacial geomorphology, neither is principally concerned with this aspect.

The writing of this book, commenced in 1964, was inspired by W. Vaughan Lewis, whose untimely death in 1961 meant such a tragic loss to glaciology and geomorphology. Under Vaughan Lewis, to whose memory this book is dedicated, we both received our first introduction to glaciers and glacial landforms, especially in the several expeditions which he organized or initiated to study glaciers such as Vesl-Skautbreen, Veslgjuv-breen, and Austerdalsbreen in Norway. His infectious enthusiasm and unfailing cheerfulness, his generosity and helpfulness to his students, will always be remembered.

We have attempted to concentrate on the more modern contributions to the subject. The literature available is of gigantic proportions and is accumulating at an ever-increasing rate. The need to collate and summarize the results of research, from time to time, is just as important as the need for research itself. The revised edition of the text was completed in 1974; in this revision, considerable changes and additions of new material to the text were made. References to the older literature are necessarily very selective, and those who seek a fuller treatment of historical aspects and discarded theories are referred to Professor Charlesworth's masterly summary. In compiling material for certain parts of the book, his survey of the Quaternary period has proved invaluable. We also acknowledge the help in collecting the most recent material provided by *Geographical Abstracts*, edited by K. M. Clayton.

References are listed at the end of each chapter. There is inevitably some duplication of references for this reason. Abbreviations of the titles of journals are in accordance with the recommendations given in the *World List of Scientific Periodicals* (4th edition). Place of publication, in the case of books, is given only for those not published in London or New York. All measurements used in the book are given in the metric system using standard SI abbreviations.

The authors make no apology for the occasional overlap of material between one chapter and another. Cross references are given where this occurs. Such overlap is unavoidable when a particular topic is relevant to more than one chapter, and is essential for clarity of presentation.

We are grateful to the staff of the Drawing Office in the Department of Geography, King's College, London, and to Mrs A. Rogers for help in typing the revised manuscript.

January 1975   CLIFFORD EMBLETON
                                                      CUCHLAINE A. M. KING

Glacial and Periglacial Geomorphology

Second edition
Volume 1

To the Memory of
W. Vaughan Lewis 1907-1961

who inspired in us a lasting interest in glaciers and glacial geomorphology

Glacial Geomorphology

Clifford Embleton

Reader in Geography, University of London King's College

Cuchlaine A. M. King

Professor of Physical Geography, University of Nottingham

Edward Arnold

© Clifford Embleton and Cuchlaine A. M. King 1975

First edition published 1968
Reprinted 1969, 1971, 1974
Second edition first published 1975
by Edward Arnold (Publishers) Ltd is a
fully revised version of parts I–III
of *Glacial and Periglacial Geomorphology*

ISBN: 0 7131 3791 7 Cased Edition
ISBN: 0 7131 5792 5 Paper Edition

Printed in Great Britain by
Butler & Tanner Ltd,
Frome and London

Contents

Plates

For the moving of large masses of rock, the most powerful engines without doubt which nature employs are the glaciers ... These great masses are in perpetual motion, ... impelled down the declivities on which they rest by their own enormous weight, together with that of the innumerable fragments of rock with which they are loaded. These fragments they gradually transport to their utmost boundaries ... In this manner, before the valleys were cut out in the form they are now ... huge fragments of rock may have been carried to a great distance; and it is not wonderful, if these same masses, greatly diminished in size, and reduced to gravel or sand, have reached the shores, or even the bottom of the ocean.

(J. PLAYFAIR, 1802)

Introduction

Geographers and geologists today have grown up with the idea that glaciers and ice sheets once extended much farther south in Europe and North America, so that it is difficult to appreciate the doubts and difficulties that attended the establishment of the glacial theory. The complex of forms associated with mountain glaciation, the U-shaped valleys, rock-basin lakes, cirques and other features, are so well known that they are among the most easily recognized of landforms. They are, however, by no means easy to explain in detail, and many less striking but equally important features of glaciation may even be entirely overlooked.

The extent and character of the effects of the ice age only became known gradually (F. J. North, 1943). W. Buckland, who was born in 1784, was aware of the presence of large erratics that could not have been carried to their present positions by rivers. He explained these and many other features, however, as the result of the Noachian flood. It was this idea that led to the introduction of the term 'drift'. Later, Buckland was converted to the glacial view by Louis Agassiz. The glacial theory had its beginning in the Alps, whose glaciers were larger in the late eighteenth and early nineteenth centuries, when the glacial theory was being slowly developed, than they are now.

As early as 1723, J. J. Scheuchzer (1672–1733) put forward a theory of glacier movement. He suggested that water entered crevasses in the ice and froze, causing the ice to move downhill. One of the chief initiators of the glacial theory was H. B. de Saussure (1740–99) whose views were a mixture of modern and medieval. He appreciated the formation, movement and some of the effects of glaciers, especially their ability to transport large boulders. De Saussure greatly influenced the views of J. Hutton on glaciation. In 1815, J. P. Perraudin, a Swiss guide, suggested that glaciers had formerly been more extensive, a view which was supported by J. Venetz, an engineer, in 1821. Another early advocate of the glacial theory was J. de Charpentier (1786–1855). It is interesting to note that de Charpentier credited J. W. von Goethe, the poet, with the discovery of the ice age around 1829–30. A. Bernhardi also published a paper on the glacial theory in 1832. The term 'ice age' was first coined in 1837 by K. Schimper.

The work of these pioneers and others, such as that of H. T. de la Beche in 1842, was largely ignored until the second half of the nineteenth century. It was not until Louis Agassiz turned his attention to the glacial theory that much notice was taken of the new ideas by the most influential scientists of the day. Agassiz was born in 1807. He was convinced by de Charpentier of the validity of the glacial theory. Agassiz himself, however, did not in fact add much new to the theory, but his well-known name gave greater weight to the new views. In 1837 he delivered his famous presidential address on ice ages to the Société Helvétique des Sciences Naturelles (the 'Discourse of Neuchâtel'), and published his views in 1840 under the title *Études sur les glaciers*. Agassiz converted Buckland and C. Lyell to the glacial theory when he visited Britain in 1840. After travelling to various parts of the British Isles, Agassiz came to the conclusion that all of this area had once been beneath an ice sheet. Although a great weight of evidence was presented to members of the Geological Society of London in 1840 by Agassiz, Buckland and Lyell, there was still considerable hesitation in accepting the implications of the glacial theory. In 1847 Agassiz left for North America where he also stimulated interest in the glacial theory.

In Britain, the glacial theory did not receive much further support until A. C. Ramsay started his work on glaciation (see B. Hansen (1970) for an account of the early history of the glacial theory in British geology). In 1860, he published a study of the glacial features of Switzerland and North Wales. It was mainly Ramsay who eventually overcame the iceberg theory, to which Lyell had reverted by this time and which was widely held. He worked with T. F. Jamieson, who first put forward the true explanation of the parallel roads of Glen Roy. Meanwhile, many sound glaciological observations were being made in Switzerland by J. D. Forbes, J. Tyndall and others, who were working on the mechanics of glacier movement. These early workers have been followed by countless others, through all of whose efforts light has gradually been shed on the complexities of glacial processes and on the landscapes which they produce.

The Pleistocene ice sheets and glaciers moved over a landscape that had previously been fashioned by subaerial processes and especially river action. The extent to which the ice modified these earlier landscapes varied with many factors, including the nature of the pre-glacial relief, the amount of ice accumulation and the ease with which it could flow outward. Thus in many areas, the landscape now shows an assemblage of forms that owe their character to both glacial and non-glacial processes. The fact that so many landscapes are polygenetic greatly complicates any analysis of their evolution. The situation is rendered even more complex by the geologically rapid waxing and waning of the ice sheets during the Pleistocene period, which is now thought to have covered at least the last one and a half million years.

The appearances of glacial landforms belonging to different periods of the ice age are also significantly different. There is, for instance, a striking difference between the character of the glacial landforms in the areas covered by the Newer Drift and the Older Drift in Britain. The Older Drift has subdued relief and the deposits are deeply weathered. They occur mainly on the interfluves as erosion has removed them from the valley floors. These characteristics contrast sharply with the fresh appearance of the Newer Drift, with its constructional forms and its frequent occurrence in the valley bottoms. This contrast has been referred to by K. M. Clayton (1957) as the 'attitude' of the drift.

The features produced by glacial action also develop and change with time from immature, youthful forms to mature forms. The ice age, however, was short from the viewpoint of geological time. It is unlikely, therefore, that glacial forms have developed beyond a fairly young stage in most areas. This applies particularly to those areas that lay near the limits of glaciation. The most advanced forms of glacial action described by D. L. Linton (1963) occur in the Antarctic, where they still remain partially buried beneath a thick cover of ice. Thus when glacial forms are examined, it must always be remembered that the time they have had to develop has been relatively short. The glacial processes have modified a pre-existing landscape, and the resultant landforms have also suffered modification under different conditions since their emergence from beneath the ice.

In view of the short duration of the ice age, it is apparent that in many circumstances the action of ice is very effective in producing a new landscape, which bears unmistakable evidence of its formative agent. Active glaciers are one of the most effective of all geomorphological agents, acting both destructively and constructively. Nevertheless there are also conditions under which ice can protect the land surface on which it rests.

There is still much to be learnt about the initiation of the glacial period and the factors on which the fluctuations of the ice sheets and glaciers depend. It seems more likely that we are living in an interstadial or interglacial period at present than that the Pleistocene period has finally come to an end, and the behaviour of the world's great ice sheets and glaciers is of direct relevance to mankind. Quite small changes in the quantity of ice on land, for instance, can effect very significant changes in world sea level. Melting of all the present land ice would suffice to raise sea level by more than 50 m and flood the world's major cities. Barring human intervention, such a possibility seems to be extremely remote, and much more important are the minor fluctuations of climate such as the warming-up from the 1920s to the 1950s reflected in worldwide glacier recession and sea levels rising at the order of 1 mm/year. Equally significant for mankind are any signs of cooling and renewed growth of glaciers. At present, the snow-line lies just above the top of Ben Nevis in Scotland, so that only a small lowering of the temperature would allow snow to remain throughout the summer and consolidate into small cirque glaciers.

There are few parts of the world which have not been affected directly or indirectly by the events of the ice age. A great deal of work has been done and much has been written concerning the effects of the ice age. Much of this work has been devoted to a study of the chronology of glaciation, as revealed in the stratigraphy of glacial deposits on land and in the sedimentary record in the sea. This aspect is not, however, stressed in this book, which is concerned essentially with the direct effects of glacial action on the landscape. Much of the British Isles, Europe and North America as well as large parts of Asia, South America and small parts of Australia and Africa show evidence in their landforms of the influence of glacial processes.

In the large area covered by ice sheets or glaciers a wide range of relief types, rock types and climatic conditions occurs. The range of these variables leads in turn to a great diversity of glacial forms that have been recognized and studied. This variety means that many different approaches to the problem of understanding the genesis of glacial landforms have been adopted, and many different techniques have been applied in these studies.

Present-day glaciers provide a valuable laboratory in which to examine glacial processes and glacial landforms in the making. The surface movement of glaciers can easily be observed, but it is along their sides and at their bases that erosion and most deposition occurs. For this reason, much effort has been expended in drilling boreholes through glaciers and digging tunnels into them to examine their movement relative to their beds and to study the ice-rock interface. Instruments have been developed to measure the rate of movement of the ice, its temperature, its crystal structure and other characteristics. Observations such as these have provided important evidence on the manner in which ice moves and moulds its bed. Such field-work has been supplemented by laboratory studies on ice and by theoretical studies of its behaviour under specified conditions. The theoretical studies and laboratory results can then be checked in the field by direct observation. In this way, valuable results have been obtained concerning the processes of glacial erosion and deposition in terms of the flow characteristics of the ice.

The formation of many glacial deposits and depositional features is less easy to observe directly because the processes in question are subglacial or englacial. Some marginal features, however, can be observed in the process of formation: for example, the formation of eskers has been observed on glacier margins. In depositional forms, the character of the deposit itself gives the most valuable evidence concerning their method of formation. For this reason, much work has been devoted to the study of till fabrics and the internal structure of the depositional features.

In order to understand as fully as possible the glacial landforms that have been revealed by the withdrawal of the ice from areas previously covered by the ice sheets, it is necessary to consider the studies that have been made of the existing ice masses. A study of the processes acting at present can throw light on the landforms that were created by similar processes in the past. Thus, the first section of the book is devoted to an account of the characteristics of the glaciers and ice sheets that still cover substantial portions of the earth's surface. This section is followed by studies of the forms produced by glacial and fluvioglacial erosion (Part II) and by glacial and fluvioglacial deposition (Part III).

References

AGASSIZ, L. (1840), *Études sur les glaciers* (Neuchâtel)
   (1840), 'On glaciers, and the evidence of their having once existed in Scotland, Ireland, and England', *Proc. geol. Soc. Lond.* **3**(2), 327–32
BECHE, H. T. DE LA (1832), *Geological Manual* (London)
BERNHARDI, A. (1832), 'Wie kamen die aus dem Norden stammenden Felsbruchstücke und Geschiebe ... an ihre gegenwärtigen Fundorte?' *Heidelb. Jb. Minert. Geogn. Petrefaktenk.* **3**, 257–67
BUCKLAND, W. (1824), *Reliquiae Diluvianae* (London)
CAROZZI, A. V. (1967), *Studies on Glaciers preceded by the Discourse of Neuchâtel, by Louis Agassiz* (London), 213 pp.
CHARPENTIER, J. DE (1835), 'Notice sur la cause probable du transport des Blocs erratiques de la Suisse', *Annls Mines, Paris* **3**, 8, 219–36
CLAYTON, K. M. (1957), 'The differentiation of the glacial drifts of the East Midlands', *E. Midld Geogr.* **7**, 31–40
FORBES, J. D. (1843), *Travels through the Alps of Savoy* (Edinburgh)
HANSEN, B. (1970), 'The early history of glacial theory in British geology' *J. Glaciol.* **9**, 135–41
HUTTON, J. (1795), *Theory of the Earth with proofs and illustrations* (Edinburgh)
JAMIESON, T. F. (1862), 'On the ice-worn rocks of Scotland', *Q.J. geol. Soc. Lond.* **18**, 164–84
LINTON, D. L. (1963), 'Some contrasts in landscapes in British Antarctic Territory', *Geogrl. J.* **129**, 274–82

NORTH, F. J. (1943), 'Centenary of the Glacial Theory', *Proc. Geol. Ass.* **46,** 1–28

PERRAUDIN, J. P. (1815), quoted in J. VENETZ (1821)

RAMSAY, A. C. (1860), *The old glaciers of Switzerland and North Wales* (London)

(1864), 'On the erosion of valleys and lakes', *Phil. Mag.* **4,** 28, 293–311

SAUSSURE, H. B. DE (1779–96), *Voyages dans les Alpes* (Paris), 4 vols

SCHEUCHZER, J. J. (1723), *Itinera per Helvetiae Regiones Alpinas* (Leyden)

SCHIMPER, K. (1837), 'Über die Eiszeit', *Mém. Soc. helv. sci. nat.* **5,** 38–51

TYNDALL, J. (1858), 'On some physical properties of ice', *Phil. Trans. R. Soc.* **148,** 211–29

(1860), *Glaciers of the Alps* (London)

VENETZ, J. (1833), 'Mémoire sur les variations de la température dans les Alpes de la Suisse', *Mém. Soc. helv. sci. nat.* **1,** pt. 2

Part I

Basic concepts of glaciation and glacier behaviour

Glacier ice is able to produce the phenomena for which we seek to account; and *that* in every place where those phenomena occur. (M. H. CLOSE, 1866)

1

Ice ages and World glaciation

Finally, two or three hitherto silent guests called to their aid a period of intense cold, with glaciers descending from the highest mountain ranges, far into the low country.... (J. W. VON GOETHE, *Wilhelm Meister*, 1829)

Throughout the geological history of the earth, long periods of warm climate have alternated with shorter phases of generally cold climate during some of which ice sheets and glaciers have appeared. There is good evidence that glaciation occurred more than once in the Pre-Cambrian and again in the Permo-Carboniferous period, as well as in the Pleistocene. Pre-Cambrian tillites and boulder-beds are known from many widely separated areas of the world, such as Scotland (C. Kilburn, W. S. Pitcher, and R. M. Shackleton, 1965) or northern Michigan (USA). There is even clearer evidence of the Permo-Carboniferous ice age. In South Africa, for instance, the Dwyka tillite at the base of the Karroo System contains scratched and faceted stones, and rests in many places on a striated rock surface; there are also varve-like sediments. The approximate world extent of the Permo-Carboniferous glaciation is known, but the same cannot be said for any of the Pre-Cambrian glaciations where the evidence is very much more fragmentary. In neither case is it known how long glaciation lasted, nor whether there was a series of glacial and interglacial episodes within one major glaciation.

The Permo-Carboniferous glaciation was followed by a lengthy span of time, through the Mesozoic era, when world temperatures were higher than those of today, and when ice sheets and glaciers were absent. In the Cenozoic era (a term embracing the Tertiary and Quaternary periods), great continental ice sheets and glaciers reappeared. Until the 1950s, most investigators looked for the onset of this the latest of the world's ice ages in the European and North American stratigraphic record of late Pliocene–early Pleistocene time (K. K. Turekian, 1971). Controversy centred around the exact date of the beginning of Pleistocene glaciation (most opinions favoured dates of about 1·5 to 2 million years BP), where to place the boundary in the stratigraphic sequence, and how many glaciations there were within the Pleistocene ice age. Two major sets of investigations have changed our whole outlook: recent discoveries from the study of deep-sea sediments and of the geophysics of the ocean floors, and new evidence on the glacial history of Antarctica. It is now known that continental ice sheets have existed since at least the Miocene in Antarctica (R. H. Rutford *et al.*, 1968), and we can no longer think in terms of a sudden onset of Pleistocene refrigeration. There is some indication from deep-sea sediments in the south Pacific (R. Gram, 1969) that ice, whether of continental proportions or in the form of local glaciers, existed in Antarctica as far back as the Eocene, but there is also clear evidence from fossil flora that, at other times in the early Tertiary, Antarctica was relatively warm. Future research may well establish a succession of glacial and interglacial periods throughout Cenozoic times, the severity of the cooling in the glacial periods increasing to a maximum in the Pleistocene.

1 The Pleistocene period

1.1 *General introduction*

'Pleistocene' was the name given by C. Lyell (1839) to the 'most recent' epoch of geological history. Until the last few years, most geologists have used the term in a slightly narrower sense. Thus, the Cenozoic era is conventionally divided into the Pleistocene and the Recent (or Holocene) epochs. The Recent period is sometimes referred to as the post-glacial and is now usually taken as beginning with the Pre-Boreal (Table 1.3) pollen zone IV, about 10,000 BP at the beginning of the Flandrian transgression (Holocene Commission, INQUA 1969). R. G. West (1968) recommends that, because of the arbitrary nature of these various subdivisions of the Cenozoic and the problems of definition, the terms Tertiary, Quaternary and Holocene should be abandoned. Instead, he suggests that the Cenozoic era be divided into six epochs (Palaeocene, Eocene, Oligocene, Miocene, Pliocene, Pleistocene) and the Pleistocene be regarded as extending to the present day.

Repeated glaciation of large land areas was the outstanding feature of the Pleistocene. At times, ice spread out over an area totalling 46 million km², more than three times the present ice cover of the world (Figs. 1.1 and 1.2). Agreement is far from being reached on the number of glacial periods in the Pleistocene, but there were certainly four or five principal ones, each of which may have included several cold phases alternating with milder intervals or 'interstadials'. Even less certainty surrounds the question of the number of pre-Pleistocene, Tertiary, glacial periods. In southern Alaska, recent work by G. H. Denton and R. L. Armstrong (1969) shows evidence of no less than sixteen glacial expansions, of which four are likely to be Pleistocene with ages of less than 2·7 million years, while the oldest known is of the order of 10 million years.

Table 1.1 shows regional examples of the sequences of glacial and interglacial cold and temperate stages in the Pleistocene. The table is by no means complete, nor are the implied correlations between different areas generally agreed except in the case of the Upper Pleistocene. It is simply intended to show some of the commonly encountered names in Pleistocene terminology, and it should be stressed that increasing uncertainty characterises the sequences and their correlation as one goes farther back in time before the Upper Pleistocene. Not all the cold periods were marked by glaciation

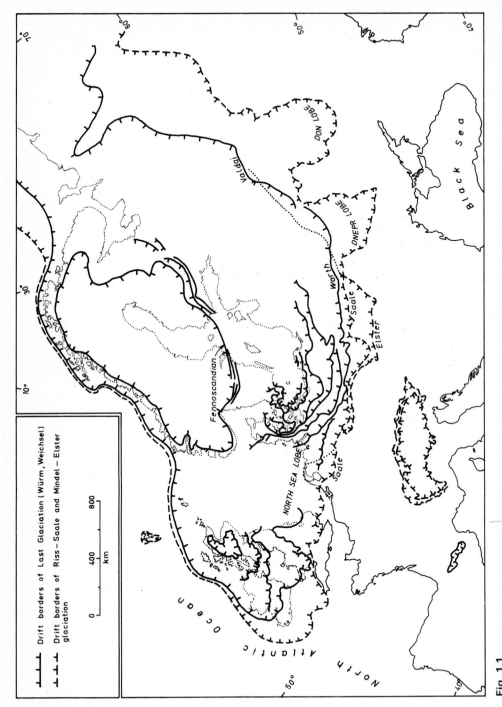

Fig. 1.1
European Pleistocene ice limits (R. F. Flint, *Glacial and Pleistocene Geology*, John Wiley & Sons, New York 1957—with minor amendments)

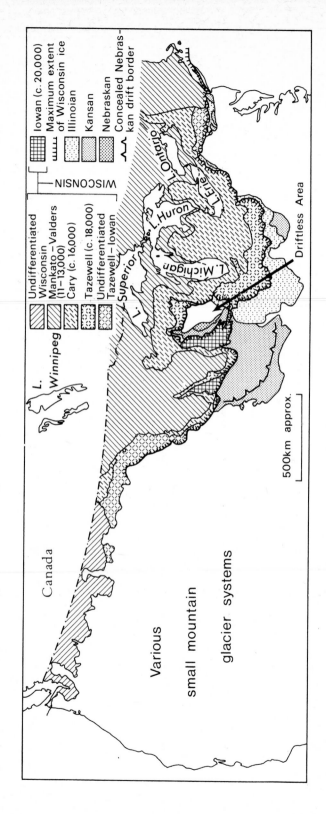

Fig. 1.2
Pleistocene ice limits in the United States and eastern Canada (R. F. Flint, 1957; W. D. Thornbury, 1965; H. E. Wright and D. G. Frey, 1965)

Table 1.1   Subdivision of the Pleistocene in the main glaciated areas

	Great Britain		Alps	Northern Europe	European Russia	North America
	Former terminology	New terminology*				
Upper Pleistocene	Newer Drift	Devensian	Würm	Weichselian	Valdai	Wisconsin
	Ipswichian†	*Ipswichian*		*Eemian*	*Mikulino*	*Sangamon*
	Gipping	Wolstonian	Riss	Saale	} Moskva } Dnepr	Illinoian
Middle Pleistocene	*Hoxnian*	*Hoxnian*		*Holstein*	*Likhvin*	*Yarmouth*
	Lowestoft { Lowestoft Till Corton Sands Cromer Till	Anglian	Mindel	Elster		*Kansan*
	Cromerian and Cromer Forest Bed Series	*Cromerian*				
		Beestonian				*Aftonian*
		Pastonian				
Lower Pleistocene	Weybourne Crag	Baventian	Günz			Nebraskan
		Antian				
		Thurnian				Pre-Nebraskan
	Norwich Crag	*Ludhamian*				
	Red Crag	Waltonian	Donau			
Pliocene	Coralline Crag and older formations					

* After B. W. Sparks and R. G. West, 1972; G. F. Mitchell, L. W. Penny, F. W. Shotton and R. G. West, 1973.
† Named interglacials are shown in italics.

in every era. There was probably no glacial ice in Britain in the Lower Pleistocene (the Crag deposits listed in the table are mainly marine) whereas in the Alps at this time, glaciers appeared in the Günz and Donau stages.

Several of the interglacial stages are thought to have been of considerable duration, longer than the glacial periods, and all of them appear to have been much longer than the post-glacial. They are represented in the stratigraphical sequence by such features as fossil soils, weathering horizons, or fluvial and lacustrine deposits containing temperate fauna. In contrast with the voluminous glacial and fluvioglacial deposits, interglacial relics are generally fragmentary and very restricted in occurrence, and this is one of the fundamental reasons why Pleistocene stratigraphy in many areas is so uncertain.

The distribution of ice was not precisely the same in each glacial period. We know very little so far about world distributions of ice in the Tertiary period; it may be that ice was then mainly restricted to Antarctica, Greenland, and high altitude areas. It seems likely that the maximum spread of ice occurred in the Middle Pleistocene glacials. The Riss/Saale glaciation in Europe and the Illinoian in North America were generally the most extensive though in part of west-central North America, the preceding Kansan ice extended even farther (Fig. 1.2). The last glaciation (Weichsel, Wisconsin, etc.)

was more restricted. On a world-wide view, however, these differences are of relatively minor importance, and it should be noted that the several Pleistocene glaciations in fact covered much the same territory, adopted similar patterns of ice flow, and were governed by very similar physical and climatic controls.

Table 1.2, mainly abbreviated from W. L. Donn *et al.* (1962) and R. F. Flint (1971), summarizes the major distributions of Pleistocene ice according to recent evidence.

Table 1.2   Areas covered by ice in the last glaciation and at the Pleistocene maximum

Area	Maximum extent	Extent in last glaciation
	km² × 10⁶	
Antarctica	13·80	13·80
Laurentide ice sheet	13·39	12·34
Scandinavian ice sheet	6·67	4·09
Siberian ice sheet	2·71	1·56
N. American Cordilleran ice	2·83	2·53
Greenland	2·30	2·30
All others in N. Hemisphere	1·79	1·17
Southern hemisphere excluding Antarctica and including all South America	0·91	0·80
Totals	44·40	38·59

cf. *Present-day ice-covered area:*   14·90

Both the Antarctic and Greenland ice sheets differed little in extent from their present sizes, for they were limited as now by calving into deep water. They were, however, very much thicker. The great Laurentide and Scandinavian ice sheets are now represented by quite insignificant and fragmented masses: over 99 per cent of their bulk has melted away. Relatively little is known about the thicknesses of ice in the Pleistocene, except for a few small areas, and consequently estimates of the volume of ice vary widely. H. Valentin (1952) computed a figure of 60·9 million km³; W. L. Donn and others (1962) put the figure of the Riss/Illinoian glaciation at 84–99 million km³, and 71–84 million km³ for the last glaciation. R. F. Flint (1971) gives 77 million km³ for a glacial age (unspecified). These figures represent about three times as much ice as remains today.

1.2 *Climatology*

The climatology of the Cenozoic glacial ages is very speculative and inadequately understood: some hypotheses to explain the changing climate will be summarized later. There is an enormous literature concerned with probable climatic conditions at different stages, yet opinions are so divergent (except in regard to the last 100,000 years) that it is unsafe to generalize beyond such statements that the average world temperature in the Pleistocene glacials was possibly 5–7°C cooler than at present while the interglacials may have been somewhat warmer than at present in middle to high latitudes. The snow-line in the glacials was depressed everywhere, but not uniformly. In the Alps, it was reduced in elevation by as much as 1300 m, in Britain by up to 1200 m, and in the Western Cordillera of North America by about 800 m.

The most detailed long-term climatic record so far obtained is that revealed from deep ice coring at Camp Century on the Greenland ice sheet (W. Dansgaard *et al.*, 1969; W. Dansgaard *et al.*, 1971). The borehole, in latitude 77° 10′N, longitude 61° 08′W, reached bedrock at 1387 m, and samples of ice were analysed for variations in oxygen isotope ratios. The index of concentration of O^{18}, δ, is related, in the case of high-latitude snow, to the temperature at the time of snow/ice formation. There are thus seasonal variations of δ in accumulated snow and ice, and also long-term variations owing to climatic change. The interpretation is simpler for polar ice than for temperate ice where refreezing meltwater and isotopic exchange between the water and the snow complicate matters. In theory, for polar ice, it is a question of counting

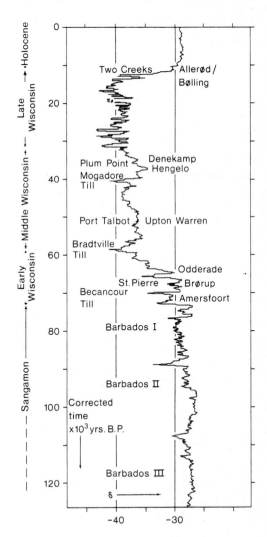

Fig. 1.3
Interpretation of climatic record over the last 120,000 years from the Camp Century drill hole, Greenland. δ in 200-year intervals is plotted on a corrected time-scale (W. Dansgaard *et al.*, 1971)

annual layers, δ varying between a summer maximum and winter minimum, to obtain a chronology, against which the longer-term climatic oscillations of δ can be studied. However, with increasing age, the annual layers become blurred by molecular diffusion, so other methods of establishing age have to be tried. Dansgaard *et al.* (1971) calculate a theoretical age-depth scale from glacier flow theory applied to ice sheets, and show that the probable age at a depth of 1140 m, for instance, is 10,000 years, and, at the base of the borehold, about 120,000 years. Fig. 1.3 shows the complete climatic and stratigraphic interpretation for this deep ice core. The 2164 m borehole at Byrd Station, Antarctica (latitude 80° 01′S, longitude 119° 31′W; S. Epstein *et al.*, 1971), broadly confirms the Camp Century record. It reveals that the Wisconsin glaciation in Antarctica began about 75,000 years ago, that the temperature fluctuations since that time were essentially synchronous as between Greenland and Antarctica, and that pre-Wisconsin temperatures near the base of the core were warmer than at present. It seems that the world's great ice sheets provide sensitive indicators of late Pleistocene climate at least, and further analyses of deep ice cores promise to add greatly to our knowledge of recent climatic change.

1.3   *Stratigraphy and correlation*

Because of the relative youthfulness and therefore often well-preserved nature of its record in terms of deposits, landforms, and fossil plants and animals, the Pleistocene is the best documented of all geological periods. This very wealth of data, and the multiplicity of changes within a short time span have led to great difficulties of interpretation and wide differences of opinion concerning sequences of events and their correlation. Pleistocene stratigraphy is complicated at the outset by the enormous range of environments in which deposits accumulated—marine, fluviatile, lacustrine, glacial, arid and volcanic sequences are included. Even more formidable from the point of view of correlation and attempts to establish successions is the fact that the great bulk of the Pleistocene sediments now visible consist of terrestrial and particularly glacial accumulations, characterized by rapid lateral and vertical variation, often poorly understood mode of deposition, and lack of significant fossil content. Exposures, moreover, are often inadequate, and lack clarity because of the unconsolidated nature of the sediments. Nor was the Pleistocene period one of crustal stability; not only were large areas affected by glacio-isostatic deformation, but tectonic disturbances resulting in faulting, tilting and warping were widespread in certain areas (see, for example, P. B. King, 1965).

The principal bases of correlation of Pleistocene successions may be summarized as follows:

(*a*)   *Fossils*   The stratigraphy of the Pleistocene, unlike that of the older geological periods (whose time span was individually vastly longer), does not depend to any great extent on fossil zonation, whether by plants, molluscs, or vertebrates. This is fundamentally because of the short duration of the Pleistocene, the frequent absence or rarity of fossil remains, and the time-transgressive nature of the fossils that do exist. The fossil content of deep-sea cores has, however, been of value in estimating Pleistocene changes of temperature (see p. 24).

(b) *Lithological units* The tracing and correlation of lithological units is a basic tool (and one of the oldest) for elucidating Pleistocene successions. The mapping of particular drift sheets, solifluction sheets, loess horizons and spreads of fluvioglacial gravels (including terraces) has been the foundation of Pleistocene stratigraphy in North America and Europe. The method is not without its hazards, especially the result of rapid lateral variations in the unit being used for correlation.

(c) *Palaeosols and weathering zones* Weathered layers and fossil soils have long been used stratigraphically to separate Pleistocene deposits, especially in the Mid-West of North America (R. V. Ruhe, 1965). Pioneer workers such as F. Leverett (1930) and G. F. Kay (1931) recognized different depths of weathering associated with till sheets of different ages, and from these observations were able to make tentative estimates of absolute age, which in fact were not too badly in error. Subsequent workers continued to 'use soils and weathering horizons as the primary basis for separation of the four principal drift sheets; they have also been used to correlate pluvial events in the Great Basin with glacial successions in the Sierra Nevada (R. B. Morrison, 1964, 1968) and elsewhere. Much work has also been done on fossil soils, often associated with loess sheets, in Europe.

(d) *Prehistoric cultures* Archaeological investigations have helped (though in some cases also served to hinder) the unravelling of Pleistocene successions and their approximate dating. Such methods are now recognized as rather unreliable, because of the way in which the cultures are strongly time-transgressive. J. K. Charlesworth (1957, pp. 1028–39) gives a detailed summary of some relevant literature while B. W. Sparks and R. G. West (1972, pp. 225–50) describe the archaeology of the British Pleistocene (see also K. P. Oakley, 1964).

(e) *Pollen stratigraphy* The study of pollen sequences has proved to be one of the most useful of all methods in elucidating the climatic changes of the last 15,000 years (see, for example, H. Godwin, 1956; R. G. West, 1968). A series of stratigraphical units and pollen zones has been established in the British Isles and northern Europe, and is also being worked out in North America. Table 1.3 gives the principal time divisions now widely adopted in north-west Europe, though there are minor differences between the schemes formulated independently in the several countries. The names in column 2 are based on the classical work of A. Blytt (1876) and R. Sernander (1908) in Scandinavia.

For each zone in a particular region, a characteristic flora is reflected in the pollen assemblage. There are, of course, variations from one region to another within any one zone and, taking this into account, it is possible broadly to reconstruct vegetation maps for each stage which throw light on the climatic conditions prevailing, and to learn about the distribution of glacial ice in particular stages. Zone III, for instance, witnessed the last phase in Britain when glaciers reappeared, mainly in Scotland, the Lake District and North Wales; because the stratigraphical evidence for the age of this phase is particularly clear in the Loch Lomond area, it is often referred to as the Loch Lomond readvance.

Table 1.3 Pollen zones of the Late- and Post-glacial stages in north-western Europe

Stage		Zone	Time, radiocarbon years BP
Post-Glacial	VIII	Sub-Atlantic	Up to 2,500
	VII b	Sub-Boreal	2,500– 5,000
	VII a	Atlantic	5,000– 7,000
	VI	Late Boreal	7,000– 9,000
	V	Early Boreal	9,000– 9,500
	IV	Pre-Boreal	9,500–10,300
Late-Glacial	III	Younger Dryas	10,300–10,800
	II	Allerød	10,800–12,000
	I c	Older Dryas	
	b	Bølling	12,000–13,500
	c	Oldest Dryas	

Note: In North America, Zone III probably corresponds to the Valders advance, Zone II to the Two Creeks interval, Zone 1 to the Mankato advance.

1.4 *Pleistocene chronology*

The last twenty years have seen the development and application of entirely new methods to the elucidation of stratigraphical problems and correlation in the Pleistocene. Prior to about 1950, Pleistocene chronology rested on very uncertain foundations or on techniques of very limited application; it is well, however, to outline these older techniques first, before considering the new methods.

(*a*) *Rates of operation of geomorphological processes*   Such assessments were the basis of the earliest attempts to devise a time-scale. A well-known example is the measurement by various workers of the rate and distance of recession of Niagara Falls; other workers have investigated rates of delta-building since an area was uncovered by ice, rates of weathering of drift sheets, or rates of limestone solution around erratics deposited originally on the limestone surface but now found resting on a limestone pedestal (for a summary, see Charlesworth, 1957, pp. 1517–22). All such measurements were subject to grave difficulties of interpretation, and many were later shown to be invalid. In the case of Niagara, for instance, it is certain that rates of water discharge over the Falls must have varied with climatic and glacio-isostatic changes, thus affecting the rates of fall recession. Another complication was the discovery that, in part, the Niagara gorge had been cut in more than one deglacial period and that the river has been eroding not only bedrock but also older infills. There are so many possible variables that the simple equation (time=length of gorge divided by rate of recession) is, for Niagara, meaningless. As a general principle, extrapolation of rates of denudation backwards in time should not rest on the assumption that the rates have been constant.

(*b*) *Dendrochronology*   The study of annual growth rings of trees can provide an accurate clock within particular regions (H. C. Fritts, 1965; M. A. Stokes and T. L. Smiley, 1968). The method involves counting and measuring the spacing of the rings on one or more radii of the trunk. Variations in thickness of rings is probably mainly related to climatic events and may thus permit correlations over small regions. It is important to use only trees with clear annual rings related to annual growth cycles – the method is inapplicable in climates lacking strong seasonal rhythm – and 'false' rings of non-annular character, caused by a period of unusual drought, for instance, can be troublesome. Nevertheless, the method can be extremely accurate. Its main limitation is of course, the maximum age of available trees growing in useful stratigraphical positions. Research on the bristlecone pine (*Pinus aristata*) in the south-western USA has shown that the method here can be extended to about 7000 BP, though this is unusual. In the Alps, V. C. Lamarche and H. C. Fritts (1971) have shown the possibilities of using *Pinus cembra L.*, correlating variations in ring spacing with the known glacial history from 1800 AD, and suggest that the method might be applicable over the last 1000 years if suitable sites are discovered. J. Alestalo (1971) has reviewed the application of tree-ring studies to the dating of a range of glacial and other geomorphological features. The method has also been used to check radiocarbon dates; the small discrepancies that are found arise partly from the fact that the tree-ring dates are calendar years whereas the radiocarbon dates are not necessarily so.

(*c*) *Lichenometry*   Lichen dating was pioneered by R. E. Beschel (1950) who later published a comprehensive summary describing the technique (Beschel, 1961). The basic idea is that the diameter of the largest lichen thallus growing on a surface is proportional to the length of time that that surface has been exposed. However, it must not be assumed that growth rates are linear with time since climate (and especially the length of growing season) is not constant, and, particularly, it must not be assumed that the relationships developed in one area will apply to other environmentally different areas. It is also obviously unsound to extrapolate growth rates to surfaces hundreds or thousands of years old from rates measured over only a few decades. *Rhizocarpon geographicum* has been the most widely used of lichen species: it has a broad distribution, a long life, and a nearly constant age-size relationship. It also tends to produce roughly circular thalli, a help in measurement. Fig. 1.4 shows a growth curve

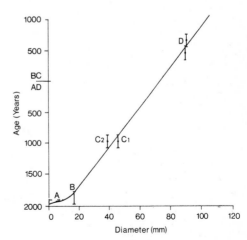

Fig. 1.4
Growth curve for *Rhizocarpon geographicum* in the Indian Peaks region, Colorado Front Range. Lichen diameters were measured on (A) twenty-four historically dated surfaces, (B) mudflow levees, (C) an Indian wall and (D) a moraine of the earliest Little Ice Age. Vertical lines show the statistical errors of the radiocarbon dates (J. B. Benedict, *J. Glaciol.*, 1967, by permission of the International Glaciological Society)

established by J. B. Benedict (1967) in the Colorado Front Range. In this area, there is seen to be an initial phase of rapid growth (*c.* 0·14 mm/yr) over about a century, subsequently slowing off to an average rate of *c.* 0·03 mm/yr. Such low values reflect moisture deficiency in this area and may be compared with the rates established by D. N. Mottershead and I. D. White (1972) for Tunsbergdal, southern Norway (initially 0·85 mm/yr, reducing to *c.* 0·2 mm/yr) with a less severe climate. The age range of the technique is similar to that of dendrochronology. Benedict (1967) notes the largest *R. geographicum* thallus in the Indian Peaks region of the Colorado Front Range as 220 mm in diameter, which must make it about 6000 years old.

An important use of lichenometry has been in dating moraines (Chapter 15). Fig. 1.5 is a map of Midtdalsbreen, southern Norway, prepared by J. L. Andersen and J. L. Sollid (1971), on which moraine ridges and glacier limits have been dated by lichenometry, back to 1750 AD. J. A. Matthews (1973) has looked at the problems of lichen growth interpretation on active moraines, and has shown that lichens can survive on boulders incorporated into pre-existing moraines or even on boulders deposited in pro-

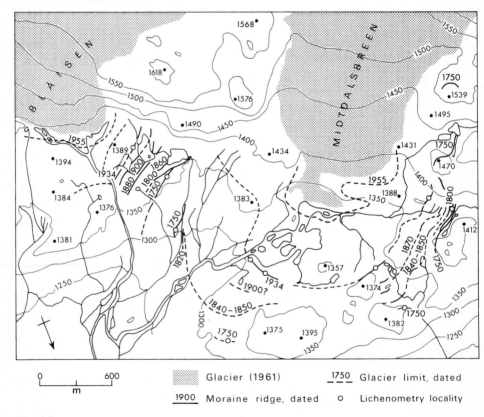

Fig. 1.5
Proglacial areas of Midtdalsbreen and Blåisen, Norway, showing moraines dated by lichenometry. The 1955 moraine was dated from air photographs (J. L. Andersen and J. L. Sollid, *Norsk geogr. Tidsskr.*, 1971, Universitetsforlaget, Oslo)

glacial outwash areas. Such circumstances indicate the need for caution in using lichenometric techniques, as with all methods of dating.

(*d*) *Rhythmite studies*   The use of rhythmites or varves in glacial lake deposits for chronological purposes was pioneered by De Geer and his associates in Sweden in the early part of this century, and later applied by E. Antevs in eastern North America. The method is discussed in Chapter 19. Until the advent of radiometric dating, this was the most reliable basis of absolute dating of late Pleistocene events in Scandinavia and North America. Much valuable information was accumulated, though the validity of some correlations has been questioned.

(*e*) *Fluorine-phosphate ratios in fossil bones*   The fluorine content of bones affected by percolating ground water containing minute traces of this element increases with time. The method excited public interest with the exposure of the Piltdown forgery, but the finding of suitable naturally emplaced bones in Pleistocene successions is a relatively rare event.

(*f*) *Dating by radioactive isotopes*   The introduction of these techniques has revolutionized the whole subject of Pleistocene chronology. For the first time, relatively accurate dating of horizons in Pleistocene successions is becoming possible, and the next decades are likely to witness greater progress in the unravelling of Pleistocene stratigraphy and the correlation of events than has been made in the whole of the last century or so of investigation. Space does not permit more than a brief indication of the methods so far established and their potentialities: for a fuller recent discussion, see W. S. Broecker (1965) and F. W. Shotton (1967).

Carbon 14   The first of the methods to be explored was based on the isotope carbon 14 (W. F. Libby, 1955), and can be applied to organic and carbonate materials. The rather short half-life of C^{14} (5730 years) limits its use to about ten half-lives, or about 60,000 years; beyond about 35,000 years, however, pre-enrichment techniques and sensitive detectors are needed to measure the very small amounts of radioactive carbon surviving. The method nevertheless is thought to be capable of spanning most of the last glaciation (Weichsel/Wisconsin), and has shown that this episode contained two major extensions of glaciers (the Hauptwürm of Europe or the Late Wisconsin of North America, *c*. 10–25,000 BP, and an earlier stadial, corresponding to the Early Wisconsin, *c*. 40–70,000 BP). The method is not without its problems and sources of error, of which the most serious are errors introduced by contamination. Artificial contamination can be guarded against, but natural contamination (e.g. by more modern plant roots or by percolating groundwater) can only be assessed and allowed for so far as possible. Contamination by 'modern' carbon introduces the greatest errors (Table 1.4) and may even give an apparent age to a sample too old to have any measurable activity of its own. The table also indicates why C^{14} dates greater than about 40,000 years for organic materials are considered rather unreliable. In the case of carbonate materials (shells, etc.), a range of 25,000 years might be a safer maximum.

Table 1.4 Errors resulting from contamination by 'modern' carbon

True age (years)	Apparent age (years) resulting from contamination levels of:	
	1%	10%
1,000	910	160
10,000	9,730	7,620
25,000	23,400	15,500
60,000	36,600	18,400

Potassium-argon The second method depends on the decay of the isotope potassium 40 to argon 40. It can yield reliable ages for certain minerals in which the argon produced has not been able to escape and where there has been no likelihood of contamination by atmospheric argon. It has so far been applied only to volcanic rocks in the Pleistocene, and the main problem is thus of finding suitable rocks in critical stratigraphical positions. The method is rather uncertain for dates younger than 20,000 years, but even this gives a possible overlap with radiocarbon dating. A most interesting result recently obtained is the dating of the Bishop Tuff in western North America at 710,000 years (R. P. Sharp, 1968), for the Bishop Tuff overlies glacial till, possibly the earliest in North America. In Europe, some very significant results have been obtained from K-Ar dating of lavas in the Eifel region associated with the Rhine terraces (and therefore with Penck and Bruckner's classical Alpine stages). J. von Frechen and H. J. Lippolt (1965) suggest ages of early glacial periods as follows:

Mindel-Riss interglacial	140–150,000 years
Mindel II glacial	*c.* 220,000 years
Mindel I glacial	*c.* 300,000 years
Günz-Mindel interglacial	*c.* 320,000 years
Günz II glacial	340–350,000 years
Günz I glacial	410,000 years

Such a chronology, it may be noted, agrees approximately with the 'short' time-scales for the Pleistocene proposed by C. Emiliani (1957, 1966, etc.) (see pp. 24–5), but poses the difficult question as to what events occupied the early Pleistocene, for other K-Ar dates suggest that the beginning of the Pleistocene should be placed at around 1 or even 2 million years ago.

The K-Ar method has also been used in attempts to date pre-Pleistocene ice ages. In Antarctica, some lava flows separating glacial tills are as old as 3·7 million years, and we conclude that a major ice sheet existed here in the Pliocene (G. H. Denton, R. L. Armstrong and M. Stuiver, 1970). W. E. le Masurier (1972) shows that subglacial volcanic eruptions in Antarctica occurred probably in the intervals 6–12 million years and 18–25 million years BP, during which times an ice sheet several hundred metres thick must have been present.

Uranium-series methods Thirdly, there are the uranium series methods based on uranium 234, thorium 230 and protactinium 231 which have recently been applied to some Pleistocene marine and other carbonate sediments. Modern deep-ocean sediments contain a constant proportion of Th^{230} precipitated from sea-water. As Th^{230} decays with time (eventually to lead, as with other unstable isotopes in the uranium series), the amount occurring in older sediments will be less and will provide a measure of age so long as there has been a constant rate of deposition of this isotope. Pa^{231} is formed in deep-sea sediments similarly, but at a different rate; it has been used on its own for dating, and also in the form of the ratio Th^{230} to Pa^{231}. The great interest of uranium-series methods is that potentially they might fill the zone of uncertainty between C^{14} and K-Ar dating, and that they can be applied to sediments, unlike K-Ar. However, uranium-series methods are still in an exploratory stage, and as many of the dates obtained so far appear to conflict with K-Ar dates, their reliability is still questionable. Th^{230}-U^{234} dating has been applied to shelly faunas of Monastirian (last interglacial) beaches in the Mediterranean and coastal Morocco (C. E. Stearns and D. L. Thurber, 1967), yielding ages from 85,000 to 135,000 years BP, generally consistent with expectations. Provisional Pa^{231}/Th^{230} dating of the Sangamon (Riss-Würm) interglacial suggests an age around 100,000–150,000 years BP.

The methods so far mentioned are summarized in Table 1.5. Others, not yet properly

Table 1.5   Principal radioactive isotopes used in the dating of Pleistocene events

	Half-life (years)	Approximate range of method (years)
C^{14}	5730 ± 40	0–70,000
K^{40}	$1\cdot3\times10^{9}$	20,000+
U^{234}	250,000	50,000–1,000,000
Th^{230}	75,000	0–400,000
Pa^{231}	32,000	5000–120,000

evaluated, are based on the isotopes helium 4, chlorine 36, and beryllium 10. With the aid of radioactive isotopes, a reliable chronology for the last 150,000 years is being steadily built up, and there is considerable promise that its extension farther back in the Pleistocene will be feasible.

(g) *Tephrochronology*   Introduced by S. Thorarinsson (1944, 1949) in Iceland, and developed by S. Kaizuka (1961) and others in Japan, this method of dating and correlation depends on the fact that a mantle of volcanic ash, deposited over a wide region in a brief moment of geological time and later buried, forms an ideal stratigraphical marker. Its age may be ascertained by the K-Ar method, or by C^{14} dating of associated carbonaceous material, if any. An example is the Pearlette ash layer of the central-western USA, emanating from a still unlocated source. It forms an important link between Kansan glacial deposits in the Great Plains, the Cedar Ridge glaciation in the Rockies, an ancient subaerial stage of Lake Bonneville, and alluvial deposits in the unglaciated part of the Great Plains. Its detailed distribution has yet to be systematically studied, but palaeomagnetic dating of a core in the Bonneville lake basin of Utah suggests a date of about 600,000 years (A. J. Eardley *et al.* 1973).

(h) *Palaeomagnetic studies* Palaeomagnetic correlation of igneous and some sedimentary rocks is possible because the direction of the earth's magnetic field is thought to have suddenly and periodically reversed at certain points in geological time, and because certain rocks become permanently magnetized in the earth's field at the time they are formed. The reversals are necessarily world-wide and synchronous, and are now broadly dated by the K-Ar method. Consequently, correlations and further dates can now be obtained by simple palaeomagnetic measurements. Four epochs are recognized (A. Cox, 1969):

Table 1.6   Late Cenozoic epochs of normal and reversed magnetization

Age (10^6 years)	Name	Magnetization
0·69	Brunhes epoch	Normal
0·69–2·43	Matuyama epoch	Reversed
2·43–3·32	Gauss epoch	Normal
3·32–>4·50	Gilbert epoch	Reversed

Within these epochs, short-period 'events' represented temporary changes of polarity (for example, the Mammoth event of reversed polarity, 2·94–3·06 million years BP, within the normal Gauss period). Attempts to extend the scale back beyond about 4–5 million years have been unsuccessful because possible errors in the K-Ar dates become larger than the duration of many polarity events. Nevertheless, this range covers the whole of the Pleistocene and part of Pliocene time, and the method is a very important adjunct to K-Ar dating (see, for instance, M. G. Rutten and H. Wensink, 1960; I. McDougall and H. Wensink, 1966). The main problem hindering further progress is that, in common with the K-Ar method, of finding juxtapositions, on the one hand, of rocks suitable for palaeomagnetic dating, and, on the other hand, of glacial or fossiliferous deposits (A. Cox, R. R. Doell, and G. B. Dalrymple, 1965).

(i) *Studies of deep-sea sediments* Development of deep-sea core-sampling techniques has made possible the stratigraphical study of Pleistocene deep-sea sediments. Changes in the type and abundance of fossils in various layers can be used to reveal changes of sea-water temperature, which to some extent must reflect changes in Pleistocene climate. C. Emiliani (1955) developed an additional technique based on oxygen isotope ratios for pelagic foraminiferal shells, enabling actual temperature curves to be deduced with some degree of confidence, and suggesting that Pleistocene temperatures of tropical sea-surfaces fluctuated within a range 23–29°C, and that the lowest temperatures of presumed glacial maxima were much the same, as were the highest temperatures of the interglacials. Having obtained temperature curves, the problem remains of dating the events shown on them. If the most recent layers of a core are dated by such means as carbon 14, extrapolations of age can be made to the older layers (more than 30,000 years old) by assuming uniform rates of sediment accumulation. This is, of course, very uncertain, and herein lies the weakness of the system until absolute dates for the older layers are available to act as independent checks. Meanwhile, Emiliani suggests that the maxima of the Riss-Saale glaciation occurred at about 110,000 BP,

of the Mindel-Elster at about 180,000 BP, and of the Günz at about 280,000 BP;
W. S. Broecker and others (1958) extended Emiliani's time scale by about 50 per cent,
by allowing for variable rates of sedimentation, but Emiliani (1970) claims that his
time-scale is correct to within a few percent at least back to 175,000 years. Completely
different correlations between the glacial sequence and the deep-sea sediment record
have been made by D. B. Ericson, M. Ewing and G. Wollin (1964). According to
them, for instance, the first glacial period of the Pleistocene (? Günz) is dated at about
1,500,000 years, and there is only agreement with Emiliani over the last 120,000 years.
A. Holmes (1966) also favoured a 'long' time-scale placing the Donau glaciation
between 1·8 and 1·37 million years, and the Günz between 1·15 and 0·9 million
years.

2 Present-day glacierization

About 15 million km², or 10 per cent of the earth's land area, are ice-covered at the
present day. Table 1.7 shows the principal glacierized regions, ranked in order of size.
It should be noted that the two largest, Antarctica and Greenland, account for 96
per cent of the total ice-covered area. The figures are those quoted by R. F. Flint (1971),
rounded to the nearest 100 km², and they should be regarded, like all measurements
of large areas, as approximations only. Rather larger figures were given by P. A. Shumskiy *et al.* (1964), totalling 16,215,000 km² compared with Flint's total of approximately
14,898,000 km². Table 1.8 shows Shumskiy's figures for the Antarctic. Many ice-
covered regions are still imperfectly mapped, as for instance in Central Asia (M. G.
Grosval'd and V. M. Kotlyakov, 1969), but new techniques of remote sensing, both
from aircraft and satellites, are rapidly providing more accurate data (see *Symposium
on glacier mapping*, 1966, for example).

Table 1.7 demonstrates how varied are the climatic circumstances under which

Table 1.7   Present-day ice-covered areas (in
km²)

Antarctic	12,588,000
Greenland	1,802,600
North-east Canada	153,200
Central Asian ranges	115,000
Spitsbergen group	58,000
Soviet Arctic islands	55,700
Alaska	51,500
South American ranges	26,500
West Canadian ranges	24,900
Iceland	12,200
Scandinavia	3,800
Alps	3,600
Caucasus	1,800
New Zealand	1,000
USA (excluding Alaska)	500
Others	about 100

(Total volume of present ice: 28 to 35 million
km³)

Table 1.8   The Antarctic ice sheet (P. A. Shumskiy, 1970)

A. Area

| | Area km² × 10³ | As a percentage of the | | |
		Antarctic ice-sheet area	Antarctic surface area	Earth's glacierized area today
Ice-sheet total	13,779	100·0	98·6	85·7
Land-based portion	12,150	88·2	87·0	75·6
Floating ice	1,460	10·6	10·4	9·1
Ice islands	169	1·2	1·2	1·0

B. Volume

| | Volume km³ × 10⁶ | As a percentage of the | |
		Antarctic ice-sheet volume	Earth's total ice volume today
Ice-sheet total	23·7 to 31·3	100	88·5 to 91·2
Land-based portion	23·0 to 30·4	97·1 to 97·8	86·2 to 89·1
Floating ice	0·6	2·5 to 1·9	2·2 to 1·8
Ice islands	0·08	0·04 to 0·3	0·4 to 0·3

glacier ice exists in the world today. By far the greatest part of the world's ice, in the Antarctic and northern Greenland especially, is found in high latitudes characterized by very low winter temperatures (the lowest temperature yet recorded is −88°C at Vostok near to the south geomagnetic pole at 3500 m elevation), low summer temperatures, small annual precipitation (about half of Antarctica has a net accumulation of less than 10 cm/year), and minimal ablation. These ice masses as a result are relatively inactive and stable. Greater activity is noticeable in areas of less extreme cold and moderate precipitation, as in southern Greenland for example. The most active glaciers, generally speaking, are those in regions where winter temperatures are low but summer temperatures high, where there is abundant precipitation in the form of snow, but at the same time where ablation losses are often heavy. Such conditions prevail in the Alps, in New Zealand, and in many other regions of middle latitudes.

The vertical distribution of ice, now as in the past, is controlled primarily by the altitude of the firn-line on the ice, for this affects the size of the accumulation area above it (see Chapter 2). The snow-line in ice-free areas varies considerably in altitude both in any given area and seasonally; it is often highly irregular, and strongly influenced by such considerations as shade and snow-drifting as well as by precipitation and ablation. For this reason, it is more useful usually to deal with the average 'regional' rather than the 'local' snow-line, the regional snow line, sometimes termed the 'climatic' snow-line, being defined as the average lower limit of snow persisting throughout the year. The height of the regional snow-line varies at present from sea level to over 6000 m in the Central Andes. In northern Britain, its theoretical height is put at about

1500 m. Its height is controlled both by snowfall and by summer temperatures, which determine the extent to which winter snow can survive throughout the summer season. Any rain falling in summer, too, is a potent instrument aiding the dissipation of winter snow.

The distribution of ice in the world is a constantly changing one. Glacier fluctuations are obvious in terms of the retreat or advance of glacier snouts over periods of a few years or decades, and are readily measurable by comparisons of maps of different dates, or more precisely by ground survey or photogrammetry. Less obvious, less easily measured, but of even greater significance are changes in the thickness of glaciers, which in turn cause changes in area or length. Trim-lines, representing abrupt changes in vegetation, sometimes mark the former higher margin of a glacier (Chapter 15). It is well known that, since the mid- or late-nineteenth century, the world's glaciers have in general been receding—the tongue of the Rhône glacier for instance now hangs high above the village of Gletsch, to which it closely approached in 1920. Many glaciers this century have stagnated; some, as in the Pyrenees, have vanished entirely. Alpine glaciers overall have lost more than one-third of their area since the 1870s, though there have been interruptions in this recession. G. Patzelt (1970), reviewing the glaciers of the eastern Alps, shows that here the main period of retreat was from 1928 to 1964, while between 1965 and 1969, most were still retreating but many fronts were stationary and up to 25 per cent were beginning to advance again. Data on glacier fluctuations relating mainly to the period 1850–1950 are extensively summarized in Charlesworth (1957, pp. 142–4).

Glaciers vary greatly in their rates of response to climatic fluctuations (see Chapter 2). The least sensitive are the largest and most slowly moving ice sheets, lying in high-altitude regions of severe climate. Much better indicators of climatic change are glaciers of middle latitudes with a critical mass balance, where slight changes in rates of accumulation or ablation may have very profound consequences. It is quite possible that only small changes of climatic régime were necessary to induce the rapid expansion of middle-latitude ice sheets in the Pleistocene.

3 Theories of climatic change

Any modern theory of the causes of ice ages must be able to satisfy the following requirements, most of which are now well established:

1 There were Ice Ages at various geological epochs before the Cenozoic.
2 Cenozoic glaciation began as far back as the Miocene and possibly even earlier in the Eocene, and there is evidence that the degree of cooling intensified through the Tertiary.
3 The glacial periods were cold climatically, and temperatures over the whole earth were less than those now experienced.
4 There were interglacial periods, some of which at least were relatively warm.
5 The difference of mean annual air temperature between glacial and interglacial periods was of the order of 6°C in the Tropics (as shown by deep-sea sediment studies) and possibly 8° or 9°C in higher latitudes.

6 Precipitation in and around ice-covered areas in the glacial periods was slightly less than in these areas today (G. Manley, 1951, for instance, suggests 80 per cent of present precipitation for British Highlands), but more of it fell as snow.

7 The snow-line rose and fell through a vertical interval of as much as 1300 m in places.

8 The onset of a glacial stage and the disappearance of ice at the end of it apparently occurred over periods of the order of only 10,000 years (G. de Q. Robin, 1966).

9 There were minor fluctuations of climate superimposed on the general rhythm of glacial and interglacial episodes.

10 The climatic fluctuations in the northern and southern hemispheres were probably synchronous (see, for example, J. T. Hollin, 1962): this is fairly certain for the last 100–120,000 years from comparisons of the Camp Century and Byrd Station records, but has been disputed for earlier stages.

Two further conditions apply to the Pleistocene, but not to the older glacial epochs:

11 Land and water had no *radically* different distribution from that of today though changes in the detailed patterns of land and sea (e.g. associated with fluctuating relative sea-levels) and of highland and lowland have certainly occurred.

12 The earth's poles were in much the same positions as they are today.

The past century and a half has seen innumerable theories expressed on the cause of Ice Ages ranging 'from the remotely possible to the mutually contradictory and the palpably inadequate' (Charlesworth, 1957, p. 1532). Few, indeed, can claim to be able to meet more than a small number of the conditions given above. The older theories are adequately reviewed elsewhere (C. E. P. Brooks, 1949, and J. K. Charlesworth, 1957, for instance), while J. C. Crowell and L. A. Frakes (1970) and J. M. Mitchell (1968) provide reviews concentrating on more modern theories of climatic change. The chief purpose of the next section is to classify, as concisely as possible, the main groups of the more plausible theories. Attention will be focused on those that have emerged in the last twenty years, and both primary and secondary mechanisms will

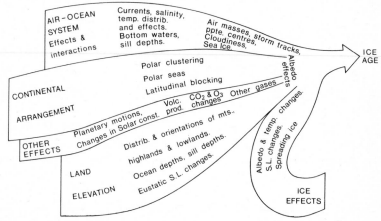

Fig. 1.6
The causes of ice ages (J. C. Crowell and L. A. Frakes, *Am. J. Sci.*, 1970)

be considered. Fig. 1.6 shows a diagrammatic summary of the principal mechanisms and changes which may be relevant in any attempt to explain ice ages.

3.1 *Geophysical Theories*

(*a*) *Continental drift*   As a hypothesis which, in modified form, has returned to favour amongst geologists in recent years as a result of palaeomagnetic and other studies, it has been used as a means of bringing various regions into near-polar positions at various times. Its principal application is in substantially helping to solve the riddle of the onset of the Permo-Carboniferous glaciation, but its bearing on Cenozoic ice ages is far from clear. Differential movement of the continents during the $1\frac{1}{2}$ million years or so of the Pleistocene seems unlikely to have exceeded a few tens of kilometres, which is hardly significant climatically. Moreover, the fact that successive Pleistocene glaciations covered mainly similar areas is strong evidence against large relative displacements. But since early Cenozoic times, continental movement is probably of a larger order of magnitude—perhaps even several hundred kilometres. Seafloor spreading has caused a widening of the Atlantic and, along with other changes such as the closure of the Tethys geosyncline and the creation of the Panama isthmus, is likely to have led to some fundamental changes in oceanic circulation. There may well be a connection, therefore, between the onset of early Cenozoic glaciation and movement of the continental plates. A cautionary note should be sounded, however; it is a mistakenly-held belief that polar positions provide the optimum conditions for glaciation, whereas in fact much more favourable positions are those of middle-latitude highlands experiencing heavy snowfall and cool summers. Nevertheless, the onset of glaciation in Antarctica has been attributed to the arrival of that continent in polar latitudes in the early Cenozoic (M. J. Selby, 1973).

(*b*) *Polar wandering*   This is really another means of achieving the same aim as continental drift or plate tectonics. For earlier geological epochs, there is no longer doubt that movement of the poles occurred, relative to the land masses. But certainly for the Pleistocene, there is no evidence of any significant polar wandering. Despite this, the idea was used (M. Ewing and W. L. Donn, 1956–8) as a means of explaining the supposed onset of glaciation at the beginning of the Pleistocene, it being postulated that at this time the earth's poles first took up roughly their present positions. Other factors were then brought into play to account for the glacial-interglacial sequence. The theory was proposed before the extent of pre-Pleistocene glaciation was known.

3.2 *Land-water changes*

(*a*) *Oceanic changes*   Since some major ocean drifts and currents serve to transfer considerable quantities of heat from tropical to polar latitudes, changes in these water movements could provoke changes in polar climates. The most critical point in regard to the Arctic is the Faroe-Icelandic sill, shallowing of which could result in less tropical water entering the Arctic Ocean. Glacio-eustatic lowering of sea level (Chapter 5) could in itself produce this effect and it thereby becomes a secondary mechanism of promoting cooling and ice extension in the Arctic once a glacial period is under way.

Ewing and Donn (1956–66) utilize this idea. Earlier, C. E. P. Brooks (1949) also postulated rearrangements of oceanic circulation caused by tectonically-created land bridges to cause climatic cooling of certain areas. J. T. Hollin (1962) also points out that sea-level lowering in the glacial periods would result in northward displacement of the 'grounding-line' in Antarctica—the line along which the ice-shelves began to float. This was an important factor aiding extension of the Antarctic ice and the implications are discussed more fully in Chapter 2.

(b) *Changes of land-mass altitude*   The effect of elevating the land is to bring larger areas above the snow-line. This was a view long held to account for Pleistocene glaciation following on the widespread orogenic activity of Tertiary time, until the fact of the glacial-interglacial succession became established. Although there is no evidence for rapidly alternating uplifts and subsidences to explain climatic fluctuations within the Cenozoic, the basic idea of the Cenozoic ice ages being associated with a period of tectonic uplift is a sound one. There is abundant geological and geomorphological evidence of late Cenozoic uplift in many areas. The Alps, for instance, are thought to have been raised some 2000 m from the Pliocene to the mid-Pleistocene; the Sierra Nevada of California may have risen 2000 m or more in the Pleistocene (see P. B. King, 1965); the Himalayas were possibly uplifted 3000 m in the Pliocene-early Pleistocene. It is well known that quite small altitudinal changes can have a disproportionately great effect in increasing snowfall and lowering temperatures, and the widespread nature of Cenozoic uplift undoubtedly helped to promote refrigeration, even though it cannot explain the fact of multiple glaciation. The recognition of extensive glaciation in the Miocene and Pliocene as well as the Pleistocene has greatly strengthened the links between mountain building and ice ages. The idea is also applicable, it should be noted, to the Permo-Carboniferous glaciation which also succeeded an epoch of major orogenesis in many parts of the world.

A further point in connection with land-mass altitude and glaciation concerns glacial isostasy. Once ice on a land area reaches a certain thickness, isostatic depression of the crust commences, and may eventually amount to one-quarter to one-third of the ice thickness. Thus the accumulation of great thicknesses of ice is encouraged, and this in turn leads to more prolonged survival of the ice.

3.3   *Atmospheric changes*

(a) *Variations in carbon dioxide content*   By reducing the amount of CO_2 in the atmosphere, the rate of absorption by the latter of long-wave radiation will also be reduced, and thus the atmosphere cooled. The essence of this theory was stated as long ago as 1899 by T. C. Chamberlin. Modern calculations suggest that even to lower surface air temperatures by 3°C would necessitate halving the present atmospheric CO_2 content, and it is practically impossible to devise a way of upsetting the earth's carbon cycle to such an extent. The earth's carbon cycle is very complex, involving plants, animals, volcanic gases, chemical weathering, carbonate fixation in rocks, oceanic concentrations, and so on; simple cyclic fluctuations seem highly improbable. However, CO_2 may be an important secondary factor in the glacial-interglacial alternation, inasmuch as with the formation of ice sheets (containing little carbonate matter), the atmosphere's content of CO_2 would be indirectly increased, causing some warming of the atmosphere and helping to bring glaciation to an end.

(*b*) *Variations in the amount of suspended volcanic dust* Volcanic (or other) dust in the atmosphere serves to cut out some solar radiation, but the screening effect is essentially regional rather than world-wide, and would have to be maintained for long periods. Evidence of volcanic activity on a sufficient scale throughout the Pleistocene is lacking. H. H. Lamb (1969) has investigated the records of volcanic eruptions since 1500 AD. He finds a correlation between atmospheric temperature and great eruptions, but the temperature effect fades after 2 to 7 years (perhaps as long as 15 years in high latitudes). This suggests that volcanic dust concentrations are unlikely to be a cause of any longer-term climatic change.

(*c*) *Other atmospheric changes* involving ozone and water vapour content, for example, have been considered, but the causes of such changes remain speculative (see section 3.7) and it does not seem probable that they could explain the glacial-interglacial succession.

3.4   *Meteorological theories*

These consist essentially of secondary mechanisms for promoting glaciation once started, or for bringing a glacial period temporarily to an end. The basic idea is that quite small changes in climate, or in the atmospheric or oceanic circulatory systems, might have far-reaching effects, especially on ice-masses in a critical condition. C. E. P. Brooks (1949) argued, for instance, that quite a small drop in temperature of an ocean already near freezing-point would cause it to freeze over, and that the increased albedo of ice compared with ocean water would result in further heat loss and still lower temperatures. Donn and Ewing (1966) have recently elaborated such ideas. Starting with an open Arctic sea to supply abundant precipitation to adjacent land masses, they argue that, with cooling as the earth's poles moved to their present positions (see section 3.1), more of this precipitation would fall as snow, leading to the formation of ice caps on the land. The albedo effect would begin to operate, as already described, and the Polar sea would freeze, still further increasing the cooling. Simultaneously, other changes would be occurring. First, with the growth of glacial centres near the Arctic, the cyclonic storm belts would be pushed equatorward, nourishing the equatorward side of the ice masses at the expense of the polar sides. If this process continued, the storm belts could be shifted into latitudes where more precipitation fell as rain, checking or even reversing the process of glacial expansion. Secondly, freezing of the Polar sea, helped by a lowered sea level (see section 3.2), would also cut off the source of moisture for the polar sides of the ice masses, again helping to bring the ice age to an end. Many of these and other assumptions in the Ewing-Donn and similar theories need to be more rigorously tested. Data on the present Antarctic and Greenland ice sheets, for instance, show that these are remarkably stable. Their high albedo helps to create their own favourable temperature conditions and it is argued that very considerable rises in temperature (10°–15°C according to G. de Q. Robin, 1966) would be required to dissipate them. They also exist in what is virtually an 'ice desert'—and disprove the popular fallacy that high precipitation is needed to cause glaciation. All these ideas are still speculative. We know very little about the precise effects, and the ensuing feedbacks, of change in any one of the components of the complex air-ocean system, but it certainly seems possible that the explanation of the glacial-interglacial oscillations may be found here.

3.5   *Glaciological theories*

Under this heading it is convenient to consider A. T. Wilson's theory of ice ages (Wilson 1964, 1969). Wilson's idea is that periodic basal melting (see pp. 89–90) of the Antarctic ice sheet could cause massive surging of the ice, creating a great ice shelf around Antarctica. Such an ice shelf would displace sea water causing a sudden rise in world sea level. J. T. Hollin (1965, 1969) has estimated the probable magnitude of the rise at 10–30 m, assuming that the surge captured the ice drainage from half the ice sheet and lowered that half by 50 per cent, and that the resulting ice shelves would be approximately as thick as at present (*c.* 200 m). The large area of new ice shelf would increase the albedo, leading to global cooling, the formation of more ice in the northern hemisphere (when sea level would fall) and further cooling. The end of such a glacial stage would be brought about by disintegration of the ice shelf, reduction of albedo, and consequent warming. Hollin (1965) has discussed possible stratigraphical evidence for the very rapid rise of sea level marking an Antarctic surge and just preceding the falling sea level as world glaciation develops. There are possible examples of sudden marine transgressions late in an interglacial in eastern England, but the evidence is not decisive.

Wilson's theory provides another possible way of initiating a glacial-interglacial oscillation within an epoch of general cooling due to other mechanisms, but lacks confirmatory evidence. It is also doubtful whether such a vast ice shelf continuous around Antarctica, but not bonded laterally to land (as are the present relatively small Antarctic ice shelves such as the Ross and the Filchner), would be stable for any significant length of time, and whether ice-sheet surges could have cooled the Earth enough to cause ice ages.

Recently, some attention has been given to the question of the rates of growth or shrinkage of the Pleistocene ice sheets. J. Weertman (1964) has analysed the time required to build up or destroy an ice sheet of given size, using the model of perfect plasticity for ice flow. If net accumulation averaged 20 to 60 cm a year, the time required to form an ice-age ice sheet would be of the order of 15,000 to 30,000 years, possibly the length of a glacial sub-stage. On the other hand, if during deglaciation the average ablation rate over large areas were substantially larger than the average accumulation rate, and ice movement ceased, the ice sheets would disappear comparatively rapidly.

The time required for the build-up of the ice sheets can be considerably lessened, however, by bringing in another factor, that of changes in the temperature of the basal ice (A. T. Wilson, 1964; J. Weertman, 1966). If we begin with a cold ice cap frozen to bedrock, outward movement will be slow (see Chapter 4). As increased accumulation causes it to thicken, there will be a gradual warming of the basal ice layer by geothermal heat, owing to the blanketing effect of the ice. Eventually, the basal ice layer will reach pressure melting point, and substantial sliding over bedrock will be possible, facilitated by basal meltwater. Spreading of the ice will thus result, but at the same time, as Weertman (1966) argues, these changes may in fact reduce the equilibrium thickness of the ice sheet by a factor of 1·4 to 2 times, since the basal shear stress will be very much less. Thus after a phase of relatively rapid expansion, shrinkage may set in until the ice sheet either disappears or approaches a new equilibrium size. These suggestions considerably help with one of the most difficult problems that many ice-age theories encounter—that of bringing glaciation to an end once it has started. Weertman's theory considerably reduces the size of any change in climate required to make an ice sheet disappear.

3.6   Changes in the Earth's orbit and axis

There are three main variables in the geometrical relationships between Earth and Sun (for a detailed discussion, see W. Munk and G. J. F. Macdonald, 1960):

1   The precession of the equinoxes, with a periodicity of about 21,000 years. The distribution of solar radiation over the Earth's surface varies slightly according to whether the equinox, for instance, occurs in aphelion or perihelion.
2   The obliquity of the plane of the ecliptic varies from 21° 58′ to 24° 36′, with a periodicity of about 40,400 years.
3   The eccentricity of the Earth's orbit; periodicity about 91,800 years.

All these, singly and in combination, affect the receipt of solar radiation and its distribution over the globe, and are indisputable. J. Croll (1875) was one of the first to base a detailed theory of climatic fluctuations on varying Earth–Sun relationships. He used the precessional period as a means of altering the distribution of heat over the Earth's surface, and his theory required that glaciation should alternate between the northern and southern hemispheres. This, of course, is no longer tenable, nor does the supposed periodicity of 21,000 years conform to what we now know of the glacial-interglacial chronology. Other workers such as W. Köppen and A. Wegener developed theories taking other Earth–Sun relationships into account, but the principal discussion of the so-called 'astronomical theory' was by M. Milankovitch (1920 *et seq.*). His curves combining the various geometrical relationships showed four main groups of low temperatures, possibly ice ages, within the last 600,000 years. This was a notable improvement over Croll's hypothesis; the fluctuations were of more reasonable duration and, moreover, did not have to alternate rigidly between opposite hemispheres—but neither did they coincide.

The main problem with the astronomical theory is not the reality of the variations but the smallness of their probable effect, which is likely to be no more than 1° or 2°C, yet on the other hand, the periodicities can be fitted to certain known Pleistocene fluctuations. The idea has been used by C. Emiliani and J. Geiss (1957) in a complex theory involving also changes in solar radiation and topographical-climatic changes.

Planetary changes cannot be used alone in formulating any theory of climatic change as they offer no explanation for the non-occurrence of glacial-interglacial periods between the Permian and the Cenozoic, and more specifically, for the gradual cooling throughout the Tertiary. It seems certain that the planetary changes must be superimposed on changes resulting from other causes. They thus provide 'background noise' but not a basic cause. Their periodicity and the magnitude of their effects are still being disputed (K. J. Mesolelia *et al.*, 1969; Donn and D. M. Shaw, 1967; D. B. Ericson and G. Wollin, 1968); C. Emiliani (1966) claims that analysis of deep-sea sediment cores strongly suggests a causal relationship between core periodicities of 40–50,000 years and astronomically calculated ages of summer insolation minima at latitude 65°N. He expects renewed glaciation in a few thousand years, reaching a peak about 15,000 years from now.

3.7 *Changes in solar emission*

In our present state of knowledge, this seems to be the most likely first cause of major climatic fluctuations (E. J. Öpik, 1968), with certain other changes (especially 3.6 and 3.2) superimposed as additional mechanisms. The Sun is the basic energy source for all atmospheric processes. Measurements over the last 40 years, however, have shown that variations in the 'solar constant' are less than 0·1 per cent (M. Jerzykiewicz and K. Serkowski, 1968). Such variations, associated with sunspot cycles of 11, 22, 44 years, etc., are probably much too small to explain ice ages, and the often-suggested correlations between sunspot activity and climatic change over the last 200 years may have little statistical significance. Although changes in the quantity of radiation can therefore probably be eliminated from the list of possible first causes of ice ages, changes in type of radiation and its components are much more promising to look at in the search for a cause. There are now known to be sizeable variations in the ultra-violet and X-ray wavelengths at times of solar flares, linked to sunspot cycles. These may cause changes in the production of upper atmosphere ozone, leading to short-term climatic changes (R. W. Fairbridge, 1970), which may in turn trigger or reinforce longer-term changes.

More speculative, but possibly of vital importance to the solution of the Ice Age problem, are longer-term changes in solar emission, such as Sir George Simpson (1934, 1957) has postulated, with a periodicity of, for instance, 200,000 or 400,000 years. Certain aspects of Simpson's theory are unacceptable in the light of recent evidence showing colder conditions in the tropics during glacial periods, but the basic mechanism is still available. It is used by R. F. Flint (1957, 1971) in his 'Solar-topographic concept', which appeals to changes in solar emission in the first place, and secondly to Cenozoic mountain-building and uplift. Flint contends that solar variations have occurred throughout geological time, but that they were only effective in causing a glacial-interglacial succession on land-areas provided with high-altitude elements: highlands are said to be the pre-requisite for extensive glaciation. With the growing evidence of earlier Cenozoic ice ages strengthening the link between mountain building and ice ages, Flint contends (1971) that his theory is now even more plausible than it was when first formulated in 1947. Substantially similar is the theory offered by R. P. Beckinsale (1965).

This postulates long-term changes in land-mass altitude through the Pleistocene: a considerable increase in mean land altitude is held to be essential for the formation of an ice age. Superimposed on this are shorter-term oscillations in the receipt of solar radiation due to changes in planetary geometry, and variations in solar emission. The answer to the question of the causation of ice ages seems likely to reside in compound theories such as these.

4 Conclusion

Major ice ages have occurred in the Pre-Cambrian, Permo-Carboniferous and Cenozoic periods of earth history. Cenozoic ice ages, possibly as far back as the Eocene, gradually developed in intensity up to the beginning of the Pleistocene about 1·5 to 2 million years ago. The maximum extent of ice in the Pleistocene was about three times its present spread and volume, and was associated not with the last (Würm, Weichsel, Wisconsinan) glaciation, but with earlier glacial periods, possibly the Riss (Saale, Illinoian).

Pleistocene stratigraphy is based on the fossil content of marine and some terrestrial formations, on the correlation of lithological units such as till sheets, on the existence of weathering zones or palaesols, on archaeological evidence and on palynology. The absolute chronology of the period is now founded mainly on radiometric methods, using particularly the isotopes carbon 14 and potassium 40, on palaeomagnetic, tephrochronological and deep-sea sediment studies, and to a less extent on glacial lake rhythmites, dendrochronology and lichenometry.

Glacier ice today covers about 10 per cent of the world's land surface, and exists in a great variety of climatic and topographical conditions. Fluctuations in its extent, vertically and areally, are often well documented for the last 200 years, and fairly well established in the Late- and Post-glacial periods.

The causes of ice ages are still the subject of speculation and controversy. Most probably, several different factors are involved, and the most plausible theories are those which are based on a combination of changes in land-mass altitude, short-term changes in the receipts of solar radiation resulting from the varying geometrical relationships of Earth and Sun, and longer-term changes in the quality of solar emission.

5 References

ALESTALO, J. (1971), 'Dendrochronological interpretation of geomorphic processes', *Fennia* **105**, 140 pp.

ANDERSEN, J. L. and SOLLID, J. L. (1971), 'Glacial chronology and glacial geomorphology in the marginal zones of the glaciers Midtdalsbreen and Nigardsbreen, south Norway', *Norsk geogr. Tidsskr.* **25**, 1–38

BECKINSALE, R. P. (1965), 'Climatic change: a critique of modern theories' in *Essays in Geography for Austin Miller* (ed. WHITTOW, J. B. and WOOD, P. D.), 1–38

BENEDICT, J. B. (1967), 'Recent glacial history of an alpine area in the Colorado Front Range, U.S.A. I. Establishing a lichen-growth curve', *J. Glaciol.* **6**, 817–832

BESCHEL, R. E. (1950), 'Flechten als Altersmassstab rezenter Moränen', *Z. Gletscherk. Glazialgeol.* **1**(2), 152–61

    (1961), 'Dating rock surfaces by lichen growth and its application to glaciology and physiography (lichenometry)' in *Geology of the Arctic (Proc. 1st int. Symp. Arctic Geology, Calgary 1960*; ed. RAASCH, G. O.) **2**, 1044–62

BLYTT, A. (1876), *Essay on the immigration of the Norwegian flora during alternating rainy and dry periods* (Christiania)

BROECKER, W. S. (1965), 'Isotope geochemistry and the Pleistocene climatic record' in *The Quaternary of the United States* (ed. WRIGHT, H. E. and FREY, D. G.), 737–754

BROECKER, W. S., TUREKIAN, K. K. and HEEZEN, B. C. (1958), 'The relation of deep-sea sedimentation rates to variations in climate', *Am. J. Sci.* **256**, 503–17

BROOKS, C. E. P. (1949), *Climate through the Ages*

CHAMBERLIN, T. C. (1899), 'An attempt to frame a working hypothesis of the cause of glacial periods on an atmospheric basis', *J. Geol.* **7**, 545–84

CHARLESWORTH, J. K. (1957), *The Quaternary Era*

COX, A. (1969), 'Geomagnetic reversals', *Science* **163**, 237–45

COX, A., DOELL, R. R. and DALRYMPLE, G. B. (1965), 'Quaternary paleomagnetic stratigraphy' in *The Quaternary of the United States* (ed. WRIGHT, H. E. and FREY, D. G.), 817–30

CROLL, J. (1875), *Climate and time in their geological relations. A theory of secular changes of the Earth's climate*

CROWELL, J. C. and FRAKES, L. A. (1970), 'Phanerozoic glaciation and the causes of ice ages', *Am. J. Sci.* **268**, 193–224

DANSGAARD, W. and JOHNSEN, S. J. (1969), 'A flow model and a time scale for the ice core from Camp Century, Greenland', *J. Glaciol.* **8**, 215–23

DANSGAARD, W., JOHNSEN, S. J., CLAUSEN, H. B. and LANGWAY, C. C. (1971), 'Climatic record revealed by the Camp Century ice core' in TUREKIAN, K. K. (op. cit., 1971), 37–56

DANSGAARD, W., JOHNSEN, S. J., MØLLER, J. and LANGWAY, C. (1969), 'One thousand centuries of climatic record from Camp Century on the Greenland ice-sheet', *Science* **166**, 377–81

DENTON, G. H. and ARMSTRONG, R. L. (1969), 'Miocene-Pliocene glaciations in southern Alaska', *Am. J. Sci.* **267**, 1121–42

DENTON, G. H., ARMSTRONG, R. L. and STUIVER, M. (1970), 'Late Cenozoic glaciation in Antarctica: the record in the McMurdo Sound region', *Antarct. J.U.S.* **5**, 15–21

DONN, W. L. and EWING, M. (1966), 'A theory of ice ages, III', *Science* **152**, 1706–12

DONN, W. L., FARRAND, W. R. and EWING, M. (1962), 'Pleistocene ice volumes and sea-level lowering', *J. Geol.* **70**, 206–14

DONN, W. L. and SHAW, D. M. (1967), 'The generalized temperature curve for the past 425,000 years: a discussion', *J. Geol.* **75**, 497–504

EARDLEY, A. J. *et al.* (1973), 'Lake cycles in the Bonneville basin, Utah', *Bull. geol. Soc. Am.* **84**, 211–16

EMILIANI, C. (1955), 'Pleistocene temperatures', *J. Geol.* **63**, 538–78

(1957), 'Temperature and age analysis of deep-sea cores', *Science* **125**, 383–7

(1966), 'Isotopic palaeotemperatures', *Science* **154**, 851–7

(1970), 'Pleistocene palaeotemperatures', *Science* **168**, 822–5

EMILIANI, C. and GEISS, J. (1957), 'On glaciations and their causes', *Geol. Rdsch.* **46**, 576–601

EPSTEIN, S., SHARP, R. P. and GOW, A. J. (1971), 'Climatological implications of stable isotope variations in deep ice cores, Byrd Station, Antarctica', *Antarct. J.U.S.* **6**, 18–20

ERICSON, D. B. and WOLLIN, G. (1968), 'Pleistocene climates and chronology in deep-sea sediments', *Science* **162**, 1227–34

ERICSON, D. B., EWING, M. and WOLLIN, G. (1964), 'The Pleistocene epoch in deep-sea sediments', *Science* **146**, 723–32

ERICSON, D. B. and WOLLIN, G. (1964), *The Deep and the Past*

EWING, M. and DONN, W. L. (1956–8), 'A theory of Ice Ages', *Science* **123**, 1061–6; **127**, 1159–62

FAIRBRIDGE, R. W. (1970), 'World paleoclimatology of the Quaternary', *Rev. Géogr. phys. Géol. dyn.* **12**, 97–104

FLINT, R. F. (1957), *Glacial and Pleistocene geology*

(1971), *Glacial and Quaternary geology*

FRECHEN, J. VON and LIPPOLT, H. J. (1965), 'Kalium-Argon-Daten zum Alter des Laacher Vulkanismus, der Rheinterrassen und der Eiszeiten', *Eiszeitalter Gegenw.* **16**, 5–30

FRITTS, H. C. (1965), 'Dendrochronology' in *The Quaternary of the United States* (ed. WRIGHT, H. E. and FREY, D. G.), 871–80

GENTILLI, J. (1948), 'Present-day volcanicity and climatic change', *Geol. Mag.* **85**, 172–5

GODWIN, H. (1956), *The history of the British flora*

GRAM, R. (1969), 'Grain surface features in *Eltanin* cores and Antarctic glaciation', *Antarct. J.U.S.* **4**, 174–5

GROSVAL'D, M. G. and KOTLYAKOV, V. M. (1969), 'Present-day glaciers in the U.S.S.R. and some data on their mass balance', *J. Glaciol.* **8**, 9–22

HOLLIN, J. T. (1962), 'On the glacial history of Antarctica', *J. Glaciol.* **4**, 173–95

(1965), 'Wilson's theory of ice ages', *Nature, Lond.* **208**, 12–16

(1969), 'Ice sheet surges and the geological record', *Can. J. Earth Sci.* **6**, 903–10

JERZYKIEWICZ, M. and SERKOWSKI, K. (1968), 'A search for solar variability' in MITCHELL, J. M. (op. cit., 1968)

KAIZUKA, S. (1961), 'Geochronology based on volcanic ejecta and its contributions to archeology in Japan', *Asian Perspective (Far east. prehist. Ass.)* **5**, 193–5

KAY, G. F. (1931), 'Classification and duration of the Pleistocene period', *Bull. geol. Soc. Am.* **42**, 425–66

KILBURN, C., PITCHER, W. S. and SHACKLETON, R. M. (1965), 'The stratigraphy and origin of the Portaskaig Boulder Bed Series (Dalradian)', *Geol. J.* **4**, 343–360

KING, P. B. (1965), 'Tectonics of Quaternary time in middle North America' in *The Quaternary of the United States* (ed. WRIGHT, H. E. and FREY, D. G.), 831–70

LAMARCHE, V. C. and FRITTS, H. C. (1971), 'Tree rings, glacial advance and climate in the Alps', *Z. Gletscherk. Glazialgeol.* **7**, 125–31

LAMB, H. H. (1969), 'Activité volcanique et climat', *Rev. Géogr. phys. Géol. dyn.* **11**, 363–80

LAMB, H. H. and JOHNSON, A. I. (1959), 'Climatic variation and observed changes in the general circulation', *Geogr. Annlr* **41**, 94–134

LE MASURIER, W. E. (1972), 'Volcanic record of Antarctic glacial history: implications with regard to Cenozoic sea levels', *Inst. Br. Geogr. Spec. Publ.* **5**, 59–74

LEVERETT, F. (1930), 'Relative length of Pleistocene glacial and interglacial stages', *Science* **72**, 193–5

LIBBY, W. F. (1955), *Radiocarbon dating* (Chicago, 2nd Ed.)

LYELL, C. (1839), *Elements of Geology* (French translation), 621. See also *Ann. Mag. nat. Hist.* **3** (1839), 323

MCDOUGALL, I. and WENSINK, H. (1966), 'Paleomagnetism and geochronology of the Pliocene-Pleistocene lavas in Iceland', *Earth planet. Sci. Letters* **1**, 232–6

MANLEY, G. (1951), 'The range of variation of the British climate', *Geogr. J.* **117**, 43–68

MATTHEWS, J. A. (1973), 'Lichen growth on an active medial moraine, Jotunheimen, Norway', *J. Glaciol.* **12**, 305–13

MESOLELIA, K. J., MATTHEWS, R. K., BROECKER, W. S. and THURBER, D. L. (1969), 'The astronomical theory of climatic change: Barbados data', *J. Geol.* **77**, 250–74

MILANKOVITCH, M. (1920), *Théorie mathématique des phénomènes thermiques produits par la radiation solair* (Paris)
  (1930), 'Mathematische Klimalehre' in *Handbuch der Klimatologie* by KÖPPEN, W. and GEIGER, R. (Berlin)

MITCHELL, G. F., PENNY, L. F., SHOTTON, F. W. and WEST, R. G. (1973), 'A correlation of Quaternary deposits in the British Isles', *Geol. Soc. Lond. Spec. Rep.* **4**, 99 pp.

MITCHELL, J. M. (ed.) (1968), 'Causes of climatic change', *Am. met. Soc., met. Monogr.* **8**(30), 159 pp.

MORRISON, R. B. (1964), 'Lake Lahontan—geology of southern Carson desert, Nevada', *U.S. geol. Surv. Prof. Pap.* **401**
  (1968), *Means of correlation of Quaternary successions* (Salt Lake City)

MOTTERSHEAD, D. N. and WHITE, I. D. (1972), 'The lichenometric dating of glacier recession, Tunsbergdal, southern Norway', *Geogr. Annlr* **54**A, 47–52

MUNK, W. and MACDONALD, G. J. F. (1960), *The rotation of the Earth* (Cambridge)

OAKLEY, K. P. (1964), *Frameworks for dating fossil man*

ÖPIK, E. J. (1968), 'Ice ages' in *The planet Earth* (ed. BATES, D. R. 2nd ed.), 164–94

PATZELT, G. (1970), 'Die Längernmessungen an den Gletschern der Österreichischen Ostalpen', *Z. Gletscherk. Glazialgeol.* **6**, 151–9

ROBIN, G. DE Q. (1966), 'Origin of the Ice Ages', *Sci. J.* **2**, 53–8

RUHE, R. V. (1965), 'Quaternary paleopedology' in *The Quaternary of the United States* (ed. WRIGHT, H. E. and FREY, D. G.), 755–64

RUTFORD, R. H., CRADDOCK, C. and BASTIEN, T. W. (1968), 'Late Tertiary glaciation and sea-level changes in Antarctica', *Palaeogeogr. Palaeoclim. Palaeoecol.* **5**, 15–39

RUTTEN, M. G. and WENSINK, H. (1960), 'Paleomagnetic dating, glaciations and the chronology of the Plio-Pleistocene in Iceland', *Rep. 21st int. geol. Congr. (Copenhagen)* **4**, 62–70

SELBY, M. J. (1973), 'Antarctica: the key to the Ice Age', *N.Z. Geogr.* **29**, 134–50

SERNANDER, R. (1908), 'On the evidence of Postglacial changes of climate furnished by the peat-mosses of northern Europe', *Geol. För. Stockh. Förh.* **30**, 465–78

SHARP, R. P. (1968), 'Sherwin till–Bishop tuff geological relationships, Sierra Nevada, California', *Bull. geol. Soc. Am.* **79**, 351–63

SHOTTON, F. W. (1967), 'The problems and contributions of methods of absolute dating within the Pleistocene period', *Q.J. geol. Soc. Lond.* **122**, 356–83

SHUMSKIY, P. A. (1970), 'The Antarctic ice sheet', *Int. Symp. Antarct. Glaciol. Exploration* (Hanover, N.H., 1968), *Int. Ass. scient. Hydrol., Publ.* **86**

SHUMSKIY, P. A., KRENKE, A. N. and ZOTIKOV, I. A. (1964), 'Ice and its changes', *Res. Geophys.* **2**, 425–60

SIMPSON, G. C. (1934), 'World climate during the Quaternary period', *Q.J.R. met. Soc.* **60**, 425–78

(1957), 'Further studies in world climate', *Q.J.R. met. Soc.* **83**, 459–81

(1959), 'World temperatures during the Pleistocene', *Q.J.R. met. Soc.* **85**, 332–49

SPARKS, B. W. and WEST, R. G. (1972), *The Ice Age in Britain*

STEARNS, C. E. and THURBER, D. L. (1967), 'Th^{230}/U^{234} dates of late Pleistocene marine fossils from the Mediterranean and Moroccan littorals', *Progr. Oceanogr.* **4**, 293–305

STOKES, M. A. and SMILEY, T. L. (1968), *An introduction to tree-ring dating* (Chicago)

Symposium of glacier mapping, Can. J. Earth Sci. **3**(6), 737–901

THORARINSSON, S. (1944), 'Tefrokronologiska studier på Island', *Geogr. Annlr* **26**, 1–203

(1949), 'Some tephrochronological contributions to the vulcanology and glaciology of Iceland', *Geogr. Annlr* **31**, 239–56

TUREKIAN, K. K. (ed.) (1971), *The Late Cenozoic glacial ages* (New Haven)

VALENTIN, H. (1952), 'Die Küsten der Erde', *Petermanns Mitt., Ergänz.* **246**, 118 pp.

WEERTMAN, J. (1964), 'Rate of growth or shrinkage of non-equilibrium ice sheets', *J. Glaciol.* **5**, 145–58

(1966), 'Effect of a basal water layer on the dimensions of ice-sheets', *J. Glaciol.* **6**, 191–207

WEST, R. G. (1968), *Pleistocene geology and biology*

WILCOX, R. E. (1965), 'Volcanic-ash chronology' in *The Quaternary of the United States* (ed. WRIGHT, H. E. and FREY, D. G.), 807–16

WILSON, A. T. (1964), 'Origin of ice ages: an ice shelf theory for Pleistocene glaciation', *Nature, Lond.* **201**, 147–9

(1966), 'Variation in solar insolation to the South Polar region as a trigger which induces instability in the Antarctic Ice Sheet', *Nature, Lond.* **210**, 477–8

(1969), 'The climatic effects of large-scale surges of ice-sheets', *Can. J. Earth Sci.* **6**, 911–18

WOLDSTEDT, P. (1954), 'Die Klimakurve des Tertiärs und Quartärs in Mitteleuropa', *Eiszeitalter Gegenw.* **5**, 5–9

2

Glacier régimes

Consider, then, the glaciers as the outlets of the vast reservoirs of snow...—as icy streams moving downwards, and continually supplying their own waste in the lower valleys, into which they intrude themselves like unwelcome guests, in the midst of vegetation, and to the very threshold of habitations.... (J. D. FORBES, 1843)

Glacier régimes are concerned with the loss and gain of snow in a glacier. When the glacier or ice sheet has a positive régime, it is gaining ice and will advance and thicken, while the reverse applies when the régime is negative. Glacier régimes are intimately related to climate and climatic changes, and therefore to changes in sea level relative to the land. They also provide important hydrological data. From the point of view of glacial geomorphology, the régimes of glaciers are important in that they affect the behaviour of the ice. They can be directly related to the formation of moraines, which may mark stages of halt and readvance, trim-lines, some meltwater channels and other features.

First, some terms will be defined and the methods of obtaining glacier régimes will be mentioned. Some characteristic glacier régimes in different climates and environments will then be discussed. Then the effect of climatic change on glaciers will be considered in order that these factors may be correctly related.

1 Definitions

The two essential factors that make up the régime or mass balance of a glacier system are the accumulation and the ablation, the gain and loss of mass respectively. (For a standard terminology and definitions of mass-balance terms, see IASH, 1969.) The term *accumulation* includes all those processes by which solid ice and snow are added to a glacier or ice sheet. *Ablation* refers to all the processes by which ice and snow are lost from the glacier. Accumulation on the glacier is brought about mainly by the precipitation of snow. Rain, if it freezes on impact, can also be included. Refreezing of liquid water and condensation of ice direct from vapour (sublimation) also cause accumulation, and material may be added to the glacier by windblown snow or avalanches. The processes that cause ablation include melting, evaporation, calving, wind erosion, and removal of snow or ice by avalanches. One of the problems of measuring accumulation and ablation is that the processes operate not only on or near the surface but also within and beneath the ice. Changes deep within and beneath the ice are usually ignored as they are small compared with the surface and near-surface changes, except on floating ice shelves or glacier tongues.

The budget year is the most important unit of time when glacier mass balance studies are being considered. The budget year runs from the time when ablation has reached its maximum extent after the summer season of one year until the similar state in the following year. This period need not be an exact calendar year and will vary over the glacier surface. The accumulation and ablation are recorded in water equivalent (in cm³ or m³) over the whole glacier surface, or in cm or m of water equivalent at a point. In order to measure the water equivalent, it is necessary to know the density of the snow or ice.

The terms used in mass balance studies include the following:

1 *Gross annual accumulation* is the total volume of water equivalent added to a glacier during the budget year.
2 The *net annual accumulation* is the amount of water equivalent added to the glacier and still present by the end of the budget year. This material is limited to the accumulation area of the glacier in which there is a surplus at the end of the budget year.
3 The *gross annual ablation* is the total amount of water equivalent of snow and ice consumed during the budget year by all processes of ablation.
4 The *net annual ablation* is the volume of water equivalent actually lost from the glacier during the budget year. The difference between the gross and the net ablation is made up by such processes as refreezing of meltwater on or within the glacier, the melting of temporary accumulation in the ablation area, and the internal storage of water.

In the ablation area all the material gained that year is lost before the end of the budget year. This loss in the case of temperate glaciers takes place largely by surface melting. The melting occurs mainly in the lower parts of the glacier during the summer ablation season, although some ablation may go on at low levels throughout the whole of the year in some glaciers. In other areas, however, such as the Antarctic ice sheet, calving is the most important process of ablation.

The line dividing the accumulation area from the ablation area is called the *equilibrium line* (Fig. 2.1). This line should not be confused with the *firn-line*, or *annual snowline*. The two lines have sometimes been confused or regarded as identical, although in fact they may not coincide. The firn-line is the uppermost line on the glacier to which the snowfall of the winter season melts during the summer ablation season. It is a clearly marked line, separating hard blue ice below from snow above in many glaciers. However, in some glaciers there may be an accumulation of dense ice between the firn-line and the equilibrium line, which lies at a lower level. This dense ice is also called *superimposed ice* and is formed by the refreezing of snow-melt water or rain. In high arctic glaciers in particular there is often a complex transition zone between the accumulation and ablation zones.

R. M. Koerner (1970) shows how superimposed ice on the Devon Island Ice cap occurs in two zones, one continuous and the other discontinuous. An annual increment of superimposed ice may consist of several layers according to the melting and freezing conditions. There is no valid method of distinguishing annual layers under these conditions. An increase in mean crystal size and in the standard deviation from the mean was found with decreasing altitude between the firn edge and the equilibrium line. This pattern was reversed below the equilibrium line, so that this characteristic can be used to determine long-period variations in the level of the equilibrium line. The zone of superimposed ice on the Devon Island ice cap mostly lies between 1300 and 1500 m elevation in the north and west of the ice cap, but is lower in the south.

The problems of the transition zone are discussed by F. Müller (1962) in a study of the glaciers on Axel Heiberg Island. A number of zones with different characteristics can be identified. The *dry snow zone* is the highest and always has a temperature below 0°C, so that there is no penetration of meltwater. The *percolation zone* lies below this zone, and is divided into two parts. In the upper one, *zone A*, percolation does not reach to the previous year's summer surface, although the temperature does rise to 0°C below the surface and meltwater can penetrate a little way. In *zone B* some of the current season's meltwater penetrates and refreezes to form ice lenses in the earlier accumulation. The 0°C line sinks below the previous summer surface. Below zone B lies the *slush zone*, from which material can be lost by slush avalanches. These avalanches are caused by the saturation of the snow above the impermeable superimposed ice layers and can cause the removal of material on gentle slopes of only 3 or 4 degrees. The upward limit of this zone is called the *slush limit*. The slush avalanches also carry away some of the superimposed ice and even the older accumulation beneath it below the equilibrium line. The older accumulation of the previous summer can be separated from the accumulation of the current season by the development of *depth hoar* between

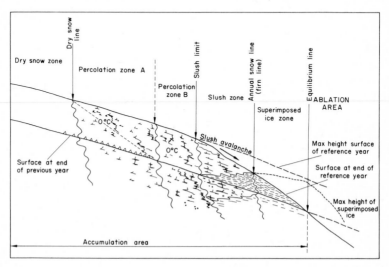

Fig. 2.1
Schematic zonation of the accumulation area   (F. Müller, *J. Glaciol.*, 1962, by permission of the International Glaciological Society)

the two layers. The depth hoar consists of large cup-shaped or bar-like crystals and results from changes of snow character with seasonal weather changes. Some of these zones are shown in Fig. 2.1.

2 Mass balance studies

The mass balance or régime of a glacier can be assessed in two ways. It is equal to the gross annual accumulation less the net ablation during the budget year. Alternatively, it is equal to the net annual accumulation in the accumulation area at the end of the budget year less the dense ice lost from the surface exposed at the end of the previous budget year in the ablation zone. The quantities required for the second method are often easier to measure. Details of the measurements necessary to establish the glacier budget are given in the Technical Note No. 1 of the Glaciological Research Sub-Committee of the Glaciological Society.

A detailed study of the problems of establishing the mass balance of a glacier or ice sheet has been given by M. F. Meier (1962). The budget quantities that he gives are measured in a vertical direction so that they can be readily related to area on a map. The results are given in water equivalent values, which means that it is necessary to observe the density of the ice or snow. In the accumulation area, where the annual accumulation is in the form of snow or firn, a known volume of material must be obtained and its weight gives the density. The water equivalent values can be given for each unit area. Meier uses the term 'specific mass budget' to define the budget at a single point. If these values are integrated over the whole area of the glacier, the total mass budget can be obtained.

Accumulation and ablation vary with both time and space over the whole glacier system. The accumulation reaches a peak at any one point during the early part of the budget year and may fall off to small or zero values during the latter part of the year, during the summer season. The ablation changes, on the other hand, have an almost opposite trend. The difference between the curves for accumulation and ablation shows the specific mass budget at any one point. The value can also be shown cumulatively. This gives the net specific budget throughout the budget year.

The change of the specific budget with altitude is important in calculating the total budget. In the lower part of a mountain glacier, the specific budget will be negative, while in the upper part it will be positive. A direct comparison of these values will not give a true idea of the total budget unless the area of the glacier at the different altitudes is taken into account. Budget values for different altitudes must be related to the areas between the appropriate contours. The total mass budget can be relatively accurately assessed by using these values. An important point on a curve showing the variation of either the specific or total budget with altitude is the point at which the curve crosses from the positive upper part to the negative lower part (Fig. 2.2). This point indicates the position of the equilibrium line, separating the ablation and accumulation areas. L. R. Mayo, M. F. Meier and W. V. Tangborn (1972) discuss two systems of mass balance calculation: the stratigraphic system and the annual or fixed-date system. The former refers to annual maxima and minima at points on the glacier, while the latter refers to values at the beginning and/or end of a hydrological year.

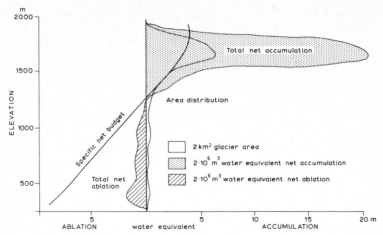

Fig. 2.2
The mass balance of Nigardsbreen for the budget year 1961–2, when the surplus accumulation amounted to 94·9×10⁶ m³ water equivalent (G. Østrem, *Norsk geogr. Tidsskr.*, 1961–2, Universitetsforlaget, Oslo)

Most of the quantities measured refer to summer surfaces, and these correspond to annual minima at the points measured. The integration of the values over the whole glacier may be meaningless, because the values refer to different times at different places. A combined system, using both methods, can provide area-average results that can be related to hydrological and meteorological data.

Measurements to summer surfaces at the beginning and end of a hydrological year are summed with proper reference to the type of material involved, for example, old firn, ice, snow and superimposed ice of the year under study, new firn formed in that year, and late snow deposited towards the end of the year. Other 'balance increment' terms relate values at the beginning and end of a hydrological year to appropriate area-average balance minima. The two types of 'net balance' are defined in this way and other terms are given precise meaning. The scheme is illustrated in Fig. 2.3 and is versatile. It can be used on any glacier, although summer surfaces cannot be defined on glaciers in which ablation or accumulation is continuous throughout the year. The key to the combination of the two systems is recognition of the type of material involved.

In a glacier system that has a high rate of accumulation and also a high rate of ablation, the rate of flow of the ice must be considerable. This applies, for example, to the Franz Josef glacier, which flows to within about 200 m of sea level in a relatively low latitude. Its accumulation area lies in an area of very heavy precipitation on the western side of the Southern Alps of New Zealand, while its snout penetrates down into the coastal rain forests. At the other extreme are some of the Antarctic glaciers that receive a very small quantity of precipitation, owing to the extreme cold, and whose main method of ablation is by calving, as temperatures rarely rise to freezing-point. These glaciers move extremely slowly (Chapter 4). The Ferrar glacier near the Ross Sea, for example, moves at only about 5 cm/day although it is a very large glacier, and probably, similar rates of motion, up to a few tens of metres a year, characterize

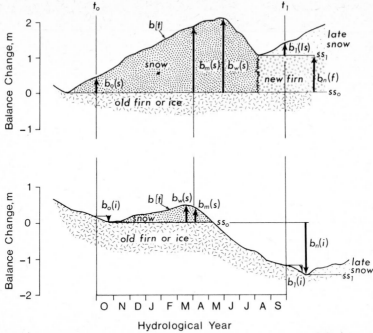

Fig. 2.3
Balance quantities as measured at specific points on a glacier using a stratigraphic system referenced
to summer surfaces. Shown here are typical balance curves $b\,[t]$ for points in the accumulation area
(top) and in the ablation zone (lower diagram). The vertical scale is in metres of water equivalent,
with an arbitrary zero. The zero may be placed in any other position depending on the needs of
the analysis. Balance terms that increase with time, or that represent an increment of mass added
to the system, are considered positive (arrows pointing up). Measurements at the beginning (t_o)
and end (t_1) of the hydrological year, in spring, and at a time after the hydrological year (perhaps
next spring) define these quantities:

$b_o\,(s)$	the initial snow balance
$b_o\,(i)$	the initial ice balance
$b_m\,(s)$	the measured winter snow balance
$b_1\,(ls)$	the final late snow balance
$b_1\,(i)$	the final ice balance
$b_n\,(f)$	the net firnification
$b_n\,(i)$	the net ice balance

Also shown is $b_w\,(s)$, the maximum winter snow balance, which is usually not measured, being
dependent on knowledge of $b\,(t)$ at each point; note that it occurs at different times at different
points. Note the formation of summer surfaces (ss) and how they form at different times at the two
points represented by the diagrams (L. R. Mayo *et al.*, *J. Glaciol.*, 1972, by permission of the International Glaciological Society)

about half the coastline of Antarctica (G. de Q. Robin, 1972). An improvement of the
climate, with higher temperatures, would allow the air to hold more moisture, while
still remaining below freezing-point. This would increase the snowfall and thus the
glaciers would become more active.

The size of a glacier budget relative to the glacier area gives a good indication of
the state of activity of the glacier. If the budget is large relative to the glacier area,

the ice will flow fast through it. The reverse will also apply. In both instances if the budget is balanced, the glacier will maintain its volume and position.

2.1   *Measurement of accumulation*

It is generally easier to measure net accumulation than gross accumulation. The former measurement also has the advantage that accumulation of several seasons can be measured at one time. This can be achieved by digging pits at suitable positions in the accumulation area. These give accurate values for the accumulation of several years at the positions and altitudes chosen. Then the total net accumulation can be reasonably accurately calculated from the different specific values obtained from the individual pits.

In analysing the results obtained from pit observations, it is clearly necessary to be able to establish the boundaries between different budget years of accumulation. Sometimes this can be done fairly easily if dirt layers have accumulated on the snow surface at the end of the budget year, but this criterion is not always accurate. The summer snow also usually hardens at the surface where the upper layers of snow have partially melted and refrozen. This gives a thin hard crust that can be identified in the pit profile. The formation of depth hoar in very cold climates has also been mentioned.

Where the accumulation is heavy, pits must be deep to penetrate through a number of budget years of accumulation. For example in Iceland an annual accumulation of 2·38 m was recorded in a pit dug near the southern margin of the Vatnajökull ice cap at 1200 m. A pit 6 m deep in this area, therefore, can only penetrate through about two complete accumulation layers, especially if the pit is dug early in the ablation season. In this area, the budget years can be fairly readily distinguished by dust layers as the pit was sited fairly close to the rock rim of the ice cap.

The reverse situation is found in the Antarctic where accumulation is very small. The individual layers are difficult to identify owing to the lack of summer melting. Here the average accumulation is about 30 to 60 cm, but this amount is gradually reduced by compression to layers 10 to 20 cm in thickness. A pit 6 m deep in this area would cover somewhere between 20 and 50 years of accumulation.

Another difference between these two areas is the density of the snow in the pits. The density of the Antarctic snow is much less than that in more temperate areas. In the Antarctic, it varies between 0·3 and 0·45 over a period of three years' accumulation, the density being least at the beginning of the budget year. In the Iceland pit, on the other hand, the density varied between 0·55 and 0·6 in snow of the previous season. This is caused by the greater pressure of overlying material and much more active melting.

The density within any one budget year does not necessarily vary in the same way with depth in all climates. For example, in the Austrian Alps the density at one point was found to be at a minimum at the bottom and reached a maximum somewhere in the centre of the annual accumulation layer. Changes in crystal size and colour allowed the beginning of the budget year to be identified in this area. One of the characteristics of very mountainous areas of this type is the very uneven nature of the accumulation. Many pits must be dug to obtain a good value for the total net accumulation in these conditions. However, where the accumulation area is broad and even, as in the case of the Vatnajökull plateau, it is only necessary to dig representative pits in one or two positions at each level.

In order to calculate the mass budget of a glacier, the net accumulation is the only accumulation value that it is essential to obtain. Methods of obtaining the gross accumulation on a glacier are more complex and will not be discussed.

2.2 *Ablation measurements*

The net accumulation must be equated with the ablation of dense ice to obtain the mass balance. There is no need to measure the ablation in the accumulation area in order to obtain the mass balance. However, when the net accumulation is measured in pits during the summer season, the figures obtained will refer to past budget years, while the measure of ablation of solid ice in the ablation area will refer to the current year. This value of ablation must be balanced against the net accumulation found in the following summer season or at the end of the current budget year. The loss of dense ice in the ablation area can be found by continuous measurements of ablation during the whole of the ablation season. These measurements must start while the glacier is still covered with winter snow by drilling stakes down into the solid ice beneath. The depth of snow can then be recorded to indicate the position of the ice surface. However, the level of the ice surface may rise as the season progresses owing to the addition of frozen meltwater beneath the snow. This must be allowed for before the loss of older ice starts and net ablation begins. Some observations of ice density must be made to obtain the water equivalent of the ice lost during the ablation season.

A large number of ablation stakes must be inserted to give an accurate value of the loss of dense ice. The stakes need frequent redrilling during the height of the ablation season when up to 15 cm or more of ice may be lost from the surface of temperate glaciers in one day near their snouts. For example, stakes drilled to a depth of 7·5 m about the middle of the ablation season in 1955 had completely melted out by the early part of the following ablation season near the snout of Austerdalsbreen in Norway. It has been estimated that about 12 m of ice are lost from near the snouts of many Icelandic, Alaskan and Norwegian glaciers during one ablation season. Ablation on Nigardsbreen in Norway has been known to exceed 20 m in one summer.

On the whole, ablation does not vary annually as much as accumulation, although it can vary considerably locally over a glacier surface. It is great near rock walls and where warm, dry katabatic winds blow down the glacier from higher levels. Where there are waves on the glacier surface, the ablation tends to be greater on the crests of the waves, as on Austerdalsbreen, Norway. The amount of ablation often decreases linearly with height and also varies greatly with the season. It rises sharply in the northern hemisphere to a maximum in the first half of July and falls off rapidly again to low values in early September, when the rate resembles that of mid-May.

Another useful observation to make in connection with ablation and mass balance studies is the measurement of the discharge of streams issuing from the glacier. This includes both meltwater and rain falling on the glacier basin during the summer where this does not freeze before reaching the glacier snout. The discharge cannot give the total ablation as ice lost by evaporation is not included in the meltwater, but for hydro-logical purposes the amount of meltwater is important. It is closely associated both with the glacier mass régime, as discussed by G. Østrem (1963), and with the river régime of the meltwater stream. A measurement of the load of the meltwater stream also gives a useful indication of the effectiveness of glacial erosion by the glacier con-cerned (Chapter 11).

Statistical analysis has shown that any particular glacier has a characteristic vertical net-budget gradient, vertical ablation gradient and equilibrium line altitude (R. E. Dugdale, 1972). The net-budget gradient is the slope of the net budget curve at the equilibrium line. The ablation gradient is the straight line that best fits the ablation curve. V. Schytt (1967) has studied the ablation gradient of Storglaciären, and has shown that there is considerable annual variation, with a complicated dependence on meteorological parameters. There is no clear relation between ablation gradient and latitude according to studies carried out in the Arctic from Scandinavia to Canada. The degree of continentality exerts a strong influence on the ablation gradient. Total ablation on Storglaciären correlates much more closely with mean summer temperature at 1130 m than does the ablation gradient. Other significant variables are the length of the ablation season and radiation conditions. In the maritime area of western Norway with a glacial limit of 1300 m, the maximum values of the ablation gradient are 100 cm/100 m, while the minimum values of about 15 cm/100 m occur in Graus-breen. Here, the glacial limit in the continental climate of the east is very high at more than 2200 m.

3 Glacier mass balance

The mass balance of the glacier depends on the balance between accumulation and ablation. Equilibrium is rarely achieved for long and the problem of the relationship between the mass balance and the movement of the glacier is a complex one. It involves changes in the volume, form and flow. Variations in régime may be short-term fluctua-tions and many areas vary very much from year to year. The glacier does not respond at once to these short-term annual changes as the variations often cancel each other out in succeeding years before the glacier can react to them. Rather more prolonged trends, such as a succession of warm or snowy years, are required before the glacier responds to these climatic variations by a change in form.

The characteristics of the glacier snout can be used to estimate the budget of an unknown glacier. An advancing glacier with a positive budget normally has a steep or vertical front. A glacier in equilibrium has a front of moderate gradient and one that has a negative budget and is retreating generally has a gently sloping front. This gentle slope often becomes buried under morainic debris. When this cover becomes thick, the ice disappears largely by slow down-wasting and the exact end of the glacier may be difficult to determine. This difference of frontal types enables an aerial survey to be used to distinguish glaciers in various states of mass balance in areas where no detailed work has been undertaken. An example of three adjacent glaciers each of which shows clearly one of the three states mentioned is the Three Congruent Glacier of the St. Elias Range in Alaska (E. LaChapelle, 1962). As this example shows, there are often interesting variations between neighbouring glaciers in their response to similar climatic changes. These depend on the characteristics of the individual glacier basins, such as the ratio of the accumulation area to ablation area and altitude distribution. Before considering the response of glaciers to changes in climate, a few examples of specific mass balance studies in different climatic areas will be considered. The examples are arranged in order of latitude, beginning with the highest latitudes.

3.1 *Antarctic*

The Antarctic continent contains by far the greatest amount of ice and snow in the world (M. Mellor, 1961). Its ice sheet covers about 12·6 million km², an area seven times that of Greenland's ice sheet and one and a half times the area of the United States. Soviet exploration has shown that the surface of the ice rises in a dome to just over 4000 m at latitude 81°S, longitude 78°E (Robin, 1972). The mean thickness of the ice is about 2000 m, or possibly as much as 2500 m. If all the ice were to melt, sea level would rise by about 61 m. This amount would be reduced to about 40 m if isostatic recovery of the continent were allowed for.

In the last decade, observations of accumulation, ablation and flow in the Antarctic have increased considerably, but there is still some doubt concerning its mass balance. The accumulation over most of the ice sheet is very low and about half its area receives net accumulation of less than 10 cm of ice a year, a value which, in terms of its water equivalent, is only equalled in the arid sub-tropics. Density only increases slowly with depth in the ice sheet. Near the surface, the firn has a density of between 0·3 and 0·4. The density increases to 0·82 at a depth between 50 and 100 m. This density increase is accompanied by a downward increase in the size of the crystals, while the enclosed bubbles of air become correspondingly smaller. At a depth of 300 m, it is estimated that in one area the ice is 1600 years old.

A new method has had to be developed to establish the seasonal layers of accumulation both in the Greenland and Antarctic ice sheets, because in these areas dirt layers are largely absent and changes in the texture of the ice are small. The method depends upon the proportion of the isotopes oxygen 18 and oxygen 16 as explained on p. 15. These measurements can also give some indication of actual temperatures when the snow, now found as ice deeply buried in the ice sheet, was originally deposited on the surface. Dating of ice layers has also been carried out by C^{14} analysis of the carbon-dioxide bubbles trapped in the ice. One result gives an age of 3100 years for a sample of ice in Greenland.

Only one borehole has yet penetrated the full thickness of the inland Antarctic ice. It has yielded some very significant results, and it is worth summarizing the main findings at this point (A. J. Gow *et al.*, 1968; Gow, 1970; H. T. Ueda and D. E. Garfield, 1970). The location of the borehole, at Byrd Station, was 80° 01′S, 119° 31′W, elevation 1530 m. The ice was 2164 m thick and more than 99 per cent of the core was recovered. The lowest temperature recorded was −28·8°C at 800 m, and at 1800 m the value was −13°C. Liquid water indicative of pressure melting was encountered at the bottom, possibly as a basal water layer of the order of 1 mm thick. The base of the borehole reached bedrock. Dirt was abundant in the lowest 4–5 m of the core, including bands of silt, sand and pebbles, all interspersed with ice. Some large fragments up to 5 cm across were found, including granite, the nearest outcrop of which is 360 km away. Between 1300 and 1700 m several thin layers of dirt (possibly volcanic ash) up to 0·5 mm overlain by up to 10 mm of refrozen meltwater occurred. The melting probably resulted from radiational heating associated with the dirt, and no other signs of melting were found in the core. At 906 m the ice density was 0·9198, and at 56 m it was 0·8400, a brittle layer occurring between 400 and 900 m. The age of the lowermost layers of the ice is estimated to be greater than 50,000 years, and may well be as old as the last interglacial.

Accumulation in the Antarctic is difficult to measure owing to the large amount of drifting in the high winds characteristic of the area. Nevertheless there are many estimates on this positive side of the budget. M. Mellor (1959) has discussed the accumulation in the Mawson area. The results suggest very uneven addition of material, varying from a minimum of 2 cm water equivalent a year 160 km from the coast to a maximum of 12 cm water equivalent at 130 and 230 km inland. These are very low values and estimates from other areas give rather higher figures. Values will be given in cm water equivalent per annum. P. A. Shumskiy (in M. Mellor, 1959), for example, gave a value of 85 cm within 60 km from the coast at Mirny, falling to 20 to 30 cm at 70 to 450 km inland and 8 cm at 850 km inland. In other parts of the continent, a value of 15 cm for the area from 100 to 600 km inland has been given. Results of the Norwegian-British-Swedish expedition give 36·5 cm for the shelf, 25 cm for the plateau slopes and 12 cm for the edge of the high plateau of Dronning Maud Land. Ice-shelf accumulation in Wilkes Land is estimated at 35 to 40 cm. An estimate by P. A. Siple (1958) for the South Pole gives 5 to 6 cm for 10 months February through November. A. J. Gow (1965) gives 20 cm/year firn or 7·4 cm water equivalent at the Pole.

J. T. Hollin (1962) has assembled the available data on accumulation in the Antarctic. For the Indian Ocean sector, he suggests a rapid increase in the rate of gross accumulation from zero actually at the coast to a maximum of nearly 30 g/cm^2 near the coast. From this point, the value falls gradually to a latitude of 70°S and thereafter more rapidly to very low values of less than 10 g/cm^2 around 86°S in the centre of the continent. More recently, accumulation rates recorded by L. D. Taylor (1971) gave 6·8 g/cm^2 at the South Pole and 10·8 g/cm^2 near the Queen Maud Mountains. R. M. Koerner (1971) recorded only 2·8 g/cm^2 for the Plateau Station at 79° 15′S, 40° 30′E, and showed that there has been little variation in accumulation over the last 127 years. The central sector of the ice cap is one where accumulation is reduced by strong and persistent katabatic winds, so that the figures are conservative. F. Loewe (1960) gives a tentative value of accumulation for the area without shelves ranging from a minimum of 10 cm/year to a maximum of 12 cm/year. When the ice shelves are included, he gives a minimum value of 10·4 cm and a maximum of 13 cm/year. On the shelves, the minimum is 15 cm and the maximum 25 cm/year. The total positive side of the budget using these figures and taking the area of the ice sheet into account is a minimum without shelves of 127×10^{10} t (metric tonnes) and a maximum with shelves of 178×10^{10} t.

Against these rather variable estimates of the accumulation on the Antarctic continent and its ice shelves must be set the estimated loss of material by the various processes of ablation. The Antarctic ice sheet differs from others in that a considerable part of its loss is provided by calving of the large ice shelves around the coast. The largest is the Ross ice shelf (Fig. 3.13), and the Filchner ice shelf in the Weddell Sea is also very extensive. From these floating shelves, pieces break off to form large tabular icebergs. Melting, on the other hand, is one of the least important methods of loss in the Antarctic, where temperatures are low and ablation is restricted to favourable localities. One method by which snow is removed from the land is by wind drifting. Mellor (1959) has estimated that the annual mass transport by the wind at Mawson is $0·22 \times 10^{14}$g/ km/year. The estimates of Loewe (1960) of the loss of snow by this process range between a minimum value of -2×10^{10} t and -50×10^{10} t. Ablation also takes place by evaporation throughout the year. Melting only occurs locally during December and January. Measurement of ablation rates of snow during three years at 60 m elevation in the Mawson area gives figures varying between 53·5 and 79 cm/year. Values fall to lower rates with increasing altitude, for example 22 cm at 365 m. Hollin (1962) suggests that ablation as a whole is extremely small away from the coast, but that at the coast it may be double the value given for accumulation.

The mass balance of the small Meserve Glacier ($9·9$ km^2) has been discussed by C. Bull and C. R. Carmein (1970). Here, the annual loss of ice amounted to 60×10^6 kg or $0·61$ g/cm^2, while the snow accumulation was less than normal in the year observed, so that the budget is probably usually positive. In the year under discussion the balance was nearly in equilibrium. The loss by ablation on the tongue, which has an area of $1·8$ km^2, was 34 g/cm^2 at 440 m near the snout. Only 2 to 3 per cent of the total loss was by meltwater runoff, dry calving from marginal cliffs accounted for $1·5$ per cent of the total mass loss, while 40 per cent was lost by evaporation and the rest by sublimation.

The most effective form of loss from the main ice sheet is by flow offshore followed by calving. Against this loss must be set the possible accretion of ice below the shelves. There is, however, doubt as to whether the shelves lose or gain underneath. It seems likely that melting from below dominates (R. H. Thomas and P. H. Coslett, 1970) except beneath the thinner parts of the ice shelves. F. Debenham in 1948 showed that in a few places the shelves gain by freezing on their lower side. Such a gain would help to account for the headless fish and mirabilite deposits which have been found resting on the surface of the shelf. In these areas of unusual relief conditions, where rock outcrops occur, the surface is melting and the underside of the shelf is freezing.

The removal of mass by iceberg formation, both from the continental ice sheet and the glaciers, but excluding the shelves, has been estimated by Loewe to be between a minimum of -14 and a maximum of -30×10^{10} t. The greater part of this loss is supplied by the glaciers, because some of these move with speeds of several hundred metres a year, while the ice sheet as a whole only moves at a velocity of about 30 m/ year. The marginal ice shelves, however, supply the greatest amount of loss by this process of calving. Loewe estimates a loss of between -36 and -65×10^{10} t from the ice shelves. These move at 300 to 500 m/year (other estimates are as high as 2000 m/ year; Robin, 1972), and have a thickness of between 190 and 200 m; calving takes place from an estimated length of about 7000 km.

The mass balances produced by various workers vary as much as the constituent ingredients. On the whole, however, most of the estimates indicate that the ice sheet may have a positive mass balance. Mellor (1959) suggests an annual gain of 1.7×10^{12} t and a loss of 0.69×10^{12} t. Loewe's (1960) budget varies between $+88 \times 10^{10}$ t for minimum values of all the different items to $+30 \times 10^{10}$ t for a maximum value for all the items. Of eight values of the mass balance, one is negative, two suggest equilibrium, and the four most recent all show a fairly large positive budget. The average value of these four estimates is $+1.10 \times 10^{12}$ t/year.

In view of the probable dominance of calving on the negative side of the budget, it is important to note that the movement of the Antarctic ice is probably not as small as was thought at one time. Some velocities have already been mentioned and Hollin (1962) is of the opinion that the mean velocity around the periphery of the ice sheet is of the order of hundreds of metres a year. In the case of a few discharging glaciers, movement may approach 1 km/year, while the two major ice-shelf areas, as noted above, may move at even greater rates.

The ice régime of the eastern part of the Ross ice-shelf drainage system has been described by M. B. Giovinetto and J. H. Zumberge (1967). The net accumulation is estimated at $219 \pm 46 \times 10^{15}$ g/year. The ice discharge from the shelf is about $168 \pm 18 \times 10^{15}$ g/year, and the total ice régime is estimated at $133 \pm 50 \times 10^{15}$ g/year, which indicates a probably positive net budget. It seems that a positive budget is likely for the whole of the Ross ice-shelf system, and also for the inland ice of west and east Antarctica. Errors in the estimated thickness and velocity of the ice may be substantial, particularly in the case of the ice discharge value. Lack of information on subglacial melting and freezing under the shelf is another source of uncertainty. The input of glacier ice to the shelf is $48 \pm 15 \times 10^{15}$ g/year: the large error term is because of uncertainty concerning the ice thickness. The net freezing is estimated at 20×10^{15} g/year under the shelf, which is a minimum estimate.

An important point concerning the fluctuations of the ice sheet is the grounding line beyond which the ice moving seaward starts to float (p. 30). There is reason to believe that the Antarctic ice sheet is thinning slightly at present in phase with the northern hemisphere retreat. However, it can be argued that warmer conditions in the Antarctic are liable to lead to increasing accumulation, but little change in ablation would be expected. This is the basis of some hypotheses that have suggested that the advances and retreats of the northern and southern hemispheres could be out of phase. But there is evidence that the changes are in fact in phase, as Hollin (1962) shows.

He relates this correspondence in phases of retreat and advance to changes in sea level. When the northern glaciers and ice sheets fluctuate, which they can do much more readily than the Antarctic owing to their ending on land and their more temperate environment, sea level fluctuates with them. A fall of sea level accompanying increasing northern hemisphere glaciation causes the grounding line of the Antarctic ice sheet to advance seawards. Thus the whole profile of the Antarctic ice sheet can shift seaward. A considerable change in thickness, amounting to 1230 m, would occur at the original grounding line if the grounding line advanced 90 km. This distance depends on the slope of the sea floor and the extent of the lowering of sea level. To obtain the result given above, the slope was taken to be 0·1 degrees and the fall of sea level 150 m.

The reverse would occur with a rise of sea level. The profile of the ice sheet would retreat inland, causing a lowering of ice level near the margin. The change in thickness is rapidly reduced inland and depends on the existence of a stable profile. It seems likely that this process keeps the Antarctic in phase with northern hemisphere fluctuations. These changes must be initiated in the north and are then followed in the Antarctic because of the resulting changes of sea level. Causes of major glacial fluctuations must, therefore, be sought in the northern hemisphere.

It seems likely that the Antarctic has been ice-covered throughout the Pleistocene, and there is indisputable evidence that a major ice cap existed here, from time to time, during the Tertiary era (see pp. 10 and 22). It may have become fragmented at times in the earlier Tertiary, or even disappeared temporarily in milder intervals, but though it may have fluctuated in size, the ice cap is unlikely to have entirely disappeared in later Tertiary time (J. T. Hollin, 1970). There is no prospect of its disappearance in the foreseeable future, owing to the low rates of ablation. It is likely that the Antarctic ice, although frozen to its bed in the peripheral zones, is at pressure melting-point in parts of the interior. This partly arises from the small accumulation, which prevents any rapid downward transference of cold from the surface. More detailed discussion of the temperature distribution will be given in the next chapter (see, for example, Fig. 3.7).

The possibility has been considered by several workers that, if the thickness of the ice sheet were to increase substantially, the base of the ice might attain pressure melting-point over a much wider area. The ice might then be able to surge forward and extend rapidly, perhaps as far as 50°S, in extensive shelves, causing a sudden rise in world sea level because of the displacement of ocean water. As discussed on page 32, A. T. Wilson (1964) has utilized this idea of massive surging of the Antarctic ice in his theory of the causes of ice ages.

3.2 *Greenland*

The mass balance of the Greenland ice sheet has been worked out by A. Bauer (1955). The total area of the ice sheet is about 1.7264×10^6 km². The greatest area, 18 per cent, lies between 2440 and 2740 m and 6·5 per cent lies between 3050 and 3390 m. The firn-line is at about 1400 m and 83 per cent of the total area lies in the accumulation zone above this elevation. The estimated budget consists, on the positive side, of accumulation at a mean rate of 0·31 m of water equivalent, giving a total of 446 km³ of water equivalent a year. The ablation is estimated at 1·1 m water equivalent or 315 km³ of water equivalent. The discharge from glaciers represents 240 km³ of ice and is divided between 10 km³ of ice in north Greenland, 90 km³ of ice on the west coast, 120 km³ on the east coast and 20 km³ of ice in Melville Bay. The total deficit is about 100 km³ water equivalent for the whole ice sheet. The accuracy of these values is, on the whole, low and other estimates suggest a balanced budget. In 1966, a borehole at Camp Century in north-west Greenland was completed to bedrock at a depth of 1387 m (B. L. Hansen and C. C. Langway, 1966). Temperature records in the borehole (Fig. 3.6B) are discussed in the next chapter (p. 88); but they also provide an insight into accumulation rates and their recent variations. Weertman (1968) has computed the theoretically expected form of the temperature curve for the drill-hole and finds that there are small discrepancies, up to 2°C, from the actual measured temperatures. These differences, he claims, may be explained by postulating an accumulation rate about 40 per cent smaller than now in the past 10–15,000 years, and by small (<0·5°C) variations in the mean annual surface temperature over the past 1000 years. The accumulation rate that fits the measured curve is 22·4 cm³/cm²/year.

S. J. Mock (1968) reports accumulation measurements made on the Thule peninsula in Greenland. Several different measuring methods were employed, including pit studies, stake and board measurements, elevation and levelling, and ice thickness profiles. The accumulation in general decreases towards the interior, but there are wide fluctuations between adjacent stations in some areas. The accumulation rate is associated with surface relief and, in turn, with sub-surface relief. In this area there is no zone where accumulation increases with altitude and no abrupt change occurs where the mountain crest is crossed. Variations in precipitation are caused by variations in surface relief, high accumulation being associated with topographic lows. The relief consists of a series of waves, which are the expression of bedrock relief. The mean accumulation decreases from 37 g/cm² in the south-east to less than 29 g/cm² in the north-west.

The same author (S. J. Mock, 1967) has also carried out a study of the accumulation over the whole ice sheet. The analysis is in terms of latitude, longitude and elevation. 127 stations were used and separate analyses were carried out for the northern and southern domes. The results show that, in northern Greenland, the accumulation falls from a maximum of 50 g/cm² of water equivalent in the south-west to about 15 g/cm² in the north-east. In the south the highest values occur near the south-east coast and here exceed 90 g/cm² water equivalent, falling towards the west. Asymmetry in accumulation is a feature of both the northern and southern domes.

3.3 The Canadian Arctic

The Barnes ice cap in Baffin Island (Fig. 2.4) is an example of a rather unusual arctic type of glacier. The ice cap has an area of 5900 km² and is elliptical in shape, its maximum length being 150 km and its maximum width 62 km. It rests on a slightly dissected plateau at about 610 m. The centre of the ice cap forms a broad dome rising to 1128 m at its maximum. The margins are cut by deep surface meltwater streams. The area of the ice cap is fairly static at present and its movement is very sluggish or negligible. The ice cap is unusual in that it all lies below the firn-line and has no firn accumulation zone. Its balance is maintained by the addition of superimposed ice.

The total accumulation for the years 1961–2, 1962–3, and 1963–4 is estimated by R. B. Sagar (1966) to be 960×10^6 m³, 1020×10^6 m³, and 1060×10^6 m³ respectively. This yearly increment amounts to about 0·12 per cent of the total mass of ice, which is estimated to be about 910×10^9 m³. The greatest addition of mass occurs in the higher parts of the ice cap. Except for 1962 when ablation occurred throughout nearly all the ice cap, ablation was limited to the lower levels. The accumulation area of the glacier lies above the equilibrium line and has the characteristics of the percolation zone B and the slush zone. In these zones, percolating water refreezes, adding to the accumulation and densifying the snow at depth.

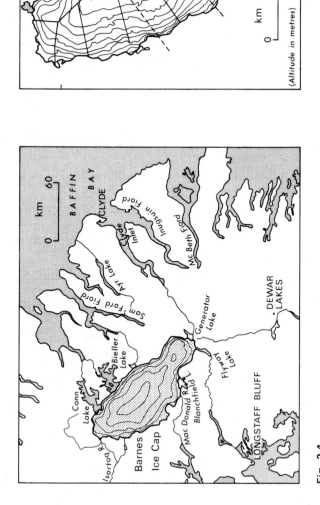

Fig. 2.4
Location of the Barnes ice cap in Baffin Island; the detailed map of the ice cap (right) shows surface contours (m), zone boundaries (broken lines) and stake lines (dots) for mass balance measurements (after O. H. Løken and R. B. Sagar, *Int. Ass. Scient. Hydrol.,* 1967)

Plate I

Aerial view in South Baffin Island. Cirques, arêtes and horns rise above a series of glaciers in which the contributory ice streams can be distinguished. The lake on the lower left has been dammed by morainic and fluvioglacial debris, but the meltwater from the glacier turns away from the lake to flow to the west (right). (Surveys and Mapping Branch, Department of Energy, Mines and Resources; *Exposure no.* A16817–102. Canadian Government Copyright.)

Plate II
Aerial view in North Baffin Island. A former transfluent glacier linking the two main glacier systems is now only represented by a short tongue projecting into the col, where a lake is now impounded. Marginal moraines and supraglacial meltwater channels can be traced, as well as various crevasse patterns. (Surveys and Mapping Branch, Department of Energy, Mines and Resources; *Exposure no*. A16350–34. Canadian Government Copyright.)

As melting proceeds in summer, the snow becomes isothermal, slush forms and streams flow. The surface is lowered by these ablation processes at the margin of the ice cap. The ablation season lasts for about two months. Ice layers form within the snow during the season and these form the superimposed ice. When this is allowed for, the ablation is about 20 to 21 cm of water equivalent.

The total net budget was found to be a loss of 1.9×10^9 m³ in 1961–2, which was an abnormally warm and dry year. For 1962–3 and 1963–4, the values were -0.12×10^9 m³ and $+0.13 \times 10^9$ m³, giving a total loss for the three seasons of -0.23 per cent of the total mass. This caused some wasting of the northern part of the ice cap.

Further data on mass balance are given by O. H. Løken and R. B. Sagar (1967). The southern part of the ice cap receives more accumulation in the winter. Near the ice cap centre, winter accumulation is three to four times that of nearby weather stations. In 1965–6 the mean mass balance was -93 cm, a loss equal to two to three times the mean annual winter accumulation. The 1964–5 mass balance was near zero at the northern end, where the equilibrium line was at 800 m above sea level. The north-east part of the ice cap has a more positive balance than the south-west, and this pattern appears to have persisted for at least 5000 years on the evidence of moraines. The ice cap is aligned north-west to south-east, and Table 2.1 gives mass balance values for zones from north to south on either side of a central line (Fig. 2.4).

Table 2.1   Barnes ice cap: mass balance values for 1964–5 and 1965–6, cm water equivalent

Zone	Year	Total	South-west area	North-east area
I	1964–5	−17	−31	+2
II		+4	−4	+15
I	1965–6	−108	−118	−94
II		−90	−92	−86
III		−85	−93	−71
IV		−78	−84	−69
V		−100	−111	−78

The mass balance of the Devon Island ice cap has been assessed by R. M. Koerner (1970). This sub-polar type of ice cap has an area of 15,570 km², reaching a maximum elevation of 1885 m. Its mean level is 600 m and the ice reaches sea level, particularly in the better nourished south-eastern part. The ice temperature is $-23°C$ at a depth of 12 m at 1800 m, $-18°C$ at 1300 m and $-13°C$ at 300 m in the ablation zone. Measurements were made along traverses by means of stakes, dye and percolation trays in the accumulation zone. The superimposed ice was measured by means of dye, while percolation trays were used in the firn zone. Stakes and wires were used in the ablation zone. On the north-western part of the ice cap the mass balance for the years 1961–1966 was -76 kg/m²/year. In some years there was a positive balance. The accumulation was two to six times greater and the ablation twice as much in the south-east as in the north-western part of the ice cap. In the south-east, calving is more active whereas it is negligible in the north-west. Part of the difference in the two zones is accounted for by variations in height-area relationships; 41 per cent of the south-east lies below 800 m, compared with only 16 per cent of the north-west. Such a difference must have taken a long time to establish, possibly hundreds if not thousands of years. A great deal of variation was shown over the 6 years of observations. There is a tendency to show a decrease in net balance between 1934 and 1950, followed by an increase, on the evidence derived from pits. The overall balance has probably been negative since 1934, but if summers like those of 1963–5 continue, the balance may become positive.

3.4   USSR and Central Asia

Some information on the mass balance of the ice masses of the USSR has been reported by M. G. Grosval'd and V. M. Kotlyakov (1969). There are four major glacier areas in the USSR, the Atlantic-Arctic, the Atlantic-Eurasian, East Siberian and Pacific-Asian areas. These zones can be divided into nineteen separate glacier areas, fourteen of which are of mountain type and five occurring on Arctic islands. The total glacier area is estimated to be 81,900 km² and the volume of water stored is about 13,750 km³. The glaciers of the East Siberian area were only discovered in the 1940s and 1950s, and they are very continental in type. Cirque glaciers predominate in the northern Urals and east Siberia, valley glaciers in the Caucasus and the Tien Shan, and transection glaciers in the Central Asian mountains. Glacier mass balance studies are being undertaken, the longest term one being on the Lednik IGAN in the northern Urals. Following a period of positive budget in the last part of the nineteenth century, this glacier has had a negative budget since the beginning of this century. It fluctuates with a wave period of 22 years, similar to that of the Grosser Aletschgletscher, but in the opposite phase. Seven selected glacier basins are now being studied in connection with the surveys for the International Hydrological Decade.

The régime of a glacier in Afghanistan has been described by O. Gilbert *et al.* (1969). In this area the climatic snow line is only a little below the mountain tops of the Hindu Kush. The Mir Samir glacier has peaks rising to 5809 m above it, but only the north-facing glaciers are of any size. The firn line on the glacier stands at about 4900 m. Between 1957 and 1965 the mean accumulation was estimated at 1300 kg/m², and the mean gross ablation was between 10 and 20 kg/m²/day. A small lowering of the firn line would lead to a substantial advance owing to the distribution of area with height. At present, however, on the evidence of lichen measurements, the slow retreat of the glacier has accelerated.

3.5   Northern Sweden

The small glacier, Storglaciären, at Kebnekajse (lat. 68°N) in Swedish Lapland has been studied in detail since 1946 by V. Schytt (1962). This small glacier has an area of 3·1 km². The average accumulation is $4 \cdot 0 \times 10^6$ m³ of water equivalent and the average ablation is $6 \cdot 2 \times 10^6$ m³ per budget year. Thus the average net loss is about 55 per cent more than the total net accumulation. This results in rapid glacier wastage and retreat. Of the fourteen years of detailed observations, twelve years showed a deficit, one year the budget was balanced, but in only one year, 1948–9, was there a surplus. The years of greatest loss were 1946–7 and 1959–60, both characterized by exceptionally hot dry summers with an excess of ablation. The budget varies very much from year to year so that the mean over a considerable number of years is required to give a good indication of the trend. The glacier will respond to the longer-term trends, but not to short-term variation from year to year.

3.6   *Iceland*

The mass balances of many of the outlet glaciers of Vatnajökull vary considerably from year to year. Estimates based on limited observations for the small outlet glacier of Morsárjökull from the years 1951–2, 1952–3, and 1953–4 suggest a total negative balance of approximately $6 \times 10^6 \, \text{m}^3$ and $26 \times 10^6 \, \text{m}^3$ of water equivalent for the first two years respectively, while in the third year there was a net gain of about $2 \times 10^6 \, \text{m}^3$. In general, however, the balance has been negative during this period. This is despite a relatively large proportion of the glacier lying above the equilibrium line. The whole accumulation area, however, lies at a fairly low altitude. The generally negative budget has been operating for several decades and the glacier has retreated about 1 km during the period from 1904 to 1953.

This glacier may be compared with two nearly adjacent glaciers, Skaftafellsjökull and Svinafellsjökull (Fig. 3.9). These two glaciers, and particularly the latter, differ from Morsárjökull in having a smaller accumulation area relative to their ablation area. On the other hand, their accumulation areas attain a considerably greater altitude. This means that winter snowfall is greater and, more important, the loss of snow from the upper part of the accumulation area is much less. Indeed, snow can fall throughout the whole year at the higher levels. The greater net accumulation has resulted in a smaller amount of retreat, particularly in Svinafellsjökull. One part of this glacier, which is fed from the highest part of the accumulation area, has not retreated at all during the 50 years before 1954. Fifty per cent of Svinafellsjökull lies above 1400 m, while the figures for the area above 1400 m of Skaftafellsjökull and Morsárjökull are 22 per cent and zero respectively.

H. W. Ahlmann (1948) has reported the results of work on the budget of Hoffellsjökull, situated a little farther east, for the period 1935–8. The results gave a budget of $-364 \times 10^6 \, \text{m}^3$, $+191 \times 10^6 \, \text{m}^3$ and $-111 \times 10^6 \, \text{m}^3$ for the three successive years. These figures indicate the variability of these outlet glaciers and the generally negative budget at this time is also apparent. The variations in this instance were the result of changes in both accumulation and ablation. Lower ablation accounted for the positive balance in 1936–7. The mass balance had the high total mean value of $1415 \times 10^6 \, \text{m}^3$ water equivalent for a total glacier area of $312 \, \text{km}^2$. This type of régime is characteristic of a very maritime climate with heavy precipitation of rain and snow, and relatively mild temperatures at all seasons. Ablation rates reach 6·5 cm/day on Morsárjökull, 5·6 cm/day on Skaftafellsjökull and 5·85 cm/day on Svinafellsjökull. For Skeidarárjökull (Fig. 3.9), G. Wojcik (1970) recorded a maximum of 2·86 mm/hour in the period 21 July–22 August, 1968.

3.7 Southern Norway

The detailed study by Østrem (1961–2) of Nigardsbreen may be taken as an example of the mass balance of a glacier in southern Norway. This glacier drains from the mountain ice cap of Jostedalsbreen. The district has a climate rather similar to that of Iceland, and snowfall is high, exceeding 5 m in places. Nigardsbreen has a very large accumulation area on the plateau ice cap and a narrow outlet glacier tongue. Most of its area lies between 1600 and 1700 m.

In the year 1961–2, when detailed mass balance studies were made, the specific net balance at different altitudes was such that the maximum accumulation took place between 1700 and 1900 m. The equilibrium line was at 1300 m. Thus the accumulation area occupied that part of the altitudinal range of the glacier where the area was great. The total net accumulation was very high owing to the near coincidence of the maximum area and the maximum specific accumulation. The very high ablation rate over the narrow glacier trunk at low elevation was not such an important item in the balance because the area involved was small. As a result, in this particular year, there was a strongly positive mass balance. The accumulation was 122×10^6 m³ and the ablation was 27×10^6 m³, giving a balance of $+95 \times 10^6$ m³. This positive budget was caused by unusually low ablation values throughout the whole area as the precipitation was normal (Fig. 2.2).

In this particular year, the mass balance was unusual, because during the last few decades Norwegian glaciers have been retreating rapidly. However, since the mid-1950s there has been some indication of more positive mass balances. This has resulted in the snouts of some of the shorter glaciers advancing during the period 1950 to 1960. Briksdalsbreen, which is a very steep glacier draining from the north side of Jostedalsbreen, has advanced during several years. The larger glaciers, Nigardsbreen and Austerdalsbreen, have continued to retreat until 1962. These two glaciers, however, advanced during the first decade of this century and again during the 1920s. Both these advances were associated with terminal moraine formation. The glacier fluctuations have been correlated with climatic variations, periods of advance being associated with spells of cooler and wetter conditions. Temperatures rose rapidly and precipitation fell after 1930, giving conditions associated with very rapid glacier recession during the next two decades.

3.8 Alaska

The Alaskan glaciers are of considerable interest as they show some irregularities. Some glaciers undergo rapid advance while neighbouring ones are retreating. The Black Rapids glacier suddenly advanced several kilometres down its valley over a period of 5 months in 1936–7, and then started to retreat again. The advance was probably the result of a succession of very snowy years which affected the glacier after a time lag of 7 years. This type of sudden advance is created by a surge in the glacier, a process that will be considered in more detail at the end of the chapter.

The Malaspina glacier system is one of the larger Alaskan glaciers, covering 4200 km² and containing about 1700 km³ of ice. Its accumulation area lies around the high peaks

of the Mount St. Elias range, forming the Upper Seward glacier, which covers 1318 km²
at 1520–2130 m. The firn limit is at about 915 m and lies across the outlet glacier, the
Lower Seward Glacier, which feeds the Malaspina Glacier. The Lower Seward glacier
is 5 to 6·5 km wide and 32 km long, leading down into the piedmont lobe of the Malaspina Glacier. This lobe spreads out over 2640 km² and rises to 610 m. It therefore
lies wholly in the ablation area. The outer 8 km are debris-covered and consist of dead
ice, on which spruce trees 100 years old are growing.

The annual accumulation of the Upper Seward glacier averages 75 cm of water equivalent, but is variable. It was 65 cm in 1947–8 and 175 cm in 1948–9. The normal
budget of the glacier is a deficit of 13×10^6 m³, but during 1948–9, a favourable year
for the glacier, the balance was positive. The glacier is, however, slowly wasting away
as a negative budget is much more common than the occasional positive one (R. P.
Sharp, 1958).

3.9   *Alpine glaciers*

The Hintereisferner has been studied in detail by H. Hoinkes and R. Rudolf (1962)
and may be taken as an example of an Alpine glacier. The observations cover the
period 1952–61. The main glacier had an area of nearly 10 km² in 1959, its mean height
being 2981 m. The mass balance, like that of the other glaciers mentioned, is very variable from year to year. With one exception, however, it was negative. The maximum
negative balance of −98 cm or $−9·83 \times 10^6$ m³ of water equivalent over the whole
glacier occurred in 1957–8. The only year with a positive balance was 1954–5, when
the glacier gained $0·77 \times 10^6$ m³. The mean value throughout the nine-year period was
$−3·60 \times 10^6$ m³, with an accuracy of $\pm 0·54 \times 10^6$ m³.

These results are similar to those given for the other areas mentioned in sections
3.5–3.8. They show that strongly negative budgets have been widespread in the last two
decades in many temperate northern areas. In this Alpine glacier, the accumulation
area is about 47 per cent larger than the ablation area, the percentage varying from
season to season. Because of the prolonged negative budget, the glacier snout has been
retreating on average about 25 m a year, but the retreat has been variable. At times,
about 100 m of dead ice become detached from the snout and gradually melt away.

3.10   *Southern Alps, New Zealand*

The highest mountains of the Southern Alps of New Zealand (43–45°S) support a
number of glaciers. There is an interesting contrast in mass balance between the glaciers
on the wetter slopes and those of the drier eastern side. The longest glacier is the Tasman, draining south-east and south on the east side of the watershed. This glacier flows
considerably more slowly than the Franz Josef on the western side and the former has
a much smaller mass balance. The rate of flow of the Tasman falls off from 0·5 m/
day at 19 km from the snout, to 0·35 m/day at 10 km from the snout. The ice in this
glacier has been wasting away slowly during the last few decades and its lower 8 km
are covered with a thick layer of ablation moraine. This down-wasting has resulted

in a thinning of the glacier by about 60 m, although the debris-covered snout has not noticeably changed position. The snout lies at a height of about 760 m.

On the western side of the Alps, the Franz Josef and Fox glaciers differ greatly from the Tasman. Until about a decade ago the Franz Josef glacier descended to a level of about 200 m above sea level, reaching into an area of dense rain forest. The glacier moves actively right to its snout, which consists of clean ice with only a thin layer of debris. Records of its velocity at the end of the last century suggest that it flowed at 525 cm/day about 3 km from the snout and 60 cm/day only 0·5 km from the snout. The glacier descends from a very high accumulation area where the precipitation probably exceeds 500 mm/year and is evenly distributed. On the low ground near the glacier snout, the annual precipitation averages between 340 and 286 mm. Records of the movement of the snout show that the glacier retreated slowly from 1910 to 1922, then advanced by a few tens of metres until 1933. A rapid retreat of nearly 1000 m then started and continued until 1946, when a renewed advance of a little over 200 m lasted until 1951 (Fig. 2.5). This was followed by renewed and accelerating retreat of 600 m until the early 1960s. This phase has come to an end with a sudden and vigorous advance.

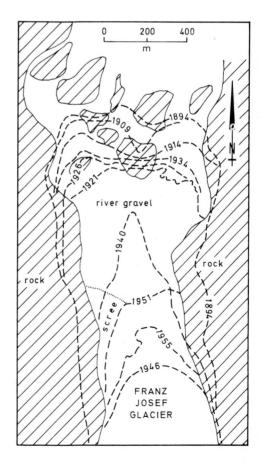

Fig. 2.5
Variations in the position of the snout of the Franz Josef glacier on the western side of the Southern Alps, New Zealand (R. P. Suggate, *N.Z. J. Sci. Technol.*, 1951; Department of Scientific and Industrial Research, Wellington)

Changes in the mass balance of the glacier appear to be related mainly to the precipitation. Variations in the accumulation are, therefore, mainly responsible for the changes in the snout position. There is a close correlation between the measured precipitation near the snout and the movement of the snout. The response of the snout lags five years behind the change in precipitation. The glacier starts to advance five years after the wetter period begins. In general, the periods of heavier precipitation were also periods of lower temperature and less sun. The mass balance was affected both by a rise in accumulation and a fall in ablation.

3.11   *Equatorial Africa*

At the highest levels on the Equatorial mountains of Africa there are still some small glaciers. They are of interest as their régimes depend on different seasonal changes from those of the temperate and high-latitude glaciers already mentioned. On Kilimanjaro, the Penck glacier is the longest. It falls from 5800 m to 4600 m in 2·4 km. No details of the mass balance are available but it seems likely that there are two ablation seasons, the longer occurring from July to September and the shorter in January and February. The total ablation in 1957 at the snout of the glacier probably did not exceed 35 cm and the accumulation was probably of the same order. The main accumulation season is March to June. This glacier differs from those in other climates in that the maximum snowfall probably occurs near the snout of the glacier and decreases upward. The main accumulation zone at present is probably in the lower part of the glacier and melting takes place at the top of the mountain. Clearly this régime cannot be stable for long.

On the volcanic peak of Mount Kenya, there are twelve small glaciers of which the Lewis glacier is the largest. This glacier has an area of 0·36 km² and is the lowest, descending to 4480 m. During the period from 1934 to 1958, the glacier receded up to 60 m. A total loss of about 18×10^5 m³ of ice has taken place, though the mass balance of the glacier is not accurately known. Accumulation occurs at two seasons, March to May and November to December, and is very variable. Glaciers can only survive where they are sheltered from the sun and exposed to the precipitation. C. M. Platt (1966) suggests that the annual amounts of accumulation and ablation on this glacier are small compared with those of temperate glaciers.

3 The response of glaciers to climatic change

Any positive change in the net budget of a glacier arising from climatic or other events will cause corresponding perturbations in glacier thickness, length and rate of flow. The response of glaciers to such events has been considered theoretically by J. F. Nye (1958, 1960, 1963, 1965a and 1965b), making use of the kinematic wave theory of Lighthill and Whitham. They applied this theory to river floods and traffic flow on roads, both of which are group phenomena. The same theory can be applied to the movement of a wave through a glacier at a speed greater than that of the ice velocity.

Nye's theory of the influence of climatic change on glacier variation considers the effects of seasonal as well as longer-term changes of climate. These changes give rise to variations in accumulation and ablation. His results are based on the assumption that there are no changes in temperature in the ice and that the ice is incompressible. His other basic assumption is that the discharge at any cross-section in the glacier is dependent upon the level of the ice and the slope of the upper glacier surface. Nye's theory is applicable to a valley glacier of varying width, but the simpler case of a valley of uniform rectangular cross-section of unit width may be considered first. Two quantities must be considered, the discharge, q, and the ice thickness, h. The relationship between these two variables is approximately a fourth power one, such that $q \propto h^4$. The term 'kinematic wave' is used to refer to a moving point for which q is constant. The velocity of the moving point is c and this varies down the glacier. The velocity of the kinematic wave under the conditions given is $c = \delta q/\delta h$ and is four times that of the ice if the former relationship holds. Because of this, the wave will reach the glacier snout long before any new material could actually be transported that far. Arriving at the snout, the kinematic wave velocity is approximately equal to the ice velocity and gives rise to a sudden forward surge. A wave passing down the Nisqually glacier (Washington) has been closely studied since 1945 (M. F. Meier and A. Johnson, 1962), and its velocity has been found to vary between two and six times the mean ice velocity.

The kinematic wave is the mechanism by which a glacier responds to changes in mass balance. If accumulation and ablation are taken into account, the value of q is no longer constant, but changes with a, the accumulation (a is negative for ablation). The effects of changes in accumulation and ablation are stable in those parts of the glacier where the flow is extending (that is, where the velocity of ice flow is increasing down-glacier). But in regions where the ice flow is compressing, or decelerating down-glacier, any change will tend to be accentuated and the glacier becomes inherently unstable.

It is assumed that the glacier accelerates steadily to a point P (Fig. 2.6) and then decelerates. If there is a uniform and shortlived addition of snow throughout the glacier, the upper part will increase uniformly in thickness. The lower part on the other hand will thicken at an increasing rate as c_0, the steady-state kinematic wave velocity, moves down the glacier at a decelerating rate. The kinematic wave is initiated at P, where a step occurs. This step appears because the upper part is falling to its original level while the lower part is still rising. This increase in elevation or step is magnified as it moves down the glacier, so that the effect near the snout is much greater than that elsewhere. In the case of the Nisqually glacier referred to above, the increase in surface level with the arrival of the wave amounted to more than 30 m in places. After the kinematic wave has passed, the glacier surface subsides downstream of P.

The passage of a kinematic wave can sometimes be traced by photogrammetric methods. In front of the wave, there is a region where the ice is compressing and becoming thicker; behind the wave, the ice is in tension, it is intensely crevassed, and its surface level drops as noted. The tension-compression boundary can sometimes be picked out visually on photographs, and successive photogrammetric surveys may enable the down-glacier velocity of the kinematic wave to be determined. A. E. Harrison

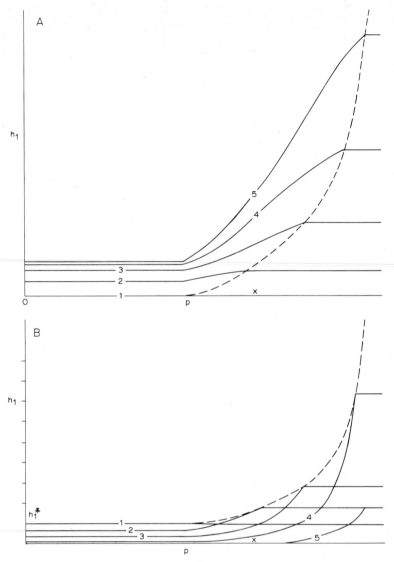

Fig. 2.6
Diagrams to show the effect of a kinematic wave in a glacier. The graph plots h_1, increase in thickness of the ice, against x, distance down glacier. Curves 1 to 5 refer to five successive time intervals, at 0, 5, 10, 20 and 40 years respectively if the strain rate is +3% a year above P and −3% a year below P, giving a time constant of 10 years. The broken line curves show the progress of the kinematic wave from P. The upper Figure (A) has different initial conditions, with the glacier in a steady state. There is then a sudden, permanent and uniform increase in the rate of accumulation. In the lower Figure (B), there is only one increase on the glacier (J. F. Nye, *Int. Ass. Scient. Hydrol., Snow and Ice Commission*, 54, 1961)

(1964) in this way found the wave velocity on the Muldrow glacier (Alaska) in 1956
to be 350 m/day.

The response time of smaller glaciers to climatic change varies between 3 and 30
years, but it may amount to hundreds or even thousands of years for very large ice
masses. For the Antarctic ice sheet, it is probably about 5000 years. This means that
some apparently different ice-sheet advances in the Pleistocene may not be of world-
wide chronological significance, but may merely reflect different frequency responses
to the same climatic event. The seasonal changes in accumulation and ablation are
of high frequency compared with the response time. The resulting thickness changes
in the glacier lag about three months behind the greatest seasonal accumulation. For
longer-term climatic changes, the response of the glacier comes into phase with the
climatic change in its upper parts, but in the lower parts the response is more complex.
There are two separate effects. First, there is the direct response, which may not be
in phase with the cause. Secondly, there is the delayed, indirect, response caused by
the travelling down-glacier of the wave form. The size of the wave varies down-glacier,
instability tending to cause an increase while diffusion effects cause a decrease. The
direct and indirect effects may cause one glacier to behave differently from its
neighbour if it is of different dimensions. Thus, it is possible to explain the great variety
of glacier response to changes in accumulation and ablation.

The response of each glacier can be calculated if the relevant data are available.
These data include q, the discharge, B, the breadth, h, the height above an arbitrary
datum line, a, the rate of accumulation averaged along a transverse line, and α, the
slope of the ice surface. These values are required for different points down the glacier.
Using these values, the changes in the glacier can be assessed by calculating c, the
velocity, and D, the diffusion co-efficient for the kinematic waves, if observed rates
of change of accumulation are known. The formulae for c and D are

$$c = \frac{1}{B} \cdot \frac{\delta q}{\delta h} \qquad\qquad D = \frac{1}{B} \cdot \frac{\delta q}{\delta a}$$

Nye (1965) has applied his theoretical analysis to actual glaciers, using a computer
program to calculate the frequency response of South Cascade glacier (Washington)
and Storglaciären (north Sweden). The latter glacier moves at about half the speed
of the former and its response time is double that of the Cascade glacier, the values
being 55 years and 26 years respectively. The amplitude of change is twice as great
in Storglaciären, owing to its lower velocity of flow. Nye (1965) has also shown that
it is possible to work back from the known changes in the position of the glacier snout
to obtain the glacier budget in the past. In the more remote past, the evidence of glacier
fluctuation is largely geomorphological. This evidence can then be more accurately
related to climatic change, through calculated values of accumulation and ablation.
The annual budget changes are not recorded by snout changes, but the observed
changes related to the mean of ten years' budget values agree well with theoretical
changes at the snout.

One of the problems of relating glacier changes to climate is the effect of local influences on the weather, which at times causes neighbouring glaciers to respond differently. The different climatic elements exert different effects at different times of the

year; for example in the Arctic, the winter temperature has little effect on glacier régime, but small variations in summer temperature can have major results. One year
may also influence the next, in that if much snow is left over from one season this will
influence the rate of melting the following season. There is also a feedback relationship
between the glacier and the climate, at least on a local scale, in the short term, and
there are similar larger-scale effects associated with longer time scales (W. S. B. Paterson, 1969).

H. C. Hoinkes (1968) has studied the relationships between glacier variations and
weather conditions by means of Alpine weather station records. He found that five-
yearly overlapping mean deviations agreed well with glacier fluctuations. The summer-
winter range was large around 1890 and small around 1920, resulting in a decrease
of continentality, which began in the late eighteenth century. This change is related
to increased zonal circulation: glacier advance is thought to be in inverse relationship
to the intensity of zonal circulation. The view that glaciers are most advanced during
the weakest circulation is, however, not entirely supported by the evidence. Seasonal
effects are important. There is a correlation between glacier variations and the average
height of the 500 mb surface. Low summer indices for the circulation, described by
deviation profiles along the 10°E meridian, are associated with positive budgets, and
high ones with negative budgets. The deviations can thus be related to Grosswetterlagen and these in turn to glacier budget responses. The anticyclonic Grosswetterlagen
of the last four decades are related to glacier retreat. This type of weather analysis
includes both an analysis of zonal, meridional and mixed types of circulation.

4 Glacier surges

There are some glaciers in the world that from time to time experience short phases
of exceptionally rapid advance, separated by longer intervals of quiescence or even
stagnation. The behaviour of such 'surging' glaciers, which should not be confused
with the passage of kinematic waves through glaciers, has long attracted the attention
of glaciologists and others; and it raises such fascinating questions that it is useful to
consider the phenomenon at some length. Research into surging promises to yield valuable information about many inter-relationships in the glacier system and about glacier
movement.

The most complete inventory of surging glaciers so far is for North America. E. V.
Horvath and W. O. Field (1969) have collected references to and dates for surges from
about 1900 onwards; A. Post (1969) and M. F. Meier and Post (1969) have identified
204 surging glaciers from recent air photographs. Some surges have been minor and
brief, but others have been major events in the history of a glacier, and for a time
have completely transformed the terminal area. The year 1966, for instance, was a
remarkable one for surges in Alaska. A dozen major glaciers surged forward, including
the Bering glacier, the largest in North America, which advanced spectacularly along
a front 42 km wide, the advance amounting to up to 1200 m since 1963. On the Walsh
glacier, whose lower half had been stagnant since 1918, a surge began in the upper
reaches in late 1960, and in 4 years the central portion of the glacier moved about

10 km (A. S. Post, 1966). In the upper reaches since 1961, the ice surface has fallen by as much as 150 m.

Glacier surges have long been known and recorded from many other parts of the world. S. Thorarinsson (1969) describes glacier surges in Iceland, referring to the impressive 1963 surge of Brúarjökull whose frontal advance at up to 5 m/hour (= 120 m/day) could be felt, seen and heard. Equally rapid advances are reported from the Karakoram Himalaya (for a recent review see K. Hewitt, 1969). In 1904–5, the Hassanabad glacier in the Karakoram is said to have advanced some 10 km in two and a half months (= 130 m/day), and in 1953, the Kutiah glacier in the same region moved forward at an estimated rate of 113 m/day (W. Kick, 1958). These rates are almost unbelievably high, and the figures may not be entirely reliable. In 1930, P. C. Visser (1938) observed and photographed the rapid advance of the Sultan Chusku glacier in the Khumdan valley. With the passage of the wave, the surface of the firn dropped suddenly by 100 m, leaving a rim of sheared-off tributary glaciers and old avalanche cones. The lower glacier surged forward over its old moraines, translating in effect up to 300 million m³ of ice from the upper glacier. Then it virtually stagnated.

Glaciers that have experienced surges in the past can be recognized by several distinctive features. Of these, the most striking and most reliable indicator is the presence of folded moraines (Fig. 2.7), especially the loops and folds in medial moraines. Following surges in 1957–60 and 1965–6, the Bering glacier in Alaska displayed 'accordiontype' folds in its medial moraines (Post, 1972). The folding is related to situations where

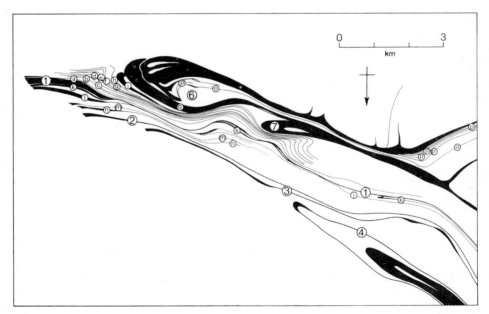

Fig. 2.7
Tongue of the Bjørnbo glacier, east Greenland, showing folded moraines (H. Rutishauser, *J. Glaciol.*, 1971, by permission of the International Glaciological Society)

tributary glaciers with near-constant flow push out into the trunk glacier between surges on the latter, forming loops in the medial moraines which are then deformed and displaced several kilometres down-glacier during surges. Since a surging glacier may move at 100 times its normal velocity, the surface becomes highly crevassed and at times great transverse cracks develop, as on the Medvezhii Glacier where one fracture was 8 km long and 10 m wide. Waves were also observed to form with 2 km lengths and heights of 50 to 70 m, which travelled along the whole glacier tongue. Some other effects of surges are illustrated by P. G. Johnson (1972) for the Donjek glacier, Yukon: he notes push-structures in older ice-cored moraines, deformation of till and outwash, and truncation of older crevasse fillings and alluvial fans. Older surges may have resulted in drainage diversion in this area. An interesting possibility for distinguishing between the tills formed by surging and 'normal' glaciers is one based on till fabric studies. N. W. Rutter (1972) finds that till from a surging glacier in Alaska shows no consistent fabric pattern, in contrast to a nearby normal glacier till. This differentiation could be a useful indicator if supported by further studies.

Although numerous hypotheses have been put forward to explain surges, we still lack sufficient quantitative data even to be able to define or describe surges accurately. We do, however, know enough about them to be able to state certain characteristics (M. F. Meier and A. Post, 1969); and E. Brückl (1972) has developed a mathematical model of a surging glacier.

1 All glaciers that surge apparently do so cyclically, at regular intervals of time (Robin and Weertman, 1973). Old moraine loops sometimes suggest periodic surging over hundreds of years.
2 The periodicity of surging differs from one glacier to another; it is therefore unlikely to be explained in terms of external events, such as earthquakes or climatic fluctuations.
3 Most surges occur over short time intervals, often two to three years, followed by a period free from surging lasting from about 15 to more than 100 years.
4 The cycles are not related to glacier dimensions or average speed of flow, nor to any particular climatic environment.
5 The velocity of ice movement in a surge may be one to two orders of magnitude greater than normal.
6 Glaciers that surge do not grow longer and longer as a result of periodic surging: each surge reaches approximately the same limits as the previous one.
7 In our present state of knowledge, surges are not generally thought to be related to peculiar bedrock configuration, though some hypotheses claim otherwise.
8 The general pattern of a surge is as follows. In a quiescent phase, the ice in the accumulation area thickens. When it becomes sufficiently thick, or the surface slope steep enough, the basal shear stress ($\tau_B = \rho gh \sin \alpha$: see p. 125) reaches a critical value and a surge begins. The rapidity of flow suggests a sudden and remarkable decoupling of the glacier from its bed. Then τ_B appears to return to more normal values, the surge ends, and the over-extended tongue stagnates.

Several possible mechanisms of surging have been suggested. J. Weertman (1962, 1969) and G. de Q. Robin (Robin and Weertman, 1973) have elaborated a basal meltwater lubrication mechanism. The basic idea in this is that, the thicker the basal water film beneath the glacier, the smoother the bed appears to the glacier. If the thickness increases to the point where it equals the controlling obstacle size (see p. 141), conditions for surging are created. Once surging begins, the increased velocity of basal sliding will generate a large further increase in basal meltwater owing to frictional heat and accelerate the process. So the initial problem is to arrange for enhanced accumulation of basal meltwater, after which the mechanism will be self-reinforcing. Let us consider conditions after a previous surge. The lower end of the glacier will be stagnating. At the upper end, the ice thickness will once more be growing and the ice becoming more potentially active as τ_B increases. In the zone between the 'active' and 'stagnant' portions, a strong gradient in τ_B will be set up, and hence also a gradient (which may become zero or even negative) in the basal water pressure (see p. 144). This may lead to quantities of meltwater accumulating in this junction or 'trigger' zone, and from then onwards, conditions are ready for renewed surging as already explained. The important part of this theory, then, is that a strong gradient in τ_B may cause water to accumulate under the glacier. There are some ancillary conditions which may help: one could be relatively impermeable bedrock, another could be an exceptionally smooth bed (e.g. controlling obstacle size of the order 1–2 mm), and another could be anomalously high subglacial temperatures related to high geothermal heat flow (Post, 1969).

L. E. Nielsen (1968,1972) has suggested that the lower stagnating portion of a glacier following a surge may act as a *relatively* immobile dam. The ice thickness behind such a dam would then increase until the resistance of the dam was overcome. Then a breakthrough would occur; the ice would become highly crevassed and broken and might be able to flow exceptionally rapidly as a suspension of ice blocks in meltwater, akin to a powder-flow process.

It has been suggested that surges could be extra-large kinematic waves passing down the glacier. The difference, however, is related to the fact that kinematic waves are associated with changes in the volume of ice, a bulge of extra thickness passing down the glacier, while a surge is usually only a redistribution of ice within the glacier. Kinematic waves due to an increase of mass balance do not appear to be the cause of surges and cannot account for the very high speeds recorded in surging glaciers. Kinematic waves may, however, be transmitted down a glacier during a surge.

Surging of cold glaciers might be possible (though data are quite inadequate to support such a hypothesis so far), if warming of the basal ice occurred. Higher temperatures would permit faster rates of ice deformation even though temperatures remained below pressure-melting point (see p. 81). More significant is the possibility of continued warming of the base of a cold glacier to pressure-melting point. Basal meltwater would then suddenly become available to lubricate the bed and sudden surging would be expected. This is the basis of A. T. Wilson's (1964, 1969) theory of Antarctic surging (see p. 32). V. Schytt (1969), discussing observations from the Nordaustlandet ice cap in Spitsbergen, suggests another relationship between surges and ice temperature distributions. Some sub-polar ice caps consist of a main central zone of ice at pressure-melting point surrounded by a ring of thinner ice (in the ablation zone) frozen to bedrock. Any increase in size of the 'warm' ice could lead to unstable conditions in which the dam of cold ice might be unable to prevent more rapid outward movement and surging would then follow.

It is evident that there are still many unanswered problems about glacier surges. Such phenomena are of great interest because of their widespread and frequent occurrence, their sometimes catastrophic nature, and the question of whether surging glaciers are capable of unusually high rates of erosion. Further research will have to attempt to answer such questions as:

1 What are the precise relationships of surging to climatic and mass-balance histories?
2 What is the role of water at the glacier bed? Does it exist in channels, cavities, or as a sheet (see p. 142)? What sort of pressure gradients exist in the basal water?
3 What sort of temperature régimes characterize surging glaciers? Can surges occur in cold glaciers, or only after the basal ice temperatures have been raised to pressure-melting point?
4 In what ways may the nature of the subglacial bed contribute to surging? Do surging glaciers often rest on exceptionally smooth or water-tight beds, and is there a relationship between surging and rates of geothermal heat flow?

Finally, a very basic question concerns the definition of surges. Are surging glaciers fundamentally different from 'normal' fast-moving glaciers such as Jakobshavn Isbræ ($c.$ 20 m/day)? Weertman (in discussion following Meier and Post, 1969, p. 816) makes the interesting suggestion that these fast-moving glaciers are really 'surging' in a steady state—because they have a large enough supply of ice to do so.

4 Conclusions

The glacier or ice-sheet régime depends upon the balance between accumulation and ablation. This balance is rarely achieved for a long period so that glaciers are continually fluctuating. The mass balance depends on the difference between accumulation and ablation. But equally important from the point of view of glacier activity are the actual amounts of accumulation and ablation, because these are related to glacier flow. Where the amount of snow added is high the glacier will be active, as in the case of the glaciers of south Iceland and Norway. But where the accumulation is very small and ablation equally low because of extremely low temperatures, glaciers will tend to move more slowly in relation to their dimensions and are geomorphologically less active. From the geomorphological point of view, therefore, both net and total budget values must be taken into account. The net budget determines the movement of the glacier snout and the variations in ice thickness. These changes in turn lead to the formation of terminal moraines, lateral moraines and trim-lines. The total budget determines in part the degree of glacial activity, and therefore is related to the erosive and depositional effects on the landscape.

The application of the kinematic wave theory to glacier fluctuations provides a useful method of linking climatic change with glacier régime and glacier response, and hence with geomorphological features. This link can work in both directions. The investigation of surging glaciers is another major research field at the present day. Many aspects of their behaviour are not yet properly understood, but basal meltwater lubrication to facilitate the rapid movements observed is one of the most likely mechanisms.

5 References

AHLMANN, H. W. (1948), 'Glaciological research on the North Atlantic coasts', *R. geogr. Soc. Res. Ser.* **1**, 83 pp.

BAUER, A. (1955), 'The balance of the Greenland ice-sheet', *J. Glaciol.* **2**, 456–62

BRÜCKL, E. (1972), 'A model of a surging glacier', *J. Glaciol.* **11**, 215–18

BUDD, W., JENSSEN, D. and RADOK, U. (1969 [1970]), 'The extent of basal melting in Antarctica', *Polarforschung* **6**, 293–306

BULL, C. (1958), 'Snow accumulation in North Greenland', *J. Glaciol.* **3**, 237–48

BULL, C. and CARNEIN, C. R. (1970), 'The mass balance of a cold glacier: Meserve glacier, South Victoria Land, Antarctica', *Int. Symp. Antarct. Glaciol. Exploration* (Hanover, N.H., 1968), *Int. Ass. scient. Hydrol. Publ.* **86**, 429–46

CHARNLEY, F. E. (1959), 'Some observations on the glaciers of Mount Kenya', *J. Glaciol.* **3**, 480–92

DEBENHAM, F. (1948), 'The problem of the Great Ross Barrier', *Geogrl J.* **112**, 196–218

DESIO, A. (1954), 'An exceptional glacier advance in the Karakoram-Ladakh region', *J. Glaciol.* **2**, 383–5

DUGDALE, R. E. (1972), 'A statistical analysis of some measures of the state of a glacier's "health"', *J. Glaciol.* **11**, 73–9

GILBERT, O., JAMIESON, D., LISTER, H. and BENDLINGTON, A. (1969), 'Régime of an Afghan glacier', *J. Glaciol.* **8**, 51–65

GIOVINETTO, M. B. and ZUMBERGE, J. H. (1967), 'The ice régime of the eastern part of the Ross Ice Shelf drainage system', *Int. Ass. scient. Hydrol., Gen. Assembly Bern 1967, Publ.* **79**, 255–66

GOW, A. J. (1965), 'On the accumulation and seasonal stratification of snow at the South Pole', *J. Glaciol.* **5**, 467–77

(1970), 'Preliminary results of studies of ice cores from the 2164 m deep drill hole, Byrd Station, Antarctica', *Int. Symp. Antarct. Glaciol. Exploration* (Hanover, N.H., 1968), *Int. Ass. scient. Hydrol. Publ.* **86**, 78–90

GOW, A. J., UEDA, H. T. and GARFIELD, D. E. (1968), 'Antarctic ice sheet: preliminary results of first core hole to bedrock', *Science* **161**, 1011–13

GROSVAL'D, M. G. and KOTLYAKOV, V. M. (1969), 'Present-day glaciers in the U.S.S.R. and some data on their mass balance', *J. Glaciol.* **8**, 9–22

HANSEN, B. L. and LANGWAY, C. C. (1966), 'Deep core drilling in ice and core analysis at Camp Century, Greenland', *Antarct. J. U. S.* **1**, 207–8

HARRISON, A. E. (1964), 'Ice surges on the Muldrow glacier, Alaska', *J. Glaciol.* **5**, 365–8

HEWITT, K. (1969), 'Glacier surges in the Karakoram Himalaya (Central Asia)', *Can. J. Earth Sci.* **6**, 1009–18

HOINKES, H. C. (1968), 'Glacier variation and weather', *J. Glaciol.* **7**, 3–19

HOINKES, H. and RUDOLPH, R. (1962), 'Mass balance studies on the Hintereisferner, Ötztal Alps, 1952–61', *J. Glaciol.* **4**, 266–80

HOLLIN, J. T. (1962), 'On the glacial history of Antarctica', *J. Glaciol.* **4**, 173–95
(1970), 'Is the Antarctic ice sheet growing thicker?' *Int. Symp. Antarct. Glaciol. Exploration* (Hanover, N.H., 1968), *Int. Ass. scient. Hydrol. Publ.* **86**, 363–74

HORVATH, E. V. and FIELD, W. O. (1969), 'References to glacier surges in North America', *Can. J. Earth Sci.* **6**, 845–51

HUMPHRIES, D. W. (1959), 'Preliminary notes on the glaciology of Kilimanjaro', *J. Glaciol.* **3**, 475–9

I.A.S.H. (*Int. Ass. scient. Hydrol.*) (1969), 'Mass-balance terms', *J. Glaciol.* **8**, 3–7

JOHNSON, P. G. (1972), 'The morphological effects of surges of the Donjek glacier, St Elias Mountains, Yukon Territory, Canada', *J. Glaciol.* **11**, 227–34

KICK, W. (1958), 'Exceptional glacier advances in the Karakoram', *J. Glaciol.* **3**, 229

KING, C. A. M. and IVES, J. D. (1955), 'Glaciological observations on some of the outlet glaciers of South-West Vatnajökull, Iceland', *J. Glaciol.* **2**, 563–9

KOERNER, R. M. (1970), 'Some observations on superimposition of ice on the Devon Island ice cap, N.W.T., Canada', *Geogr. Annlr* **52**A, 57–67
(1971), 'A stratigraphic method of determining the snow accumulation rate at Plateau Station, Antarctica, and application to South Pole-Queen Maud Land Traverse 2, 1965–66' in *Antarctic snow and ice studies. II* (ed. A. P. CRARY), *Am. geophys. Un., Antarct. Res. Ser.* **16**, 225–38

LACHAPELLE, E. (1962), 'Assessing glacier mass budgets by reconnaissance aerial photography', *J. Glaciol.* **4**, 290–7

LOEWE, F. (1960), 'Notes concerning the mass budget of the Antarctic inland ice', *Antarctic Meteorology*, 361–9

LØKEN, O. H. and SAGAR, R. B. (1967), 'Mass balance observations on the Barnes ice cap, Baffin Island, Canada', *Int. Ass. scient. Hydrol., Gen. Assembly Bern 1967, Publ.* **79**, 282–91

MAYO, L. R., MEIER, M. F. and TANGBORN, W. V. (1972), 'A system to combine stratigraphic and annual mass-balance systems: a contribution to the International Hydrological Decade', *J. Glaciol.* **11**, 3–14

MEIER, M. F. (1962), 'Proposed definitions for glacier mass budget terms', *J. Glaciol.* **4**, 252–65
(1965), 'Glaciers and climate' in *The Quaternary of the United States* (ed. WRIGHT, H. E. and FREY, D. G.), 795–805

MEIER, M. F. and JOHNSON, A. 1962, 'The kinematic wave on Nisqually Glacier, Washington', *J. geophys. Res.* **67**, 886

MEIER, M. F. and POST, A., (1969), 'What are glacier surges?', *Can. J. Earth Sci.* **6**, 807–17

MELLOR, M. (1959), 'Mass balance studies in Antarctica', *J. Glaciol.* **3**, 522–33
(1961), 'The Antarctic ice sheet', *U.S. Army Cold Regions Res. Engng Lab.* (Hanover, N.H.), Part **1**, sect. B, no. 1, 50 pp.

MOCK, S. J. (1967), 'Calculated patterns of accumulation on the Greenland ice sheet', *J. Glaciol.* **6**, 795–803

(1968), 'Some accumulation studies on the Thule peninsula, Greenland', *J. Glaciol.* **7**, 59–76

MÜLLER, F. (1962), 'Zonation in the accumulation areas of the glaciers of Axel Heiberg Island, N.W.T., Canada', *J. Glaciol.* **4**, 302–11

NIELSEN, L. E. (1968), 'Some hypotheses on surging glaciers', *Bull. geol. Soc. Am.* **79**, 1195–1201

(1972), 'The ice-dam, powder-flow theory of glacier surge', *Icefield Ranges Research Project, Scientific Results* (*Am. geogr. Soc.& Arct. Inst. N. Am.*), ed. BUSHNELL, V. C. and RAGLE, R. H., **3**, 71–4

NYE, J. F. (1960), 'The response of glaciers and ice-sheets to seasonal and climatic changes', *Proc. R. Soc.*, **256**A, 559–84

(1963), 'The response of a glacier to changes in the rate of nourishment and wastage', *Proc. R. Soc.*, **275**A, 87–112

(1965), 'The frequency response of glaciers', *J. Glaciol.* **5**, 567–87; 'A numerical method of inferring the budget history of a glacier from its advance and retreat', ibid., 589–607

ØSTREM, G. (1961–62), 'Nigardsbreen hydrologi', *Norsk geogr. Tidsskr.* **18**, 156–202

PATERSON, W. S. B. (1969), *The physics of glaciers* (Oxford)

PLATT, C. M. (1966), 'Some observations on the climate of Lewis Glacier, Mount Kenya, during the rainy season', *J. Glaciol.* **6**, 267–87

POST, A. S. (1960), 'The exceptional advances of the Muldrow, Black Rapids and Susitna glaciers', *J. geophys. Res.* **65**, 3703–12

(1966), 'The recent surge of Walsh glacier, Yukon and Alaska', *J. Glaciol.* **6**, 375–81

(1969), 'Distribution of surging glaciers in western North America', *J. Glaciol.* **8**, 229–40

(1972), 'Periodic surge origin of folded medial moraines on Bering Piedmont Glacier, Alaska', *J. Glaciol.* **11**, 219–26

ROBIN, G. DE Q. (1962), 'The ice of the Antarctic', *Scient. Am.* **861**

(1969), 'Initiation of glacier surges', *Can. J. Earth Sci.* **6**, 919–28

(1972), 'Polar ice sheets: a review', *Polar Rec.* **16**, 5–22

ROBIN, G. DE Q. and WEERTMAN, J. (1973), 'Cyclic surging of glaciers', *J. Glaciol.* **12**, 3–18

RUTISHAUSER, H. (1971), 'Observations on a surging glacier in east Greenland', *J. Glaciol.* **10**, 227–36

RUTTER, N. W. (1972), 'Comparison of moraines formed by surging and normal glaciers', *Icefield Ranges Research Project, Scientific Results* (*Am. Geogr. Soc.& Arct. Inst. N. Am.*), ed. BUSHNELL, V. C. and RAGLE, R. H., **3**, 39–46

SAGAR, R. B. (1966), 'Glaciological and climatological studies on the Barnes Ice Cap', *Geogr. Bull.* **8**, 3–47

SCHYTT, V. (1962), 'Mass balance studies in Kebnekajse', *J. Glaciol.* **4**, 281–8

(1967), 'A study of "ablation" gradient', *Geogr. Annlr* **49**A, 327–32

(1969), 'Some comments on glacier surges in eastern Svalbard', *Can. J. Earth Sci.* **6**, 867–73

SHARP, R. P. (1958), 'The Malaspina glacier, Alaska', *Bull. geol. Soc. Am.* **69**, 617–47

SHUMSKIY, P. A. quoted by MELLOR, M. (1959), 525

SIPLE, P. A. (1958), 'Man's first winter at the South Pole', *Natn. geogr. Mag.* **113**, 439–78

SUGGATE, R. P. (1950), 'Franz Joseph and other glaciers of the Southern Alps, New Zealand', *J. Glaciol.* **1**, 422–9

TAYLOR, L. D. (1971), 'Glaciological studies on the South Pole Traverse, 1962–63' in *Antarctic snow and ice studies. II* (ed. CRARY, A. P.), *Am. geophys. Un., Antarct. Res. Ser.* **16**, 209–24

THOMAS, R. H. and COSLETT, P. H. (1970), 'Bottom melting of ice shelves and the mass balance of Antarctica', *Nature, Lond.* **228**, 47–9

THORARINSSON, S. (1969), 'Glacier surges in Iceland, with special reference to the surges of Brúarjökull', *Can. J. Earth Sci.* **6**, 875–82

UEDA, H. T. and GARFIELD, D. E. (1970), 'Deep core drilling at Byrd Station, Antarctica', *Int. Symp. Antarct. Glaciol. Exploration* (Hanover, N.H., 1968), *Int. Ass. scient. Hydrol. Publ.* **86**, 363–74

VISSER, P. C. (1938), *Wissenschaftliche Ergebnisse der Niederländischen Expeditionen in den Karakorum und die angrenzenden Gebiete in den Jahren 1922, 1925, 1929–30 und 1935*, Bd. II, Glaziologie (Leiden)

WEERTMAN, J. (1961), 'Stability of ice-age ice-sheets', *J. geophys. Res.* **66**, 3783–92

(1962), 'Catastrophic glacier advances', *Int. Ass. scient. Hydrol., Commn Snow Ice (Obergurgl 1962)*, 31–9

(1968), 'Comparison between measured and theoretical temperature profiles of the Camp Century, Greenland, borehole', *J. geophys. Res.* **73**, 2691–700

(1969), 'Water lubrication mechanism of glacier surges', *Can. J. Earth Sci.* **6**, 929–42

WEXLER, H. (1961), 'Ice budgets for Antarctica and changes in sea-level', *J. Glaciol.* **3**, 867–72

WILSON, A. T. (1964), 'Origin of ice ages: an ice shelf theory for Pleistocene glaciation', *Nature, Lond.* **201**, 147–9

(1969), 'The climatic effects of large-scale surges of ice sheets', *Can. J. Earth Sci.* **6**, 911–18

WOJCIK, G. (1970), 'Ablational processes on the Skeiðarárjökull (Iceland)', *Bull. Acad. pol. Sci. Sér. chim. géol. géogr.* **18**, 251–8

3

The physical properties of ice and types of glacier

A glacier is not coherent ice, but is a granular compound of ice and water, possessing, under certain circumstances, especially when much saturated with moisture, a rude flexibility sensible even to the hand. (J. D. FORBES, 1843)

Most of the ice that makes up glaciers and ice sheets originates as newly fallen light snow flakes. The newly fallen snow has a low density as much air is trapped between the hexagonal snow crystals, but as the delicate points of the crystals melt, the snow settles and increases in density. Melting is not the only cause of increase in density, however; other factors include the temperature and the original form of the snow crystals. The crystals themselves are smaller when the temperature is low. As the temperature increases many crystals stick together and fall as large snow flakes. Heavy snowfall only occurs when the temperature is fairly close to freezing-point. When the air is very cold, it can hold little moisture and the snowfalls are lighter.

The newly fallen snow alters in several stages to become glacier ice. The first stage is the change from snow crystals to granular snow. Since the vapour pressure of a snow crystal is greatest at its points, these melt first and the grain becomes more rounded. This process is much more rapid in places where temperatures are near zero centigrade; in such places, the newly fallen snow becomes coarse and granular in a few days. In very cold polar climates such as those of central Greenland or Antarctica, the process may be delayed for years. In intertropical latitudes, on the other hand, the snow often falls as soft hail. It then partially melts under the powerful sun during the day and refreezes at night. By this process, the snow is rapidly transformed into the coarse granular form. Climate, therefore, plays an important part in the change to granular snow.

Another important factor in accounting for the change from fresh powdery snow to ice is the effect of compression. Compression will also tend to be more rapid where the temperature is close to zero, because the individual snowfalls may be heavy. Where the pressure of overlying layers is considerable, the snow has small grains but is compact and very resistant. The density of newly fallen snow is about 0·06 to 0·08, but after two days on a slope it increases to 0·2 in a temperate climate.

The second stage is the conversion of granular snow to firn. After one winter, the snow reaches a density of 0·4 to 0·55. At this stage it may be called 'firn'. There has been some discussion concerning the use of the terms 'firn' and 'névé', which are the German and French respectively for material that could be described in English as 'consolidated, granular snow not yet changed to glacier ice' (M. M. Miller, 1952). Miller suggests that the material be called firn or firn-snow, and the area where it occurs be called the névé. The term 'firn' means literally 'of last year' and this is the simplest way to use the term, to refer to snow that has survived one summer season.

The lower boundary of the firn is not so easily defined. But M. F. Meier (1962) supports the suggestion that it should lie at the point at which the densifying material becomes impermeable to water. This boundary sometimes occurs at a density of 0·55, a density at which Benson and Anderson (in Meier, 1962) have found that the rate and mechanism of densification changes. An older suggestion was that the lower limit of firn should be placed at a density of 0·4. This limit does not seem so satisfactory as greater densities than this have been found in temperate glaciers in snow that is less than a year old. For instance, snow on the Cascade glacier in Washington had a density exceeding even 0·55. The definition of firn as material that has survived one ablation season but is not yet impermeable to water seems a satisfactory definition at least for temperate glaciers. A somewhat different definition may be needed for polar ice sheets. It is important to bear in mind the distinction between temperate and polar glaciers, a distinction discussed later in this chapter.

The process of change from snow to ice through firn is a continuous one. As the firn increases in age, it also increases in density and crystal size (Table 3.1).

Table 3.1   Relationship between age, density and crystal size of firn on the Claridenferner, Switzerland

Age (years)	1	4	10	12
Density	0·59	0·60	0·68	0·70
Crystal length (mm)	0·7	1·5	2·5	4·5

The physical properties of snow and firn are important from several points of view. Snow is a bad conductor of heat, and it therefore prevents the penetration of cold into the ground when the snow layer is thick. This is important in considering permafrost and periglacial forms. On the other hand, snow can hold a large quantity of water, up to 40 per cent by volume or 75 per cent by weight. Snow in this waterlogged condition may give rise to dangerous and geomorphologically effective avalanches. At low temperatures, the snow has the properties of an elastic body, but at temperatures around zero it shows viscous properties. Thus the snow is able to creep as friction releases heat; the heat melts the crystal points and allows movement between the grains.

The third stage is the conversion of firn to ice, and the time taken for this is very variable. In the Claridenferner, the material is still firn after 12 years, and about 25 to 40 years are necessary before it becomes ice. The rate is slower in Greenland, where ice of a density of 0·8 or greater is not found at depths less than 100 m. This represents an age of about 150 to 200 years. These figures provide reasonable values for temperate and polar glaciers respectively.

As firn turns into ice, the crystal pattern changes and the bubbles of air are slowly transformed and reduced in size. The material becomes hard blue ice and the crystals increase in size until at the end of a glacier they may reach about 10 cm in length. The air bubbles are only expelled very slowly from the ice, but finally they become very small and the ice reaches a density of over 0·9. In the Mer de Glace, for example, a density of 0·88 only increases to 0·91 after the glacier has been flowing for 50 years. J. C. Behrendt (1965) has shown that density in the Antarctic ice sheet increases rapidly at first with depth. It reaches a value of 0·9 at depths of about 120 m, as shown on Fig. 3.1A. At a depth of 40 m, there is a correlation of density with accumulation rate as shown on Fig. 3.1B.

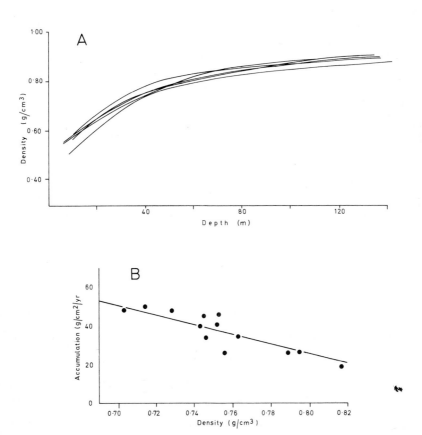

Fig. 3.1
A Graph relating density with depth in the Antarctic ice sheet.
B Graph relating density with accumulation at 40 m depth in the Antarctic ice sheet (John C. Behrendt, *J. Glaciol.*, 1965, by permission of the International Glaciological Society)

1 Polycrystalline glacier ice

Glaciers are formed of polycrystalline ice derived mainly from the original snow crystals. This material can be studied experimentally and its properties defined. Experiments on single ice crystals show that they can deform plastically by gliding of one layer over another parallel to the basal plane. Therefore, as J. W. Glen (1958a) has shown, the orientation of the individual crystals in polycrystalline ice may be expected to play an important part in its deformation. The ice of glaciers often shows a preferred crystal orientation, which is not so marked, however, as that in ice which has formed on a free-water surface. In ice of this type, all the basal planes tend to be parallel. Therefore, in using ice for tests in a laboratory, it is important to know the source of the ice. Experiments in which grain structure is studied under polarized light have been made with lake ice, glacier ice and icicles. The results show that all types of ice deform plastically both in tension and compression. It has also been found that the deformation of ice is analogous to that of metals at high temperatures, an analogy which led Glen to carry out experiments similar to those carried out on metals. P. Barnes, D. Tabor and J. C. F. Walker (1971) show that, at moderate stresses, three main processes occur depending on temperature. Below $-8°C$, the behaviour is dominated by basal gliding; between $-8°$ and $-1°C$, creep is associated with a liquid phase at the grain boundaries and grain-boundary sliding; while near $0°C$, pressure melting and regelation are the chief processes if the pressure applied is high enough.

Tests have been made over a wide range of controlled temperatures and stresses to assess the effects of both these variables on the deformation of polycrystalline ice. As more data have been collected, so it has become apparent that the behaviour of ice under stress is more complex than was originally thought. Early theories of ice behaving as a Newtonian viscous substance, or as a perfectly plastic substance (Fig. 3.2), have been abandoned. In respect of the theory of perfect plasticity, continued

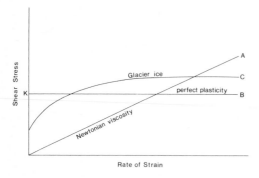

Fig. 3.2
Idealized diagram comparing the behaviour of glacier ice under stress with perfectly plastic and Newtonian viscous substances (after R. P. Sharp, 1960, *Condon Lecture Publications, Eugene, Oregon; Univ. of Oregon Press*)

experiments have shown that there is no 'yield stress' below which ice ceases to deform, but, on the other hand, as the rates of deformation are very small below stresses of about 1 bar (10^5 Nm^{-2}), the theory of perfect plasticity is a useful simple approximation, provided it is realized that it is no more than an approximation.

When a specimen of ice is first of all subjected to stress, the strain rate (rate of deformation) decreases initially with time, but after a few hours (provided the stress level is not much below 1 bar), the strain rate settles to a steady value. As stresses in actual glaciers act over long periods, the early stage of transient creep is unimportant and will not be considered further. When the stress is increased beyond about 4 bar, the creep rate falls initially and then accelerates to a higher rate of creep, a change probably associated with recrystallization as the crystals become aligned in a more favourable pattern for creep. The crystal sizes of samples subjected to higher stresses are found to be smaller than those deformed by lower stresses.

About 1955, J. W. Glen showed in a series of experiments that a power flow law was better able to describe the behaviour of ice under stress than viscous or plastic flow laws. The relationship proposed was of the form $\dot{\varepsilon}=k\tau^n$, where $\dot{\varepsilon}$ is the strain rate and τ the applied shear stress, k is a constant depending on temperature, and the exponent n is greater than unity. (Note that in a Newtonian viscous relationship, $n=1$, while in the theory of perfect plasticity, $n=\infty$.) In his experiments, Glen (1955, 1958b) used a range of stresses from 1 to 11 bar, and temperatures from $-0.02°C$ to $-12.8°C$. It should be noted that the range of stresses includes values much higher than those encountered in glaciers, that the tests were performed at temperatures slightly below to well below pressure melting-point, and that for the lower stress levels, the tests were not continued long enough to reach a steady state. Over the range that he tested, Glen found values for the exponent n of about 3.2. S. Steinemann (1958) used a stress range of 0.7 to 20 bar, a temperature range of $-1.9°C$ to $-21.5°C$, and suggested that values of n increased with increasing stress, varying from 1.85 to 4.2. These and other experiments failed to give consistent values either for k (a constant sensitive to temperature) or for n. Part of this failure was because a steady state of deformation was not always achieved in the tests, and part because the samples of ice used in different experiments were structurally different. Nevertheless, it has become apparent that the simple power flow law, although a better approximation than a perfectly plastic flow law, is not adequate to describe the rheology of ice over a large stress range. M. F. Meier (1960) and Mellor and Smith (1967) proposed a two-term flow law of the form $\dot{\varepsilon}=A\tau+B\tau^n$ (see p. 127) to fit a range of data including not only laboratory tests but also measurements on actual glaciers. Some workers have recently looked more closely at the response of polycrystalline ice to very low stresses. All emphasize the great difficulty of testing the behaviour of ice under small stresses because of the time-consuming nature of the experiments, especially at low temperatures. At $-10°C$, for instance, a stress of 0.5 bar produces a strain rate of less than 0.001 year^{-1} (see p. 114) and it will be apparent that even experiments extending over as long as a year may be much too short to give reliable results. Yet the vast bulk of glacier ice is being subjected to shear stresses of 1 bar or less. Mellor and Testa (1969) conducted tests at 0.43 and 0.093 bar, with a temperature of $-2.06°C$ and found that a power law with $n=1.8$ represented a good fit. S. C. Colbeck and R. J. Evans (1973) reported the results of ninety experiments using compressive stresses of less than 1 bar, in which large blocks of glacier ice ($0.52\times0.13\times0.13$ m) were tested in a tunnel in the glacier from which the ice was obtained. This enabled the tests to be conducted at pressure melting-point. After the initial phase of 'rapid' transient creep lasting 20–30 hours, a steady state was attained and the tests continued for up to 100 hours, some up to 200 hours. Assuming a power flow law, the best fit of the data by least squares gives $k=0.33$ and $n=1.3$, but there is in fact a wide scatter of points (Fig. 3.3A). A better form of flow law to fit the data

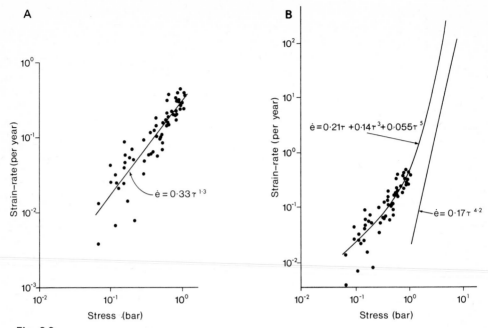

Fig. 3.3
A A polynomial flow law (Colbeck and Evans, 1973) for temperate glacier ice.
B Comparison with Glen's power flow law with $n=4\cdot2$. Also shown is a best-fit curve of polynomial type. (S. C. Colbeck and R. J. Evans, *J. Glaciol.*, 1973, by permission of the International Glaciological Society)

is of the polynomial type suggested by Lliboutry (1969), in which $\dot{\varepsilon}=A\tau+B\tau^2+C\tau^4$. Colbeck and Evans find a good fit in the following expression: $\dot{\varepsilon}=0\cdot21\tau+0\cdot14\tau^3+0\cdot055\tau^5$ (Fig. 3.3B). They argue that this form of flow law is superior to Glen's for glacier ice at pressure melting-point and at realistic stress levels.

Laboratory observations show, therefore, that ice is not a substance of constant viscosity, but a pseudo-plastic substance which can deform by slow creep under stress. Field observations of the way in which ice takes on the form of a rock surface across which it moves have confirmed this. Examples are given by J. G. McCall (1960) from observations in the upper tunnel cut through Vesl-Skautbreen, Norway. This small cirque glacier is moving across a head-wall gap, the ice forming the roof of the gap. Grooves impressed upon the ice as it slides across its rock bed above the gap remain imprinted on the ice-roof of the cave for a distance of 50 m. As the glacier only moves at 3 m/year at this point, the grooves are apparently able to survive for about 15 years. The ice, therefore, behaves plastically, moulding itself to the indentations of its floor, but then remaining rigid when the pressure against the floor is relieved. H. Carol (1947) has also studied the plastic deformation of ice in a cave beneath the Ober Grindelwald glacier at a depth of 50 m. The ice was compressed against a rock knob and in close proximity to the rock surface the velocity was increased and its consistency was softened. Carol described it as being like cheese. A plane separated the more rapidly moving ice (72 cm/day) from the slower moving ice above (36·8 cm/day). W. H. Theakstone

(1966) has described rather similar plastic contortions in ice caves beneath Østerdal-sisen in northern Norway.

The ability of ice to spread under its own weight is a property well exemplified in the character of the ice shelves that float in the Antarctic bays. This spreading illustrates the response of ice by plastic deformation to stresses within it. Evidence from Antarctic ice shelves analysed by R. H. Thomas (1971) supports a generalized power flow law in which $n=3$ over a stress range of 0.4–10 bar. It is possible to calculate the theoretical profile of an ice sheet or glacier if the properties and behaviour of ice are taken into account. A very simple demonstration may be made using the model of perfect plasticity (see p. 124) as an approximation to reality. Orowan (1949) and J. F. Nye (1952) have shown that the profile of an ice cap in a steady state resting on a horizontal base of even roughness, with mass balance and temperature held constant, would be defined by the equation $h=\sqrt{2h_o(R-r)}$ where $h_o=\tau_o/\rho g$ and the other notation is as given in Fig. 3.4A. The profile is assumed to follow a flow line in the ice, and is of parabolic form. τ_o is the maximum basal shear stress given by the 'yield stress' (about 1 bar) in the theory of perfect plasticity, ρ is the density of ice, and g the acceleration of gravity. If $\tau_o=1$ bar, then $H=4.76\sqrt{R}$, giving a direct relationship between ice-cap thickness and radius.

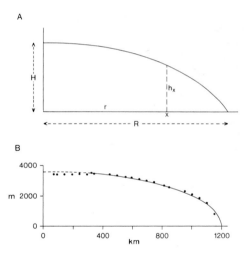

Fig. 3.4
A Notation and co-ordinates for idealized ice-sheet profile
B Surface profile of the Antarctic ice sheet, Mirny-Vostok. The dots show the field observations while the continuous line is the theoretical profile (see text) (S. S. Vialov, *Int. Ass. scient. Hydrol.*, 1958)

When it is considered how many assumptions and generalizations are involved in this simplified model, it is surprising that in fact the results of such theoretical calcula-tions agree quite well with surveyed profiles of the Antarctic ice sheet, for instance. Attempts to improve the agreement by using a power flow law create much more diffi-cult problems since the constant k is very sensitive to temperature and the ice sheets of polar regions cannot be assumed to be isothermal. But since it is likely that tempera-tures rise towards the base (see pp. 49 and 87), and because of the strain rate–tem-perature relationship, it can be reasonably argued that the basal layers will experience most deformation. Further simplifying, let us assume that all shear takes place in an infinitely thin layer next to bedrock, giving a uniform vertical velocity profile above, then the sliding velocity $V=k\tau_b{}^n$ where τ_b is the basal shear stress. Since the ice sheet is assumed to rest on a horizontal base, τ_b must be a function of ice surface slope:

$$\tau_b = -\rho g h \left(\frac{dh}{dx}\right)$$

To satisfy these several conditions, it can be shown (W. S. B. Paterson, 1969) that the surface profile would be given by an equation of the form

$$\left(\frac{h}{H}\right)^i + \left(\frac{x}{R}\right)^j = 1$$

If the exponents i and j are equal to 2, the equation defines an elliptical profile (compare the parabolic form derived on the theory of perfect plasticity). Paterson derives values of $i=2\cdot5$ and $j=1\cdot5$; Vialov (1958) gives values of $i=2\cdot6$, $j=1\cdot3$. Fig. 3.4B compares an actual ice-sheet profile with the theoretical profile: the close agreement confirms the idea that the shape of an ice sheet is largely controlled by the properties of the ice. But for ice caps not in a steady state, or affected by irregular bedrock relief, agreement is rather poor as would be expected. Buried bedrock ridges are reflected by inflections in the ice surface. Nye (1959) has shown how the heights of such ridges can be theoretically estimated from the change of ice surface slope, and radar sounding now permits such ideas to be checked.

2 Classification of glacier types

H. W. Ahlmann (1948) has suggested various methods of classifying ice masses. He gives a morphological classification based on the size and form of the ice, a dynamic classification based on the degree of activity of the ice, and finally, a thermal classification. These three types of classification all have a bearing on the geomorphological action of moving ice on the landscape. The three classifications will be considered in the reverse order, as this is the order of increasing subdivision.

2.1   Thermal classification

The temperature of an ice mass plays a fundamental part in its morphological activity. This has already been noted in connection with glacier surges and it will be stressed in later chapters. There are two basic types of glacier from this point of view: temperate and cold (or polar) glaciers (Ahlmann, 1935).

(a) *Temperate glaciers*   Temperate glaciers are approximately at pressure melting-point throughout their thickness, except for the uppermost few metres (maximum about 15–20 m) which may become temporarily colder in winter. Below the depth of penetration of this winter cold wave, the ice temperature will decrease, because of increasing pressure of ice overburden, at about 6×10^{-4} degrees C per metre (R. L. Shreve and R. P. Sharp, 1970). Temperate glaciers are not therefore strictly isothermal, though this term has been applied to them in a general way because temperatures are relatively close to 0°C. At the base of the temperate glacier, melting must occur because, with the reversed temperature gradient of pressure melting-point (the ice becomes warmer upwards), geothermal heat and any heat generated by friction as the ice drags over its bed cannot escape upwards through the ice. Fig. 3.5 shows actual temperatures recorded in bore-holes in the Athabasca glacier, and the small departures from the

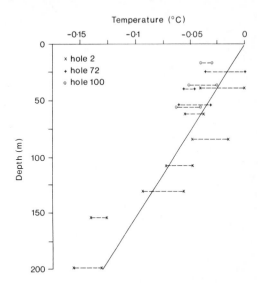

Fig. 3.5
Measured temperatures in Athabasca glacier, Alberta. The broken lines join the readings of the two thermistors at each depth. The solid line indicates theoretical pressure melting point (W. S. B. Paterson, *J. Glaciol.*, 1971, by permission of the International Glaciological Society)

theoretical straight-line plot of pressure melting-point. Such small departures arise from two main groups of factors (Paterson, 1971): first, impurities in the ice (including salt content and air bubbles), and secondly, heat sources within the glacier (mainly the result of deformation by movement, and the water content of the ice, most of which probably represents water trapped in the firn). Falling snow may have quite a significant salt content; concentrations of sodium ions up to 5000 μg/kg and of chlorine ions up to 5500 μg/kg have been reported (Lliboutry, 1971). In the light of these facts, Lliboutry proposes the following definition: 'Temperate ice is ice which contains within it a liquid phase (whether the liquid inclusions communicate or not) and which is in local equilibrium with it.' He adds a rider that the concentration of salts in the liquid must not be too high.

The majority of temperate glaciers can be generally recognized in the field from the presence (and, often, abundance) of meltwater throughout their thickness. Even if no detailed temperature measurements can be made, the temperate character can often be identified from the fact that meltwater issues from beneath the edge of the ice. Ice-dammed lakes and crevasses can remain full of water up to a certain depth (see Chapter 19). In truly temperate glaciers, meltwater formed during thermal boring should not, in theory, refreeze—and in fact drill pipes in some such glaciers have remained free for days—but owing to winter cold-wave penetration, conduction down metal pipe linings, and the existence of chilled layers at depth in the glacier (which can arise, for instance, from inclusion of sub-freezing snow into deep crevasses), many such drill holes do in fact tend to become blocked with ice. Such an occurrence is not incompatible with a general temperate character. The precise definition and recognition of temperate glaciers is thus fraught with problems, and it should be borne in mind that this category includes, in practice, many departures from the ideal model.

Despite the fact that meltwater may be encountered in fissures and channels through-out the thickness of a temperate glacier, the permeability of unfissured temperate ice is small. As far back as the time of Agassiz, it has been known that dye can penetrate the upper few metres of a temperate glacier, and especially the relatively bubble-free blue ice. Air bubbles inhibit percolation because of surface tension effects, but the main cause of impermeability of temperate ice, as Lliboutry (1971) observes, is deformation and recrystallization taking place in the moving ice which blocks the capillary channels. Melting occurs at the grain boundaries because of the high local stresses continually recurring there, followed by refreezing. In the absence of fissures or channels, temper-ate ice is thus permeable at any given time for a distance of several grains only.

Many of the glaciers of the Alps and southern Scandinavia are temperate in charac-ter. In this type of glacier, meltwater at the bed helps the ice to slide over the wet rock floor (Chapter 4), and thus rates of movement are, other things being equal, greater than in the case of cold glaciers.

(b) *Cold glaciers*   *Cold or polar glaciers* differ in very important respects from temperate glaciers. Ahlmann has subdivided this type into two, the *sub-polar* type and the *high-polar* type. He differentiates the two mainly on their firn characteristics. In the accumu-lation area of the sub-polar type, the material consists of crystalline firn down to a depth of 10 to 20 m. In summer, the surface can melt and water can be present. In the high-polar type, the temperature in the firn remains well below freezing-point and even in summer there is no melting on the surface in the accumulation area. In this type, firnification is a very slow process and does not take place above a depth of 75 m. The ablation zones of high-polar glaciers may be permanently dry; ablation takes place solely by sublimation with, possibly, a minor amount of loss by mechanical deflation, as in some glaciers in the Transantarctic Mountains described by J. H. Mercer (1971).

From the geomorphological point of view, one of the most important properties of a cold ice sheet is the absence of meltwater at depth in the ice. Melt-streams, where they occur, for example on the margins of the Greenland ice sheet, flow only on the surface of the ice. They may reach very large dimensions during the short ablation season. A large slush zone also indicates the effectiveness of surface melting in the sub-polar type of ice mass. At the base of the glacier, the ice is well below pressure melting-point because any sources of basal heat can be conducted upwards through the ice (contrast temperate glaciers). The ice is thus frozen to the bedrock on which it is resting,

Fig. 3.6 (*opposite*)
A Cross-section of inferred temperature distribution across central Greenland (G. de Q. Robin, *Polar Rec.*, 1972)
B Measured and theoretical temperature profiles for the Camp Century drill hole, Greenland. Theoreti-cal curve 1 is based on the local value of accumulation rate gradient; curve 2 uses an average accumu-lation rate gradient (J. Weertman, *J. geophys. Res.*, 1968)
C Temperature profile for the Byrd Station drill hole, Antarctica. Circles, joined by a solid line, indicate measured values; extrapolation of lower part of graph (broken line) reaches pressure melting point ($-1\cdot6°C$) at base of drill hole, 2164 m (A. J. Gow *et al.*, *Science*, 1968)

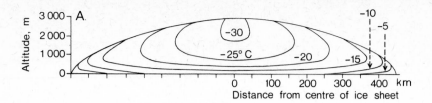

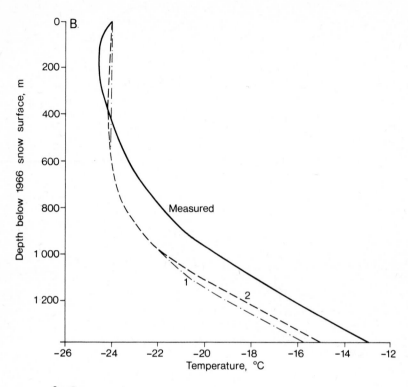

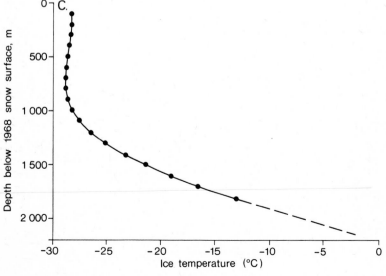

which directly affects the way in which the ice moves over its bed (Chapter 4). It seems probable that ice masses in this condition cannot achieve as much erosion as the temperate type.

The temperature distribution in glaciers and ice sheets depends on several factors, including the surface temperature fluctuations, both seasonally and through a longer period, the thickness of the ice and its conductivity, and the geothermal heat flow from below the ice. The temperature distribution for a thick polar ice sheet has been discussed by G. de Q. Robin (1955). He shows how rates of accumulation on the surface of the ice sheet affect its internal temperature. He considers the problem theoretically, assuming a stable ice sheet of the form described by J. F. Nye (1952), and calculates the temperature at various depths in the ice sheet. The geothermal heat flow, assumed to be 38 cal/cm²/year, gives a gradient of 1°C/44 m at the base of the ice sheet. Another source of heat at the base of the ice sheet is that resulting from shearing: Nye has shown that most of the movement takes place by shearing in the lowest layers of the ice sheet (Chapter 4). Robin has calculated a shear stress of 0·88 bar at the base of the Greenland ice sheet, and if the basal movement is 18 m/year, a heat supply could be generated equal to that of the geothermal heat flow. For a movement of 10 m/year, the additional heat supply would be 21 cal/cm²/year. This source of heat must be taken into account away from the centre of the ice sheet. Ice must move downward in the centre to compensate for the outward movement of ice from the centre, and this will also generate heat. Another consideration is the surface heat supply. The temperatures are lowest at the highest elevations on the surface, but as the ice moves outward and downward, it comes to be overlain by layers of firn that are slightly less cold. Thus there should be an increase in the negative temperature gradient close to the surface outward from the centre of the ice sheet, provided that no heat reaches the surface from below.

The Greenland ice sheet illustrates these points. It has a maximum thickness of about 3000 m and an average accumulation in its central parts of about 30 cm/year. Fig. 3.6A shows an idealized cross-section and the probable theoretical distribution of temperature in the ice. This temperature pattern has been confirmed by such observations as those made in the Camp Century borehole in north-west Greenland. The drilling here (B. L. Hansen and C. C. Langway, 1966) reached bedrock at a depth of 1387 m; the temperature at the ice-rock interface was −13°C, well below pressure melting-point, compared with −24°C at the surface (Fig. 3.6B) (W. F. Budd, D. Jenssen and U. Radok, 1971). Seismic observations support the idea that, generally, the central and southern interior parts of the Greenland ice sheet are also frozen to bedrock; on the other hand, the failure of echoes to return from the base of the ice sheet in interior northern Greenland has suggested that there may be areas here where the basal ice is at pressure melting-point, a possibility that has been utilized by J. Weertman (1961) in a hypothesis to explain the formation of cold-ice moraines (Chapter 15). Robin has argued that, if accumulation on the interior parts of the Greenland ice sheet were to fall to 10 cm/yr or thereabouts, basal melting would be expected over considerable areas, with important consequences for flow rates and bedrock erosion. Recent data discussed by Robin et al. (1969) suggest the possibility of a basal layer of ice at pressure melting-point 100–200 m thick in part of Greenland.

In Antarctica, patterns of temperature distribution in the ice have also been

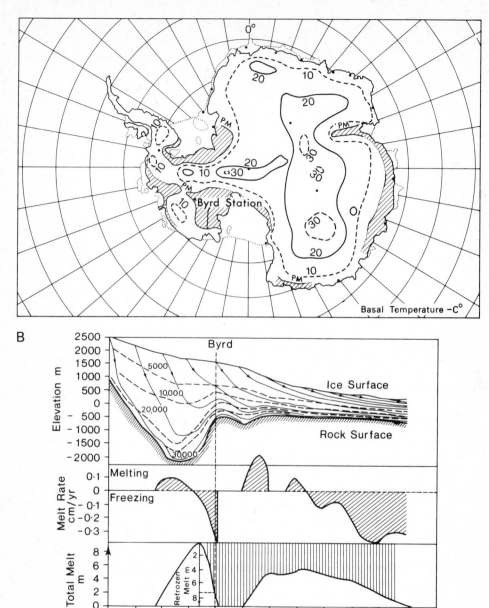

Fig. 3.7
A Calculated basal temperatures for the Antarctic ice sheet, contoured for each 10°C below pressure melting-point. The shaded areas are at pressure melting-point.
B Flow-line profile in west Antarctica through Byrd Station. Ice-particle paths and ages of ice in years are shown at the top; the middle graph shows calculated rates of melting or freezing; the lower graph shows the total amount of ice melted or refrozen assuming no migration of meltwater (W. F. Budd *et al., Polarforschung,* 1969 [1970])

examined both theoretically and from data yielded by the Byrd Station borehole. Budd, Jensen and Radok (1969[1970]) have computed the probable pattern of basal temperature (Fig. 3.7) from available data on surface and basal relief, accumulation rates and surface temperatures, assuming a steady state. The most extensive area where basal melting is predicted is in the region of thick ice with comparatively low surface elevations in Byrd Land, extending to the Ross ice shelf. Flow rates are also relatively high in this sector. The 2164 m deep borehole at Byrd Station confirmed basal melting; when completed in January 1968, water rose 60 m up the drill hole, and extrapolation of the curve of temperatures measured down to 1800 m depth in the hole (Fig. 3.6C) suggests a basal temperature equivalent to pressure melting-point, here $-1 \cdot 6°C$ (A. J. Gow et al., 1968).

It is clear, then, that many complications in the distribution of temperature within ice masses have come to light since Ahlmann first proposed his simple thermal classification. In addition, the choice of terminology has been criticized and alternatives suggested. A. Court (1957) proposed that the terms *permelting, refreezing* and *nonmelting* be used instead of temperate, sub-polar and high polar, respectively. Court's terms are meant to signify that glaciers in the first category are permeated by meltwater throughout. The second type has a layer in which water refreezes below a surface layer of melting. The third type never experiences melting on the surface. It is important to note that the whole of one ice mass need not fall solely into one category. Ahlmann used the surface conditions as the criteria for classification, but the internal temperature distribution is a more valid basis for classification. However, as just discussed, this can lead to complications because high-polar ice sheets and even parts of the Antarctic ice sheet can be at the pressure melting-point at their bases. On the whole, however, it seems best to continue to use the widely adopted terms proposed by Ahlmann, as long as it is realized that the whole ice mass need not necessarily fall into a single category.

An example of this fact has been mentioned by F. Loewe (1966) in connection with the Sukkertoppen ice cap in Greenland. The upper part of this glacier above the firnline has a temperate character, the temperature at a depth of 4 m being close to melting-point because meltwater sinking into the firn releases heat as it refreezes. In the ablation zone, however, where meltwater runs off, the ice attains the mean annual temperature of the air. This is well below the freezing-point and the ice here has polar characteristics. Inversions of this type have also been reported from Greenland and Nordaustlandet, Svalbard (V. Schytt, 1964).

(c) *Thermal régime and glacier erosion, transport and deposition*   It has already been indicated that the distribution and range of temperature in a glacier can have important consequences for conditions at the ice-bedrock interface. Under temperate glaciers, basal meltwater must be present and substantial rates of basal sliding may occur. Under cold glaciers frozen to bedrock, there can be no meltwater present, by definition, and all ice movement must occur by deformation in the ice (see Chapter 4). Many glaciers and ice sheets, as we have seen, however, do not fall simply into one or other of these categories. In accumulation zones of many thick sub-polar glaciers, for example, the upper layers may be cold while the lower layers may be at pressure melting-point, but in the ablation zones of the same glaciers where the ice is thinner, the whole thickness of ice may be cold and frozen to bedrock. It is also possible for a glacier to be completely cold in a high-altitude accumulation zone, and to flow to much lower levels with a more temperate climate where the glacier may warm up to pressure melting-point throughout. Fig. 3.8, after G. S. Boulton (1972), summarizes some of these poss-

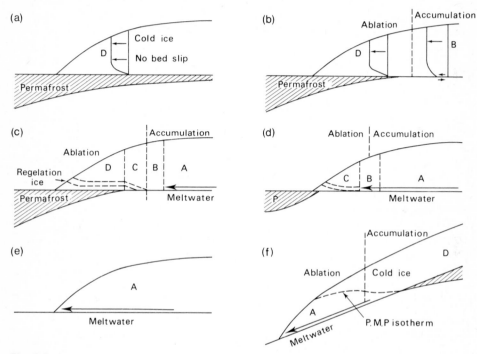

Fig. 3.8
Idealized examples of possible thermal zonation in glaciers and ice sheets (for explanation and discussion, see text) (after G. S. Boulton, *Institute of British Geographers, Spec. Publ.*, 1972)

ible relationships. In zone A, geothermal heat (H_{geo}) and heat from sliding friction (S) cannot escape upwards from the glacier sole because of the reversed temperature gradient of pressure melting-point, and basal melting therefore occurs. In zone B, only the actual base of the glacier is at pressure melting-point, and $H_{geo} + S = T$ (the rate at which heat is conducted upwards through the cold ice). In zone D, $H_{geo} + S < T$ and the ice is frozen to bedrock. C represents a zone where basal meltwater produced in A may move towards D to refreeze.

Let us now briefly look at the geomorphological implications of the varying thermal conditions in these zones.

Zone A: Mechanical crushing and shearing of bedrock obstacles may occur because the ice is able to slide over the rock floor. Any significant freeze–thaw action at the glacier bed is unlikely because temperatures overall will be at or above pressure melt-ing-point. Subglacial streams can exist and their morphological activity may be con-siderable. Relatively little debris may be entrained in the ice, except in the glacier sole (see p. 138) where regelation occurs due to local pressure variations. Deposition of lodgement till may be favoured if the ice is thick, the bed rough or the bedrock permeable (reducing the basal water pressure and thus increasing the resistance of the bed to glacier slip over it).

Zone B: Little meltwater will be produced and subglacial streams, in theory, absent. Freeze–thaw action at the bed due to pressure variations may be a significant process. Mechanical abrasion as in Zone A is possible but more limited because of slower sliding speeds—bedrock protruberances will project into cold ice.

Zone C: Basal sliding will take place as meltwater from Zones A or B moves to refreeze on the sole in this zone. Circumstances will thus be favourable for freeze–thaw action and entrainment of debris in considerable quantity. Lodgement till is unlikely to form: this will be dominantly a zone of erosion and transport.

Zone D: No slip can occur at the ice–rock contact. Basal abrasion will therefore be limited to places where larger particles of rock embedded in the moving ice a few centi-metres or more above the contact strike the bedrock. Boulton (1972) remarks that extraction of large frozen erratics, even down to the permafrost base, is feasible in this zone. At the junction with Zone C, compression of the moving ice must occur because the ice is sliding over bedrock in Zone C but not in Zone D. Thus it is likely that the basal ice of C will be thrust over that in Zone D, and debris therefore carried into higher levels of Zone D. This is a very important process of entrainment of debris in cold glaciers.

   Boulton (1970) has suggested that cold glaciers such as those of Spitsbergen, Green-land and Baffin Island may therefore carry larger loads of basally derived debris than their temperate counterparts in Iceland and the Alps, but this generalization has been contested by J. T. Andrews (1971, 1972), noting that Baffin ice seems relatively clean. The argument is handicapped by lack of sufficient and accurate field data—assessing the relative abundance of debris in different glacier systems is extremely difficult in practice, not least because chance exposures of the glacier in section may not be repre-sentative. Moreover, thermal régime is by no means the only factor controlling the efficacy of erosion, transport and deposition; lithology and roughness of the bed, and the degree of activity of the ice both past and present (see next section), are also impor-tant considerations. Nevertheless, there are fundamental differences between the morphological activities of glaciers, and differences of thermal régime are undoubtedly a main part of the explanation for this. Further quantitative data are urgently needed to test the hypothesized relationships between thermal régime and glacial sedimenta-tion and transport.

2.2 Dynamic classification

The activity of glaciers is influenced to a certain extent by their thermal characteristics: a cold-based glacier requires a larger shear stress to induce movement than a temperate glacier. However, the dynamic activity of a glacier is also closely associated with its mass balance, as mentioned in the last chapter. A classification based on the glacier's dynamic activity consists of three main types: 1 active,   2 passive or inactive,   3 dead glaciers. The importance of these three types from the point of view of glacial deposits is considered in more detail in Part III.

Active glaciers are normally fed by a continuous ice stream from an accumulation zone that may lie in a cirque basin or on a plateau. Glaciers fed by accumulation on a plateau may be termed outlet glaciers. Another group, fed entirely by ice avalanches falling from an upper accumulation area on to lower ground, may be termed 'regenerated'. The Glacier de Nantes in the French Alps is an example of a small regenerated active glacier. It does not move far from its avalanche fan head, but it has formed large terminal moraines that demonstrate its activity. The eastern side of the outlet glacier of Morsárjökull in Iceland is now fed entirely by avalanches from Vatnajökull above (Fig. 3.9). The glacier receives sufficient material to continue down the valley as a normal glacier for several kilometres. It is joined to its western portion which was still connected to the ice cap above by a continuous ice-fall in 1954. The eastern part of the glacier was severed from direct contact with the ice cap above by thinning in about 1937 (Plate IV).

The dynamic characteristics of glaciers are not directly dependent upon a positive mass balance. Some glaciers can maintain active dynamic movement even with a negative budget, but it is obvious that this state cannot continue indefinitely, as the velocity of a glacier is partly dependent upon its thickness and this in turn on its mass balance. The total budget is more important in determining the relative activity of a glacier. Active glaciers tend to have a large total budget. A glacier in this state can flow actively right down to its snout even if this is retreating up-glacier as a result of a temporarily negative budget.

The glaciers on the west side of the Southern Alps of New Zealand provide good examples of active glaciers. In this area, the precipitation is very high and ablation fast; the slope is steep and the glaciers very active. The flow rates measured at the end of the last century are much faster than those measured recently by B. M. Gunn (1964). Suggate (1950) records a maximum rate of 525 cm/day in 1894, 3·2 km from the snout of the Franz Josef glacier. This value compares with 66 cm/day in 1956 measured at about the same distance from the snout. The minimum values given are not so different, but nevertheless indicate a considerable slowing of the glacier over the last 50 years. During the period from 1935 to 1960, the glacier retreated 1260 m, 600 m of this retreat taking place since 1956. Thus although the glacier can still be described as active, and shows characteristics of this state in its lower reaches, its velocity has decreased as it has thinned and retreated over the last 50 years. A dynamically active glacier is one which is flowing fast, whether it is retreating or advancing at its snout. But it will be more active when it is thicker than when it is thinning.

Where the supply of snow to feed the glacier is small, for example on the lee side

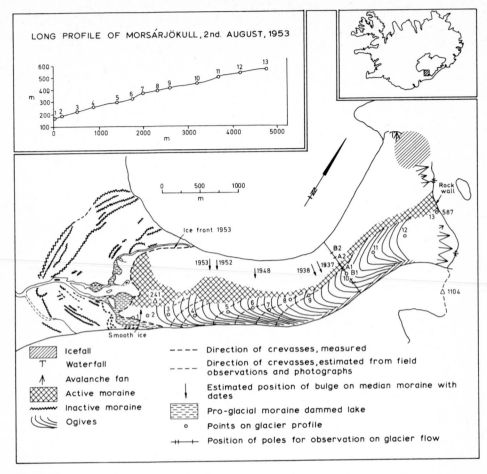

Fig. 3.9

Map of Morsárjökull (J. D. Ives and C. A. M. King, *J. Glaciol.*, 1954, by permission of the International Glaciological Society)

of a mountain range, the ice may become passive. This may also occur where the slopes are gentle. The lowlands to the east of the Scandinavian mountains illustrate the type of area where passive ice may be expected to occur. Similarly on flat areas of high ground, where lack of slope inhibits movement, the ice may be passive. The ice under these conditions still receives nourishment in the form of snow accumulation. But the ice is not dynamically active, nor can it be described as climatically dead.

Dead ice is not necessarily immobile according to Ahlmann's definition of the term. He suggests that dead ice is that which no longer receives a supply from an accumulation area. Its movement is restricted to that dependent on the slope. It is better to qualify this form of dead ice as 'climatically dead'. A further category of dead ice may be described as 'climatically and dynamically dead'; in this situation, it is not only not being nourished but has also ceased to move. It will survive for a time only on its own mass, which is a slowly wasting asset under these conditions. There may be a considerable time lag between the time at which an ice mass becomes climatically dead and the time at which it has become both climatically and dynamically dead. In the case of large ice sheets this difference in time may be several thousand years.

Lobes of dead ice in areas where the ice was at one time very thick may become isolated from their original source. It is sometimes necessary to postulate the former presence of dead-ice lobes to account for particular geomorphological features. For example, a dead-ice lobe in the Vale of Belvoir has been proposed to account for the present course of the middle Trent. The river flows in a deep trench, cut anomalously and obliquely to the strike of the rocks between Nottingham and Newark. The ice in the Vale of Belvoir, which lies just east of the Trent trench, had accomplished in its active phase considerable erosion in the soft clays exposed in this area. It would, therefore, have been relatively thick here when the ice sheet started to decay. It is reasonable to suppose that a dead-ice mass would survive in this particular low-lying area because of its relatively great thickness and because it would exist in a position likely to cause it to be covered with debris such as outwash, further helping to protect it and preserve it. Meltwater flowing along the margin of the dead ice initiated the Trent trench. Dead ice was also very important in the deglaciation of parts of Sweden, such as Norrland, which is an area far from the accumulation zone, particularly during the later stages of deglaciation when the ice centre had again moved to the west of the mountains (C. M. Mannerfelt, 1945). At present, dead ice occurs on the outer margin of the Malaspina glacier in Alaska. The geomorphological significance of dead ice in accounting for a variety of glacial depositional landforms is considered in Chapter 17.

2.3 *Morphological classification of glaciers and ice masses*

The classification of morphologically distinct ice masses is based essentially on their size and the characteristics of their environment. Altitude in relation to the areal distribution of the ice is an important factor in this classification. Some glaciers have large areas at high altitudes and others large areas at low elevations. The following classification is suggested:

1 Niche, wall-sided or cliff glacier
2 Cirque glacier
3 Valley glacier—Alpine type
4 Valley glacier—outlet type
5 Transection glacier
6 Piedmont glacier
7 Floating glacier tongues and ice shelves
8 Mountain ice cap
9 Glacier cap or ice cap
10 Continental ice sheet.

1 Niche glaciers in Spitsbergen have been described by G. E. Groom (1959). They consist of a triangular wedge of ice, often with a slightly convex surface, lying in a shallow funnel-shaped hollow in the upper part of the hillside. They develop on steep slopes (up to 42°) and are often associated with rock benches, formed where harder rocks outcrop. A gully runs from the lower end of the glacier to an alluvial fan below. At times, these small ice accumulations are connected to ice on the plateau above, but in other instances they are isolated features. It is thought that these niche glaciers develop independently of the plateau ice cap above, but this raises the problem of how snow can accumulate on slopes as steep as 42°. The glaciers probably originated as snow patches, which rest between the steep rock slope and the scree beneath. The rock slopes become notched by gullies and it is in these longitudinal gullies that snow drifts can accumulate and develop into niche glaciers. The rock benches give rise to steps on which snow can accumulate in the gullies. The gullies can then be enlarged by nivation processes until the snow can accumulate and consolidate to a sufficient thickness to form a small glacier. The niche glacier is thus genetically similar to the rather larger cirque glacier and owes its particular features both to structure and climate. It forms an early stage in cirque glacier development.

2　　The detailed work carried out by W. V. Lewis (1960) and those working with him on the cirque glaciers of the Jotunheim in Norway provides a valuable analysis of the characteristics of this type of glacier. Vesl-Skautbreen is a small cirque glacier that has been studied from many points of view and may be taken as an example of this type (Fig. 7.2A). The glacier is less than 1 km in diameter and rests against a steep rock wall. It has a mean surface slope of 26° and lies on a rock floor whose profile is arcuate. The radius of curvature of the bed and back-wall is 240 m and the front end of the glacier rests against a snow-bank, which lies between it and the moraine at its foot. The moraine in turn is separated from the rock bar at the lip of the cirque by a lake. The cirque glacier, therefore, lies in a true rock basin, which is characteristic of many cirques. Further details of the character and movement of this particular glacier are discussed in Chapter 7 in connection with the formation of cirques.

3　　When the snow-line is falling in elevation, the ice of the cirque can move out of the cirque basin and down the valley to form a valley glacier. Sometimes the ice of several cirques combines to form the valley glacier, as is common in the Alps. This has given rise to the term 'Alpine type', applicable to this form of valley glacier.

Ahlmann has further subdivided the valley type of glacier according to the area-height relationships, giving four subdivisions. His first type of valley glacier is exemplified by the Rhône glacier or Hintereisferner. These glaciers have a considerable proportion of their area a little above their median height. The areas at the highest and lowest levels are reduced to small amounts. The second type is shown by the Grosser Aletsch-gletscher (Fig. 3.10). This glacier has a large high-level névé basin, so that the greatest area occurs in the upper part of the glacier basin. The third type consists of glaciers that have most of their area at low levels and are largely fed by avalanching from higher snowfields. These include many of the glaciers of Central Asia, originating in high mountains. The Styggedalsbreen in Norway is another example. The fourth group includes glaciers such as those of north-west Spitsbergen which have large lateral tributaries. These have the greatest proportion of their area at levels a little below the mean elevation.

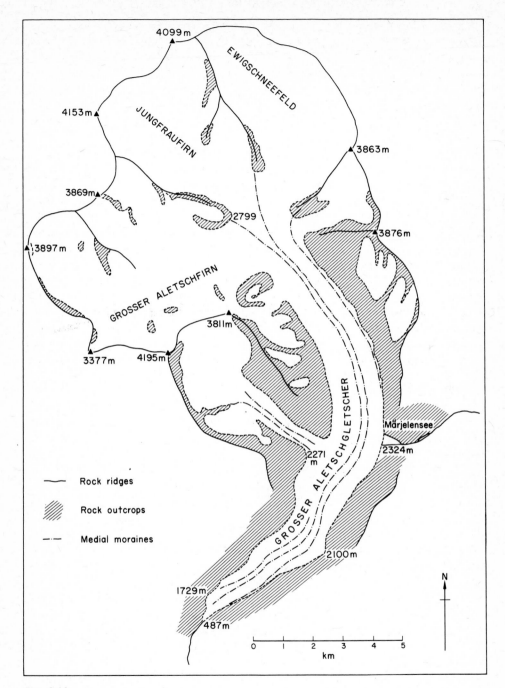

Fig. 3.10
Map of the Grosser Aletschgletscher

The importance of the relative area of a valley glacier at different levels has already been mentioned. The different behaviour of neighbouring glaciers can be accounted for in terms of these glacier types. Those with a large area in the zone of maximum accumulation will be in a more favourable position than those with only a small area at this level. One essential characteristic of all the glaciers in this category is that they are confined within valley walls throughout their length and terminate in a narrow tongue.

4 The outlet type of valley glacier is similar in its lower reaches to the former type, but it is fed in its upper reaches from an ice cap and not from an individual cirque or series of cirque basins. The outlet glaciers draining from the ice caps of Iceland and Norway are good examples of this type. Larger glaciers of essentially the same type drain from the large Greenland and Antarctic ice sheets. These are wider and may flow with considerable rapidity on steep slopes; they often end in the sea, in which case they must be partly classed in category 7.

Outlet glaciers are very susceptible to the position of the equilibrium line in relation to its source region. This is demonstrated, for example, in Morsárjökull in Iceland and Nigardsbreen and Austerdalsbreen in Norway. At the present time, their equilibrium lines lie near the top of the valley glacier part of the system. But if they were to rise a little higher, then the whole system would become rapidly moribund, as the equilibrium line would rise above the plateau surface. This situation is not so likely to arise in the Alpine type, in which the accumulation zone covers a wider height range. Any movement of the equilibrium line in these glaciers involves only a relatively small area of the whole glacier system.

5 Transection glaciers are those which occupy much of a mountain group, from which glaciers flow down in several directions into a system of radiating valleys. The accumulation area at a high elevation is not, however, large enough to be called a mountain ice cap. In other cases, the mountains may be too deeply dissected to allow an ice cap to form. The Löwenskjold glacier in Spitsbergen is an example. The glaciers flow down about 10 to 15 km from the central area of snowfields. Glacier systems of this type are responsible for forming transfluent glacial breaches and cols, which are discussed in Part II. The proportion of the area at the higher elevations is rather larger in these glacier systems, although the area at the highest levels falls off more than it does in ice caps because of the steeper relief.

6 Piedmont glaciers form when the valley glacier advances out from the containing mountain walls into a lowland beyond. The classic example of a piedmont glacier is the Malaspina glacier in Alaska, which spreads out as a broad lobe on the Pacific coast lowlands. The ice in the lobe is about 600 m thick and it lies in a basin at least 250 m below sea level. As the glacier spreads out, it maintains a sufficient thickness and surface slope to enable it to flow uphill across the hollow in the coastal foreland (R. P. Sharp, 1958). The dimensions and régime of the glacier have already been commented upon and its stagnant marginal zone noted. This type of glacier is characterized by its relatively large area at the lowest altitude. Nearly one-quarter of the area covers only one-tenth of the height range in another typical example, the Murray glacier in Spitsbergen.

Skeiðarárjökull is another piedmont glacier (Fig. 3.11). It is an outlet glacier flowing south from Vatnajökull in Iceland and spreads out on to the sandur (outwash plain) at the foot of the mountains. The glacier is only 8 km wide where it breaks through the mountain wall but it expands into a broad lobe about 25 km along the front. The

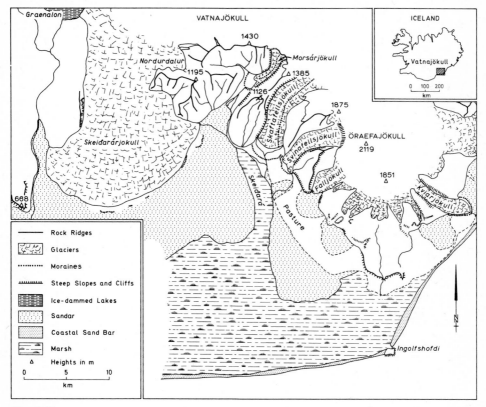

Fig. 3.11
Map of Skeiðarárjökull and other outlet glaciers on the southern margin of Vatnajökull in Iceland (J. D. Ives and C. A. M. King, *J. Glaciol.*, 1955, by permission of the International Glaciological Society)

ice is not dead at its margin as the whole system is fairly active. The accumulation zone of the glacier is on the mountain ice cap of Vatnajökull but only a relatively small part at a low elevation drains through this outlet glacier. The activity of Skeiðarárjökull is evident in the very large sandur that it has built out in front of its broad snout.

7 Floating ice tongues are at present restricted to high latitudes where glaciers can reach to sea level. The form that the floating part of the glacier takes depends on the surrounding coastal relief. Where the glacier is confined within a valley, the floating part of the glacier will be no wider than the grounded part. The glacier will lose mass

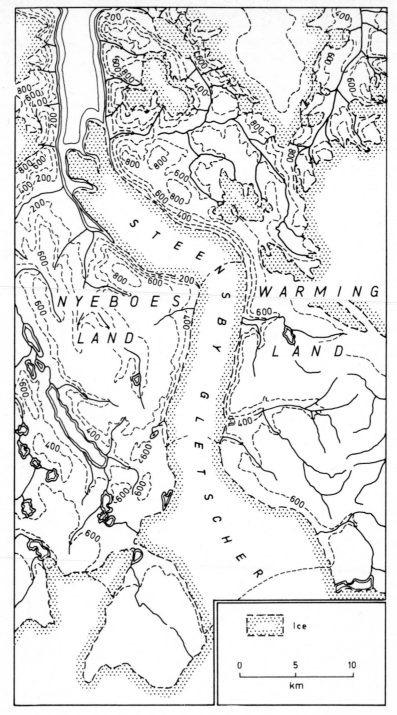

Fig. 3.12
Map of the Steensby Gletscher, north Greenland (F. Ahnert, *J. Glaciol.*, 1963, by permission of the International Glaciological Society)

by calving, thus creating icebergs. An example of a glacier ending in a fjord with a floating ice tongue is the Steensby Gletscher in north Greenland (Fig. 3.12). Glaciers of this type are relatively rare in the Arctic, because the glaciers are so heavily crevassed that they usually calve off before their tongues float. The floating part of the Steensby Gletscher tends to disintegrate into separate lobes. The glacier is an outlet glacier draining the inland ice and is 48 km long and 11 km wide near the outlet. It has a low gradient of 1·4 per cent as do the other floating ice tongues in this area. The outermost 8 km of the Steensby Gletscher are floating, and its thickness in this section is only about 75 to 100 m, judging from the 15 to 18 m high ice cliffs. The disintegration is thought to be caused by a bend in the fjord, which forces part of the glacier to ground while the rest continues to move. The movements causes rotation and splitting up of the floating ice. Two other examples of floating ice tongues in north Greenland are the Petermanns Gletscher and Ryder Gletscher, both of which are flat glaciers with little crevassing.

These narrow floating glacier tongues differ greatly from the floating ice shelves in size. Ice shelves are more characteristic of the Antarctic: a good example is the Ross shelf. The material of the shelf is partly derived from the outward flow of the inland ice sheet, but it is mainly supplied by the accumulation of snow on the upper shelf surface. This form of nourishment is particularly important in the Antarctic where the accumulation is much greater at the edge of the continent. The bulk of the Ross shelf ice is, therefore, formed of firn rather than true glacier ice. The floating ice is about 300 to 400 m thick and covers an area of 550,000 km², or more than the area of France (Fig. 3.13). The ice cliffs along the shelf edge are fairly high because the shelf consists mainly of not very dense firn. The density at a depth of 6 m is about 0·5. The calving of the barrier, as the ice shelf is called, produces very large tabular icebergs that only occur in the southern hemisphere, as ice shelves of this type are confined to the Antarctic. The barrier spreads out and moves under its own weight and its rate of movement is very variable. Observations record rates of from 4 m/year to a maximum of 844 m/year (Robin (1972) estimates maximum rates of the order of 2000 m/year; see p. 51). The shelf melts from below over most of its area, a process that is assisted by brine soaking. Along much of the Antarctic coastline, the ice sheet reaches the sea in ice cliffs.

8   Mountain ice caps are accumulation areas from which outlet glaciers flow to lower levels. They usually rest on upland plateau surfaces. Two good examples of this type of ice cap are Vatnajökull in Iceland and Jostedalsbreen in Norway. The former ice cap is about 600 to 760 m thick, resting on a flattish plateau at about 1000 to 1220 m in elevation. Jostedalsbreen also occupies a high plateau area that was uplifted in the late Tertiary. It has been dissected mainly by the vigorous outlet glaciers that drain the mountain ice cap. The Jostedalsbreen ice cap occupies the area from which the great Scandinavian ice sheet originally emanated in the Pleistocene, and to which it eventually shrank again. However, the present ice cap is probably not the direct remnant of the large Pleistocene ice sheet, but a renewed growth of ice since the climatic optimum. There are still several vigorous outlet glaciers descending from this elevated

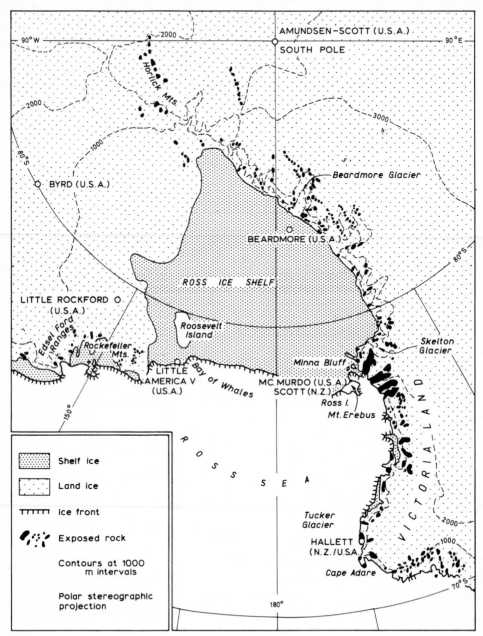

Fig. 3.13
Map of the Ross Ice Shelf, Antarctica

plateau ice cap which lies at about 1370 to 1500 m. The ice thickness is small and probably does not exceed 300 to 600 m.

9   Glacier cap or lowland ice cap are terms that can be used to describe the small ice masses that develop at fairly low levels on flattish country in the high arctic. There are good examples of this type of ice cap on the islands of the arctic regions of northern Canada. The Barnes ice cap on Baffin Island may be taken as an example. Its régime has already been discussed and its dimensions noted on page 54. It lies on a relatively low plateau surface and has no vigorous outlet glaciers. The ice cap is not active glaciologically, partly because of its arctic environment and partly because of the gentle relief around it. The Barnes ice cap is probably frozen to its base at least at its margins, although there is reason to believe that in the central part its base is at pressure melting-point (Holdsworth, 1973). There are large ice-dammed lakes on the north-eastern edge of the ice cap.

10   There are only two of the largest type of ice mass at present, namely, the ice sheets of Greenland and Antarctica. Some details of the dimensions of these ice sheets have already been given in connection with their régimes. The overwhelming size of the Antarctic ice sheet compared with any other ice mass is clearly apparent. Its presence alters the whole climate and life of the southern hemisphere. The other two major ice sheets of the Pleistocene ice age that rivalled it in size also exerted a similar widespread effect when they were at their maximum extent. At present, there is about 90 per cent of the world's ice in the Antarctic and about 9 per cent in Greenland (Chapter 1). The Laurentide ice sheet at its largest covered much the same area as the present Antarctic ice sheet, but the European one covered only about half this area during the maximum of glaciation.

A relationship between the surface gradients of ice sheets and glaciers on the one

Table 3.2   Regression analysis of glacier length against glacier gradient

Groups	Alaska				Canada			
	A	B	C	D	A	B	C	D
a value	6·15	7·95	3·67	7·81	11·92	15·08	—	3·61
b value	0·44	0·45	0·48	0·35	0·50	0·44	—	0·64
r value	0·856	0·770	0·935	0·747	0·781	0·736	—	0·976
s value	3·84	5·15	3·12	2·95	5·97	6·08	—	1·14
number	26	8	13	5	41	35	1	5

Groups	Antarctica				Greenland			
	A	B	C	D	A	B	C	D
a value	4·68	13·54	5·35	—	5·5	5·52	—	7·58
b value	0·52	0·47	0·49	—	0·76	0·78	—	0·62
r value	0·930	0·799	0·909	—	0·721	0·731	—	0·922
s value	11·54	18·52	7·68	—	18·26	19·08	—	6·28
number	37	11	26	—	34	76	3	5

x=glacier length in km, *y*=glacier gradient expressed as a fraction (1:25, 1:30, e.g.); *y*=*a*+*bx*
r=correlation coefficient, *s*=standard error

hand, and their size (length of glacier, for example) on the other hand, has been deduced by J. T. Buckley (1969). Although such a correlation must not be pressed too far in individual cases, and the data involved are clearly of a general nature only, the idea of such a relationship is useful in reconstructing the shapes of former (Pleistocene) ice sheets and glaciers. Buckley's data were obtained from four areas, (a) coastal Alaska, (b) coastal Greenland, (c) coastal Antarctica in the Ross ice-shelf area, and (d) the north-east Canadian Arctic Islands. The glaciers were divided into four groups A—all glaciers, B—glaciers flowing from ice caps into fjords, C—glaciers flowing from mountain headwalls to fjords or shelves, D—glaciers flowing from ice caps to fjord heads or land less than 200 m above sea level. The longer glaciers consistently have lower gradients; outlet glaciers have concave profiles, with a steep up-valley section, and a gentle middle section. Table 3.2 gives the complete results of the analysis.

4 Conclusions

The character of glacier ice movement depends upon the properties of the ice. The ice in turn depends upon the snow of which it is formed. The external factors that govern its change into ice and the forces exerted on the ice must also be considered. The physical properties of ice have been studied in the laboratory, and a simple power flow law relating strain rate to stress provides an approximate model of ice flow. More complex flow laws have recently been suggested.

Glaciers can be classified in various ways. The thermal classification is particularly significant in connection with the nature of ice flow. The differentiation between temperate ice, which is, in theory, at pressure melting-point throughout, and cold or polar ice is very important. The lack of meltwater at the base of the cold type is important and results in considerably slower movement. An ice sheet need not belong to one type over its whole area, and even large polar ice sheets are known to have basal areas at pressure melting-point. Glaciers can also be classified, according to their dynamic character, as active, passive, and dead. Dead ice need not necessarily be immobile, but the term implies that the ice no longer has a source of supply. The ice mass can live on capital for up to thousands of years in large ice sheets. Morphological characteristics provide a third criterion of classification, ranging from small niche glaciers to continental ice sheets. Only two of the latter type now exist, but in the Pleistocene period the Laurentide and north European ice sheets rivalled the largest existing ice sheet in Antarctica.

5 References

AHLMANN, H. W. (1935), 'Contribution to the physics of glaciers', *Geogrl J.* **86**, 97–113
  (1948), 'Glaciological research on the North Atlantic coasts', *R. geogr. Soc. Res. Ser.* **1**, 83 p.
AHNERT, F. (1963), 'The terminal disintegration of Steensby Gletscher, North Greenland', *J. Glaciol.* **4**, 537–45

ANDREWS, J. T. (1971), 'Englacial debris in glaciers', *J. Glaciol.* **10**, 410
(1972), 'Englacial debris in glaciers', *J. Glaciol.* **11**, 155

BARNES, P., TABOR, D. and WALKER, J. C. F. (1971), 'The friction and creep of polycrystalline ice', *Proc. R. Soc.* **324**A, 127–55

BEHRENDT, J. C. (1965), 'Densification of snow on the ice sheet of Ellsworth Land and South Antarctic Peninsula', *J. Glaciol.* **5**, 451–60

BENSON, C. S. and ANDERSON, quoted by MEIER, M. F. (1962), 260

BOULTON, G. S. (1970), 'On the origin and transport of englacial debris in Svalbard glaciers', *J. Glaciol.* **9**, 213–29
(1972), 'The role of thermal régime in glacial sedimentation', *Inst. Br. Geogr. Spec. Publ.* **5**, 1–19

BUCKLEY, J. T. (1969), 'Gradients of past and present outlet glaciers', *Geol. Surv. Pap. Can.* **69-29**, 13 pp.

BUDD, W. F., JENSSEN, D. and RADOK, U. (1969 [1970]), 'The extent of basal melting in Antarctica', *Polarforschung* **6**, 293–306
(1971), 'Re-interpretation of deep ice temperatures', *Nature, Lond. (Phys. Sci.)* **232**, 84–5

BULL, C. (1957), 'Observations in North Greenland relating to theories of the properties of ice', *J. Glaciol.* **3**, 67–72

CAROL, H. (1947), 'The formation of *Roches Moutonnées*', *J. Glaciol.* **1**, 57–9

COLBECK, S. C. and EVANS, R. J. (1973), 'A flow law for temperate glacier ice', *J. Glaciol.* **12**, 71–86

COURT, A. *et al.* (1957), 'The classification of glaciers', *J. Glaciol.* **3**, 2–7

GLEN, J. W. (1955), 'The creep of polycrystalline ice', *Proc. R. Soc.*, **228**A, 519–38
(1958a), 'Mechanical properties of ice. I. The plastic properties of ice', *Phil. Mag.* Suppl. 7, 254–65
(1958b), 'The flow law of ice', *Int. Ass. scient. Hydrol. Symposium Chamonix 1958, Publ.* **47**, 171–83

GOW, A. J., UEDA, H. T. and GARFIELD, D. E. (1968), 'Antarctic ice sheet: preliminary results of first core hole to bedrock', *Science* **161**, 1011–13

GROOM, G. E. (1959), 'Niche glaciers in Bünsow-land, Vestspitsbergen', *J. Glaciol.* **3**, 369–76

GUNN, B. M. (1964), 'Flow rates and secondary structures of the Fox and Franz Joseph Glaciers, New Zealand', *J. Glaciol.* **5**, 173–90

HAEFELI, R. (1952), 'Observations on the quasi-viscous behaviour of ice in a tunnel in the Z'Mutt Glacier', *J. Glaciol.* **2**, 94–9

HANSEN, B. L. and LANGWAY, C. C. (1966), 'Deep core drilling in ice and core analysis at Camp Century, Greenland', *Antarct. J.U.S.* **1**, 207–8

HOLDSWORTH, G. (1973), 'Barnes ice cap and englacial debris in glaciers', *J. Glaciol.* **12**, 147–8

LANDAUER, J. K. (1959), 'Some preliminary observations on the plasticity of Greenland glaciers', *J. Glaciol.* **3**, 468–74

LEWIS, W. V. (ed.) (1960), 'Norwegian cirque glaciers', *R. geogr. Soc. Res. Ser.* **4**, 104 pp.

LLIBOUTRY, L. (1969), 'The dynamics of temperate glaciers from the detailed viewpoint', *J. Glaciol.* **8**, 185–205

LLIBOUTRY, L. (1971), 'Permeability, brine content and temperature of temperate ice', *J. Glaciol.* **10**, 15–29

LOEWE, F. (1966), 'The temperature of the Sukkertoppen ice cap', *J. Glaciol.* **6**, 179

MANNERFELT, C. M. (1945), 'Några glacialmorfologiska formelement', *Geogr. Annlr* **27**, 1–239

MCCALL, J. G. (1952), 'The internal structure of a cirque glacier: report on studies of englacial movements and temperatures', *J. Glaciol.* **2**, 122–30
  (1960), 'The flow characteristics of a cirque glacier and their effect on glacial structure and cirque formation' in *Norwegian cirque glaciers* (ed. W. V. LEWIS), *R. geogr. Soc. Res. Ser.* **4**, 39–62

MEIER, M. F. (1960), 'Mode of flow of Saskatchewan glacier, Alberta, Canada', *U.S. geol. Surv. Prof. Pap.* **351**
  (1962), 'Proposed definitions for glacier mass budget terms', *J. Glaciol.* **4**, 252–265

MELLOR, M. (1959), 'Creep tests on Antarctic glacier ice', *Nature, Lond.* **184**, 717

MELLOR, M. and SMITH, J. H. (1967), 'Creep of ice and snow' in *Physics of snow and ice: International Conference on Low Temperature Science, Sapporo 1966* (ed. H. ŌURA), **1**(2), 843–55

MELLOR, M. and TESTA, R. (1969), 'Creep of ice under low stress' and 'Effect of temperature on the creep of ice', *J. Glaciol.* **8**, 131–52

MERCER, J. H. (1971), 'Cold glaciers in the central Transantarctic Mountains, Antarctica: dry ablation areas and subglacial erosion', *J. Glaciol.* **10**, 319–21

MILLER, M. M. (1952), 'The terms "névé" and "firn"', *J. Glaciol.* **2**, 150–1

NYE, J. F. (1952), 'The mechanics of glacier flow', *J. Glaciol.* **2**, 82–93
  (1959), 'The motion of ice sheets and glaciers', *J. Glaciol.* **3**, 493–507

OROWAN, E. (1949), Discussion on 'The flow of ice and of other solids', *J. Glaciol.* **1**, 231–6 and 238–40

PATERSON, W. S. B. (1969), *The physics of glaciers* (Oxford)
  (1971), 'Temperature measurements in Athabasca glacier, Alberta, Canada', *J. Glaciol.* **10**, 339–49

ROBIN, G. DE Q. (1955), 'Ice movement and temperature distribution in glaciers and ice sheets', *J. Glaciol.* **2**, 523–32
  (1972), 'Polar ice sheets: a review', *Polar Rec.* **16**, 5–22

ROBIN, G. DE Q., EVANS, S. and BAILEY, J. T. (1969), 'Interpretation of radio-echo sounding in polar ice sheets', *Phil. Trans. R. Soc.* **265**, 437–505

SCHYTT, V. (1964), 'Scientific results of the Swedish expedition to Nordaustlandet, Spitsbergen, 1957 and 1958', *Geogr. Annlr* **46**, 243–81

SHARP, R. P. (1958), 'Malaspina Glacier, Alaska', *Bull. geol. Soc. Am.* **69**, 617–46

SHREVE, R. L. and SHARP, R. P. (1970), 'Internal deformation and thermal anomalies in Lower Blue Glacier, Mount Olympus, Washington, U.S.A.', *J. Glaciol.* **9**, 65–86

STEINEMANN, S. (1954), 'Results of preliminary experiments on the plasticity of ice crystals', *J. Glaciol.* **2**, 404–12
  (1958), 'Experimentelle Untersuchungen zur Plastizität von Eis', *Beitr. Geol. Schweiz (Hydrologie)* **10**, 72 pp.

STUART, A. W. and BULL, C. (1963), 'Glaciological observations on the Ross ice shelf near Scott Base, Antarctica', *J. Glaciol.* **4**, 399–414

SUGGATE, R. P. (1950), 'Franz Josef and other glaciers of the Southern Alps, New Zealand', *J. Glaciol.* **1**, 422–9

THEAKSTONE, W. H. (1966), 'Deformed ice at the bottom of Østerdalsisen, Norway', *J. Glaciol.* **6**, 19–21

THOMAS, R. H. (1971), 'Flow law for Antarctic ice shelves', *Nature, Lond.* **232**, 85–7

VIALOV, S. S. (1958), 'Regularities of glacial shields movement and the theory of plastic viscous flow', *Int. Ass. scient. Hydrol., Symposium Chamonix 1958, Publ.* **47**, 266–75

WEERTMAN, J. (1961), 'Mechanism for the formation of inner moraines found near the edge of cold ice caps and ice sheets', *J. Glaciol.* **3**, 965–78

    (1964), 'The theory of glacier sliding', *J. Glaciol.* **5**, 287–303

4

Ice motion

My theory of Glacier Motion then is this: a Glacier is an imperfect fluid, or a viscous body, which is urged down slopes of a certain inclination by the mutual pressure of its parts.

(J. D. FORBES, 1843)

The movement of glaciers has been known to the inhabitants of Alpine regions since at least the sixteenth century, from the records of ice invading formerly uncovered ground, and from the changes in position of the most favourable crossings of the glaciers. In 1653, prayers were offered and a procession led by the Bishop in an attempt to stop the advance of the Aletsch glacier over cultivated land. Actual measurements of the rates of motion were undertaken in the eighteenth century and subsequently: G. J. Hugi recorded that his hut on the Unteraar glacier shifted 1·4 km between 1827 and 1840. An early theory of glacier motion, that of H. B. de Saussure (1760), held that ice moved downhill as a rigid body, sliding on bedrock, but by 1773, an element of plasticity or viscosity in glacier movement had been detected by A. C. Bordier.

Further observations and measurements of ice motion were undertaken by a number of pioneer workers in the mid-nineteenth century—L. Agassiz on the Unteraar glacier, J. D. Forbes on the Mer de Glace, and the Schlagintweit brothers on the Pasterze glacier. Agassiz (1840) established that the centre of a glacier flows faster than its sides, and that the rate of movement is slower in the source regions and in its terminal parts than in between. In the years 1845 and 1846, Agassiz and his co-workers succeeded in measured velocities at two sites on the Unteraar glacier for every month, finding

Table 4.1 Displacement of surface markers on the Mer de Glace near Montan-vert measured by J. D. Forbes in 1842

Marker	Distance (yards)		Mean daily motion in summer (inches)
	From left bank	From right bank	
D_2	100	(615)	16·3
D_5	230	(485)	20·8
D_6	305	(410)	21·2
D_3	365	350	23·1

rates varying from 48·5 m/yr in the winter of 1845–6 to 138 m/yr for the summer of 1846. This confirmed Forbes' earlier prediction (1843) that the velocity of flow is greater in warmer weather. In addition, Forbes found that glaciers often flow faster where their valleys narrow, and that since the central portions flow faster than the sides (Table 4.1), it seemed to him likely that the surface layers flow faster than the basal layers. He further argued that glacier ice is plastic under stress, and that whereas some glaciers may be frozen to their beds, others may not be, allowing them greater freedom of movement. It is remarkable that these assertions, made over a century ago, have proved to be substantially correct.

The latter half of the nineteenth century witnessed the emergence of various theories of partial or complete plastic deformation of ice under stress (G. Seligman, 1949). F. Pfaff (1874) claimed that plasticity decreased with a lowering of ice temperature, and also, incidentally, was the first to notice, in the firn regions, that vertical as well as horizontal components of surface motion existed. A classic summary of the facts of glacier motion then known, and of the various theories of ice motion propounded, is contained in A. Heim's *Handbuch der Gletscherkunde* (1885), which gives a full exposition of the concept of glacier yielding by intergranular liquefaction and regelation. Theories of plastic flow by the turn of the century were numerous and varied, though none had yet been tested in the laboratory; they included theories of differential movements of ice crystals, deformation of the ice crystals themselves including gliding along crystallographic planes, and rearrangement of molecules within the ice crystals (H. Hess, 1904). On the other hand, many were still convinced that glacier yielding was more a matter of differential flow along internal shear planes, possibly numerous and closely spaced. Such was the opinion, particularly, of H. Philipp (1920). Forbes had long ago demonstrated the existence of such shear planes, and R. T. Chamberlin (1928) successfully measured differential motion along them in the terminal parts of the Brenva glacier.

Studies of glacier flow as a mathematical and hydrodynamic problem began to emerge in the 1920s. Outstanding for the clearness of its exposition was the work of C. Somigliana; it was based, however, on the erroneous assumption that ice is a substance of constant viscosity, as was the work of M. Lagally (1929, 1934, 1939). Lagally arrived at a figure of 10^{14} poises for the approximate viscosity of glacier ice, and deduced a relation between viscosity (μ), ice thickness (h), ice density (ρ), bedrock slope (α) and velocity (v) of a point on the glacier centre line:

$$v = \frac{h^2 \rho g \sin \alpha}{2\mu} \quad \text{(where } g = \text{gravitational acceleration)}$$

The relationship was checked by observations of v and seismic determinations of h on the Pasterze glacier, and it was concluded that ice behaved as a viscous liquid, a conclusion that has since been firmly refuted.

Improvements in techniques for measuring glacier flow continued, and European glaciers were quite intensively surveyed by the turn of century (see, for instance, P. L. Mercanton's (1916) classic study of the flow of the Rhône glacier, 1874–1915). In 1928, R. Finsterwalder (1931) was the first to adapt photogrammetric techniques to the precise measurement of ice motion. Equally important to theories of glacier flow is a knowledge of ice thickness and bedrock configuration; in this field too, developments in seismic (and later, gravimetric) techniques made possible much more reliable determinations of the form of subglacial rock surfaces. The year 1937 saw the beginning of some very important glaciological investigations of the Jungfraujoch area in Switzerland. M. F. Perutz and others found evidence both of plastic deformation of moving ice, and of laminar flow involving some thrusting along discrete shear planes. Plastic deformation was associated with evidence of strong preferred orientation of the ice crystals, whose basal planes were found to be aligned parallel to the direction of ice flow (Perutz, 1940).

With the Jungfraujoch investigations, we enter the modern period of research into the deformation of ice and the mechanism of glacier flow. Since 1937, field research and laboratory experimentation have led to great advances in our knowledge of the behaviour of ice under stress, and the following sections of this chapter will present the most important results of this work as it bears on the problem of ice motion. There are still, however, considerable gaps in our understanding of this problem, for field measurement has lagged substantially behind the development of theoretical concepts. Moreover, most laboratory and field research on glacier flow has until recently been carried out in connection with temperate glaciers, which yield more readily to stress and flow more quickly than cold glaciers. We still need to know much more about the internal movements and temperatures of cold ice sheets and glaciers. We also need a great deal more information about the mechanism by which ice slides over an uneven surface of bedrock, for this is now known to be the most important contributor, in the case of temperate glaciers, to total glacier movement and ice discharge. The mechanism of basal sliding is much less well understood than the internal deformation of a moving glacier but it is of fundamental importance geomorphologically since it is the key to understanding glacial erosion.

1 Surface velocity observations

Maximum rates of ice motion vary enormously from one glacier to another, and over periods of time. The majority of valley glaciers move at speeds of less than one metre a day. For brief periods, however, some glaciers experience surges (Chapter 2) in which rates of flow become abnormally high: during part of 1937, for instance, the Black Rapids glacier in Alaska probably attained a speed of 75 m/day, while in the Alps, the Vernagtferner (whose normal maximum velocity is about 5 cm/day) in 1845 reached a speed of 11 m/day. Over longer periods, the most rapidly moving glaciers in the world are those descending steep slopes and fed by large accumulation areas: of the Greenland outflow glaciers, the Jakobshavn Isbræ moves at up to 20 m/day (A. Helland gave such an estimate for this glacier's movement as long ago as 1877), as does the Great Qarajaq glacier (L. A. Lliboutry, 1969a), and the Rinks Isbræ up to 28 m/day (W. Kick, 1957). The rates of discharge of these glaciers are also immense: Jakobshavn Isbræ pours out 14 km³/year. The fact that the lower portions of these glaciers float and calve into sea inlets also helps them to attain these unusually high rates of movement. Only a few Antarctic glaciers attain such high velocities and discharges. Although difficult to measure because of the absence of bedrock outcrops, most of the inland ice moves sluggishly. About a quarter of it is discharged in relatively narrow ice streams, some of which attain velocities of 20 m/day where they reach the sea and float. The high velocities of parts of the floating ice shelves have already been noted (p. 101).

The methods used to measure surface velocities consist of trigonometrical survey by theodolite of stakes bored into the ice to sufficient depth to offset surface ablation, of photogrammetric survey (Finsterwalder, 1931), and of mechanical devices connecting ice to bedrock (R. W. Galloway, 1956, and the 'cavitometer' described by R. Vivian and G. Bocquet, 1973). The first, though laborious, is by far the most accurate, and can detect vertical as well as horizontal components of motion. A problem for movement studies of ice caps and ice sheets is the lack or rarity of bedrock outcrops for trigonometrical or photogrammetric control. In such cases, traversing may be the only practicable form of survey. The traverse should follow flow lines so far as these can be estimated in advance and new techniques of electronic distance measurement are replacing conventional methods using steel tape. Where no bedrock outcrops are available, or the traverse becomes unacceptably long, the only recourse is to astronomical fixes repeated over time, but such determinations of ice movement, even over periods of several years, are relatively inaccurate.

An analysis of the flow of the Saskatchewan glacier (M. F. Meier, 1960) exemplifies the level of precision attainable by modern theodolite survey in the case of a valley glacier. The probable errors in the determinations of flow were less than 0·06 cm/day for horizontal motion and 0·11 cm/day for vertical movements. Maximum velocities, up to 32 cm/day, occurred at the firn-line, diminishing to less than 1 cm/day near the snout. This reflects the fact that, in a glacier flowing down a channel of uniform cross-section and slope, velocity will be greatest at the firn-line where the volume of ice being discharged reaches a maximum; down-valley, the volume of ice decreases by ablation. Actual glaciers do not flow in channels of uniform cross-section or slope, but nevertheless a longitudinal diminution of velocity below the firn-line is often apparent. Fig. 4.1 shows the variation in rate of flow along Austerdalsbreen, Norway, from the point where it leaves the plateau ice-cap to its snout.

Transverse surface velocity profiles for the Saskatchewan glacier show greatest speeds in the central portion and a four- or five-fold decrease within 50 m of the glacier margin. The form of the profiles is parabolic. Fig. 4.2A plots the ratio of velocity down-glacier (V_x) to centreline velocity down-glacier (V_c) as a function of distance from centre to margin. The distance S represents side-slip of the glacier; the results represent the average of four transverse velocity profiles, and are typical of most valley glaciers. Fig. 4.2B shows some plots of transverse variations of surface velocity on Athabasca glacier for comparison.

Precise surveys also often reveal actual transverse components of flow in valley glaciers. Flow of part of the Saskatchewan glacier, for instance, is directed obliquely towards the southern valley side, a movement that represents ice taking the place of that lost by increased ablation near the rock side. Transverse components of flow also often appear where valleys and their glaciers widen.

Vertical components of flow (those resulting from ice motion, not accumulation or ablation) are more difficult to measure, but must not be neglected. The general

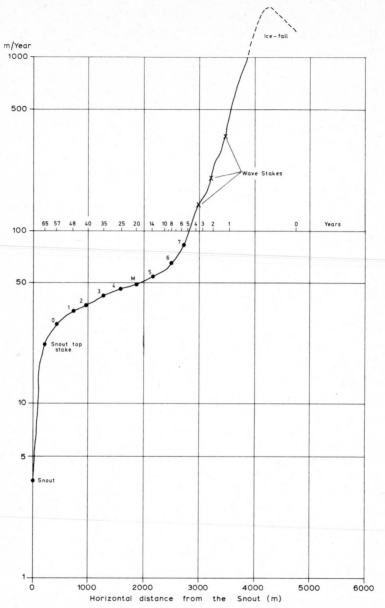

Fig. 4.1
Variation in the rate of flow of Austerdalsbreen, Norway, from the head of the glacier (ice-fall) to its snout (C.A.M. King and W.V. Lewis, *J. Glaciol.*, 1961, by permission of the International Glaciological Society)

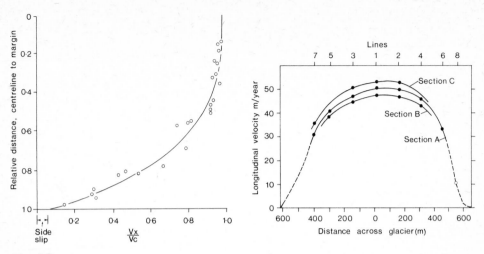

Fig. 4.2
A Ratio of down-glacier velocity (V_x) to centreline down-glacier velocity (V_c) as a function of relative distance from the centreline. Data obtained from four transverse profiles, Saskatchewan glacier, Alberta (M. F. Meier, *U.S. geol. Surv. Prof. Pap.* 351, 1960)
B Lateral variation of surface velocity for three cross-profiles on Athabasca glacier, Alberta (C. F. Raymond, *J. Glaciol.*, 1971, by permission of the International Glaciological Society)

tendency is for the ice to have a downward component of motion in the firn region, and an upward component near the snout. On the Saskatchewan glacier, velocity vectors plunge beneath the surface from above the firn-line to 2 km below it; farther down-glacier, the vectors gradually begin to rise out of the surface (Fig. 4.10). Near the terminus, vertical components of motion reach values of 8 mm/day, nearly as much as the horizontal component.

The Saskatchewan glacier represents a single ice-stream; some large valley glaciers, however, consist of several ice-streams flowing side-by-side, so that transverse surface velocity profiles may be more complicated. Fig. 4.3 shows the Fröya glacier in north-east Greenland (W. R. B. Battle, 1951), where three ice-streams (A, C, E) with velocities of 4·5, 10, and 10·5 cm/day flow together, separated by zones (B, D) of more slowly moving ice. Hidden rock ridges beneath B and D may explain the unusually slow movements in these two zones.

Even in single valley glaciers, however, the transverse velocity profiles are not always of the parabolic form illustrated by the Saskatchewan glacier. Some glaciers exhibit a form of flow termed plug flow or Blockschollen movement (R. Finsterwalder, 1950), in which there is a very rapid increase in velocity close to the valley sides and little differential surface movement in the main central mass of ice. Finsterwalder claimed that Blockschollen flow is mainly associated with fast-moving glaciers in which the ratio of mean velocity to width is greater than about 0·16. Blockschollen movement is typical of many surging glaciers, and often results, if the bed is irregular, in the ice surface breaking up into a series of seracs (as in an ice-fall). An example of Blockschollen flow for a sub-polar glacier, the Kongsvegen glacier in Vest-Spitsbergen, is given by W. Pillewizer (1969). The high velocities attained near to the bedrock in Blockschollen

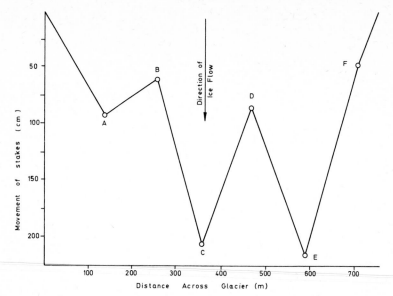

Fig. 4.3
Transverse velocity distribution of the Fröya Gletscher, Greenland, as shown by the movement of
six stakes (A to F) over a period of three weeks, August 1949 (W. R. B. Battle, *J. Glaciol.*, 1955,
by permission of the International Glaciological Society)

flow have important consequences for glacial erosion (see pp. 251–3). Blockschollen
flow is an extreme form of transverse velocity distribution; there are many glaciers
that exhibit a U-shaped transverse velocity profile, in between the plug form of flow
(Blockschollen) and the parabolic transverse velocity profile.

1.1 *Strain-rate measurements*

Changes in velocity along or across a glacier surface are often more significant than
absolute velocity values. Changes in velocity ('velocity gradients') are expressed as
strain rates, which may be extending or compressing. Strain rates may be given as
per cent change in length over a unit of time; commonly, the unit of time is a year
and the amount of extension or compression is written simply as a dimensionless number.
Thus a strain rate of $0 \cdot 001 \, yr^{-1}$ signifies a change in length of 1 part in 1000 over a
year. Longitudinal strain rates on a glacier surface can be measured with reference
to a line of stakes along a flow-line of the glacier, but this assumes knowledge of the
direction of a flow-line (although this is approximately given by the form of the channel
in the case of a valley glacier, it will not be reliably known for an ice sheet or ice cap).
A line of stakes will therefore only give an approximate picture at best, and it will
not give information about transverse strain rate. Much more satisfactory is to determine
all the strain-rate components. The standard method of performing this is to set up
a strain-rate 'diamond', consisting of a diamond pattern of four stakes, with one diagonal
lying approximately along the flow line and a stake where the two diagonals cross.
The sides and diagonals are measured with steel tape and the measurements repeated

after any desired interval of time (a few months or a year, commonly). J. F. Nye (1959c) gives details of the computation of strain rate components from these data.

1.2   *Time-variation of velocity*

Many workers since the time of Agassiz and Forbes have noticed that glaciers do not flow at absolutely constant rates. In the ablation zone, there is a tendency for the ice to move faster in summer than in winter, while in the névé region, the reverse is often the case. Short-period fluctuations also occur, but are much more difficult to measure.

(*a*) *Seasonal variations*   More rapid summer movement of ice in the ablation zone is definitely established, for instance, on the Hintereisferner (R. Haefeli, 1948), on the Saskatchewan glacier (Meier, 1960), on the Gornergletscher (G. R. Elliston, 1963), on the Athabasca glacier (W. S. B. Paterson, 1964), and on Østerdalsisen, Norway (W. H. Theakstone, 1967). Elliston found that velocities were 20–80 per cent greater than the annual average in summer, and 20–50 per cent less in winter. Paterson showed

Table 4.2   Movement of White Glacier, Axel Heiberg Island, Canada (F. Müller and A. Iken, 1973)

Profile position	Year	Annual movement cm/day	Increase of summer velocity over the mean annual movement
			%
8·5 km from	1960–61	10·1	0·2
snout (near	1961–62	10·5	28
mean equilibrium	1962–63	9·9	—
line)	1963–64	10·3	—
	1964–65	9·9	—
	1965–66	10·0	9
	1966–67	10·2	10
	1967–68	10·4	15
3·2 km from	1960–61	8·3	49
snout	1961–62	8·9	46
	1962–63	7·9	57
	1963–64	7·9	—
	1964–65	8·5	—
	1965–66	8·8	38
	1966–67	8·8	58
	1967–68	8·8	36
0·7 km from	1960–61	3·0	51
snout	1961–62	3·1	61
	1962–63	2·7	32
	1963–64	2·7	—
	1964–65	2·7	—
	1965–66	2·9	15
	1966–67	2·9	49
	1967–68	2·8	37

that mean velocities on the Athabasca glacier were 15 per cent greater for the period April–July than for the year. These variations are not confined to temperate glaciers. Paterson (1961) notes that the Sefstrøm Gletscher in north Greenland flowed faster at the beginning of August than at the end of that month when surface melting had ceased. Table 4.2 lists some data for White Glacier on Axel Heiberg Island which is of sub-polar type—it is probably not frozen to bedrock except for its marginal parts, which may only be loosened in late summer. Data from cold ice sheets and glaciers, frozen to bedrock, are still inadequate, but recent work in Antarctica (e.g., C. Swithin-bank, 1966) has not yet revealed seasonal variations in flow greater than 1 per cent.

There are several possible causes of more rapid summer flow of temperate and sub-polar glaciers. Higher air temperatures are prevalent in summer, yet of course the ice itself in a temperate glacier cannot fluctuate in temperature (other than in the surface layer affected by winter cold) as it is already at pressure melt-point. But summer is the season of surface melting, and meltwater is thought to play an important part in lubricating the bed of a glacier and thereby enhancing the sliding process (see p. 142). The problem is to what extent surface meltwater may penetrate to the bed of a thick glacier, and whether it could reach the bed of a sub-polar glacier. As truly cold glaciers are frozen to bedrock, the lack of seasonal variation in the flow of Antarctic outlet glaciers is consistent with the meltwater lubrication hypothesis.

The situation is less clear in respect of sub-polar glaciers, for which data on seasonal variations have been rather inadequate until recently. For the Bersækerbræ in east Greenland (T. W. Friese-Greene and G. J. Pert, 1965), there is a possible correlation between velocity of ice flow and ablation rates, so that some surface meltwater might be finding its way to the glacier bed in summer at least, possibly down the sides of the glacier. The Kongsvegen glacier in Spitsbergen (Pillewizer, 1969) has low rates of flow (1·35–1·50 m/day) for about 10 months of the year but then suddenly accelerates (to 3·50 m/day) for the two summer months, the timing of the change in flow rate correlating with variations in discharge of meltwater which reaches a marked peak in July followed by a rapid decline in August. Fig. 4.4 shows details of the movement and rate of ice melting for the White Glacier, Axel Heiberg Island, through part of 1968. The spring increase in velocity occurred almost simultaneously at all stations on the glacier, lagging a few days behind the earlier rapid increase in surface melting. Muller and Iken (1973) suggest that it takes a little time, in the case of a sub-polar glacier, for the englacial and subglacial drainage systems to be restored after the winter and for meltwater to raise the temperature in the basal ice to pressure melting-point. There is thus a considerable body of evidence to suggest that

1 the greater availability of meltwater in summer (from both ice melt and precipitation as rain) enhances the basal sliding of temperate glaciers;
2 in the case of sub-polar glaciers, basal sliding is restricted except in summer when meltwater may penetrate to the bed in quantity and cause a sudden acceleration in total movement;
3 cold glaciers frozen to bedrock all year do not exhibit any marked seasonal variations in rates of flow.

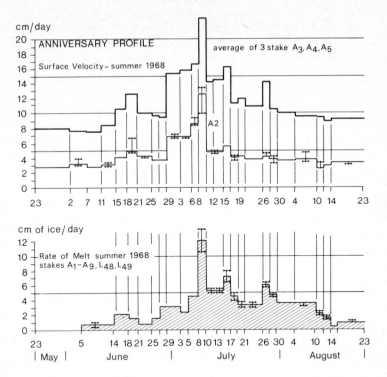

Fig. 4.4
Summer velocity changes on the White glacier, Axel Heiberg Island, and comparison with changes in ablation rates, May–August 1968 (F. Müller and A. Iken, *Int. Ass. scient. Hydrol., Symposium on the hydrology of glaciers 1969*, 1973, by permission of the authors)

The faster movements sometimes noted in winter time for the firn regions of temperate glaciers are less clearly understood. In 1942, determination of the velocity of a point on the Jungfraufirn at 3350 m revealed a winter velocity twice as great as that of the summer (Haefeli, 1948). One possible explanation is that the greater thickness of the firn-snow blanket increases the shear stress and therefore raises the flow rate. E. Orowan (1949) suggested that even a 1 per cent weight increase might raise the flow rate substantially, in spite of the lower temperatures in the upper layers. However, surface velocity measurements in firn are reliable only if the marker stakes are drilled to depths greater than the new snow layer (and preferably into ice), for new snow can exhibit considerable movement itself arising from creep and compaction (J. G. McCall, 1952).

(*b*) *Short-interval variations*   Variations in rate of flow over periods of hours, days or weeks have been recorded by a variety of methods. Repeated theodolite surveys can be used to detect variations in surface movement, though extreme precautions must be taken to minimize instrumental error as the movements in question are often minute. J. E. Jackson (1953) has drawn attention to the striking effects of atmospheric refraction changes on theodolite observations, and the theodolite must be protected from the sun. Other techniques have included continuous recording at the end of invar wire anchored to bedrock, a method of most use near to valley side-walls or in subglacial cavities, and the use of strain gauges to test basal ice movement variations directly. The first method, theodolite survey, is laborious, but has given some useful results. Meier (1960), for instance, reports short-interval variations in flow for the surface of the Saskatchewan glacier. These results were considered to be accurate to within 15 mm/day, and revealed sudden jerky movements of the ice, involving sudden increases in velocity of up to 170 per cent, and sudden decreases up to 230 per cent (giving actual reversals of motion). The fluctuations were irregular and not synchronous from one marker to another. J. W. Glen and W. V. Lewis (1961) also found rapid and irregular changes in the rate of side-slip for Austerdalsbreen (Norway). They suggested that the uneven movement recorded might result from sidewall irregularities or boulders jammed in the ice-rock gap, causing stress to build up until the ice suddenly freed itself with a jerk. But an equally important factor seemed to be weather conditions, for rapid movement often followed on heavy rain and warm air temperatures (one marker moved 5 cm in 8 hours, seven times its normal speed, following a period of rain), suggesting again that melt-water lubrication may be very important in the slip of ice over rock. On Morsárjökull in Iceland, J. D. Ives and C. A. M. King (1955) also noted a possible correlation between rapid ice movement and periods of heavy rainfall (Fig. 4.5). Meier (1960) for the Saskatchewan glacier suggested a relation between velocity and meteorological conditions, but noted that overshadowing such a correlation were other causes of differ-ential motion, namely, localized differential shearing in the ice, and irregular move-ments of inter-crevasse blocks. Paterson (1964), discussing weekly variations of flow of the Athabasca glacier, found that increased meltwater discharge from the glacier snout tended to occur 3 or 4 days after an increase in velocity had been measured about 2·5 km up-glacier. This suggested that, if an increase in basal meltwater lubrica-tion was responsible for the acceleration of ice flow, the thicker basal water film travelled down-glacier at about 0·8 km/day. J. Weertman (1962) arrived at a similar figure on theoretical grounds, postulating that the thicker zone of basal meltwater is propagated down-valley under the ice as a kinematic wave. The agreement between theory and observation is striking but may be fortuitous, and many more observa-tions are needed before the relationship, which may be very important for the basal sliding of temperate glaciers, can be regarded as more than very tentative.

Variations in meltwater lubrication probably do not, however, explain all short-interval variations in flow. R. P. Goldthwait (1973) has reported studies of irregular glacier motion for seven glaciers in widely differing environments, including both temperate and cold glaciers. Using various techniques, and over periods ranging from 15 minutes to 12 hours, he records jerky motion at the surfaces of all glaciers, though not at all positions on those glaciers. Apparently instantaneous movements occurred, from less than 1 mm to 20–30 mm. But no jerky motion was measured in the basal ice of any of the glaciers. The sample of seven glaciers is clearly small and does not reflect adequately the range of glacier behaviour, for jerky motion has been measured

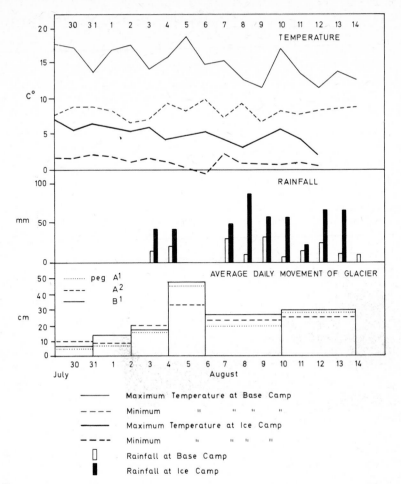

Fig. 4.5
The relationship between precipitation and ice velocity on Morsárjökull, Iceland (J. D. Ives and C. A. M. King, *J. Glaciol.*, 1955, by permission of the International Glaciological Society)

in basal ice elsewhere (e.g. Theakstone, 1967; Vivian and Bocquet, 1973). But Goldthwait's finding of jerky motion at the surfaces of cold glaciers shows that factors other than basal meltwater lubrication must be sought. He suggests tentatively that variations in solar radiation may affect the availability of interstitial water in the surface layers, since the time of greatest slip release showed some relation with the incidence of solar radiation.

2 Englacial velocity measurements

Measurements of velocity inside a glacier or ice sheet can only be made by means of artificial tunnels or boreholes (or in exceptional cases by utilizing subglacial stream tunnels). Tunnelling is an arduous task, though it provides a wealth of information not only on englacial movements but also on glacier structure. Deep boring is best carried out by electrically heated hotpoints, and the borehole must be immediately lined with aluminium or plastic to prevent its closure by refrozen meltwater. In the case of cold glaciers, water produced by the drill hot-point must be removed as it forms; otherwise it will refreeze behind the drill and block the hole. Thus drilling in cold ice is more difficult and costly. Drilling of boreholes in glaciers dates back to Agassiz' attempts on the Unteraar glacier about 1840; while between 1889 and 1909, some deep borings were made in the Hintereisferner by A. Blümcke and H. Hess, to depths of up to 224 m (Blümcke and Hess, 1909; Hess, 1933).

There is always a problem with boreholes in knowing whether they have reached bedrock, or whether they have reached a chance boulder in the ice. This uncertainty can be reduced by other methods of determining ice thickness (see p. 123) as checks (seismic, accurate to a few metres; gravimetric, accurate to about 10 m; electrical resistivity, less reliable), or by drilling a series of boreholes along a cross-profile to see if the supposed bedrock profile 'looks' reasonable. But none of these checks enables one to know whether the borehole has stopped in basal debris close to, but not actually reaching, bedrock. Since in many glaciers (especially cold glaciers) the lowermost layers may be a zone of intense deformation, the base of the borehole ending in basal debris will not give a true measure of basal sliding.

Another disadvantage of boreholes is that the hole and its casing may affect the flow of the ice, albeit to a small extent, and because the hole must change in length as it deforms with the ice flow and the casing cannot, the casing must slip lengthwise in it. So while measurements of deformation of the hole can give the two components of ice motion perpendicular to the hole, there is no means of measuring the component parallel to the hole. Thus velocity vectors are not completely determined.

The deformation of boreholes is measured by inclinometer. Such instruments have been greatly improved in recent years (for a new design, see C. F. Raymond, 1971a), but their accuracy is far less than that of surface glacier surveys. Graphs of borehole deformation always show 'waviness' and this is partly, at least, due to errors in inclino-metry.

The first reliable measurements of the vertical velocity distribution in a glacier were made in 1948–9. In August 1948, a near-vertical tube was emplaced in the Jungfraufirn (Perutz, 1950) to a depth of 136 m, where an obstruction, probably solid rock, was met. Distortion and displacement of the tube by October 1949 showed a surface movement of 38 m (in 14 months) and a bottom movement of 10 m. In the top 50 m of the borehole, there was no perceptible differential movement, and from −50 to −136 m there was a continuous retardation of flow, the rate of retardation increasing nearer the bottom (Fig. 4.6A).

A few years later, a deeper borehole was successfully sunk in the centre of the Mala-spina glacier (R. P. Sharp, 1953) where seismic data suggested an ice thickness of 595 m. The borehole reached 305 m before the hotpoint failed. Deformation of the hole after one year is shown in Fig. 4.6B where net displacement from a vertical straight line is plotted. There was no perceptible differential flow to a depth of 90 m, and only slight

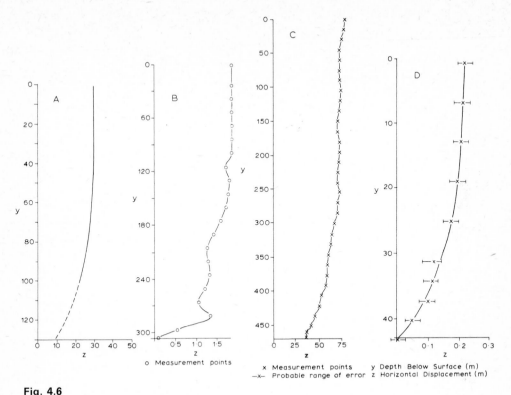

o Measurement points

x Measurement points   y Depth Below Surface (m)
—x— Probable range of error z Horizontal Displacement (m)

Fig. 4.6
The deformation of vertical boreholes in glaciers:
A Jungfraufirn (M. F. Perutz, 1950): August 1948—October 1949;
B Malaspina glacier (R. P. Sharp, 1953): July 1951—August 1952;
C Salmon glacier (W. H. Mathews, 1959): displacements computed for a one-year period;
D Saskatchewan glacier (M. F. Meier, 1960): August 1952—August 1954.
Note that the vertical and horizontal scales differ in each case. Direction of ice flow left to right.
Many of the minor irregularities in the curves for **B** and **C** may have no significance, being smaller
than the possible errors of measurement. In **D**, the probable errors are indicated for each measurement,
and the smooth curve represents a line of best fit passing within the range of error for each data
point.

variations in velocity to 285 m, but a marked retardation in flow below that level.

An unusually complete record of englacial velocity distribution was provided by boreholes in Salmon glacier, British Columbia (W. H. Mathews, 1959). About one-third of the distance across the glacier surface, a pipe was sunk to −481·5 m, ending within probably 8·5 m of the rock floor. The velocity at the surface was 22·5 cm/day, 20·7 cm/day at −240 m, and 12·4 cm/day at the bottom; the vertical velocity profile (Fig. 4.6C) consisted of a smooth curve corresponding closely to J. F. Nye's (1952) theoretical curve for laminar flow of ice resting on a smoothly inclined plane (p. 128).

The team investigating the Saskatchewan glacier (Meier, 1960) made many unsuccessful attempts to sink deep boreholes; in the end, one pipe was inserted to a depth of 50 m (glacier thickness here about 500 m). The interest of this experiment is that very precise measurements of the displacement of the pipe over 2 years revealed differential movement throughout the depth of 50 m and even in the top 10 m (Fig. 4.6D); this conflicts with the previous suggestions of many glaciologists (such as M. Demorest, 1938) that the uppermost 30 m or so of a glacier remain in a rigid 'brittle' condition. The deformation of this pipe showed no sudden change of velocity at any special depth; the velocity simply decreased exponentially.

Two of the most detailed recent surveys of englacial velocity have been carried out on Athabasca glacier, Alberta (1966–7) and Lower Blue glacier, Washington State (1957–61). Fig. 4.7A shows the deformation of nine boreholes sunk along three cross-sections of Athabasca glacier (1, 5, and 3 holes respectively), all except one penetrating 98 per cent or more of the glacier thickness (Raymond, 1971b; B. Kamb and R. L. Shreve, 1966). The relative proportions of internal ice deformation and basal sliding revealed by this survey are of great interest and will be discussed in later sections of this chapter.

The data from all these boreholes bear out Forbes' contention that, in a vertical profile, velocities generally decrease from glacier surface to bedrock. It seems unlikely that further borehole experiments will substantially alter this picture, save possibly in special locations. Further boreholes are urgently needed, however, to test theories of rates of ice deformation in different conditions, and especially in cold glaciers and ice sheets, for it should be noted that all the boreholes described have been in temperate ice.

Just as variations of strain rate can be measured at the glacier surface (p. 114),

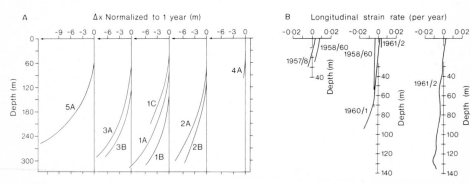

Fig. 4.7
A Longitudinal displacement with respect to the surface as measured in several boreholes in Athabasca glacier, Alberta (C. F. Raymond, *J. Glaciol.*, 1971, by permission of the International Glaciological Society)
B Variation of longitudinal strain rate with depth, for several boreholes on Lower Blue glacier, Washington State. Extensional strain rates are positive (R. L. Shreve and R. P. Sharp, *J. Glaciol.*, 1970, by permission of the International Glaciological Society)

so pairs of boreholes can be used to study variations of strain rate with depth. Fig. 4.7B illustrates, for a site on the Lower Blue glacier, how, near the surface at this particular site, the longitudinal strain rate is extensional, whereas at depth it is compressional (Shreve and Sharp, 1970). Since ice is virtually incompressible and the vertical strain rate is virtually zero, the compression at depth here must be compensated by transverse extension. This agrees with the fact that the cross-section of the Lower Blue glacier becomes wider and its sidewalls slightly steeper down-glacier from this site.

Information on englacial velocities from tunnels is limited because of the fact that tunnels in temperate glaciers must be constructed horizontally or sloping outwards to allow meltwater drainage; consequently, such tunnels have been largely confined to ice-falls or relatively steeply sloping cirque glaciers. The velocity distribution given by the Vesl-Skautbreen cirque glacier tunnel is described in Chapter 7. The data obtained from some other tunnels in Norway, Switzerland and France will be considered later in this chapter. Tunnelling in cold ice can present fewer problems in some respects, since there is no meltwater to contend with, and the termini of land-based cold glaciers often take the form of ice cliffs into the base of which tunnelling operations can be conducted. Several cold polar glaciers have been studied in this way in recent years, and later sections of this chapter will consider the results obtained, for example, on Meserve glacier in Antarctica and a glacier in north-west Greenland near Thule.

In rare cases, natural subglacial cavities and meltwater stream tunnels can be utilized for making englacial velocity observations. H. Carol (1947) was able in this way to penetrate to a depth of 50 m below the surface of the Grindelwald glacier in Switzerland. In the subglacial cavity, he was able to witness ice being forced locally over a bedrock obstruction, causing a sudden increase in the stress applied to the lower layers of ice which resulted in their being squeezed forwards at an abnormally high rate of flow (see also Chapter 3, p. 82). Within 1 m of the bedrock, speeds of flow reached 72 cm/day, whereas higher up in the ice, a more normal velocity of 37 cm/day prevailed. This is one of the few recorded cases where a basal ice layer has been shown to flow faster than the ice above, and is relevant to the discussion of the extrusion flow theory on page 133. It is important to note that this example was related to a particular bedrock obstruction and does not imply that the basal ice was travelling more rapidly over the whole width of the glacier (W. V. Lewis, 1947a).

Quite extensive exploration of subglacial cavities beneath the Glacier d'Argentière, France, has been reported by Vivian and Bocquet (1973) and others. Velocity measurements were made by cavitometer (see *Ice* **35** (1971), 6–7). Such direct observations of basal ice movement, possible only in tunnels or natural cavities, yield data of a far higher order of accuracy than those obtained from boreholes and inclinometry.

3 Measurements of ice thickness

Data on ice thickness are essential for testing any theories of ice motion, and at this point it is useful to summarize the techniques for measuring ice thickness. The most accurate and direct is to sink boreholes or drive tunnels into a glacier, and these have already been mentioned in the last section when englacial velocity measurements were described. A method giving ice depths accurate to within a few metres under good conditions is seismic survey. Somewhat less accurate are gravimetric methods, generally reliable only to the nearest 10 m. D. J. Crossley and G. K. C. Clarke (1970) compare gravity survey data with borehole data on the Fox glacier, Yukon Territory; the data from three sites are given in Table 4.3 (overleaf).

Table 4.3   Comparison of data from boreholes and gravity survey

Station	Gravity survey: ice thickness (m)	95% confidence limits (m)	Boreholes: ice thickness (m)
12	29·0	±12·3	31·4
16	30·6	±12·5	27·0
20	50·6	±14·7	47·9

Still less accurate but potentially useful in some circumstances are electrical resistivity techniques (see the series of papers in *Journal of Glaciology* **6** (1967), 599–650). Finally there is radio echo sounding which has been developed since the early 1960s. It is a technique which is not only rapid (the equipment can be installed in aircraft) but may surpass the conventional geophysical techniques just described in both accuracy and reliability of interpretation. Moreover, it yields a continuous profile of the subglacial bedrock surface. For polar ice sheets, this method has now superseded all others for determining ice thicknesses. D. J. Drewry (1972) assesses some results obtained by this method in part of Antarctica where ice thicknesses exceed 3000 m in places.

4 The internal flow of glaciers

The movement of glaciers is made up of two components, internal deformation and basal sliding. These will be considered in turn. The earliest theories of internal flow, such as those of Somigliana and Lagally, were based on the premise that ice behaves as a Newtonian viscous material, in which there is a linear relationship between shear stress and strain rate. This was found to be incorrect, and glaciologists then turned to a plastic flow law (e.g. J. F. Nye, 1951). As already outlined on p. 80, this assumed that, up to a certain shear stress level (τ_0), no deformation occurred, but at greater shear stresses, an infinite amount of deformation resulted. In other words, the ice was assumed not to be capable of supporting a shear stress $>\tau_0$. The theory of perfect plasticity showed better agreement with observational data than the viscous-type flow law, and it was used, for instance, to calculate the longitudinal profile of the Unteraar glacier

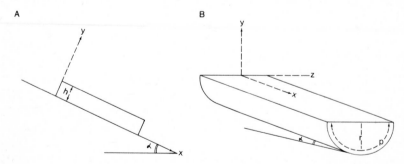

Fig. 4.8
A Notation and co-ordinates for a parallel-sided slab of ice resting on an inclined plane
B Notation and co-ordinates for ice flowing down inclined channel of uniform semi-circular cross-section.

(Nye, 1952c) which showed remarkable agreement with its actual measured profile.

Because it is a useful and simple approximation to the actual behaviour of ice under stress, it is worth considering some of the implications of this theory a little further. Let us consider first (Fig. 4.8A) a parallel-sided slab of ice of infinite width resting on, but not sliding down, an inclined plane; that is, its tendency to slide is opposed by a basal shear stress τ_b. Then

$$\tau_b = \rho g h \sin \alpha$$

where ρ is the density of ice, g the acceleration of gravity, and other notation is as in Fig. 4.8A. Thus values of τ_b can be found knowing ice thickness and angle of slope. If the slab of ice is 300 m thick and the slope is 1°, and we take $\rho = 0.9$ g/cm³, $g = 980$ cm/s², then $\tau_b \simeq 0.46$ bar. Now in the theory of perfect plasticity, the yield stress τ_o has a value of about 1 bar. Referring to the equation above, it will be seen that

$$\frac{\tau_o}{\rho g} = h \sin \alpha = \text{a constant.}$$

In other words, any change in h must be compensated by a change in α, and we have a simple (but approximate) explanation of the readily observed fact in nature that glaciers become thinner on steeper slopes and thicken on gentler ones, assuming no longitudinal change in accumulation or ablation. Furthermore, knowing α from surface measurements, the theory of perfect plasticity yields approximate maximum values of h, ice thickness. In practice, α must be taken as average slope over a distance not less than the estimated thickness of the ice.

Let us now consider ice in a sloping channel of regular cross-section, where the side-walls as well as the bed are opposing ice motion. Fig. 4.8B shows a channel of semi-circular cross-section (Nye, 1952b). Then

$$\tau_b = \rho g \frac{A}{p} \sin \alpha$$

where A is the cross-sectional area taken perpendicular to the bed, and p is the bed perimeter of this cross-section. Using estimates of A and p, applying this simple formula to sixteen Alpine glaciers, Nye found values of τ_b ranging from 0·49 bar for the Unteraar glacier at 2100 m in 1945–7, to 1·51 bar for the Grenz glacier at 2700 m in 1948–9. W. H. Ward (1955) found values of 0·42 to 0·92 bar for the Highway glacier (Baffin Island), with the lower values applying to the sluggishly-moving parts of the tongue. W. H. Mathews (1959) gave 1·42 bar for part of the Salmon glacier, British Columbia and Paterson (1970) finds a range from 0·5 to 1·4 bar for Athabasca glacier. S. Orvig (1953) found values of 0·4 bar for the Barnes ice cap, where the ice is barely moving (maximum measured velocity = 28 m/year; a recent 10 km traverse from central divide to margin encountered no velocity greater than 6 m/year). Away from the retarding effect of the bed and sidewalls, shear stresses within the ice decline towards zero at the centre of a glacier surface. From these and other assessments, it appears that τ_b for moving glaciers is usually in the range 0·5–1·5 bar, in order that motion of the glacier may take place. J. T. Andrews (1972), using an average value of $\tau_b = 1$ bar, shows that theoretical total glacier power ($W_t = \tau_b \cdot V$, where V is the average velocity through the glacier cross-section at the equilibrium line) falls in the range 0·2–1·5 W for the majority of non-surging glaciers with speeds up to 100 m/year.

The theory of perfect plasticity was a convenient approximation, but only an approximation, and in order to understand glacier flow more exactly, it was obviously necessary to determine more precisely how ice deforms under various shear stresses. This can best be done by controlled laboratory experiments. The first such experiments of value were those undertaken by Perutz (1948) in connection with the wartime 'Bergship' project. Perutz showed that ice responds to stress by creep, fairly rapidly at pressure melt-point, and that creep occurred down to temperatures as low as −20°C. Further controlled laboratory experiments, as described on p. 81, were carried out by J. W. Glen (1952, 1955) who found that when a load is applied, ice deforms almost instantaneously to a certain strain and then deforms more slowly as time proceeds. This slow deformation is termed 'creep', and it is this that allows glacier ice to flow continuously under stress. The actual mechanisms of creep, mainly intra-granular slip and recrystallization, have been described in Chapter 3. Glen showed that the relation between shear stress (τ) and strain rate ($\dot{\varepsilon}$) is of the general form

$$\dot{\varepsilon} = k\tau^n$$

where $n > 1$ and k is a function of temperature. This power law gives a much closer approximation to the way in which glacier ice deforms than the theory of perfect plasticity.

It is useful to recall some of the salient points made in Chapter 3 with regard to the power flow law. A range of values for n between about 2 and 4 has been determined experimentally, depending on shear stress and temperature. Values of n are below 2 at very low shear stresses. The constant k in the equation is very sensitive to temperature, of which it is an exponential function. Increasing the temperature from −15°C to pressure melting-point increases the flow rate by an order of magnitude. Cold ice can support greater shear stresses than ice at pressure melt-point and therefore deforms less readily. This is one reason why cold glaciers flow more slowly, and it is the reason why crevasses in cold glaciers tend to be slightly deeper than in the case of temperate glaciers. It was also pointed out in Chapter 3 that, at low shear stresses, experiments have to be continued over long periods before a steady rate of creep is achieved. Laboratory experiments have been continued over periods of months and with shear stresses as low as 0·1 bar, but in actual glaciers, ice is subjected to shear stresses over hundreds of years and the stresses may approach zero in the central surface zones. Such conditions obviously cannot be duplicated in the laboratory.

Glen's power flow law has been tested in studies of ice deformation in actual glaciers, particularly by studying the rates of closure of tunnels. The closure rate depends on the flow law provided that there are no other stresses within the ice. The tunnels on the Z'Mutt glacier (Switzerland) and Vesl-Skautbreen (Norway) gave values of $n=3$ in close agreement with the laboratory tests. However, two tunnels cut in more actively moving ice, in the Arolla glacier (Switzerland) and Austerdalsbreen (Norway), closed at a greater rate than that given by the flow law because the tunnels were near the bases of ice falls where the ice was subject to large longitudinal stresses. Meier (1960) has used the data of tunnel closure, the rates of borehole deformation, and various

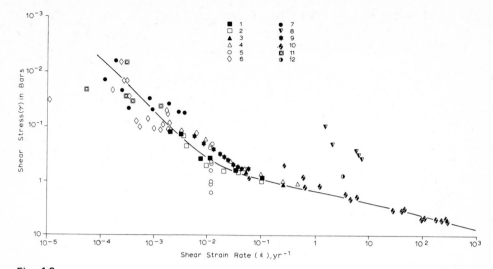

Fig. 4.9
Strain rate as a function of shear stress. Data from various glaciers and laboratory experiments on ice (M. F. Meier, 1960, *op. cit.*)

1 Surface transverse profile, Saskatchewan glacier; laminar flow analysis
2 Surface transverse profile, Saskatchewan glacier; complete analysis by Nye's theory
3 Z'mutt tunnel (J. F. Nye, 1953)
4 Vesl-Skautbreen tunnel (J. F. Nye, 1953)
5 Saskatchewan glacier borehole; complete analysis by Nye's theory
6 Malaspina glacier borehole; laminar flow analysis (R. P. Sharp, 1951–2)
7 Saskatchewan glacier borehole; laminar flow analysis
8 Austerdalsbreen tunnel (J. W. Glen, 1956)
9 Jungfraufirn borehole; laminar flow analysis
10 Laboratory experiments (J. W. Glen, 1955)
11 Malaspina glacier borehole; laminar flow analysis (R. P. Sharp, 1951–4)
12 Arolla tunnel (J. F. Nye, 1953)

laboratory experiments of Glen and others, to revise Glen's flow law, which he contends does not apply at very low shear stresses. Fig. 4.9 plots $\log \dot{\varepsilon}$ as a function of $\log \tau$; the curve which best fits all the data just mentioned is almost a straight line at stresses greater than about 0·7 bar, but curves more strongly at lower stresses. The revised flow law given by Meier is

$$\dot{\varepsilon} = k_1\tau + k_2\tau^n$$

and the best fit is obtained when $n = 4·5$, $k_1 = 0·018$, and $k_2 = 0·13$. The data are for temperate glaciers only. S. C. Colbeck and R. J. Evans (1973) and L. A. Lliboutry (1969b) have more recently proposed a flow law of polynomial type which they consider provides a better fit to experimental data (see p. 82) for temperate glaciers at the normal range of stresses

$$\dot{\varepsilon} = k_1\tau + k_2\tau^2 + k_3\tau^4$$

It is possible to make estimates of the velocities of movement expected at any given depth within a glacier from the flow law, provided that surface velocities are known and that simplifying assumptions are made about the cross-sectional form of the glacier.

For a semi-circular channel, using Glen's power flow law

$$v_0 - v = \frac{x}{n+1} \sin^n \alpha \cdot d^{n+1}$$

where v_0 is the surface velocity, v is the velocity at depth d, and $x = \rho g.k^n$. Thus vertical velocity profiles may be constructed (though clearly one cannot work out basal veloci-ties unless the thickness of the glacier is also known). Mathews (1959) showed that, for the Salmon glacier (p. 121), there is very satisfactory agreement between the measured deformation of a borehole and the calculated velocity profile based on the formula above, if $x = 1.25 \times 10^{-4}$ and $n = 2.8$. Fig. 4.10 shows the theoretical velocity

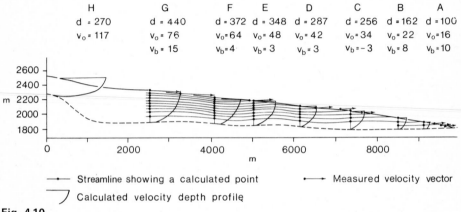

H	G	F	E	D	C	B	A
d = 270	d = 440	d = 372	d = 348	d = 287	d = 256	d = 162	d = 100
V_0 = 117	V_0 = 76	V_0 = 64	V_0 = 48	V_0 = 42	V_0 = 34	V_0 = 22	V_0 = 16
	V_b = 15	V_b = 4	V_b = 3	V_b = 3	V_b = -3	V_b = 8	V_b = 10

—•— Streamline showing a calculated point   •—→ Measured velocity vector

⌐ Calculated velocity depth profile

Fig. 4.10
Velocity vectors and calculated profiles and streamlines in a longitudinal section of the Saskatchewan glacier (M. F. Meier, 1960, *op. cit.*)

d thickness of glacier (metres) V_o surface velocity (m/year) V_b calculated basal velocity (m/year)

distributions computed by Meier (1960) for the Saskatchewan glacier; knowing the approximate thicknesses of the glacier from seismic data, it is also possible to estimate the basal sliding component V_b, which is found to vary between zero and 15 m/year. These values may well be too small, and the fact that in one case the value was negative suggested that the seismic data might be giving too great thicknesses for the ice. The diagram also shows calculated streamlines, representing the theoretical paths taken by individual ice particles. These paralleled the bedrock profile surprisingly closely.

For Athabasca glacier, Paterson (1970) has proposed a more certain method of deter-mining V_b for glaciers where boreholes penetrating to bedrock give some measured values of V_b. Given measurements of V_o, d (from seismic sounding) and α, it is possible by successive approximation to find V_b at sites where there are no boreholes. V_b for Athabasca glacier computed in this way ranges from zero to 32 m/year; it shows no clear correlation with values of τ_b computed from Nye's formula (p. 125), which may suggest that the latter is not giving sufficiently realistic values, and that bed roughness is unlikely to be uniform.

Theoretical internal velocity distributions for glaciers whose cross-sectional form is not assumed to be semi-circular are more difficult to compute. Nye (1965) has investigated glacier flow in channels of rectangular, elliptical and parabolic cross-section, assuming that there is no basal slip and that the ice is isothermal. With these limitations in mind, Fig. 4.11A shows the theoretical velocity distribution for a parabolic channel whose width is four times its depth, and Fig. 4.11B the calculated shear stress distribu-

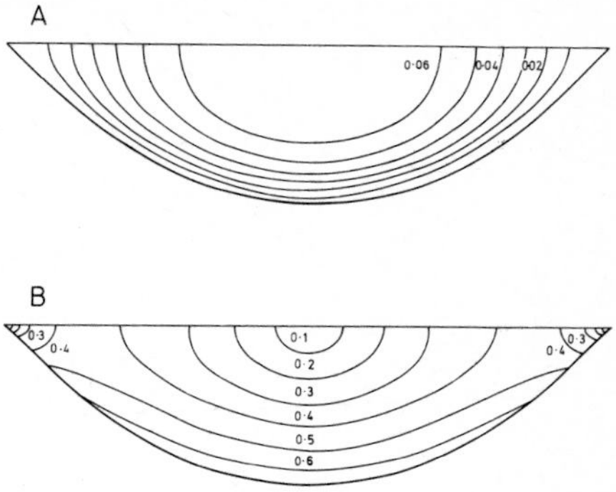

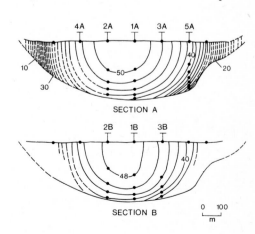

Fig. 4.11
Theoretical velocity distribution (A) and shear stress distribution (B) for a parabolic channel whose width is four times its depth. Velocity and shear stress given in dimensionless form (J. F. Nye, *J. Glaciol.*, 1960, by permission of the International Glaciological Society)

tion (in dimensionless form). Glen's simple power flow law is adopted, with $n=3$. An interesting result is the prediction that the zone of maximum shear stress at the glacier surface occurs not at the very edge of the glacier but a little way in. If the bedrock sides are steeply sloping, however, it occurs very close to the edge. This theoretically-derived result may not be unrelated to observations of crevasse patterns, where the zone of maximum marginal crevassing has often been seen to lie a small distance in from the actual ice edge.

To test the validity of theoretical patterns of velocity distribution such as are shown in Fig. 4.11, measurements in a series of closely spaced boreholes are needed. So far, only data from Athabasca glacier are really adequate for this purpose, and even here, with nine deep boreholes in three cross-profiles, extrapolation of the field data is necessary to complete the picture. Figs. 4.12 and 4.13 show the velocity distribution on 2 of the 3 cross-sections (C. F. Raymond, 1971b), and it is clear that in several respects

Fig. 4.12
Distribution of longitudinal velocity (m/year) in two cross-sections of Athabasca glacier, Alberta (C. F. Raymond, *J. Glaciol.*, 1971, by permission of the International Glaciological Society)

there are striking discrepancies between Nye's model (Fig. 4.11) and actual behaviour. Nye's model assumed zero basal slip, which is unrealistic and contrasts with the high basal sliding velocities found in parts of Athabasca glacier: up to 81 per cent and 87 per cent (maximum) of total surface velocity for the two cross-sections respectively. At one of the more marginally sited boreholes, basal slip was still 70 per cent, but there is a very rapid decrease in basal slip to small amounts at the actual margins. Measured shear strain rates also contrast with theory, being as little as $0·06\,\mathrm{yr^{-1}}$ at the bottom of deep boreholes (where so much of the movement is here taking place by basal sliding), but $0·3\,\mathrm{yr^{-1}}$ at the sides. Raymond (1971b) concludes that the glacier appears to be supported more strongly by friction near its margins as compared with basal friction at the centre. This means that such a simple approximation for estimating basal shear stress as $\tau_b = \rho g h\,A/P\sin\alpha$ is incorrect. The rapid lateral variations in the contribution of basal sliding to total glacier movement are incompatible with the assumptions of Nye's model and require physical explanation. Two possible explanations, variations in bed roughness and variations in basal water pressure, will be discussed in the section of this chapter dealing with basal sliding. Meanwhile, an improved theoretical model of the flow of Athabasca glacier will be described (L. Reynaud, 1973). Reynaud allows for basal sliding, and assumes movement in a steady state parallel with the channel long axis with velocity V_o at the surface centre. He adopts a value of friction between ice and bedrock proportional to the normal (hydrostatic) pressure of the ice less an allowance for subglacial water pressure. Using the same width–depth ratio as Nye, a maximum ice thickness of 310 m and surface slope 3° 30′, he computes the theoretical values depicted in Fig. 4.14. Table 4.4 compares some measured and

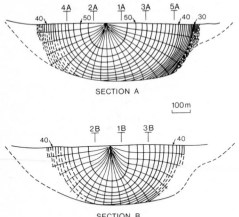

Fig. 4.13

Contours of constant longitudinal velocity (m/year) for two cross-sections of Athabasca glacier, Alberta (C. F. Raymond, *J. Glaciol.*, 1973, by permission of the International Glaciological Society)

theoretical data. The closer agreement between theory and reality is striking (compare Figs. 4.12 and 4.14A). The theoretical pattern of shear stress distribution (Fig. 4.14B; compare Fig. 4.11B) has interesting geomorphological implications. Basal shear stress is estimated to rise to a maximum nearer the glacier edges, not along the axis of the channel, which may help to explain why glacial troughs can widen into a U-shape. For further comparison, Fig. 4.15 shows the distribution of shear stress calculated from velocity data by Raymond (1973) for the two cross-sections of Athabasca glacier (Fig. 4.13).

Table 4.4   Measured and theoretical velocities (at surface centre) and ice discharge, for a cross-section of Athabasca glacier (Reynaud, 1973)

	Velocity m/yr	Total outflow hm³/yr
Measured:	52	10·9
Nye (1965):	17	2·9
Reynaud (1973):	52	11·0

In temperate glaciers, the effect of temperature on shear-stress distributions can be ignored since the ice is nearly isothermal. In cold ice sheets and glaciers of substantial thickness, the ice is not isothermal, and this will affect internal velocity and shear-stress distribution. Nye (1959b) analysed some data from the Greenland ice sheet, and argued that the basal layers are likely to be warmer than the upper layers, because of geothermal heat and the heat of friction generated by shearing in the basal layers. Since strain rates are very sensitive to ice temperature it is probable that most relative shear motion is concentrated in the basal ice layers, and that, above these, there may be little differential movement, the upper ice being carried forward by the shearing in the basal layers.

Data to support these ideas are now available from tunnels and boreholes in some cold glaciers. R. P. Goldthwait (1961) reports on observations in a tunnel in a glacier near Thule, north Greenland. The tunnel penetrated along the bed of the glacier, for

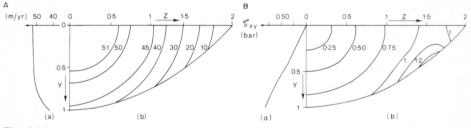

Fig. 4.14
A Distribution of longitudinal velocity for a parabolic channel whose width is four times its depth. Maximum ice thickness=310 m, ice surface slope=3°30'; $n=3$ and $k=0·17\,\text{bar}^{-3}\,\text{year}^{-1}$ in the flow law; the ratio of depth of water table below glacier surface to total ice thickness=0·33.
   (a) down the Y axis   (b) in cross-section
B Theoretical distribution of shear stress τ for a channel as defined in (A);
   (a) down the Y axis   (b) in cross-section
(L. Reynaud, *J. Glaciol.*, 1973, by permission of the International Glaciological Society)

30 m back from the terminal ice cliff. The ice above was up to 41 m thick and its maximum rate of movement at the surface was about 13 mm/day. On a vertical shaft running up from the inner end of the tunnel, ice motion was measured as follows:

At   5 m above the rock floor: 2·72 mm/day
At   1 m above the rock floor: 1·25 mm/day
At 0·1 m above the rock floor: 0·16 mm/day

At the bedrock surface, no motion of the ice was detected: the ice was frozen to a lichen- and moss-covered surface. Mosses and grass, surviving from an earlier non-glacial period (*c.* 200 years ago), projected into the basal ice for a few millimetres and had not been sheared off. A little higher, air bubbles in the ice indicated motion. Shear planes exposed in the inner tunnel dipped up-glacier at 11°, compared with 20–30° in the terminal ice cliff: they curve so as to meet the bedrock floor at a tangent. They are a typical feature of many ice margins and affect the interpretation of the ice move-ment data, for as well as processes of creep in the ice, discrete shearing is taking place in this rather special environment near an ice margin.

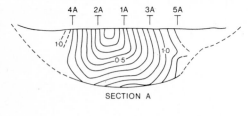

SECTION A

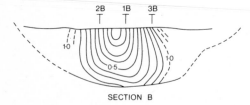

SECTION B

Fig. 4.15
Theoretical distribution of shear stress τ (bar) for the two cross-sections shown in Fig. 4.13 (C. F. Raymond, *J. Glaciol.*, 1973, by permission of the International Glacio-logical Society)

Investigations on Meserve glacier, Antarctica (Holdsworth and Bull, 1970) illus-trate the behaviour of another slow-moving (up to 15 mm/day) glacier, frozen to bedrock. The ice was up to 80 m thick and virtually isothermal at $-18°C$ below 8 m depth. A tunnel 55 m long was excavated along the bed and vertical boreholes sunk from the glacier surface to intersect it. Basal ice velocity at at least one point was zero, but in a subglacial cavity the basal ice had a finite velocity. Maximum rates of deforma-tion in the ice were associated with an amber-coloured layer of ice, up to 0·85 m thick, containing debris ranging from clay to small pebbles and a concentration of salts up to 8 ppm. (The salts may have been derived from weathered bedrock, from a former period of warmer wetter climate.) The layer of amber ice rested on ice-cemented rock debris with boulders up to 2 m in size, and was overlain by normal glacier ice with little salt content and virtually devoid of debris. For the upper part of the glacier, where shear stresses were less than 0·5 bar, a simple power flow law with $n = 1·6 \pm 0·1$ provided a good fit to the data from borehole deformation, but in the amber-ice layer, where shear stress is rapidly increasing, higher values of n ($= 4·3 \pm 0·5$) were needed to explain tunnel closure rates. The higher n values and higher creep rate in this layer are possibly related to salt content, as the ice crystal structure did not differ from that of the normal glacier ice above.

Over the past twenty years, considerable advances in our understanding of the in-ternal deformation of moving ice have been made, based on extensive laboratory study as described in Chapter 3, and on field data from a range of environments, as described in this chapter. Theoretical models of the behaviour of ice in temperate glaciers are now showing good agreement with observations from tunnels and boreholes. Understanding of the flow of cold glaciers still lags behind that of the flow of temperate glaciers, however, because of the continuing relative paucity of field data. Overall, it is still true that Glen's power flow law provides a good model of ice behaviour under a limited range of stress, though values of n in the formula have been found to vary from near 1 to more than 4. Whether more complex flow laws of polynomial type will provide a still better approximation to reality is not yet known with certainty.

4.1   Other theories of internal flow

(*a*) *The theory of extrusion flow*   Early workers such as Forbes demonstrated that the centre of a glacier flows more quickly than its marginal portions, and deduced therefrom that the surface also moves more quickly than the basal layers. Suggestions that, in certain conditions, the deeper layers of ice might be capable of movement more rapid than the upper layers have often been made, but the idea was not taken up seriously until the years 1937–8, when, independently, two glaciologists developed a more complete theory of extrusion flow and produced apparent evidence in support of it. M. Demorest (1937) suggested that, because of the plasticity of ice, ice was forced out from below the centres of accumulation of west Greenland; at the same time, higher ice moved down to take its place. In 1938, he produced evidence from striated surfaces that basal ice movements were controlled by pressure (see Chapter 6). His theory of extrusion flow was fully expounded in 1942–3; extrusion flow, contrasted with 'gravity flow', resulted from differences in ice pressure, and maximum rates of movement were postulated in the deeper layers, which were said to be more plastic under pressure.

R. Streiff-Becker (1938) put forward evidence from the Claridenfirn. Here the average accumulation over the firn basin was 3·45 million m³/year. Since the surface of the firn was not rising, this amount must be discharged from the basin, and according to Streiff-Becker it passed through a cross-section estimated at 68,000 m². Measurements of surface velocity at this cross-section (average 14 m/year) would allow for the discharge of only about 1 million m³/year; in order to discharge three and a half times this amount, either more ice must be discharged at greater velocity at depth, or the estimate of maximum depth for the cross-section would have to be increased from 110 m to about 300 m. Streiff-Becker chose the former explanation as more likely, and claimed that extrusion flow was taking place.

There is serious doubt that Streiff-Becker's estimates of the cross-section are reliable, and it has also been suggested (W. V. Lewis, in discussion following G. Seligman, 1947) that perhaps not all the ice from the Claridenfirn was escaping solely through the critical cross-section. Extrusion flow has never been encountered in any deep boreholes (a minor instance noted by Carol has already been mentioned) and the concept appears to be physically unsound. W. V. Lewis (1948) asked the very pertinent question 'why the upper layers are not carried forward on the back of the supposedly swifter moving lower layers'. Nye (1952a) pointed out that the upper layers could only be held back by either a longitudinal tensile stress or transverse shear stresses exerted by the drag of the valley sides in the case of a valley glacier. In both cases, the stress would be many times greater than the ice could support, and the process would also be opposed by the shear stress between ice and bedrock. Geomorphologists such as C. D. Holmes (1937) and S. E. Hollingworth (1931, 1947) found the theory attractive, in that in certain conditions it was postulated by Streiff-Becker particularly that fast-moving currents of ice would be carried down close to the bedrock floor and theoretically give rise to enhanced rates of erosion at these places. The fact remains that theory is against it, and it has never been shown to occur in practice.

(b) *Extending and compressing flow*   In considering the internal deformation of glaciers, the basic case considered was one of ice flowing down channels of constant gradient with uniform rate of discharge. In actual glaciers, longitudinal extension and compression frequently occur—the ice tends to thicken or 'compress' in sections where the bedslope is reduced, for instance. The term 'compression' is not meant to indicate actual reduction of volume of the ice (ice is a virtually incompressible substance) but a local thickening of the glacier; conversely with 'extension'. Not only will changes in longitudinal gradient cause compression or extension, but also changes in rates of accumulation or ablation will have this effect. Nye (1952b) has made a theoretical investigation of the effects of these changes using the simplified flow law of perfect plasticity. If R is the radius of bed curvature, α the surface slope of the ice, and ϕ the rate of ice discharge, then extending flow will occur when the expression

$$\frac{\delta\phi}{\delta x} + \frac{\phi}{R}\cot\alpha$$

is positive, and compressive flow when it is negative, $\delta\phi/\delta x$ being the rate of addition of ice to the upper surface of the glacier over a distance x. If the bed is of even gradient, R would have infinite value, but the left-hand term would still give rise to compressing flow if negative, and extending flow if positive.

Theoretically, under conditions of compressing or extending flow, the slip-lines shown in Fig. 8.4 would represent the planes along which the ice would have greatest tendency to shear. The second situation (compressive flow) helps to explain the occurrence of the thrust-planes often seen in glacier snouts. The potential slip surfaces are likely to be brought into use when they happen to coincide with any other zones of weakness in the ice, or when bedrock irregularities or morainic obstacles prompt their formation (see Chapter 8). B. M. Gunn (1964) gives examples of the development of up-sloping thrust-planes under compression. An excellent example of extending flow is given by the upper part of the Austerdalsbreen ice-fall (Plate III) down which the ice is accelerating to speeds of 2000 m/yr. Where the ice leaves Jostedalsbreen to plunge down the ice-fall, step-faulting characteristic of extending flow is seen. As it approaches the base of the ice-fall, compression commences and increases in magnitude: the flow has decelerated to about 100 m/year at the base of the ice-fall, and continues to decelerate down-glacier to less than 20 m/year (Fig. 4.1). In the tunnel dug in the lower part of the ice-fall, J. W. Glen (1956) showed that, in the top 20-metre layer of ice, there was a compressive stress of about 3 bars acting parallel to the ice surface, confirmed by the rates of tunnel closure which were abnormally high (see p. 126).

The slip-line fields deduced by Nye in 1952 have some interesting implications for

Plate III
The twin ice-falls at the head of Austerdalsbreen, Norway. The Austerdalsbreen tunnel, excavated
in 1959, was located about one-third of the way up the left-hand ice-fall. (*C.E.*)

glacial erosion of basins and steps in glacial valleys, which are discussed more fully
in Chapter 8 (see Fig. 8.5 especially). Nye and P. C. S. Martin (1968) show that there
is theoretically a limit to the amount of concavity that can develop in the valley floor
by glacial erosion, set by the maximum curvature of the α lines in the slip-line field,
but no limit to degree of convexity. They argue, therefore, that glaciers cannot excavate
enclosed basins between, say, more resistant rock outcrops, deeper than a certain limit
related to the spacing of the outcrops and the α-line curvature. If the model is developed
by using a more realistic flow law than that of perfect plasticity, the theoretical boundaries are no longer sharp but are zones of high shear strain rate, but the erosional
implications are substantially similar.

(*c*) *Rotational movements in glacier flow*   Rotational slipping in the case of cirque glaciers is now well established by J. G. McCall and others, and will be discussed in Chapter 7. W. V. Lewis (1947b) suggested an extension of the concept to sections of valley glaciers below ice-falls or steeper gradients. Rotation of the Austerdalsbreen ice-fall tunnel excavated in 1955 was occurring so rapidly (14 minutes of arc per day) that its extension to reach bedrock was never accomplished. However, actual rotation of the ice is not the only possible cause of rotation in the tunnel: a uniform shear through-out the thickness of the ice could cause apparent rotation. The tunnel was also being bent convexly upwards. The idea of large-scale rotational movements of the ice in basins below ice-falls, as suggested by Lewis, is undoubtedly too simple, though movements along potential slip-surfaces as postulated by Nye (1952b) and Nye and Martin (1968) for extending and compressing flow may well sufficiently resemble rotational move-ments. This is one of the many reasons why further detailed studies of surface and englacial ice movements in actual glaciers are required. The implications of these con-cepts of ice movement, in regard to the development of some erosional features of glaciated valleys, will be considered in Chapter 8.

5 Basal and side-slip

Internal deformation provides one component of glacier flow; sliding over bedrock provides the other. The relative contribution that each makes varies with bed slope, ice thickness, and ice temperature. Direct measurement of side slip is often possible but, in the case of basal sliding, measurements can only be obtained from tunnels or deep boring. The mechanism of sliding, and measured rates of sliding, are of the highest interest to geomorphologists since it is likely that a glacier carrying debris and slipping past bedrock will be able to erode far more effectively than one that is stuck fast to bedrock.

   Side-slip may be measured in several ways. If the marginal ice is not too heavily crevassed, a stake may be inserted as deeply and as close to the ice margin as possible, and its motion relative to two fixed points on the adjacent rock wall measured by invar tape or by a continuously-recording mechanical device. It is unusual, owing to margi-nal crevassing, to be able to find a suitable place for such a stake closer to the ice edge than 2 or 3 m. Alternatively, a stake may be pushed at an angle through the ice edge almost to touch the rock wall, and the vertical and horizontal motion of the end of the stake can be recorded relative to its initial point of contact with the rock wall. The first of these methods gives less erratic movements than the second, since the ice more than 2 or 3 m from the actual margin flows more uniformly. Since velocities may change rapidly with distance from the ice edge (as in Blockschollen flow, for instance), side-slip must always be measured as close to the ice edge as possible. Side-slip (S_s) is often expressed as a percentage of maximum surface velocity (V_m):

$$S_s = \frac{V_m}{V_s} \times 100 \qquad (V_s = \text{velocity at the ice edge})$$

For the Fox and Franz Joseph glaciers in New Zealand, B. M. Gunn (1964) obtained values of 12 to 21 per cent, with one abnormally high measurement of 68 per cent where the enclosing rock walls were very steep. On Austerdalsbreen (Norway), Glen (1958) and Glen and Lewis (1961) found side-slip of 10 to 20 per cent rising to 65 per cent near the foot of the ice-fall (Chapter 8). They also noted that side-slip was very variable with time, and from place to place, and some of the factors responsible have already been mentioned (p. 118).

The ratio of basal sliding to maximum surface velocity is equally variable. McCall (1952) in the Vesl-Skautbreen lower tunnel found that basal sliding accounted for 90 per cent of movement—and unlike the Austerdalsbreen side-slip measurements, the basal movement was smooth and steady over 12-hour periods, possibly reflecting a relatively even (abraded?) rock floor at the site of measurement. Vesl-Skautbreen is a small cirque glacier which behaves basically, as will be detailed in Chapter 7, as a rigidly rotating mass of ice. In the Jungfraufirn borehole on the Aletsch glacier, basal sliding amounted to 26 per cent, assuming that the borehole did in fact reach bedrock, and to 55 per cent for the Salmon glacier. In ice caves below Østerdalsisen, Norway, Theakstone (1967) found basal sliding of about 65 per cent within only 60 m of the ice margin, while for Athabasca glacier (Raymond, 1971b), basal sliding reaches the values, very high for valley glaciers, of 81 and 87 per cent at the centres of two cross-section velocity profiles. Fig. 4.12 shows how these high rates diminish rapidly to very small values of side-slip at the actual glacier margin, emphasizing that measurements of side-slip may not provide any useful guide to what is happening at the glacier base.

Detailed measurements of temporal variations in basal sliding (Fig. 4.16) have been made using a mechanical device ('cavitometer') in natural subglacial cavities below the Glacier d'Argentière, France (Vivian and Bocquet, 1973). Movement occurred in jerks, with velocities ranging from zero to 9 cm/hour; and there was one period of complete halting for 3 days. The possibility of basal sliding being very irregular should

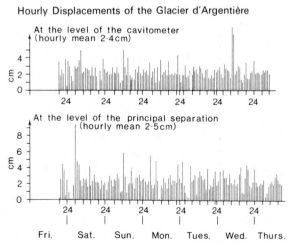

Fig. 4.16
Experimental study of velocities of ice displacement and deformation in a cavity beneath the Glacier d'Argentière, France (R. Vivian and G. Bocquet, *J. Glaciol.*, 1973, by permission of the International Glaciological Society)

not be forgotten when average figures of the percentage contribution of this process
to total glacier flow are quoted.

There is some evidence (and much speculation) that, in the case of cold glaciers
and ice sheets, basal sliding does not occur at all since the ice is frozen to bedrock.
The shear strength of a cold ice-solid interface is of the order of 10 bar if the solid
is stronger than the cold ice and if it is wettable, whereas normal basal shear stresses
of temperate glaciers are 0·5 to 1·5 bar. Observations by R. P. Goldthwait (1956, 1961)
in a tunnel in the Thule area of Greenland showed that the ice here was not sliding
over its bed, though there was measurable differential flow in the lower ice layers (see
p. 131). Cold ice tunnels on the Jungfraujoch, studied since 1950, have also shown
that cold ice here is not sliding over its frozen bed; furthermore, differential motion
within the ice is less than would be associated with temperate ice (R. Haefeli, 1963).

Theoretically, a distinction should be made between three types of 'cold' ice mass
in respect of basal sliding. First, there is the totally cold glacier frozen to bedrock (as
illustrated by the Camp Century borehole, Fig. 3.6B). No basal sliding can occur.
Secondly, there can exist cold glaciers whose base is just at pressure melting-point (Fig.
3.6C). Bedrock obstacles projecting up into the cold ice may effectively reduce basal
sliding to very limited amounts. Thirdly, there are those glaciers where cold ice overlies
a temperate basal layer, as in several parts of Antarctica probably (Fig. 3.7A), and
where basal sliding will occur just as it does beneath most active temperate glaciers.
These differences have important consequences for bed erosion. In the first category,
basal erosion will be minimal. It may not, however, be completely absent, since any
larger rock fragments held mainly by the rapidly shearing ice a few centimetres above
the actual ice-rock interface may be dragged over the latter. This is the possible
explanation of the signs of plucking and striation noted by J. H. Mercer (1971) associated with some small cold glaciers in the central Transantarctic Mountains. Glaciers
in category two will also be capable in theory of limited bedrock erosion, but the rates
of erosion are likely to be far less than in the cases of temperate glaciers and sub-polar
glaciers with high rates of basal sliding.

We thus have abundant evidence that an important part of the flow of temperate
glaciers consists of basal sliding, and more limited evidence that wholly cold glaciers
do not normally slip over their beds. The first analysis of the mechanism of sliding
based on direct observation was that of B. Kamb and E. LaChapelle (1964) who studied
the process in a tunnel in the Blue glacier, Washington. The tunnel reached bedrock
at 26 m depth, and basal sliding of 90 per cent was measured on a relatively steep
slope. If, in the tunnel, the ice was quickly removed from bedrock by excavation, it
came away easily, suggesting the presence under natural conditions of a basal water
film. The structure and texture of the lowest ice layer (up to about 3 cm thick) was
distinctive, and thought to represent a layer in which alternate freezing and melting
under the influence of changing pressure conditions—the process termed regelation—
had been taking place. This regelation layer also contained abundant rock debris (compare the 'sole' of the Vesl-Skautbreen glacier, p. 226) picked up presumably during
the regelation process. In some places, cavities separated the ice from bedrock, the
largest being 10 m wide and 20 cm high, and in some cavities, regelation ice spicules
were seen weakly attached to the under-surface of the ice. The observations supported

J. Weertman's theory (1957) that *regulation slip* is an important mechanism of basal sliding.

Kamb and LaChapelle demonstrated the process of regelation slip experimentally by hanging a weight attached by wire to a cube frozen into an ice block. Cubes of rock (dunite), plexiglass, and aluminium were used. Under load, the cubes moved slowly through the ice by melting in front (i.e. below them) and refreezing behind (i.e. above them). The shape and size of the ice block were not disturbed, nor were the texture and structure of the ice except along the path of travel of the cubes, where changes in crystal form were apparent, giving features similar to those encountered in the regelation layer at the base of the glacier. Hence, it is thought that, when glacier ice is forced to pass over a bedrock obstacle, melting of the ice occurs around the upstream face where compressive stress is greatest, the meltwater flows around the obstacle and refreezes on the low-pressure downstream side (where there may be a subglacial cavity), and there is a flow of heat from the freezing to the melting places. The energy to keep the cycle going is supplied by the basal shear stress.

Kamb and LaChapelle's field observations have been reinforced by others subsequently. R. A. Souchez *et al.* (1973) and Vivian and Bocquet (1973) have reported similar findings in cavities below the Glacier d'Argentière (Fig. 4.17). These natural

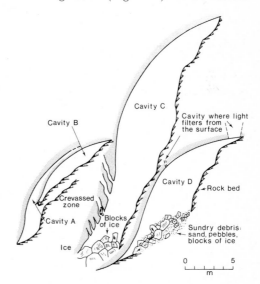

Fig. 4.17
Longitudinal profiles of subglacial cavities, Glacier d'Argentière, France (R. Vivian and G. Bocquet, *J. Glaciol.*, 1973, by permission of the International Glaciological Society)

cavities occur on the down-glacier side of a subglacial rock bar, beneath 100 m of ice. A basal regelation layer had a total thickness of 1·5 m, in which there were marked laminations, and beds of pure ice alternating with sandy ice sometimes containing pebbles up to 5 or 6 cm and also small water pockets (0·5–1 m³). Observations over 3 years in one cavity showed refrozen ice some centimetres thick on some visits, and melting (helping to concentrate the sandy material) at others. Rock debris was being expelled from the basal regelation layer as the flowing ice reached the cavity, was falling to the floor·and being taken back into the glacier at the lower end. Spicules and efflorescences of ice crystals were attached to the ice roof, some up to 10 cm long, and especially at the upper end of the cavity where the pressure on the ice was locally released. These observations, unique in their extent because of the opportunities afforded by hydro-electric engineering work to visit many subglacial localities over a period of years, definitely confirm that interstitial water can and does refreeze at the base of a temperate glacier in accordance with local changes of pressure; and that subglacial cavities may be more numerous than expected.

The rate of bulk movement of temperate ice made possible by the regulation mechanism was theoretically examined by Weertman in 1957. He proposed a highly simplified model of a glacier bed, in which bedrock protuberances consisted of cubes of dimension L and equal spacing L'. Then, assuming that all heat flow occurs through the obstacle, the speed of regelation sliding (S_a) is given by:

$$S_a = \frac{CD}{\rho HL} \cdot \tau_b \cdot r^2$$

where D is the coefficient of conductivity of the rock, H is the heat of fusion (80 cal/g), ρ is the density of ice, r is a roughness factor ($=L'/L$) and C is a constant. It will be seen that as the obstacle size becomes larger or the spacing decreases, so S_a becomes smaller. Smaller obstacles allow the establishment of greater temperature gradients, and therefore allow greater regelation slip. Conversely, obstacles larger than 100 cm would reduce S_a to negligible amounts.

Since S_a becomes negligible for larger obstacles, and since the regelation mechanism is inapplicable in the case of cold glaciers, Weertman introduced an additional sliding mechanism, that of creep around obstacles arising from concentrations of stress, on the basis of Glen's power flow law. As stress increases around obstacles, the strain rate rises, and if the hydrostatic pressure at the base of the ice is sufficiently large to prevent cavity formation on the downstream side of obstacles, Weertman (1957) showed that the sliding velocity S_b by this process is given by:

$$S_b = BL\left(\frac{\tau_b r^2}{2}\right)^n$$

where B is the constant in the power flow law. Increasing the obstacle size on this mechanism actually gives increased rates of sliding. Weertman considered that these two mechanisms together provided a satisfactory explanation of glacier sliding over obstacles of a variety of sizes and showed that the speed of sliding, S, when $S_a = S_b$ is

$$2^{1-(n/2)} \tau_b^{(n+1)/2} r^{n+1} \sqrt{\frac{BCD}{\rho H}}$$

If $n = 3$, this simplifies to

$$\frac{1}{2} \tau_b^2 r^4 \sqrt{\frac{BCD}{H\rho}}$$

Thus S is proportional to the square of the basal shear stress and to the 4th power of the bed roughness. A major problem with this relationship is that S becomes very sensitive to the values of r, yet in practice it is very difficult to make realistic assessments of r since the glacier bed is only occasionally visible.

L. Lliboutry (1959) contended that the rate of sliding S obtainable by combining the two equations was too small, and that the model of bed form adopted by Weertman was unrealistic. Taking a modified bed form consisting of a series of parallel sine waves of amplitude a and wavelength λ, he showed that the expected velocity of sliding derived from both Weertman's processes is given by:

$$S_a + S_b = \frac{Ba}{2r^2} \left[\frac{2\tau_b}{\sqrt{3\pi r}} \right]^n$$

where r is the roughness factor $(=a/\lambda)$. Taking $n=3$, $B=0.18$ bar^{-3} year^{-1}, and $r=0.1$, Lliboutry found $S_a + S_b = 4$ m/year. This, he considered, was much too small a rate of sliding ('frottement quasi-statique'), which might give rise to glacier polish, but incapable of more significant erosion.

Lliboutry therefore postulated a modification of the sliding theory. He investigated the possibility that the glacier might partially detach itself from its bed: in the sine-wave model, it might only rest on the crests of the waves, and that cavities, normally water-filled in the case of temperate glaciers, might exist in the troughs. The rate of sliding derived was:

$$S_c = \frac{Bz}{2\tau_b^2} \left(\frac{z}{\sqrt{3\tau_b\lambda}} \right)^n \qquad \text{where } z = a\pi^2 (\rho gh - p)^2$$

(p = water pressure beneath the glacier). Taking an arbitrary value of 6.5 bar for the term $(\rho gh - p)$ (see p. 143 for a discussion of the significance of this parameter), he obtained a sliding rate of 100 m/year. It is important to note that values of S_c will increase as τ_b, the basal shear stress, becomes smaller.

In 1964, Weertman published a revised form of his sliding theory, using the same double mechanism as previously, but making modifications to allow for some heat passing through the ice surrounding the bedrock obstacles. The theory allowed for rectangular obstacles with three different dimensions, and took into account the effects of obstacles both larger and smaller than the 'controlling obstacle size' (Λ). The latter is the size of obstacle for which $S_a = S_b$ and is given by the expression

$$2^{n/2}\tau_b^{(1-n)/2}r^{1-n}\sqrt{\frac{CD}{\rho HB}}, \text{ or } 2^{1.5}\frac{1}{\tau_b . r^2}\sqrt{\frac{CD}{\rho HB}} \text{ when } n=3$$

If an obstacle is smaller than Λ, then ice flows around it primarily by regelation; if larger, the ice flows largely by creep. Fig. 4.18 plots sliding velocity S and controlling obstacle size Λ against basal shear stress τ_b, using three different values of r, the roughness factor. This model gives reasonable agreement with the observations of Kamb and LaChapelle (1964). In their case, with $\tau_b = 0.7$ bar and $r = 9$, S as measured was 5.8 m/year and S from the graph is about 2 m/year. Λ from the graph is about 4 cm, compared with 3 cm as the maximum thickness of the observed regelation layer.

5.1   The rôle of meltwater in basal sliding

The general question of the form in which meltwater exists beneath temperate glaciers is critical for the whole process of sliding. It is also a controversial question, mainly

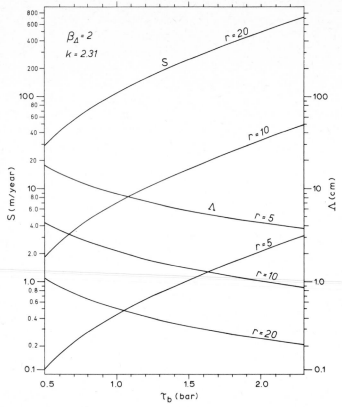

Fig. 4.18
Graph showing sliding velocity (S) and controlling obstacle size (Λ) as a function of the shear stress (τ) for three values of the roughness factor (r) (J. Weertman, *J. Glaciol.*, 1964, by permission of the International Glaciological Society)

because field data are still inadequate. There are two possibilities: one that the melt-water exists as a thin film lubricating the bedrock surface, the other that the water is mainly concentrated in subglacial cavities or channels. Cavities can form when the 'normal' overburden pressure of the ice ($=\rho g h$) is not great enough to suppress them (Nye, 1970). They are more likely to form under relatively thin ice.

Basal meltwater may accumulate from three main sources:

1 Meltwater from the glacier surface
2 Ice melted by geothermal heat (about $0 \cdot 5$ cm³/cm²/year)
3 Ice melted by basal sliding (about 3 cm³/cm²/year in the case of a sliding velocity of 80 m/year under a basal shear stress of 1 bar).

Variations in the amount of meltwater at the glacier bed may explain observed time variations in glacier flow and also the catastrophic advances or surges of some glaciers (Weertman, 1962, 1966).

Weertman in 1964 considered the theoretical effect of a basal water film on glacier sliding. Many observations have suggested that basal meltwater acting as a lubricant is an important factor in this process (p. 115). If the water layer thickness is less than

$\Lambda/100$, Weertman showed that its effect on sliding velocity will be negligible; for a water layer of thickness $\Lambda/10$, the increase in sliding velocity might be 20 per cent; while if the water layer were of the thickness (Λ) equal to the size of the controlling obstacles, then truly catastrophic increases in velocity (perhaps ten times) might be expected. Many of the observed variations in rates of glacier flow could thus be explained in terms of basal water layers whose thickness is of the order of one-tenth the height of the controlling obstacle size. For a glacier normally travelling at 100 m/ year, with a 1 bar basal shear stress, Fig. 4.18 gives $r=20$ and $\Lambda=5$ mm. A 20 per cent variation in flow could then be explained in terms of a basal water film only 0·5 mm thick.

Weertman (1966) has considered the general question of the controls on the thickness of the basal water layer. The hydraulic gradient beneath a glacier is

$$\frac{\rho g \sin \alpha}{\mu}$$

where α is the ice surface slope and μ is the viscosity of water. If the bedrock is impermeable, the thickness (H) of the water layer will be a function of hydraulic gradient and the amount of meltwater discharged in the layer (Q):

$$H = \left(\frac{Q \cdot 12\mu}{\rho g \sin \alpha} \right)^{1/3}.$$

For permeable bedrock, a correction factor must be introduced (Boulton, 1972):

$$H = \left(\frac{M \cdot 12\mu}{\rho g \sin \alpha} - 12 \, KD \right)^{1/3}$$

where M is the volume of meltwater produced up-glacier from the permeable bedrock zone, K is the bedrock permeability and D is the thickness of the permeable beds.

Lliboutry's sliding theories have considered basal water existing mainly in cavities beneath the ice (his so-called theory of subglacial cavitation, though the use of the term 'cavitation' in this sense must not be confused with the erosion process of that name (see p. 193)). As such cavities enlarge, which will happen if there is water to fill them, the area of contact between the glacier and its bed will be progressively reduced, and therefore the speed of sliding will increase. There is no doubt that such water-filled cavities exist. Apart from the recent exploration of the extensive system of subglacial cavities beneath the Glacier d'Argentière, there have been numerous instances of drill-holes suddenly piercing water-filled cavities (e.g. Paterson and J. C. Savage, 1970), tunnels encountering them (e.g. J. E. Fisher, 1953), and of sudden surges on meltwater streams which probably reflect sudden draining of such subglacial (or possibly englacial) water bodies, as was noted just before the Allalin glacier catastrophe (Vivian, 1966). The maintenance of sizeable water-filled cavities under flowing glaciers depends on the pressure of water occupying them and, as described on page 141, the effective basal pressure $(\rho gh - p)$ is an important parameter in Lliboutry's sliding theory. The problem is the paucity of field data to show what range of values this parameter might have. The term ρgh presents no problems if the thickness of ice above the cavity is known; for 300 m of ice, the value of ρgh will be approximately 27 bar.

Values of p, the water pressure in the cavities, have not often been recorded. But it is clear that, the higher the values of p, the less will be the value of $(\rho gh - p)$ (which is the effective pressure of the ice on its bed) and the greater, in theory, will be the rate of sliding. Maximum values of $(\rho gh - p)$ will be found at the level of the englacial water table; above this, the effective basal pressure will be simply ρgh. Minimum values of $(\rho gh - p)$ might be expected at the deepest part of the glacier bed. Thus there will be lateral variations in the effective basal pressure which Raymond (1971b) considers may explain the great lateral variations in basal sliding rates beneath Athabasca glacier (Fig. 4.12). With high englacial water-table levels, $(\rho gh - p)$ will be greatest relatively close to the margins of the glacier and will retard side-slip, and the largest gradients in sliding velocity, from margin to centre, will be found where valley sides are steep.

Measurements of p have given values of about 10 bar beneath the Gorner glacier (Vivian, 1970), and pressures of similar magnitude have been measured beneath the Glacier d'Argentière (Vivian and J. Zumstein, 1973). Fig. 4.19 shows the situation

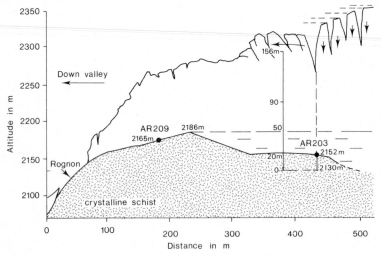

Fig. 4.19
Long profile of the lower part of the Glacier d'Argentière, France, and the subglacial rock surface. At the left may be seen the Rognon rock bar, partially exposed through the ice; also shown is the position of the borehole AR203 (R. Vivian and J. Zumstein, *Int. Ass. scient. Hydrol., Symposium on the hydrology of glaciers 1969*, 1973)

of the borehole (AR 203) beneath the Glacier d'Argentière where water pressure was monitored, and its relation with the presumed englacial water table. From the end of October to early June in the following year, a more or less constant pressure of 9–10 bar was recorded. From June to October, the main ablation season, there were daily variations in pressure of 3–4 bar, but up to 8 bar amplitude on one occasion following heavy precipitation of 104 mm (maximum $p = 16$ bar). These variations were closely matched by variations in discharge of the subglacial stream gauged at the Rognon rock bar, with a time lag of about 12 hours. The thickness of ice at the monitored borehole is less than 200 m, so that $(\rho gh - p)$ is reduced to very small values at times.

When $(\rho gh - p)$ reaches very small values, the friction between the glacier and its bed will become minimal—the glacier is being mainly supported by the pressure of subglacial water and in the ultimate case it will float. Such a situation probably explains many cases of rapid glacier advances (Lliboutry, 1964). The catastrophic fall of the Allalin glacier on 30 August 1965 (R. Vivian, 1966) is a supreme example. The tongue of this glacier lay on a 30-degree slope and over the last few decades had been gradually thinning. In the week before the disaster, the speed of ice movement had reached 4 m/day, and there had been torrential rain of 56 mm followed by rising temperatures. Crevasses had become water-filled, and sudden surges in the meltwater stream from the snout suggested the existence of numerous subglacial and englacial water bodies. The ice tongue broke away when meltwater had sufficiently reduced ice adhesion on the slope, and precipitated a million cubic metres of ice into the valley below the Mattmark Dam.

A neglected factor in discussions of basal water pressure is the permeability of the bedrock floor (Paterson, 1969; G. S. Boulton, 1972). Values of $(\rho g h - p)$ will clearly become larger where there is an increase in permeability, i.e. subglacial water is able to leak out of the subglacial cavities, or from any basal water film of the Weertman type, into the bedrock. This will retard glacier sliding and thereby affect both erosion and deposition at the glacier bed.

5.2 *Theoretical models of basal sliding*

Workers such as Lliboutry, Nye and Weertman have been striving to develop general models of glacier sliding, which is a complex process dependent on many variables that are difficult either to define or to measure. Inevitably, their models have had to be simplified compared with reality and it is not surprising that discrepancies with (rather limited) field observations have appeared. A major problem which Weertman encountered in 1957 and 1964 was how to build a simple yet acceptable model of the naturally irregular glacier bed. A spectrum of obstacle sizes had to be specified: the bed was taken to consist of rectangular obstacles whose dimension L took values of λ, 10λ, 100λ ... where λ is a constant. Nye (1970) has criticized the choice of the factor of 10, and Lliboutry (1968) claims that obstacles which are wide compared with their height are omitted. Lliboutry also considers the idea of a 'controlling obstacle size' untenable, since in reality large bumps do not exist side-by-side with small bumps, but the latter are found *on* the large bumps, and in turn still smaller bumps are superimposed on the small bumps (Lliboutry, in discussion following Weertman, 1967b). Obstacles of rectangular (Weertman) or sinusoidal (Lliboutry) form are clearly a gross over-simplification from reality, as is the assumption by both workers that the sizes or wavelengths should vary in geometrical progression. Further progress in the development of sliding theories must depend in part on defining more realistic bed forms (Lliboutry, 1968) and this needs quantitative field studies of glacier bed morphology, for instance, by illuminating at night a section of recently exposed glacier bed. The angle of illumination should be that of the direction of ice flow, so that some indication may be given of the extent of ice-rock contact on an irregular surface allowing for lee-side subglacial cavities. Spectral analysis might allow the most common wavelength bands to be determined. Subglacial localities might be used, if available, in such studies, for example the 400 m² of the bed of the Glacier Argentière recently made accessible by hydro-electric tunnels. Another group of problems faced by Lliboutry and Weertman concerns the distribution of basal meltwater and the water pressures encountered beneath actual glaciers, as discussed in the last section. Here again, further field studies are clearly necessary before improvements in this respect can be made to the sliding theories and before the different views can be reconciled.

Weertman (1971) has recently defended the comparative simplicity of his theory of glacier sliding against attempts by Nye (1969, 1970) to develop more exact solutions using a viscous flow law and different forms of wavy bedrock surface. The disadvantage of Nye's analysis lies in the viscous flow approximation, and such other assumptions as a perfectly slippery bedrock surface and no subglacial cavities. Weertman argues that the lack of precise agreement between his theoretical values and field observations is more likely to be due to deficiencies in the latter; for instance, in the regelation slip mechanism, it would be preferable to have actual measured values for D, the coefficient of thermal conductivity, instead of using average values. In short, further development in our understanding of glacier sliding is now likely to depend on more and better field data, and this is why such programmes of subglacial research as are being carried out beneath the Glacier d'Argentière are so important.

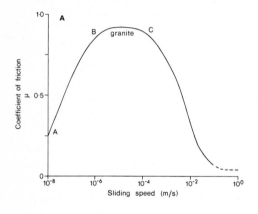

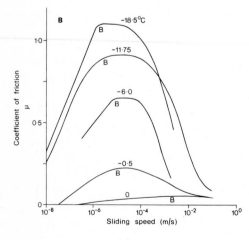

Fig. 4.20
The friction of ice over granite at different sliding speeds. The surface asperity roughness of the granite is about 1·5 μm.
A At a temperature of −11·75°C
B At a range of temperatures between 0° and −18·5°C (P. Barnes *et al., Proc. R. Soc. A*, 1971, by permission of the Royal Society)

5.3 *Experiments on the sliding of ice over rock*

Such experiments have been conducted using a range of rock types, speeds of sliding and temperatures (P. Barnes *et al.*, 1971). In Fig. 4.20, the portions of the curves falling on the left-hand parts of the graphs are of most relevance to actual glaciers; the portion *AB* of the curve for $-11 \cdot 75°C$, for instance, covers a range of speeds appropriate to most glaciers, while *BC* represents very fast-moving or surging glaciers. Friction increases as the temperature is reduced while the change in slope of the curves at *B* shifts to slower sliding speeds. The granite used in these experiments had a surface asperity roughness of about $1 \cdot 5 \ \mu m$ and its friction was about 20 times that of steel. For low or moderate sliding speeds, the high values of friction achieved below $-10°C$ show very effective adhesion between ice and rock, greater than the shear strength of ice, so that shearing takes place in the ice itself. A thin film of ice was formed on the rock at these lower temperatures, and the friction is determined by the force needed to produce creep in this film of ice. Above $-10°C$, the ice film was not observed and interfacial adhesion was less effective. R. Hope *et al.* (1972) also found that an ice-to-ice sliding surface tended to form at low temperatures and stick-slip motion then occurred. These experiments generally confirm field experience and theoretical ideas that cold glaciers do not slide over bedrock, but that as pressure melting-point is approached, the friction falls to low values. Hope *et al.* found little rock wear by cold ice, especially when ice-to-ice sliding occurred, and this also supports the general contention that cold glaciers frozen to their beds are not effective erosional agents.

6 Some structural features arising from glacier motion

6.1 *Crevasses*

The theory of the development of crevasses in terms of stress distributions in moving glaciers was first propounded by W. Hopkins in 1862. Crevasse patterns are of several varied types (Fig. 4.21):

A Chevron crevasses, forming in the marginal zones (old ones may be twisted round as shown)

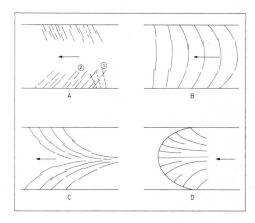

Fig. 4.21
Some examples of common crevasse patterns (R. P. Sharp, *Glaciers*, University of Oregon Press, 1960):
A Marginal (1—old rotated crevasses, 2—newly formed crevasses) **B** Transverse **C** Splaying **D** Radial. Ice flow from right to left. (By permission of author and publisher.)

B Transverse crevasses, convex up-glacier
C Splaying crevasses, beginning longitudinally and approaching the sides at 45°
D Radial splaying crevasses, characteristic of the glacier snout.

Often, these types cannot be easily distinguished: for instance, B and C have the same theoretical alignments in the marginal zone.

Crevasses may vary in width from hair-cracks to gaps many metres wide. Of greater significance to the glacial geomorphologist is the matter of crevasse depths, for crevasses are a principal means of penetration of meltwater to the deeper parts of glaciers, and in some cases may enable meltwater to reach the rock floor (as W. V. Lewis (1947) postulated in his theory of rock-step formation beneath thin glaciers). Nye (1955, 1957) has shown that, on the basis of Glen's power flow law, the rate of longitudinal extension ($\dot{\varepsilon}$) in a glacier subjected to a longitudinal tensile stress will be given by $\dot{\varepsilon} = B\sigma^n$ where σ is the yield stress in extension. Then the maximum theoretical depth of crevasses (d) will be given by

$$d = \frac{2}{\rho g} \frac{\dot{\varepsilon}^{1/n}}{B} \ (3 \sin^2 \alpha + 1)^{-1/2}$$

where ρ is the density of ice, g is the acceleration of gravity and α is the surface slope of the ice. Where α is small, the last term approximates to unity. If $n = 3 \cdot 3$, $B = 0 \cdot 1$ yr^{-1} bar^{-n}, and α is small, then $d = 22 \cdot 8 \ \dot{\varepsilon}^{1/3 \cdot 3}$ where d is in metres and the strain rate $\dot{\varepsilon}$ is given in units per year. For there to be good agreement between calculated values and measured depths, G. Holdsworth (1969a) emphasizes that regional values of $\dot{\varepsilon}$ must be used, and accurate englacial temperatures obtained to assess the value of the temperature-sensitive constant B (a change of temperature from 0°C to $-1 \cdot 5$°C, or from -1.5°C to -12°C, decreases the value of B by an order of magnitude).

Measured depths of true crevasses rarely exceed 30 m (it is important to distinguish crevasses from moulins (Chapter 6) which may be much deeper). W. Blake (1956) gives 26 m as the maximum depth in the Thule area (north-west Greenland); M. M. Miller (1954) gives 27-30 m for the Taku and Seward glaciers of Alaska, noting that it is not possible to measure by hand-line the lowest tapering part of the crack, and Holdsworth (1969b) quotes the maximum measured depth on the Kaskawulsh glacier of Alaska as 27 m. Crevasses exceeding 36 m in depth are known in the Antarctic and Greenland (Miller, 1954) and also in the Alps (F. Loewe, 1955). There are suggestions that, as cold ice can support higher shear stresses, the tensile layer and its crevasses may be deeper in polar than in temperate glaciers, and that crevasses in cold ice may be more sluggish in closing. A. H. Lachenbruch (1961) has studied the depth and spacing of tension cracks in hardware models using geological materials, which predict reasonable depths and spacings of glacier crevasses.

The presence of water in an isolated crevasse has a profound effect on the depth it can attain, because the pressure of the water will act as a wedge to cause further penetration of the glacier. Weertman (1973) shows that the depth attainable is a multiple of $\pi/2$ times the depth of a crevasse in a field of closely spaced water-free crevasses. If, however, an isolated water-free crevasse were to occur (though this is unlikely in nature), there is no reason why it should not deepen until it reached the glacier bed, though it might be pinched closed at a shallow depth.

6.2   *Foliation*

This is the name usually given to a secondary layering of glaciers produced by deformation; it is to be distinguished from primary layering of sedimentary origin inherited from the firn (see Chapter 7 for a description of sedimentary layering in cirque glaciers). Foliation layers may be up to one or more metres in thickness, and are identified not by dirt inclusions as in sedimentary layering but by differences in type of ice, especially between white bubbly ice and bluish clear ice. Foliation is most intense near to the margins of a glacier, where it often runs roughly parallel to them, near bedrock outcrops, below ice-falls and where different ice streams flow next to each other. The layers may dip at any angle, but often approach the vertical where the foliation is intense. Good examples of foliation are given by N. Untersteiner (1955) for the Pasterze glacier (Austria) and by N. W. Rutter (1965) for the Gulkana glacier (Alaska).

The exact origins of foliation are still in some doubt, but generally it is a structure resulting from deformation of ice during flow. It forms in areas subjected to intense compression or shear, as noted above, and since most of the surface area of a glacier experiences only very low shear stresses, the foliation does not normally originate where it is observed, but in the deepest layers subjected to much greater distortion. The pattern of foliation seen at the surface reflects subsequent differential down-glacier movement and surface ablation. Thus the pattern can be used as a rough measure of the

Plate IV
Morsárjökull, Southern Iceland. Showing well-developed ogives below the ice-fall and avalanche fan from the ice-cap, and strongly marked medial moraine. (*C.A.M.K.*)

relative velocity of flow. A problem not yet solved is the apparent failure of foliation in some instances to form parallel to the planes of greatest shear stress (Meier, 1960).

6.3   Ogives

These cross-glacier bands of alternating white and dark ice, curving down-glacier because of differential ice movement, were first described by Forbes (1843) on the Mer de Glace. They are not very common features, but on some glaciers (such as Austerdals-breen, or the Vaughan Lewis glacier in the Juneau ice-field of Alaska) they are exceptionally striking. They occur only on temperate glaciers below ice-falls, and their spacing (usually 50 to 200 m and decreasing down-glacier owing to compression) is approximately that of one year's flow. The white ice bands consist mainly of

Plate V
Medial moraine and ogives on Austerdalsbreen, Norway. *(C.E.)*

white bubbly ice (density c. 0·89), while the dark bands include a high proportion of blue bubble-free ice (density c. 0·91). The formation of these alternating layers is now reasonably well understood (C. A. M. King and W. V. Lewis, 1961; J. E. Fisher, 1962). Each of the dark ice bands represents that portion of the glacier that spent the summer descending the ice-fall. In the process, the ice crystals enlarge by melting and occasional refreezing, much meltwater gathers in crevasses and dirt accumulates on the stretched ice surface. At the foot of the ice-fall, the much-broken ice reconsolidates by compression into mainly blue bubble-free and dirt-stained temperate ice. In winter, on the other hand, the ice in the ice-fall remains generally frozen, crystals remain small and the surface will be mostly protected by a mantle of snow. It reconsolidates into mainly white cold bubbly ice.

On the glacier below the ice-fall, differential ablation accentuates the differences between successive bands. The white bubbly ice bands have less thermal conductivity and a greater albedo than the darker predominantly blue-ice bands and therefore tend to stand up in waves. Moreover, once formed, meltwater tends to collect and aid accumulation of dirt on the dark bands in the troughs, and the coarse-grained dark ice tends to hold dirt better than the smoother fine-grained 'winter' ice.

7 Conclusion

Glacier motion has been studied in increasing detail over the past century. Modern surveys of surface movements show that rates of flow vary along the length of the glacier and across it. In a single ice-stream, maximum velocities on a transverse profile occur in the central portions. In some glaciers, there is a very rapid change in velocity close to the valley sides, representing a form of flow termed Blockschollen movement. Variations in rates of flow with time have been investigated; in the ablation zone of temperate and sub-polar glaciers, for instance, the ice tends to flow faster in summer than in winter, probably because of the increased lubrication of the bed by meltwater. Some short-period variations may also be related to weather conditions and the availability of basal meltwater. Englacial velocity measurements by means of boreholes or tunnels show that, in a vertical profile, velocities decrease from glacier surface to bedrock, as was anticipated by Forbes in 1843. The theory of extrusion flow must be firmly abandoned.

The movement of glaciers consists of two components—internal deformation or 'creep', and basal sliding. Internal deformation approximates to a power flow law in which the exponent has a value usually between 2 and 4. Theoretical models of internal velocity distributions based on this law agree satisfactorily with observations from boreholes and tunnels. More complex forms of flow law (e.g. of polynomial type) have been more recently suggested. Longitudinal changes in rates of glacier flow are considered by Nye in his theory of extending and compressing flow, a theory that has important implications for glacial valley erosion, as will be shown in Chapter 8. The second component of glacier flow, sliding over bedrock, may be studied in the slip of a glacier past its sidewall and in the basal sliding directly observed in tunnels or deep borings. Side-slip has been found to amount to between 10 and 70 per cent of the maximum surface velocity, while basal sliding may vary from 20 to 90 per cent in the case of temperate glaciers. Cold glaciers do not exhibit basal sliding since they are frozen to bedrock, but there is relatively rapid shearing in the basal ice layers. Cold glaciers are therefore unlikely to be effective in eroding the bedrock by abrasion, though abrasion is probably not completely absent due to dragging of larger particles, mostly embedded in the rapidly shearing ice a few centimetres above the sole, over bedrock. There are also some cold glaciers whose base is just at pressure melting point, where limited sliding occurs (retarded by bedrock obstacles projecting into the cold ice above), and thick polar glaciers where cold ice overlies temperate ice, allowing rates of sliding characteristic of wholly temperate glaciers. In temperate glaciers on relatively steep slopes, basal sliding represents the principal contributor to ice discharge, and this has important consequences for bedrock erosion. The effect of higher velocities of basal sliding is to increase the number of boulders being dragged over an area of bedrock in a given time, and therefore enhance the degree of abrasion of the surface. Further details of the abrasion process will be given in Chapter 6. On flat surfaces or up reversed slopes, basal sliding will be considerably reduced. In such conditions, glacier motion is accomplished not by longitudinal pressure, which the ice would be incapable of transmitting, but by a thickening of the ice until the situation is reached at which the centre of gravity of the ice once again falls downhill sufficiently to permit movement. Movement, however, will be slow since the thicker glacier can discharge more ice for a given velocity than a thinner one.

Basal sliding beneath temperate glaciers occurs by a combination of regelation slip and creep around obstacles arising from stress concentrations. Lliboutry, Nye and Weertman have examined these processes theoretically, using highly simplified models of bed form. Weertman's revised theory (1964) shows satisfactory agreement with the observations of Kamb and LaChapelle. Both Weertman and Lliboutry consider that basal meltwater is of great importance in the sliding process. Basal meltwater may exist as a thin film lubricating bedrock and in water-filled subglacial cavities. The basal water pressure is a very important parameter in the sliding theory; by offsetting the normal overburden pressure of the ice, it affects the friction of ice on the rock. Under extreme conditions of high subglacial water pressure, catastrophic glacier advances may occur. Sudden advances may also result if warming of the basal ice of cold glaciers brings it into the temperature range where regelation slip becomes possible.

A study of glacier motion is a highly relevant preliminary to understanding the origin of many landforms resulting from glacial erosion and deposition, such as valley steps, rock basins, cirques and drumlins. The formation of these and other features will be considered in detail in Parts II and III; and the implications of both the theory and observation of glacier flow will be examined in connection with these landforms.

8 References

AGASSIZ, L. (1840), *Études sur les glaciers* (Neuchâtel)

    (1847), *Nouvelles études et expériences sur les glaciers actuels* (Paris)

ANDREWS, J. T. (1972), 'Glacier power, mass balances, velocities and erosion potential', *Z. Geomorph*. Suppl.**13**, 1–17

BARNES, P., TABOR, D. and WALKER, J. C. F. (1971), 'The friction and creep of polycrystalline ice', *Proc. R. Soc.* **324**A, 127–55

BATTLE, W. R. B. (1951), 'Glacier movement in north-east Greenland, 1949', *J. Glaciol.* **1**, 559–63

BLAKE, W. (1956), 'The depth of crevasses', *J. Glaciol.* **2**, 644–5

BLÜMCKE, A. and HESS, H. (1909), 'Tiefbohrungen am Hintereisgletscher 1909', *Z. Gletscherk.* **4**(1), 66–70

BORDIER, A. C. (1773), *Voyage pittoresque aux glacières de Savoie Suisse* (Geneva)

BOULTON, G. S. (1972), 'The role of thermal régime in glacial sedimentation', *Inst. Br. Geogr. Spec. Publ.* **4**, 1–19

CAROL, H. (1947), 'The formation of *Roches Moutonnées*', *J. Glaciol.* **1**, 57–9

CHAMBERLIN, R. T. (1928), 'Instrumental work on the nature of glacial motion', *J. Geol.* **36**, 1–30

COLBECK, S. C. and EVANS, R. J. (1973), 'A flow law for temperate glacier ice', *J. Glaciol.* **12**, 71–86

CROSSLEY, D. J. and CLARKE, G. K. C. (1970), 'Gravity measurements on Fox glacier, Yukon Territory, Canada, *J. Glaciol.* **9**, 363–74

DEMOREST, M. (1937), 'Glaciation of the Upper Nugssuak Peninsula, West Greenland', *Z. Gletscherk.* **25**, 36–56

(1938), 'Ice flowage as revealed by glacial striae', *J. Geol.* **46**, 700–25

(1942), 'Glacier regimens and ice movement within glaciers', *Am. J. Sci.* **240**, 31–66

(1943), 'Ice sheets', *Bull. geol. Soc. Am.* **54**, 363–400

DREWRY, D. J. (1972), 'The contribution of radio echo sounding to the investigation of Cenozoic tectonics and glaciation in Antarctica', *Inst. Brit. Geogr. Spec. Publ.* **4**, 43–57

ELLISTON, G. R. (1963), in discussion following J. WEERTMAN, 'Catastrophic glacier advances', *Bull. Int. Ass. scient. Hydrol.* **8**, 2, 65–6

FINSTERWALDER, R. (1931), 'Geschwindigkeitsmessungen an Gletschern mittels Photogrammetrie', *Z. Gletscherk.* **19**, 251–62

(1950), 'Some comments on glacier flow', *J. Glaciol.* **1**, 383–8

FISHER, J. E. (1953), 'Two tunnels in cold ice at 4000 m on the Breithorn', *J. Glaciol.* **2**, 513–20

(1962), 'Ogives of the Forbes type on Alpine glaciers and a study of their origins', *J. Glaciol.* **4**, 53–61

FORBES, J. D. (1843), *Travels through the Alps of Savoy* (Edinburgh)

FRIESE-GREEN, T. W. and PERT, G. J. (1965), 'Velocity fluctuations of Bersækerbræ, East Greenland', *J. Glaciol.* **5**, 739–47

GALLOWAY, R. W. (1956), 'Mechanical measurement of glacier motion', *J. Glaciol.* **2**, 642–4

GLEN, J. W. (1952), 'Experiments on the deformation of ice', *J. Glaciol.* **2**, 111–114

(1955), 'The creep of polycrystalline ice', *Proc. R. Soc.*, **228**A, 519–38

(1956), 'Measurement of the deformation of ice in a tunnel at the foot of an ice-fall', *J. Glaciol.* **2**, 735–45

GLEN, J. W. and LEWIS, W. V. (1961), 'Measurements of side-slip at Austerdalsbreen, 1959', *J. Glaciol.* **3**, 1109–22

GOLDTHWAIT, R. P. (1956), 'Formation of ice cliffs', *Tech. Rep. Snow Ice Permafrost Res. Establ.* **39**, 139–50

— (1961), 'Regimen of an ice cliff on land in north-west Greenland', *Folia geogr. dan.* **9**, 107–15

— (1973), 'Jerky glacier motion and meltwater', *Symposium on the hydrology of glaciers* (Cambridge, 1969), *Int. Ass. scient. Hydrol. Publ.* **95**, 183–8

GUNN, B. M. (1964), 'Flow rates and secondary structures of Fox and Franz Joseph glaciers, New Zealand', *J. Glaciol.* **5**, 173–90

HAEFELI, R. (1948), 'The development of snow and glacier research in Switzerland', *J. Glaciol.* **1**, 192–201

— (1951), 'Some observations on glacier flow', *J. Glaciol.* **1**, 496–500

— (1952), 'Behaviour of ice in the Z'Mutt glacier', *J. Glaciol.* **2**, 94–9

— (1963), 'Observations and measurements on the cold ice sheet on Jungfraujoch', *Bull. Int. Ass. scient. Hydrol.* **8**, 2, 122

HEIM, A. (1885), *Handbuch der Gletscherkunde* (Stuttgart)

HELLAND, A. (1877), 'On the ice-fjords of north Greenland, and on the formation of fjords, lakes, and cirques in Norway and Greenland', *Q.J. geol. Soc. Lond.* **33**, 142–76

HESS, H. (1904), *Die Gletscher* (Braunschweig)

— (1933), 'Das Eis der Erde' in GUTENBERG, B. (ed.), *Handbuch der Geophysik* (Berlin) **7**(1), 1–121

HOLDSWORTH, G. (1969a), 'Primary transverse crevasses', *J. Glaciol.* **8**, 107–29

— (1969b), 'An examination and analysis of the formation of transverse crevasses, Kaskawulsh glacier', *Icefield Ranges Research Project, Scientific Results* (ed. BUSHNELL, V. C. and RAGLE, R. H.) **1**, 109–25

HOLDSWORTH, G. and BULL, C. (1970), 'The flow law of cold ice; investigations on Meserve glacier, Antarctica', *Int. Symp. Antarct. Glaciol. Exploration* (Hanover, N.H., 1968), *Int. Ass. scient. Hydrol., Publ.* **86**, 204–16

HOLLINGWORTH, S. E. (1931), 'The glaciation of western Edenside and adjoining areas and the drumlins of Edenside and the Solway basin', *Q.J. geol. Soc. Lond.* **87**, 281–359

— (1947), in discussion following SELIGMAN, G. (1947), *J. Glaciol.* **1**, 20

HOLMES, C. D. (1937), 'Glacial erosion in a dissected plateau', *Am. J. Sci.* **29**, 41–7

HOPE, R., LISTER, H. and WHITEHOUSE, R. (1972), 'The wear of sandstone by cold sliding ice', *Inst. Br. Geogr. Spec. Publ.* **4**, 21–31

HOPKINS, W. (1862), 'On the theory of the motion of glaciers', *Phil. Trans. R. Soc.* **152**, 677–745

HUGI, G. J. (1842), *Über das Wesen der Gletscher* (Stuttgart)

IVES, J. D. and KING, C. A. M. (1955), 'Glaciological observations on Morsárjökull, S.W. Vatnajökull, Iceland', *J. Glaciol.* **2**, 477–82

JACKSON, J. E. (1953), 'Surveying on glaciers', *J. Glaciol.* **2**, 235

KAMB, B. and LACHAPELLE, E. (1964), 'Direct observation of the mechanism of glacier sliding over bedrock', *J. Glaciol.* **5**, 159–72

KAMB, B. and SHREVE, R. L. (1966), 'Results of a new method for measuring internal deformation in glaciers', *Trans. Am. geophys. Un.* **47**, 190

KICK, W. (1957), 'Exceptional glacier advances in the Karakoram', *J. Glaciol.* **3**, 229

KING, C. A. M. and LEWIS, W. V. (1961), 'A tentative theory of ogive formation', *J. Glaciol.* **3**, 913–39

LACHENBRUCH, A. H. (1961), 'Depth and spacing of tension cracks', *J. geophys. Res.* **66**, 4273–92

LAGALLY, M. (1929), 'An attempt to formulate a theory of crack formation in glaciers', *Snow Ice Permafrost Res. Establ. Translation* **47**, 18 pp.

(1934), *Mechanik und Thermodynamik des stationären Gletschers* (Leipzig)

(1939), 'Zur Mechanik eines auf seiner Sohle gleitenden stationären Gletschers', *Z. Gletscherk.* **26**, 193–8

LEWIS, W. V. (1947a), 'Some comments on Dr. H. Carol's article', *J. Glaciol.* **1**, 60–3

(1947b), 'Valley steps and glacial valley erosion', *Trans. Inst. Br. Geogr.* **13**, 19–43

(1948), in review of FLINT, R. F., *Glacial geology and the Pleistocene epoch*, *J. Glaciol.* **1**, 212

LLIBOUTRY, L. (1959), 'Une théorie du frottement du glacier sur son lit', *Annls Géophys.* **15**, 250–65

(1964), 'Subglacial "super-cavitation" as a cause of the rapid advances of glaciers', *Nature, Lond.* **202**, 77

(1964–5), *Traité de glaciologie* (Paris, 2 vols)

(1968), 'General theory of subglacial cavitation and sliding of temperate glaciers', *J. Glaciol.* **7**, 21–58

(1969a), 'How ice sheets move', *Sci. J.* **5** (March 1969), 50–5

(1969b), 'The dynamics of temperate glaciers from the detailed viewpoint', *J. Glaciol.* **8**, 185–205

LOEWE, F. (1955), 'The depth of crevasses', *J. Glaciol.* **2**, 511–12

MATHEWS, W. H. (1959), 'Vertical distribution of velocity in Salmon glacier, British Columbia', *J. Glaciol.* **3**, 448–54

MCCALL, J. G. (1952), 'The internal structure of a cirque glacier', *J. Glaciol.* **2**, 122–130

MEIER, M. F. (1960), 'Mode of flow of Saskatchewan glacier, Alberta, Canada', *U.S. geol. Surv. Prof. Pap.* **351**, 70 pp.

MERCANTON, P. L. (1916), 'Vermessungen am Rhone Gletscher, 1874–1915', *Neue Denkschr. schweiz. naturf. Ges.* **52**, 1–190

MERCER, J. H. (1971), 'Cold glaciers in the central Transantarctic Mountains, Antarctica: dry ablation areas and subglacial erosion', *J. Glaciol.* **10**, 319–21

MILLER, M. M. (1954), in discussion on 'The mechanics of glacier flow', *J. Glaciol.* **2**, 339–41

MÜLLER, F. and IKEN, A. (1973), 'Velocity fluctuations and water régime of Arctic valley glaciers', *Symposium on the hydrology of glaciers* (Cambridge, 1969), *Int. Ass. scient. Hydrol., Publ.* **95**, 165–82

NYE, J. F. (1951), 'The flow of glaciers and ice-sheets as a problem in plasticity',
   Proc. R. Soc., **207**A, 554–72

   (1952a), 'Reply to Mr. Joel E. Fisher's comments', *J. Glaciol.* **2**, 52–3

   (1952b), 'The mechanics of glacier flow', *J. Glaciol.* **2**, 82–93

   (1952c), 'A comparison between the theoretical and the measured long profiles of
   the Unteraar glacier', *J. Glaciol.* **2**, 103–7

   (1955), 'Comments on Dr Loewe's letter and notes on crevasses', *J. Glaciol.* **2**,
   512–14

   (1957), 'The distribution of stress and velocity in glaciers and ice-sheets', *Proc. R.
   Soc.*, **239**A, 113–33

   (1959a), 'The deformation of a glacier below an ice-fall', *J. Glaciol.* **3**, 387–408

   (1959b), 'The motion of ice-sheets and glaciers', *J. Glaciol.* **2**, 493–507

   (1959c), 'A method of determining the strain-rate tensor at the surface of a
   glacier', *J. Glaciol.* **3**, 409–19

   (1965), 'The flow of a glacier in a channel of rectangular, elliptic, or parabolic
   cross-section', *J. Glaciol.* **5**, 661–90

   (1969), 'A calculation on the sliding of ice over a wavy surface using a
   Newtonian viscous approximation', *Proc. R. Soc.* **311**A, 445–67

   (1970), 'Glacier sliding without cavitation in a linear viscous approximation',
   Proc. R. Soc. **315**A, 381–403

NYE, J. F. and MARTIN, P. C. S. (1968), 'Glacial erosion', *Assemblée Générale de Berne,
   Commn Snow Ice* (1967), *Int. Ass. scient. Hydrol., Publ.* **79**, 78–86

OROWAN, E. (1949), in discussion on 'The flow of ice and other solids', *J. Glaciol.* **1**,
   231–40

ORVIG, S. (1953), 'On the variation of the shear stress on the bed of an ice-cap', *J.
   Glaciol.* **2**, 242–7

PATERSON, W. S. B. (1961), 'Movements of the Sefstrøms Gletscher, north-east
   Greenland', *J. Glaciol.* **3**, 844–9

   (1964), 'Variations in velocity of Athabasca Glacier with time', *J. Glaciol.* **5**,
   277–85

   (1969), *The physics of glaciers* (Oxford)

   (1970), 'The sliding velocity of Athabasca glacier, Canada', *J. Glaciol.* **9**, 55–
   63

PATERSON, W. S. B. and SAVAGE, J. C. (1970), 'Excess pressure observed in a
   water-filled cavity in Athabasca glacier, Canada', *J. Glaciol.* **9**, 103–7

PERUTZ, M. F. (1940), 'Mechanism of glacier flow', *Proc. R. Soc.* **52**A, 132–5

   (1948), 'A description of the iceberg aircraft carrier...and some problems of
   glacier flow', *J. Glaciol.* **1**, 95–104

   (1950), 'Direct measurement of the velocity distribution in a vertical profile
   through a glacier', *J. Glaciol.* **1**, 382–3

PFAFF, F. (1874), 'Über die Bewegung und Wirkung der Gletscher', *Annls Phys.* **151**,
   325–36. See also SELIGMAN, G. (1948)

PHILIPP, H. (1920), 'Geologische Untersuchungen über den Mechanismus der
   Gletscherbewegung und die Entstehung der Gletschertextur', *Neues Jb. Miner.
   Geol. Paläont.* **43**, 439–556

PILLEWIZER, W. (1969), 'Die Bewegung der Gletscher und ihre Wirkung auf den Untergrund', *Z. Geomorph.* Suppl. **8**, 1–10

RAYMOND, C. F. (1971a), 'A new borehole inclinometer', *J. Glaciol.* **10**, 127–33

‒ (1971b), 'Flow in a transverse section of Athabasca glacier, Alberta, Canada', *J. Glaciol.* **10**, 55–84

REYNAUD, L. (1973), 'Flow of a valley glacier with a solid friction law', *J. Glaciol.* **12**, 251–8

RUTTER, N. W. (1965), 'Foliation pattern of Gulkana glacier, Alaska Range, Alaska', *J. Glaciol.* **5**, 711–18

SAUSSURE, H. B. DE (1779–96), *Voyages dans les Alpes* (Paris), 4 vols

SELIGMAN, G. (1947), 'Extrusion flow in glaciers', *J. Glaciol.* **1**, 12–21

‒ (1948), 'The movement of firn and ice in glaciers', *J. Glaciol.* **1**, 142–4

‒ (1949), 'Research on glacier flow', *Geogr. Annlr* **31**, 228–38

SHARP, R. P. (1953), 'Deformation of a vertical bore-hole in a piedmont glacier', *J. Glaciol.* **2**, 182–4

‒ (1954), 'Glacier flow: a review', *Bull. geol. Soc. Am.* **65**, 821–38

‒ (1960), *Glaciers* (Condon Lecture Publs, Univ. of Oregon Press)

SHREVE, R. L. and SHARP, R. P. (1970), 'Internal deformation and thermal anomalies in Lower Blue glacier, Mount Olympus, Washington, U.S.A.', *J. Glaciol.* **9**, 65–86

SOMIGLIANA, C. (1927), reported by FINSTERWALDER, S. (1926–7), 'Über die innere Reibung des Eises und die Bestimmung der Gletschertiefe', *Z. Gletscherk.* **15**, 55–9

SOUCHEZ, R. A., LORRAIN, R. D. and LEMMENS, M. M. (1973), 'Refreezing of interstitial water in a subglacial cavity of an alpine glacier as indicated by the chemical composition of the ice', *J. Glaciol.* **12**, 453–9

STREIFF-BECKER, R. (1938), 'Zur Dynamik des Firneises', *Z. Gletscherk.* **26**, 1–21

SWITHINBANK, C. (1966), 'A year with the Russians in Antarctica', *Geogrl J.* **132**, 463–75

THEAKSTONE, W. H. (1967), 'Basal sliding and movement near the margin of the glacier Østerdalsisen, Norway', *J. Glaciol.* **6**, 805–16

UNTERSTEINER, N. (1955), 'Some observations on the banding of glacier ice', *J. Glaciol.* **2**, 502–6

VIVIAN, R. (1966), 'La catastrophe du glacier Allalin', *Revue Géogr. alp.* **54**, 97–112

‒ (1970), 'Hydrologie et érosion sous-glaciaires', *Revue Géogr. alp.* **58**, 241–64

VIVIAN, R. and BOCQUET, G. (1973), 'Subglacial cavitation phenomena under the Glacier d'Argentière, Mont Blanc, France', *J. Glaciol.* **12**, 439–51

VIVIAN, R. and ZUMSTEIN, J. (1973), 'Hydrologie sous-glaciaire au glacier d'Argentière (Mont-Blanc, France)', *Symposium on the hydrology of glaciers* (Cambridge, 1969), *Int. Ass. scient. Hydrol., Publ.* **95**, 53–64

WARD, W. H. (1955), 'The flow of Highway Glacier', *J. Glaciol.* **2**, 592–8

WEERTMAN, J. (1957), 'On the sliding of glaciers', *J. Glaciol.* **3**, 33–8

‒ (1962), 'Catastrophic glacier advances', *Un. géod. géophys. int., Symposium at Obergurgl* (1962), *Int. Ass. scient. Hydrol., Publ.* **58**, 31–9

‒ (1964), 'The theory of glacier sliding', *J. Glaciol.* **5**, 287–303

WEERTMAN, J. (1966), 'Effect of a basal water layer on the dimensions of ice-sheets',
        J. Glaciol. **6**, 191–207

    (1967a), 'An examination of the Lliboutry theory of glacier sliding', J. Glaciol. **6**,
        489–94

    (1967b), 'Sliding of non-temperate glaciers', J. geophys. Res. **72**, 521–3

    (1971), 'In defense of a simple model of glacier sliding', J. geophys. Res. **76**, 6485-
        6487

    (1973), 'Can a water-filled crevasse reach the bottom surface of a glacier?'
        Symposium on the hydrology of glaciers (Cambridge, 1969), Int. Ass. scient. Hydrol.,
        Publ. **95**, 139–45

5

Some indirect effects of Pleistocene glaciation

Many of the Quaternary deposits in all countries ... indicate a pluvial period just as clearly as the Northern Drift indicates a glacial. (A. TYLOR, 1867)

The ice cap hypothesis advocated by Mr Croll is a probable cause of a great reduction in the level of the sea through abstraction of water from the sea, and its deposition at the poles in the form of ice. (A. TYLOR, 1868)

The geomorphological consequences of Pleistocene glaciation were by no means confined to the lands covered by ice and the adjacent periglacial zone. The radical changes of climate characteristic of the Pleistocene had far-reaching effects on the earth's water balance, so that the level of the seas rose and fell with the waning and waxing of the great ice sheets; the discharge of rivers fluctuated according to the varying rates of precipitation and evaporation; and lakes expanded and shrank, or even appeared and disappeared, adjusting themselves like the rivers to the prevailing hydrological conditions. The enormous burden of ice that accumulated in some regions caused actual deformation of the earth's crust, as, on a much smaller scale, did the weight of water represented by the largest of the temporary lakes. Changes in land, sea and lake levels, as well as changes in fluvial activity, are abundantly documented in the landforms and deposits of both glaciated and extra-glacial regions: this chapter is concerned with these changes, but does not pretend to be more than an outline of a subject whose scope is vast and whose literature is almost as voluminous as that on glacial geomorphology itself.

1 Pluvial periods in extra-glacial regions

The term 'pluvial' was introduced into geological literature by A. Tylor (1868) to refer to periods of greater rainfall and surface runoff in the Quaternary, which he supposed to have been responsible for deposition of certain sediments in England and northern France. Later, the term was transferred to other regions and drier climates; some years earlier, T. F. Jamieson (1863) had advanced the idea that the lowered temperatures of glacial times had resulted in expansion of lakes. In the hundred years that have elapsed since, the concept that present-day arid and semi-arid regions experienced pluvial and inter-pluvial periods in the Pleistocene has become firmly accepted and efforts are now directed towards establishing the chronology of such events and their correlation with climatic changes in the glaciated areas.

It is now widely agreed that the glacial periods of the Pleistocene were, in many mid-latitude areas, times of increased precipitation, lowered temperatures and reduced evaporation. The correlation between glacials and pluvials was long established in Western North America with the classic work of I. C. Russell on Lake Lahontan (1885) and the Mono Valley (1889), and of G. K. Gilbert (1890) on Lake Bonneville, but it had been anticipated in the Old World by L. Lartet (1865) who suggested that the glaciation of the Syrian highlands and former high stands of the Dead Sea had been contemporaneous. The correlations between glacials and pluvials are now known to be less simple than these pioneer workers envisaged, for the numbers of established glacial and pluvial events have multiplied, and the climatic conditions prevailing during pluvial and inter-pluvial phases have also been shown to be more complex than was originally thought.

Much of our knowledge of Pleistocene climatic fluctuations in arid and semi-arid areas is based on studies of pluvial lakes. Invaluable records of their fluctuations of level are given by their strandlines and other marginal features such as spits, bars and deltas, and by their deposits of both littoral and deep-water facies. However, it must be recognized at the outset that lake levels are determined not only by climate, and changes in level may be due to tectonic factors, erosion of the lake outlet (if any), or possible diversion of a river into the lake, as in the case of the Bear River diverted by lava to flow into the Bonneville basin. In the case of some pluvial lakes, too, entry of meltwater from nearby glacial ice has affected the level, though usually to only a minor extent. Pluvial Lake Lahontan, for example, received meltwater at times, but even if all the ice in its catchment had melted, it could not have supplied more than 20 per cent of the water accumulating there at its maximum extent.

Pluvial lakes were of wide distribution at times in the Pleistocene, and some attained impressive dimensions. The Chad basin once contained a lake ('Megachad') equivalent to four times the size of present-day Lake Victoria and whose level rose to 320 m as late as 10,000 BP (A. T. Grove and R. A. Pullan, 1963; Grove and A. Warren, 1968); the Dead Sea, whose surface now stands 400 m below sea level, rose at least once to 30 m above sea level; the Caspian Sea combined with the Aral Sea, reaching a level 115 m (possibly higher) above its present and backing up the Volga to Kazan, though meltwater from the Central Asian glaciers contributed substantially to this prodigious water body (K. W. Butzer, 1958). Since space forbids even a list of the Pleistocene pluvial lakes, three major examples from the Basin-and-Range country of the western USA are selected for further comment.

1.1   *Pluvial lakes in the Great Basin*

In the Great Basin, the general sequence of Pleistocene climatic change is thought to have been as follows:

———→Interglacial——→ ... ——→Glacial——→ ... ——→Interglacial——→
cool——→ warm——→ cool——→ cold——→ cool——→ warm
dry ————→ wet ————→ moist ————→ dry ————→

This is necessarily an oversimplification. The tectonic features of the Great Basin formed ideal situations for lake formation when climatic conditions were suitable, and over 120 former lakes have been identified. Gilbert's Lake Bonneville was by far the largest, approaching 52,000 km² or the size of present Lake Michigan, and attaining a maximum depth of 335 m. Its present-day attenuated remnants are Great Salt Lake, Provo and Sevier Lakes. Gilbert concentrated mainly on shore morphology, but though this is impressive, it does not yield as complete or unambiguous a story of the lake fluctuations as do studies of the lacustrine and inter-lacustrine deposits. The record so far deciphered in any detail covers only the 75,000 years or so of the Wisconsin (last) glaciation; Lake Bonneville undoubtedly existed at times in the pre-Wisconsin period but the data have yet to be assembled and analysed. The highest stand of the lake at the Bonneville shoreline (1550–1616 m: a precise figure cannot be quoted because of isostatic deformation) was attained on several occasions according to recent work (R. B. Morrison and J. C. Frye, 1965, and references contained therein). Other major shorelines, first demonstrated by Gilbert, occur at about 1460 m (the Provo stand) and 1370 m (the Stansbury level, distinguished by an abundance of tufa). At the highest stands, Lake Bonneville overflowed through Red Rock Pass to the Snake River, and the well-developed Provo shoreline represents the level being held by a resistant formation in this outlet. Fig. 5.1 shows the fluctuations of level postulated by R. B. Morrison and J. C. Frye during Wisconsin time. Their correlations of high-lake phases with glacial episodes should be noted. This is one of the two areas in the Great Basin where a close connection can be established, for glacial ice at times entered Lake Bonneville from the Wasatch mountains. Thus, for example, till and outwash from Little Cottonwood Canyon interfinger with lake sediments laid down at times of high water level, and both the upper and middle Bull Lake (early Wisconsin or Altonian) tills appear to have been deposited in standing water. The general synchroneity of glaciations and lake maxima is difficult to dispute. Radiometric age determinations by C^{14} and Th^{230} have given further confirmation of this correlation (see, for example, R. F. Flint and W. A. Gale, 1958; A. Kaufman and W. S. Broecker, 1965), but there still remain such problems as the fact that the highest stands of pluvial lakes were not reached in glacial maxima but during retreat (R. W. Fairbridge, 1970). The highest terrace of Lake Bonneville, for instance, was last occupied (BS, Fig. 5.1) about 13,000 BP towards the close of the Woodfordian (Flint's (1971) 'Late Wisconsin').

The lake sediments record alternating phases of deep and shallow water, and occasional intervals of complete desiccation when subaerial weathering and soil formation supervened. The most important soil recognized in the Wisconsin sequence (see Fig. 5.1) by Morrison is the Promontory Soil, correlated with the major interstadial, probably 22,000–28,000 BP, when Lake Bonneville temporarily disappeared.

Lake Lahontan (Plate VI) was the second largest Pleistocene water body in the Great Basin (about 22,000 km²; maximum depth 213 m at Pyramid Lake). Its main shorelines today stand at 1332, 1277 and 1213 m (isostatic deformation was negligible since the

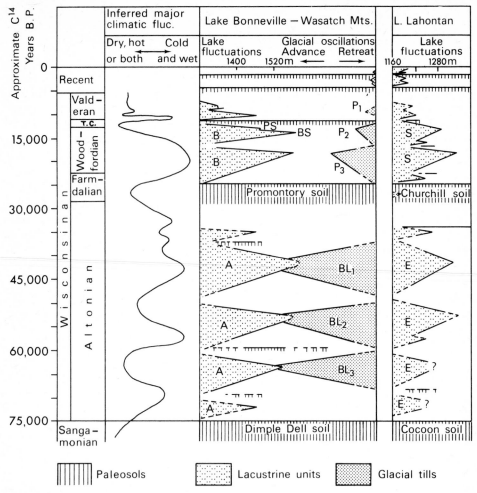

Fig. 5.1
Climatic history, lake-level fluctuations and glacier oscillations in the Bonneville and Lahontan areas, during the Wisconsin period (R. B. Morrison and J. C. Frye, *Report, Nevada Bureau of Mines,* 1965)
Lacustrine units: **B** Bonneville formation   **A** Alpine formation
                        S Sehoo formation   **E** Eetza formation
Glacial units: $P_{1,2,3}$, Upper, middle, lower Pinedale till
                    $BL_{1,2,3}$, Upper, middle, lower Bull Lake till
BS Bonneville shoreline last attained (*c.* 13,000 BP)
PS Provo shoreline

mass of water was much smaller than in the case of Lake Bonneville). Unlike Lake Bonneville, it never overflowed. The correlations proposed by Morrison are indicated on Fig. 5.1. Note the strongly developed Churchill soil, marking a lengthy interval of desiccation between the two main deep-lake periods. Pre-Wisconsin events are tentatively recognized, but not shown on Fig. 5.1; deep-lake periods have been suggested for the Kansan and Illinoian glacials.

Plate VI
Sediments (seen as a white band across the photograph) deposited in pluvial Lake Lahontan, Nevada.
(C.E.)

South from Lake Lahontan is a third pluvial lake basin of great interest, that of Mono Valley. Today, the basin contains Lake Mono, about 52 m deep and lacking an outlet. At times in the Pleistocene, water rose to 2156 m (Tioga or late Wisconsin glaciation) and 2190 m (Tahoe or early Wisconsin, when the lake was 275 m deep and overflowed briefly to the Owens River). At these times of high water, tongues of glaciers emerged from the Sierra Nevada and entered the lake ('Lake Russell'), some-times with the interposition of outwash deltas, and lake shorelines visibly cut the moraines. Again, the evidence strongly points to the broad contemporaneity of glacial and pluvial events.

The pluvial lakes of the Great Basin have been used to assess the amount of climatic change between glacial and interglacial periods. An early attempt was that of O. E. Meinzer (1922), who concluded that Arizona, for example, was about as moist in pluvial maxima as Oregon is today. More modern computations of precipitation–temperature–runoff relations needed to support a given lake level have been made by several. An example is the work of C. T. Snyder and W. B. Langbein (1962) in Spring Valley, Nevada, who postulate increases of precipitation from the present 300 mm to 500 mm in glacial times, and corresponding reductions in evaporation from 1100 to 800 mm.

Undoubtedly, the potential for further investigations of pluvial lake history both in the Great Basin and elsewhere in the world is very great, and such investigations are destined to shed much light on the events of the Pleistocene, particularly the older events that are so poorly documented in the glacial record. These studies promise eventually to reveal the total number of pluvials and glacials, their chronology and intercorrelation. The principal method of attack will be on the stratigraphy of lake sediments in enclosed basins by deep boring; such sediments must preserve a complete record of changing lacustrine and subaerial conditions since there is little if any possibility of the complete loss of any member except perhaps by wind deflation. Cores up to 200 m long have been recently obtained from the sites of some pluvial lakes in North America. One from Great Salt Lake is estimated to span the last 0·8 million years (A. J. Eardley and V. Gvosdetsky, 1960). Correlation of the core stratigraphy with the record from surface features is discussed by A. J. Eardley (1968), and its paleomagnetic dating by A. J. Eardley et al. (1973).

2 Changing stream régimes

The changing climatic conditions throughout the Pleistocene, together with the changes in the relative levels of land and sea to be described in the next sections, had a profound influence on stream activity. With fluctuations in precipitation and in rates of runoff affected, for instance, by the changing vegetation cover, river volumes rose and fell even where they were not directly fed by meltwater. Furthermore, the load of material available for river transport varied, being greatest when vegetal cover was too sparse to restrain downhill creep, when the climate favoured solifluction and major mass movement, or when the rivers were supplied with outwash or were attacking loess-covered areas. Since the hydraulic activity of streams is largely controlled by discharge and load, the streams adjusted their long profiles to the prevailing conditions, and sequences of terraces and valley fills attest to the frequency and magnitude of changes in these conditions. A study of stream channel characteristics below an active glacier in Washington (USA) by R. K. Fahnestock (1963) demonstrates that erosion or deposition of valley outwash is closely related to variations in stream discharge and debris supply. In regions not too far removed from the sea, Pleistocene changes of land and sea level also affected stream profiles, submergence causing drowning and alluviation of their lower reaches, while emergence resulted in some cases in incision of the lower reaches and the formation of knick points which receded slowly upstream.

The geomorphological evidence of these changes is clear, in terraces, buried channels, and very often also signs that the present-day streams are 'underfit' in that they now occupy valleys too large for the present rates of discharge (see, for instance G. H. Dury, 1960). What is much more difficult is to relate these features to particular climatic events in the Pleistocene, or to use them to establish the nature of climatic changes; for any such attempts to be worth while, reliable evidence on the ages of the features has to be first obtained.

The subject is too vast to elaborate, beyond referring briefly to the evidence presented in some of the features of the Thames Valley in England. The Thames received direct accessions of meltwater and outwash at certain stages (for example, via its tributary the Colne); at other times, variations in load and discharge were determined by climate. The present river is clearly underfit in relation to its floodplain and incapable of now handling the coarse material of which the floodplain is built; its present shrunken size and relatively sluggish activity attest to the changed conditions of the post-glacial period. Higher-level gravel spreads, now forming terraces or plateau cappings, are related to previous climatic episodes. The dating of many of these older gravels is fairly well established—the Boyn Hill terrace, it is generally agreed, formed in the Hoxnian (Elster-Saale) interglacial, for it overlies Elster drift at Hornchurch (Essex) and is itself overlain by solifluction deposits of Saale age (B. W. Sparks and R. G. West, 1972). The climatic conditions under which the gravels accumulated are less well known, but the impressive extent of aggradation represented by the Boyn Hill terrace, for instance (its valley-floor width approaches 20 km in the Romford area) is consistent with an early interglacial age when a river of sizeable volume was attacking deposits accumulated in its upper and middle reaches and sweeping them downstream; at the same time, base-level was rising in response to melting of the ice sheets, as will be described in the next section.

3 Glacio-eustatic sea-level changes

The accumulation of snow and ice on land-masses represents storage of moisture and is an integral part of the earth's hydrological cycle. An immediate effect of any increase in world temperatures is the liberation of meltwater from such snow and ice, most of which finds its way to the oceans and causes a necessarily world-wide rise in sea level. Conversely, any extension of ice sheets and glaciers results in a drop in sea level. Such changes of sea level are described as glacio-eustatic; the term 'eustatic' was introduced by E. Suess in the late nineteenth century to refer to world-wide sea-level changes caused by several possible factors, not only those related to ice ages. A most difficult problem, in fact, lies in attempting to distinguish glacio-eustatic from other eustatic changes of sea level, such as might arise, for instance, from changes in the capacity of the ocean basins caused by tectonic activity or deposition of sediments in those basins, from changes in the temperature of the sea water (a 1°C change would affect sea level by 1–2 m), or from changes in the level of the sub-oceanic crust as the load of sea water varies. As sea level can only be defined in terms of its relation with the land-masses, a further complication arises from the possibility that apparent changes of sea level may be the result of actual changes in the height of the land arising from crustal deformation. Crustal deformation caused by variations in ice load will be dealt with in the next section. A final consideration to be borne in mind in connection with Pleistocene sea-level changes is that even slight shifts in the earth's axis of rotation would cause changes in the shape and position of the mean sea-level (geoidal) surface relative to the land.

It is evident, then, that to disentangle the effects of purely glacio-eustatic changes of sea level from the effects of the other factors mentioned is an almost insuperable task. Nevertheless, evidence is available to show that, throughout the Quaternary and in historic times, glacio-eustatic fluctuations of sea level have taken place.

The first to recognize the possibility of glacial control of sea level was C. Maclaren (1842), who assessed the change in level resulting from Pleistocene glaciation at 107 to 213 m. A. Tylor's estimate (1872) was 180 m. In 1910, R. A. Daly expounded his hypothesis of glacial control in the evolution of coral reefs, involving the upgrowth of corals from platforms cut during ice-age low sea levels (see also Daly, 1934). The extent of glacio-eustatic sea-level change during the glacial periods still provides a formidable problem, but several lines of evidence are now converging to suggest a lowering of between 100 and 140 m in each main glacial period. The evidence comes from submerged wave-cut platforms and notches, submerged littoral or terrestrial sediments, and buried channels of coastal rivers. In all cases, however, the possibility must be continually borne in mind that the anomalies of level may be explained by crustal displacements, instead of actual sea-level changes. As examples of the data used in support of low sea stands in the last glaciation, the following may be mentioned. Many oceanic atolls possess basal platforms at about -100 m, while submarine terraces fringe many coasts at about this level. The buried channels of rivers in western Europe often grade down to levels approaching -100 m (for instance, the First Buried Channel of the Thames). H. N. Fisk and R. McFarlan (1955) in the Mississippi delta suggest a low sea level in the last glaciation of -136 m, though R. W. Fairbridge (1961) considers this figure may be too large owing to insufficient allowance being made for local subsidence and compaction. Fairbridge prefers a figure of -100 m for the Main Würm and -70 m for the Late Würm, while J. R. Curray (1965), incorporating some further data, suggests -145 m for the early Wisconsin (Main Würm) and -120 to -125 m for the late Wisconsin (Late Würm).

W. L. Donn and others (1962) approached the problem from another standpoint by estimating Pleistocene ice volumes and converting these into amounts of sea-level lowering:

Stage	Estimate A*	Estimate B*
Late Wisconsin	105 m	123 m
Early Wisconsin	114 m	134 m
Illinoian	137 m	159 m

* Two estimates of sea-level lowering are given, based on different assumed ice-thicknesses.

They support their reconstruction of Illinoian sea-level stands of -137 to -159 m with new evidence of deeply submerged and buried marine terraces (Newfoundland -145 m, West Indies -147 to -154 m, Argentina up to -158 m, etc.) on which are found shallow-water shells radiocarbon-dated at more than 30,000 years. However, estimates of ice thickness are so unreliable that the calculated figures of sea-level lowering may be widely in error. Moreover, they make no allowance for the following factors:

1 There would be semi-contemporaneous crustal adjustments due to loading and unloading of ice on the land;
2 The ocean floors would themselves respond isostatically to loading and unloading of the sea water (A. L. Bloom, 1967; N.-A. Mörner, 1971; see p. 171);
3 Changes in sea level automatically cause slight changes in oceanic areas.

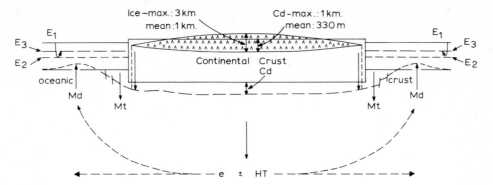

Fig. 5.2
Eustatic-isostatic relationships. The diagram shows a hypothetical ice-loaded continent bound by
simatic oceanic crust. Mean ice loading (1 km) depresses crust 330 m (Cd). Crustal displacement (Cd)
is effected by elastic compression (e) and/or horizontal transfer (HT) in the deep layers. This may
give rise to positive marginal displacement (Md).
    Sea-level E_1 is lowered glacio-eustatically to E_2, but may rise after a time-lag (for Md to occur)
to E_3. Isostatic unloading of the ocean area by water removal during transfer E_1 to E_2 will also contribute to the height of E_3. The reverse may be expected during deglaciation when, for a deglacial
eustatic rise of 100 m, a mean oceanic crustal lowering of 33 m is suggested (R. W. Fairbridge, *Physics
and Chemistry of the Earth*, ed. L. H. Ahrens, Pergamon Press, 1961)

Figure 5.2 (after R. W. Fairbridge) illustrates some of these eustatic-isostatic relationships. The final sea-level lowering resulting from ice formation would be considerably
less, perhaps one-third less, than the simple calculations performed by W. L. Donn
and others (1962) suggest. Fairbridge argues, for instance, that melting of all the
world's present ice, theoretically equivalent to a column of ocean water 95 m high
(Flint (1971): 65 m), would in the end cause a sea-level rise of little more than 50 m.
Even this, however, would flood about one-eighth of the present land area.
    It is relevant to note here the important part played by the ice of Antarctica and
Greenland in glacial/interglacial sea-level changes. These two ice masses together
account for over 98 per cent of all ice on earth today, and for perhaps 40 per cent
of the world's ice in glacial periods. Evidence is insufficient to show whether these two
ice masses remained relatively unchanged in volume throughout interglacial periods,
or whether they survived diminished in volume. Since on their own account their melting or accumulation could affect world sea level by as much as 50 m, interglacial sea
levels cannot be deduced from ice volumes. Stratigraphic evidence, however, offers
more reliable interpretations (see H. G. Richards and Fairbridge, 1965), but there are
still great differences of opinion.
    Donn and others (1962) argue that there has been approximate sea-level stability
from one interglacial to another; entirely different is the viewpoint held by F. E. Zeuner
(1959) and R. W. Fairbridge (1961) for instance, that there has been a continuous
secular lowering of sea level throughout the Quaternary, on which glacial oscillations
have been superimposed. According to this view, late Pliocene–early Pleistocene sea
level may have been 200 m or more above the present, and in the several interglacials,
maximum sea levels were attained as follows:

Eemian	+ 10 m
Hoxnian	+ 60 m
Cromerian	+100–110 m

Minimum glacial sea levels according to Fairbridge were:

Würm	−100 m
Riss	− 55 m
Mindel	− 5 m
Günz	+ 35 m

This gives an amplitude of 110–120 m for glacio-eustatic sea-level fluctuation as between one interglacial and the adjacent glacial period.

Sea-level changes over the last 20,000 years are comparatively better known because radiocarbon dates are available to establish a time-scale. But there are still considerable differences between the various reconstructed sea-level curves and their interpretations, relating to the choice of radiocarbon-dated samples and to assumptions made about

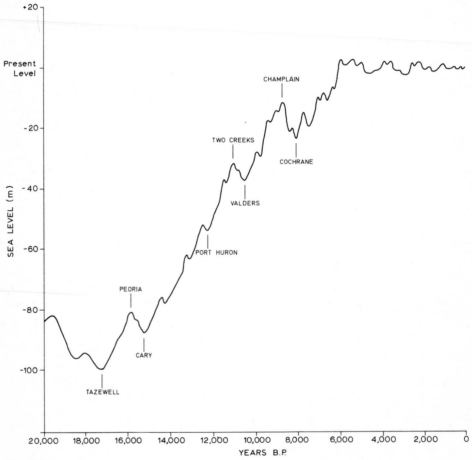

Fig. 5.3
Sea-level change since 20,000 BP (J. R. Curray, 1965; R. W. Fairbridge, 1961)

which coasts have remained tectonically stable during this period. Eventually, Th^{230} dates may help to support and extend the C^{14} time-scale. On Fig. 5.3, the portion of the curve from 20,000 to 10,000 BP is derived from Curray (1965), and from 10,000 BP to the present from Fairbridge (1961). Essentially, the overall rise from the Main Würm (late Wisconsin) minimum to about 6000 BP (the 'Flandrian transgression', with an average rate of 8 mm per year), represents the return of water to the oceans from the decaying Main Würm ice masses. The smaller additional oscillations reflect relatively minor climatic fluctuations (e.g. the Zone III deterioration at 11,000 BP). Table 5.1 compares these and some estimates of eustatic sea level over the last 10,000 years; some authorities do not accept sea levels higher than present since the last glaciation.

Table 5.1   Some estimates (m) of eustatic sea level over the last 10,000 years

Years BP:	1000	2000	3000	4000	5000	6000	7000	8000	9000	10,000
Fairbridge (1961)	+1	−2	−3	+2	+3	0	−6	−16	−14	−32
Jelgersma (1961)	−1	−2	−3	−4	−5	−7	−10	−19	−35	
Shepard (1963)	−0·5	−1	−2	−3	−4	−7	−10	−16	−22	−31
Schofield (1964)	+1	+2	+3	+5	−2	−0·5	−4	−19	−33	−36
Tooley (1974)	+1	+0·5	0	−1·5	−1	−3	−5	−14	−21	

Historic and present changes of sea level may be studied on tide-gauge records. The longest record is for Amsterdam, from 1682 onward, but this is not ideally situated owing to local subsidence. Allowing for this, however, the Amsterdam record suggests world sea level falling in the late eighteenth century and rising again from about 1880 to the present day. A correlation with world-wide glacial retreat since the mid-nineteenth century is highly probable (H. W. Ahlmann, 1953). Current sea-level rise is put at 1·2 mm a year (H. Valentin 1952); a figure of 1·1 mm a year was suggested in the first analysis of tide-gauge records in stable regions by B. Gutenberg (1941).

4 Glacio-isostatic crustal deformation

The idea that isostatic deformation of the earth's crust might arise from the weight of thick ice sheets was advanced by T. F. Jamieson over a century ago. G. K. Gilbert (1890) broadened the concept to include crustal depression by the weight of water in the largest pluvial lakes. Since Jamieson's time, the work of many in Scandinavia, Finland and North America has proved the correctness of these early ideas. It is known, for example, that glacial loading today keeps the bedrock surfaces of Greenland and Antarctica below sea level near their centres, while slow adjustment to unloading in Canada, the north-eastern USA and Fennoscandia is still in progress, even though most of the ice in these areas vanished 7000–8000 years ago. Evidence from tide-gauges, geophysical studies and tilted Quaternary shorelines is collectively unequivocal. Studies of deformed shorelines in particular have made it possible to calculate rates of uplift consequent upon ice unloading, and to delineate the areas that have been affected by glacio-isostatic deformation, the boundaries of these areas lying roughly parallel to the limits of the last glaciation. The patterns of crustal deformation are revealed in maps of isobases, or lines joining points of equal uplift on the affected strand-lines. Fig. 5.4A shows an example of isobases reconstructed for 6000 BP over north-eastern Canada by J. T. Andrews (1970), based on age and elevation data for 58 sites. At this time, deglaciation was virtually complete, and three main centres of post-glacial

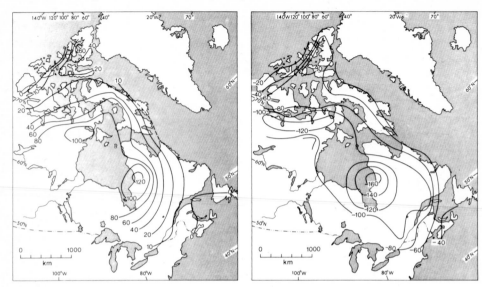

Fig. 5.4
A Isobases (in metres) for 6000 BP, northern and eastern North America
B Residual uplift (in metres) in northern and eastern North America (J. T. Andrews, *Institute of British Geographers, Spec. Publ.*, 1970)

uplift may be tentatively delineated. On isobase maps, the limits of measurable crustal deformation may be represented by so-called 'hinge-lines' (or 'zero isobases') for each shoreline; for instance, in the Great Lakes area, the southern parts of certain Late-glacial shorelines are horizontal, but beyond their respective hinge-lines, they rise northward. M. Sauramo (1955) has attempted to identify similar hinge-lines in Fenno-scandia.

The mechanics of the process of crustal deformation are incompletely understood. When the load of ice exceeds the strength of the crust, subsidence occurs, presumably displacing sub-crustal matter outward though part of the load could be taken up by an elastic reaction. There is a time-lag in the process, but it seems likely that crustal response is complete over a span of 15,000–20,000 years. With deglaciation, the crust is unloaded and gradually recovers its former level (the terms 'crustal rebound' or 'recoil' are often used in this connection). However, the ice masses were not simply loads added to or removed from particular areas of the earth's crust, because, simultane-ously, equivalent masses of water were taken from or returned to the oceans. It is, therefore, a question of load transfer (A. L. Bloom, 1967) and the model of glacio-isostasy must include the following elements:

Glaciated areas: over a time-scale of some thousands of years, loads of ice corresponding to pressures of up to 140–170 bar were added to or removed from these areas which occupy about 5 per cent of the earth's surface.

Oceanic areas: loads of water corresponding to pressures of 10–12 bar were added to or removed from areas occupying 70 per cent of the globe.

In Bloom's model, although there is still the problem, even allowing for some crustal elasticity, of how sub-crustal matter was transferred on such a huge scale from local glaciated areas to extensive oceanic areas, and vice-versa, there is no problem as to where the displaced matter was disposed. Former theories, not taking into account the simultaneous loading and unloading of oceanic areas, were forced to postulate 'peripheral bulges' around the glaciated areas to accommodate the displaced rock material, but despite intensive search, no geomorphological evidence of such bulges has been found. The minimum height and size of ice mass needed to cause crustal sinking is only approximately determinable, for crustal strength varies somewhat. Large modern reservoirs such as Lake Mead and Lake Kariba contain sufficient mass of water to cause crustal subsidence detectable in geodetic surveys. Pluvial Lake Bonneville equivalent in weight to an ice cap 330 m thick and 250 km in diameter caused 65 m of crustal subsidence (load up to about 20 bar), while smaller Lake Lahontan (equivalent to ice 180 m thick and 170 km in diameter) caused only 10 m of subsidence (load up to about 5 bar) as measured in shoreline deformation. The Bonneville data have been recently re-examined by M. D. Crittenden (1963). The Great Basin, however, may be an area of less than average crustal strength, and the amplitude of warping depends not only on ice thickness but also on the nature of the crustal and sub-crustal material. From evidence available, it seems likely that the maximum subsidence beneath a large ice sheet should be about 25–30 per cent of the maximum ice thickness, but correspondingly less beneath small ice sheets, assuming that the density of sub-crustal matter (3·0–3·3) is about 3 or 4 times that of ice.

Andrews (1970) identifies three stages in the complete process of glacio-isostatic recovery (Fig. 5.5). The first response to unloading occurs during deglaciation, beginning at the time when ice thinning commences, and is termed the phase of restrained

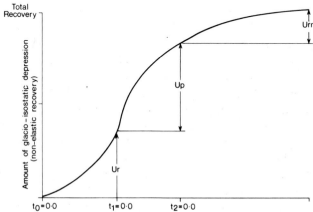

Fig. 5.5

The three periods of glacio-isostatic recovery—restrained rebound (Urr), post-glacial (Up) and residual (Ur) uplift (J. T. Andrews, *Institute of British Geographers, Spec. Publ.*, 1970)

rebound. The second stage is that of post-glacial uplift, extending from the time when ice had virtually disappeared to the present, and the third is the residual component, comprising the amount of recovery still to take place before final equilibrium is established. The first and third of these stages are relatively poorly known; most research has concentrated on the second stage because most field data relate to this episode.

4.1   *Fennoscandia*

At present, differential uplift is in progress, varying from 9 mm a year in the north of the Gulf of Bothnia to 4 mm a year near Stockholm and nil around Copenhagen. It should be remembered that these rates are determined from tide-gauge records with respect to mean sea level, which itself is currently rising at about 1 mm a year. The net emergence results in about 1000 km² of land being added to Finland each century. Studies of the crustal rebound in the Late-glacial and Post-glacial periods are based on warped strandlines, of which an enormous number has now been recognized, and their interpretation in detail is highly complex and controversial. In the centre of the region, post-glacial uplift of about 520 m is deduced, and it is thought that another 200 m or so of recovery will take place before equilibrium is achieved (E. Niskanen, 1939). Thus 700 m of isostatic depression, plus some allowance for an unknown amount of restrained rebound prior to the earliest shoreline formation (*c.* 12,000 BP), must be invoked, and if we take this as equivalent to 30 per cent of the ice thickness, as suggested above, the ice thickness is likely to have been at least 2300 m, which is in accord with evidence from other sources. Maximum rates of rebound probably attained 50 or even 75 mm a year for short periods in the neighbourhood of the Gulf of Bothnia, but they have now slowed down considerably as noted, and complete recovery may take another 10,000 years or so. M. H. P. Bott (1971) estimates the 'relaxation time' (the time required for the deviation from isostatic equilibrium to be reduced to $1/e$ of its initial value) as 5000 years. The lack of equilibrium at present is reflected in the large areas of negative gravity anomalies (up to 30 mgal).

4.2   *North America*

The story of glacio-isostatic deformation is much less well known than in Fennoscandia, for the area is vastly larger, much of it thinly populated, and inadequately mapped. Moreover, because of a faster rate of recovery and lower potential sea level, marine transgression is largely absent across much of Arctic Canada (Andrews, 1970), denying early workers the advantage that the Scandinavians had in such time-stratigraphical markers as the deposits of the Littorina transgression. Again, tide-gauge records in North America, because of the extent of unbroken land compared with Fennoscandia, reveal only a small part of the movements currently taking place. Fig. 5.6B shows the average rates of uplift in eastern North America at present, which reach a probable maximum of 13 mm/year over southern Hudson Bay, greater than the maximum rate in north-west Europe (Fig. 5.6A) partly because of the greater ice loading. In the Great Lakes today, water is rising on the southern shores and receding from the northern ones. Just in the area, however, where tide-gauge records would be of the highest in-

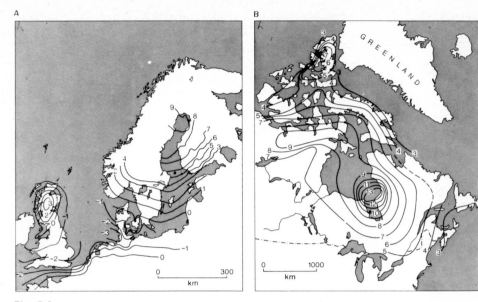

Fig. 5.6
A Contemporary glacio-isostatic uplift (mm/year) in north-west Europe. There is also subsidence, of both glacio-eustatic and non-glacial origin, in the North Sea–English Channel area. Broken and dotted isobase lines represent less certain interpolations (R. G. West, *Pleistocene geology and biology*, 1968, Longman, by permission of the publisher).
B Contemporary glacio-isostatic uplift (mm/year) in eastern and northern North America. As present sea level is thought to be rising at between 1 and 3·5 mm/year, the 3 mm/year isobase on this diagram represents an approximate boundary outside which submergence of coastal areas may be occurring (J. T. Andrews, *Institute of British Geographers, Spec. Publ.*, 1970)

terest, namely, Hudson Bay, tide-gauges are few and their records unreliable or of short duration. The longest record is for Churchill, since 1928; B. Gutenberg (1941) analysed the first 10 years and deduced a rise of 10–20 mm a year, but since then, doubt has been cast on the validity of the earliest record, and the data since 1940 have failed to show any positive trend. But on archaeological evidence, J. B. Bird (1954) thought a rise of 3 feet a century (=10 mm a year) likely. More recently, D. M. Barnett (1966) has re-examined the Churchill tide-gauge data.

Deformed and uplifted shorelines have been extensively studied, as in Fennoscandia. A line Milwaukee–Cleveland–New York roughly indicates the southernmost limit of measurable up-warping (Fig. 5.6B). Maximum post-glacial uplift is located in Richmond Gulf, Hudson Bay, where it amounts to at least 302 m (W. R. Farrand, 1962). Andrews' estimate (1970) for residual uplift in this area is about 160 m, giving a total of about 460 m since deglaciation. As in Fennoscandia, evidence on the phase of restrained rebound is slight; Andrews estimates that this might account for 50–75 per cent of total recovery. Taking the allowance as 50 per cent, the theoretical total recovery amounts to 920 m, which is approximately consistent with estimates, derived from other sources, that the Wisconsin ice here was probably over 3000 m thick. Negative gravity anomalies of around 25 mgal occur over the centre of Hudson Bay, which agrees with the estimate for residual rebound (see also I. Fischer, 1959). Rates of recovery are now much slower than in earlier post-glacial time. For example, B. Matthews (1966) in northern Ungava has deduced mean rates of uplift amounting to 60 mm a year for the period 6700–7700 BP, compared with only 3 mm a year for the period since 4000 BP. Andrews (1970) shows that, for Arctic Canada, the amount of residual rebound in metres (U_r) is given by the equation

$$U_r = 7 \cdot 16 + 125 \cdot 38 \; x$$

where x is the present rate of uplift in metres per year.

In the Great Lakes area, the history of crustal deformation and rebound has been most thoroughly investigated, beginning with the classic studies of F. Leverett and F. B. Taylor (1915) and Taylor (1927), followed by others whose work is summarized and evaluated by J. L. Hough (1958). North of the southernmost hinge-line, the former shorelines rise northward (for instance, the Algonquin shoreline from a zero isobase of 180 m in central Michigan to 335 m on the north shore of Lake Huron, rising to greater heights farther north); the oldest shorelines possess the greatest tilts. The zero isobases of successive lake stages are also displaced north through as much as 250 km from the earliest to the latest, showing that the area experiencing isostatic recovery shrank slightly over the 8000-year period since deglaciation. Many of the shorelines are remarkably clear-cut and well developed, suggesting that they were formed in pauses during intermittent isostatic recovery.

4.3   *British Isles*

Recent investigations are demonstrating that the story of glacio-isostatic rebound is here much more complex than earlier workers supposed. Indeed, the older concepts of horizontal raised beaches in Scotland, postulated by the Geological Survey and W. B. Wright (1914), for instance, at 15, 25, 50 and 100 feet (5, 8, 15 and 30 m), are plainly at variance with any idea of isostatic recovery at all, for this ought to lead to deformed rather than horizontal shorelines. Re-mapping of the raised beaches (e.g. J. B. Sissons 1962, 1967; Sissons *et al.*, 1966) has clearly shown the tilted nature of many, and the fact that many more sets of beaches exist than had been formerly thought. The Main Post-glacial Shoreline, for instance, formed about 5500 BP, stands at 14·9 m at the western end of the Carse of Stirling, reaches only 12·8 m at Stirling itself and declines to 9·2 m at Leith. In Lancashire it stands at a little over 5 m while farther south it falls below present sea level. Its appearance in Scotland and northern England demonstrates that glacio-isostatic recovery has here outpaced the eustatic rise of sea level; farther south, the latter has dominated. The highest marine limit yet recorded in Britain is about 38 m near Stirling; this is the highest point attained by the Main Perth Raised Shoreline where it merges with outwash of the Perth Readvance (*c*. 13,000 BP) (D. E. Smith *et al.*, 1969). If Fairbridge's estimate of minus 48–50 m for sea level at this time is accepted, total isostatic uplift of 86–88 m has taken place since that date. Other data in the Sterling area give amounts of uplift totalling 52 m since 10,300 BP and 12 m since 5500 BP. From this it may be inferred that rates of isostatic recovery have slowed from about 13 mm a year (13,000–10,300 BP) to 8 mm a year (10,300–5500 BP), and finally to 2 mm a year (5500 BP to the present). Tide-gauge data for Britain have been studied by such workers as H. Valentin (1953) and D. L. Gordon and C. T. Suthons (1963). Although the records of many British tide-gauges are too short for reliable trends to be established, and the number of records in Scotland, the most interesting area, are few and restricted to the east coast, the evidence suggests that Aberdeen is currently rising at 3–4 mm/year and Dunbar at 5 mm/year. Using data for the Stirling area and Gutenberg's (1941) formula for estimating residual uplift, about 50 m of iso-static recovery still remains to be achieved in this area. The probable limit of glacio-isostatic deformation lies across northern Britain in about the latitude of Lancashire, but in Britain, as elsewhere, glacio-isostatic movements are very difficult to disentangle from tectonic movements resulting from other causes in such marginal areas. An inter-esting finding from central Scotland is that the raised shorelines themselves do not always have uniform or gradually changing gradients; Sissons (1972) has shown that the Main Buried Shoreline is dislocated in two places by 1 m and 1·5 m respectively. It is possible that the stresses of glacio-isostatic deformation are locally being taken up by small movements along pre-existing faults.

5 Conclusion

Pleistocene glaciation was not only the cause of great morphological changes in the glacial and periglacial regions, but affected the whole of the earth in terms of eustatic changes in sea level, deformation of the earth's crust by the loading and unloading of ice, and variations of temperature, precipitation and evaporation. The glacial per-iods were characterized by low sea levels (probably -100 to -140 m in the last glacial period), by increased precipitation and reduced evaporation. The range of oscillation of sea level between one glacial and the adjacent interglacial was probably about 110–120 m. Pleistocene glacio-eustatic sea-level changes may have been superimposed on a continuous secular lowering of sea level throughout the Quaternary.

Pluvial lakes in many areas now arid or semi-arid testify to the wetter climatic condi-tions of the glacial periods in the extra-glacial regions, and the sequences of deposits in these lake basins are most valuable in reconstructing the chronology of the Pleisto-cene. The changing climatic conditions also had profound effects on stream activity, one of the most obvious morphological effects being the series of terraces and valley fills created in response to changes in river volume and load.

Isostatic deformation of the earth's crust by the weight of thick ice sheets has been postulated since 1863 and is confirmed by modern studies of rates of land uplift in response to glacial unloading, and by the present depression of bedrock in central Greenland and Antarctica below sea level by the weight of their ice masses. Total subsi-dence beneath large ice sheets may amount to 25–30 per cent of maximum ice thickness. The morphological effects of recovery following unloading are most clearly seen in the tilting and warping of former marine or lacustrine shorelines.

6 References

AHLMANN, H. W. (1953), 'Glacier variations and climatic fluctuations', *Am. geogr. Soc., Bowman Memorial Lectures*, Ser. III, 1–15

ANDREWS, J. T. (1970), 'A geomorphological study of post-glacial uplift with particular reference to Arctic Canada', *Inst. Br. Geogr. Spec. Publ.* **2**, 156 pp.

BARNETT, D. M. (1966), 'A re-examination and re-interpretation of tide gauge data for Churchill, Manitoba', *Can. J. Earth Sci.* **3**, 77–88

BIRD, J. B. (1954), 'Post-glacial marine submergence in Central Arctic Canada', *Bull. geol. Soc. Am.* **65**, 457–64

BLOOM, A. L. (1967), 'Pleistocene shorelines: a new test of isostasy', *Bull. geol. Soc. Am.* **78**, 1477–94

BOTT, M. H. P. (1971), *The interior of the Earth*

BUTZER, K. W. (1958), 'Quaternary stratigraphy and climate in the Near East', *Bonn. geogr. Abh.* **24**, 157 pp.

CRITTENDEN, M. D. (1963), 'New data on the isostatic deformation of Lake Bonneville', *U.S. geol. Surv. Prof. Pap.* **454** E, 1–31

CURRAY, J. R. (1965), 'Late Quaternary history, continental shelves of the United States' in *The Quaternary of the United States* (ed. WRIGHT, H. E. and FREY, D. G.), 723–35

DALY, R. A. (1910), 'Pleistocene glaciation and the coral reef problem', *Am. J. Sci.* **30**, 297–308

—— (1934), *The Changing World of the Ice Age* (Yale Univ. Press)

DONN, W. L., FARRAND, W. R. and EWING, M. (1962), 'Pleistocene ice volumes and sea-level lowering', *J. Geol.* **70**, 206–14

DURY, G. H. (1960), 'Misfit streams: problems in interpretation, discharge and distribution', *Geogrl Rev.* **50**, 219–42

EARDLEY, A. J. (1968), 'Bonneville chronology: correlation between the exposed stratigraphic record and the subsurface sedimentary succession', *Bull. geol. Soc. Am.* **79**, 907–9

EARDLEY, A. J. and GVOSDETSKY, V. (1960), 'Analysis of Pleistocene core from Great Salt Lake, Utah', *Bull. geol. Soc. Am.* **71**, 1323–44

EARDLEY, A. J. *et al.* (1973), 'Lake cycles in the Bonneville Basin, Utah', *Bull. geol. Soc. Am.* **84**, 211–16

FAHNESTOCK, R. K. (1963), 'Morphology and hydrology of a glacial stream—White River, Mount Rainier, Washington', *U.S. geol. Surv. Prof. Pap.* **442** A, 1–70

FAIRBRIDGE, R. W. (1961), 'Eustatic changes in sea level' in *Physics and Chemistry of the Earth* (ed. AHRENS, L. H. *et al.*) **4**, 99–185

—— (1970), 'World palaeoclimatology of the Quaternary', *Rev. Géogr. phys. Géol. dyn.* **12**, 97–104

FARRAND, W. R. (1962), 'Postglacial uplift in North America', *Am. J. Sci.* **260**, 181–199

FISCHER, I. (1959), 'The impact of the Ice Age on the present form of the geoid', *J. geophys. Res.* **64**, 85–7

FISK, H. N. and MCFARLAN, R. (1955), 'Late Quaternary deltaic deposits of the Mississippi River', *Geol. Soc. Am. Spec. Pap.* **62**, 279–302

FLINT, R. F. (1971), *Glacial and Quaternary geology*

FLINT, R. F. and GALE, W. A. (1958), 'Stratigraphy and radiocarbon dates at Searles Lake, California', *Am. J. Sci.* **256**, 689–714

GILBERT, G. K. (1890), 'Lake Bonneville', *U.S. geol. Surv. Monogr.* **1**, 1–438

GORDON, D. L. and SUTHONS, C. T. (1963), 'Mean sea level in the British Isles', *Admiralty mar. Sci. Publ.* **7**

GROVE, A. T. and PULLAN, R. A. (1963), 'Some aspects of the Pleistocene paleogeography of the Chad basin' in HOWELL, F. C. and BOURLIÈRE, F. (eds), *African ecology and human evolution* (Viking Fund Publs in Anthropology, Chicago), 230–45

GROVE, A. T. and WARREN, A. (1968), 'Quaternary landforms and climate on the south side of the Sahara', *Geogrl J.* **134**, 194–208

GUTENBERG, B. (1941), 'Changes in sea-level, postglacial uplift, and mobility of the earth's interior', *Bull. geol. Soc. Am.* **52**, 721–72

HOUGH, J. L. (1958), *Geology of the Great Lakes* (Univ. of Illinois Press)

JAMIESON, T. F. (1863), 'On the parallel roads of Glen Roy, and their place in the history of the glacial period', *Q.J. geol. Soc. Lond.* **19**, 235–59

(1865), 'On the history of the last geological changes in Scotland', *Q.J. geol. Soc.* **21**, 161–203

JELGERSMA, S. (1961), 'Holocene sea-level changes in the Netherlands, *Meded. geol. Sticht.*, **C-IV**, 101 pp.

KAUFMAN, A. and BROECKER, W. S. (1965), 'Radiocarbon chronology of Lake Lahontan and Lake Bonneville II, Great Basin', *Bull. geol. Soc. Am.* **76**, 537–66

KING, P. B. (1965), 'Tectonics of Quaternary time in middle North America' in *The Quaternary of the United States* (ed. WRIGHT, H. E. and FREY, D. G.), 831–70

LARTET, L. (1865), 'Sur la formation du bassin de la mer morte ou lac asphaltite, et sur les changements survenus dans le niveau de ce lac', *C.r.hebd. Séanc. Acad. Sci., Paris* **60**, 796–800

LEVERETT, F. and TAYLOR, F. B. (1915), 'The Pleistocene of Indiana and Michigan, and the history of the Great Lakes', *U.S. geol. Surv. Monogr.* **53**, 1–529

MACLAREN, C. (1842), 'The glacial theory of Professor Agassiz', *Am. J. Sci.* **42**, 346–365

MATTHEWS, B. (1966), 'Radiocarbon dated post-glacial land uplift in Northern Ungava, Canada', *Nature, Lond.* **211**, 1164–5

MEINZER, D. E. (1922), 'Map of Pleistocene lakes of the Basin-and-Range Province and its significance', *Bull. geol. Soc. Am.* **33**, 541–52

MÖRNER, N-A. (1971), 'Relations between ocean, glacial and crustal changes', *Bull. geol. Soc. Am.* **82**, 787–8

MORRISON, R. B. and FRYE, J. C. (1965), 'Correlation of the middle and late Quaternary successions of the Lake Lahontan, Lake Bonneville, Rocky Mountain (Wasatch Range), Southern Great Plains, and eastern Mid-West areas', *Nevada Bureau of Mines, Rep.* **9**, 1–45

NISKANEN, E. (1939), 'On the upheaval of land in Fennoscandia', *Ann. Acad. scient. fennicae*, A, **53**(10), 30 pp.

RUSSELL, I. C. (1885), 'Geological history of Lake Lahontan, a Quaternary lake of north-western Nevada', *U.S. geol. Surv. Monogr.* **11**, 1–288

(1889), 'Quaternary history of Mono Valley, California', *U.S. geol. Surv. 8th A. Rep.* (1886–7), 261–394

SAURAMO, M. (1955), 'Land uplift with hinge-lines in Fennoscandia', *Ann. Acad. scient. fennicae*, A-III, **44**, 25 pp.

SCHOFIELD, J. C. (1964), 'Post-glacial sea-levels and isostatic uplift', *N.Z.J. Geol. Geophys.* **7**, 359–70

SHEPARD, F. P. (1963), 'Thirty-five thousand years of sea-level' in *Essays in marine geology* (Univ. of S. California Press), 1–10

SISSONS, J. B. (1962), 'A re-interpretation of the literature on late-glacial shorelines in Scotland with particular reference to the Forth area', *Trans. Edinb. geol. Soc.* **19**, 83–99

(1967), *The evolution of Scotland's scenery.*

(1972), 'Dislocation and non-uniform uplift of raised shorelines in the western part of the Forth valley', *Trans. Inst. Br. Geogr.* **55**, 145–59

SISSONS, J. B., SMITH, D. E. and CULLINGFORD, R. A. (1966), 'Late-glacial and post-glacial shorelines in south-east Scotland', *Trans. Inst. Br. Geogr.* **39**, 9–18

SMITH, D. E., SISSONS, J. B. and CULLINGFORD, R. A. (1969), 'Isobases for the Main Perth Raised Shoreline in south-east Scotland as determined by trend-surface analysis', *Trans. Inst. Br. Geogr.* **46**, 45–52

SNYDER, C. T. and LANGBEIN, W. B. (1962), 'The Pleistocene lake in Spring Valley, Nevada, and its climatic implications', *J. geophys. Res.* **67**, 2385–94

SPARKS, B. W. and WEST, R. G. (1972), *The ice age in Britain*

TAYLOR, F. B. (1927), 'The present and recent rate of land tilting in the region of the Great Lakes', *Pap. Mich. Acad. Sci.* **7**, 145–57

TOOLEY, M. J. (1974), 'Sea-level changes during the last 9000 years in north-west England', *Geogrl J.* **140**, 18–42

TYLOR, A. (1842), 'On the formation of deltas: and on the evidence and cause of great changes in sea-level during the Glacial Period', *Geol. Mag.* **9**, 393–9 and 485–501

(1868), 'On the Amiens gravel', *Q.J. geol. Soc. Lond.* **24**, 103–25

VALENTIN, H. (1952), 'Die Küsten der Erde', *Petermanns Mitt., Ergänz.* **246**, 118 pp.

(1953), 'Present vertical movements of the British Isles', *Geogrl J.* **119**, 299–305

WALTON, K. *et al.* (1966), 'The vertical displacement of shorelines in Highland Britain', *Trans. Inst. Br. Geogr.* **39**, 1–145

WRIGHT, W. B. (1914), *The Quaternary Ice Age* (Lond., 1st ed.)

ZEUNER, F. E. (1959), *The Pleistocene period* (Lond., 2nd ed.)

Part II

Glacial and
fluvioglacial erosion

We know that the glaciers rub, wear, and polish the rocks with which they are in contact. Struggling to dilate, they follow all the sinuosities, and press and mould themselves into all the hollows and excavations they can reach, polishing even overhanging surfaces. (J. DE CHARPENTIER, 1835)

6

Small-scale features of glacial and fluvioglacial erosion

Few, few believe what I have told,
Men say that I am over bold,
What then? they sneered that Welshmen's tails
Had polished Buckland's rocks in Wales.
(A. C. RAMSAY, 1856, at the Anniversary
Dinner of the Geological Survey)

Rock surfaces in glaciated regions often display clear signs of the former passage of ice over them in the form of markings. These include the familiar striations, channels and grooves of varying dimensions and shapes, and gouges suggesting the extraction of small chips of rock by some agency. Some rock surfaces have been abraded and scoured to a highly polished form, while others, sometimes adjacent to smoothed areas, show signs of plucking. These and other features are now collectively taken as clear evidence of former glaciation, better evidence indeed than the landforms of larger scale such as hanging valleys or apparently U-shaped troughs which in some cases at least admit of other non-glacial explanations. But relatively little is known about the origins of small-scale glacial rock markings: although there is a considerable body of literature describing them, there is considerable disagreement over the exact mechanism of their formation. It is becoming increasingly apparent that moving ice alone may not be the sole agent of formation involved, but that the scouring effects of subglacial water moving under high pressure must be taken into account, and also the abrasive effects of water-saturated ground moraine being squeezed into motion by differential ice pressure. It is clear that in this context it is no longer possible to make a rigid distinction between glacial and fluvioglacial erosion.

On the question of what is to be considered small-scale, there can be no rigid answer. It is arguable that there is a gradation from the smallest to the largest scale of glacially-eroded relief. S. Rudberg (1973) has attempted to differentiate between small-scale, medium-scale and 'alpine' relief forms. His middle category includes forms with dimensions up to a few hundred metres—steep rocky faces, giant roches moutonnées and rock basins—but it is difficult to see how the upper and lower size limits of these features are to be defined. This chapter will consider forms ranging in size from a few centimetres (the length, for example, of rather short striations) to tens of metres (in the case of most roches moutonnées). Remaining chapters in Part II will consider larger-scale forms with dimensions normally exceeding hundreds of metres.

1 Striations, grooves and polished surfaces

L. Agassiz in 1838 was one of the first to ascribe polished and scratched rock surfaces in certain Alpine valleys to abrasion by moving ice whose under-surface was armed with rock fragments of various sizes. The process was vividly described by J. D. Forbes at the side of the Brenva glacier in 1843. At a point where the ice was driven hard against the rock face, he extracted a specimen of basal ice whose face was seen to be 'set all over with sharp angular fragments . . . which were so firmly fixed in the ice as to demonstrate the impossibility of such a surface being forcibly urged forward without sawing and tearing any comparatively soft body which might be below it' (p. 204). In the rock exposed by removal of the ice specimen, grooves, scratches and a general polishing were visible. Descriptions of striated and smoothed surfaces in various formerly glaciated regions culminated in T. C. Chamberlin's great monograph of 1888, dealing with *The rock scorings of the great ice invasions* in North America, a work which

Plate VII
Glacially-abraded volcanic bedrock in front of Breiðamerkurjökull, Iceland. Striations run parallel to the shaft of the hammer. (*C.E.*)

has never been surpassed in the accuracy and detail of its observation over such an extensive area.

Striations consist of lines, usually short but very occasionally a metre or more in length, engraved on suitable rock surfaces. They have been recorded in a great variety of situations—on level, sloping, vertical or curved rock surfaces, for instance—but are most prolific on gently-inclined smooth surfaces which the ice was forced to ascend. They have been recorded not only on open surfaces, but also on the risers of steps and in trenches (M. Demorest, 1938), and in recesses in vertical walls (P. Sheldon, 1926). The clarity of their development depends largely on the character of the rock, for soft, coarse-grained or brittle rocks do not take the markings well, and on exposed or easily weathered rocks, the markings are rapidly erased. Some striae begin very gradually and slowly deepen, while others begin quite abruptly; again, some fade out slowly and others disappear in a blunt or fractured end. Any individual striation can result from two diametrically opposed directions of ice motion. The larger striations are known as grooves, while at the other end of the scale, the scratches become of minute size. Rock surfaces which appear smoothed and polished to the naked eye are seen to possess innumerable small scratches under the microscope. The degree of polishing depends on the fineness of the abrading material, namely, the rock flour (N. Edelman, 1949), and in some cases a glazed appearance results. Polished rock surfaces do not usually survive long exposure to weathering. F. E. Matthes (1930) attributed the unusual abundance of glacier polish in the region above Yosemite valley to the prevalence of highly siliceous granite which weathers only slowly. Even so, the polish is largely restricted to areas affected by the last glaciation.

On the formation of striations, Chamberlin stated that 'the ice of a glacier has of itself but little abrading and practically no striating power. It can neither groove nor scratch.... The little effect that it may produce is of the nature of polishing...' (p. 208). The finer scratches are the work of grains of silt or sand, not just embedded in the ice but caught between boulders in the ice and the bedrock, for the physical properties of ice prevent it applying sufficient pressure on the very small surface area of an individual sand grain. Larger scratches and grooves are clearly etched by larger fragments: 'A block of rock that presents a large surface to the embracing ice and but a small area in contact with the rock ... floor may be held relatively firm and act as a steady graver' (ibid.). Such a situation was actually observed by H. Carol (1947) at 50 m below the Oder Grindewald glacier, where striae 4 mm deep were being carved in the bedrock. If the block is too large and has too great an area of contact with bedrock, the ice will flow over the block, and it will be the underside of the ice which is grooved. Sometimes the cutting tool appears to have rotated, thus presenting a new cutting edge, and giving rise to striations which step sideways or lie *en échelon* (Edelman, 1951).

It is relatively easy to comprehend how rock surfaces are striated and polished, but the larger grooves present quite a different problem and their origin is very obscure. Some of the largest grooves known were described by H. T. U. Smith (1948) in the Mackenzie valley of north-west Canada, where they attain lengths of as much as 12 km, depths of up to 30 m, and widths approaching 100 m (see also Chapter 9). There is no doubt that they are of glacial origin, since they are totally indifferent to structure, they accord in direction with what else is known of ice movement in this region, and they are of similar form to smaller grooves in bedrock and till elsewhere which grade downward into striations. Yet obviously they have not been gouged out by single boulders; possibly groups of boulders packed and frozen together may explain the smaller ones. Another problem is what localized the grooves, for neither structure nor relief appears to have played a part. Although the glacial origin of these particular grooves is undeniable, care must be taken to distinguish grooves of other kinds; thus Chamberlin (1888) thought that some might be merely ice-modified pre-glacial grooves, and others, especially curving and twisting grooves, might be the work of subglacial streams.

Striations and grooves have often been used as a means of determining ice motion. As will become apparent, this is by no means a simple matter, nor are the conclusions based solely on them particularly reliable. It is first necessary to appreciate that scratches on rocks or rock surfaces are not always the result of glaciation. Surface markings may result from differential chemical weathering, and a number of other nonglacial processes may produce striations. Agassiz in 1838 considered running water bearing stones along with it, and also avalanches, as possibilities. Chamberlin, at a time when the hypothesis of glaciation by land ice was still in its infancy, thought that some markings might have been made by icebergs or shore-ice during partial submergence. This possibility is in fact a very real one in Arctic regions, where it is still the reason for much uncertainty regarding directions of former ice motion. J-C. Dionne (1973) has suggested how glacial and drift-ice striations may be distinguished, in that the latter are unaccompanied by chattermarks or polished surfaces, for instance, but there is also the possibility that drift-ice striations may have beeen added later to a surface initially abraded by glacial ice.

Many observers have confirmed that mass movements can scratch rock surfaces. C. F. S. Sharpe (1938) gives an example of an alpine mudflow in the San Juan Mountains of Colorado, where blocks exceeding one cubic metre in volume had been moved, and where ledges in the channel containing the mudflow showed excellent striae orientated roughly parallel to the direction of movement. J. L. Dyson (1937) also showed conclusively that snow-slides in Glacier National Park were capable of producing fine striations. Here, fresh scratches up to 10 m long and 6 mm deep were superimposed across older (and probably glacial) striations. Their age, judged by comparing weathering of the scratches and other surface areas, was not more than a few years, and this was confirmed by the fact that in a few cases the stones responsible for the scratching still remained poised at the lower ends of the grooves. A rapidly moving recent snow-slide was suggested as the most probable mechanism. Other examples of the erosive effects of moving snow are given by A. B. Costin et al. (1964, 1973). In general, any heavy moving mass may cause scratching of the rock surface beneath, and it is important to note that the cutting tools need not be held rigidly in the moving medium (Demorest, 1938, p. 721).

The great bulk of rock scorings in regions known from other evidence to have been ice covered at some former time are, however, undoubtedly caused by glacial action. Scratched surfaces in formerly glaciated regions are often best displayed when the rock can be newly stripped of a till cover or turf and cleaned up with brush and water. Studies of such glacial striations can throw a certain amount of light on directions of

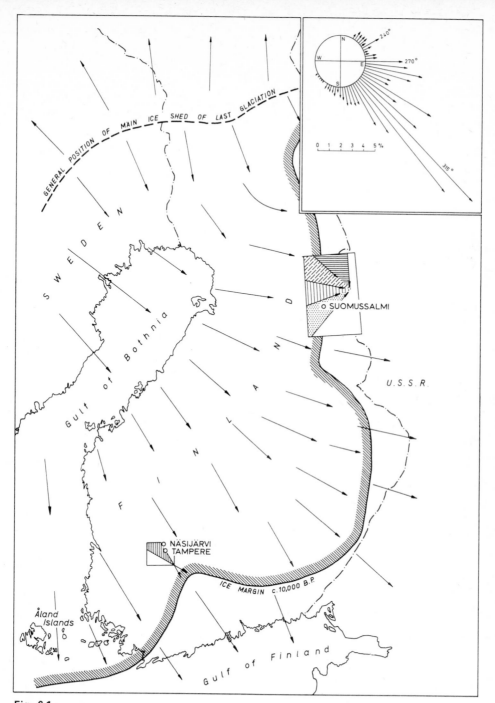

Fig. 6.1
Directions of ice motion over Finland and adjacent areas. The approximate position of the ice margin
about 10,000 years ago is indicated (the time of the Baltic Ice Lake and middle Swedish moraines).
Glacial striations in the rectangular areas around Tampere and Suomussalmi have been mapped and
analysed by K. Virkkala (1951, 1960); in these areas, the vertical line-shading represents the sector
in which most striations are orientated; while, in the Suomussalmi area, the horizontal line-shading
indicates the sector possibly representing the youngest ice movement, and the stipple indicates a
sector possibly representing ice movement locally deflected. Further detail of striae direction in the
Tampere region is given in the inset.

ice movement, and they can also provide useful information on the nature and behaviour of the basal ice.

Much detailed work on glacial abrasion marks has been carried out in Finland. As has long been known, a large number of striations must be recorded if there is to be any reliable idea gained of former directions of ice motion. An individual striation will only express ice movement at that point, and local movement will be strongly affected by topographical irregularities. K. Virkkala (1960) quotes striations on a shore cliff at Näsijärvi in southern Finland, where the orientation swings gradually from 315° on the stoss side to 345° on the lee side. Out of 1300 striae measurements around Tampere, he found that 55 per cent bore an azimuth of 300° to 330° (Fig. 6.1), the variation between these values probably reflecting purely local influences on basal ice movement. Bedrock irregularities probably exerted progressively stronger influence on ice movement as the ice thinned during deglaciation. Virkkala (1951) recorded similar local variations in striae trend in another intensive study around Suomussalmi, east Finland (Fig. 6.1), but here there was a much higher concentration of striae directions, 83 per cent trending within 15° of an east–west azimuth. Very noticeable was the parallelism between the striae and the chains of lake basins in the area. However, in both studies Virkkala found evidence of distinct sets of striations at variance with the principal sets. In the Suomussalmi area, in addition to the common east–west striae, three other trends appeared directed north, north-west and south-west. It is clear from other evidence that the main Scandinavian ice-divide lay to the north, north-west and west of Finland, and that therefore ice movement across the country was generally from these directions (Fig. 6.1). The south-westerly striae may be wholly the result of local deflections. As the north and north-westerly striae are more clearly marked than the common westerly sets, Virkkala argues that they are somewhat younger, and concludes that in the last glaciation, ice moved chiefly from the west; there followed a short phase of stagnation succeeded by renewed ice activity from different ice centres and movement from the north-west and north.

As Virkkala emphasizes, great caution must be used in interpreting striations in this way. To be of any validity, studies must be based on a very large number of striae observations over a reasonably large area, and other evidence on ice movement must be taken into consideration, particularly the dispersion of erratics. It is clear that bedrock irregularities can produce considerable local variations of basal ice movement—which may of course differ from the movement of the main bulk of the ice above—and during changes in ice thickness, the directions of ice movement as a whole may alter. It was also pointed out in Chapter 4 that, even in a valley glacier, the lines of ice flow are not solely directed parallel with the axis of the glacier. Variations in ice supply and wastage will also be accompanied by shifts in the direction of ice motion and therefore striation.

There remains the possibility that crossing sets of striations may be the work of different glacial phases with different ice centres. Sometimes one set of striae appears distinctly fresher than another. However, the question arises as to why the older striae were not erased by the later movement. Such a situation suggests that the later movement was but short-lived and that it could have caused only a very minor amount of bedrock erosion. In other words, crossed striae are unlikely to represent successive glaciations but merely shifts in ice motion during a phase of general deglaciation, as Virkkala concludes.

In the examples quoted from Finland, the question did not arise of two possible directions of ice motion differing by 180°. Even for the south-westerly set, a north-easterly source of the ice appears improbable since other evidence adequately locates the Scandinavian ice centres elsewhere as shown on Figure 6.1. But the problem has been encountered, and the form of the striae alone provides no answer to the problem (see p. 183). A particularly doubtful case was described by T. H. Clark (1937) in southern Quebec where striations run generally north–south. A northern source of ice would be expected in this area, yet there is local evidence which suggests a movement of ice from the south. This is in the form of small-scale crag-and-tail features on limestone surfaces; the 'crag' is provided by hard sand-grains (a few millimetres in size) in the limestone which possess 'tails' of limestone a few centimetres in length fading out towards the north. More recently, R. Y. Lamarche (1971) has looked again at this area, and his detailed survey of 1300 km² supports Clark's view that ice moved northward. Seventy-three measurements of 'tails' showed a strong concentration in the sector bearing between 0 and 10°. Lamarche suggests that, since the tail-like forms are dominant and cut all other striations, they must refer to the last stage of glaciation, possibly when the ice sheet had thinned and separated to such an extent that ice remaining in this area began to move northwards down the regional slope to the St Lawrence valley.

Similar tail-like forms thought to result from glacial modification of a limestone surface were noted by Edelman (1949) in south-west Finland. On the Kakskielaklobb peninsula, hard silicate knobs possess tails up to 2 m long, dying away gradually and running parallel with striae. Such tails, although relatively rarely found, were also noted by Chamberlin (1888, p. 193) and serve to eliminate one of the two possible directions of ice motion indicated by striae.

Deflections of striae caused by bedrock irregularities throw useful light on the behaviour of the basal ice. Demorest (1938) observing the vacated parts of the bed of the Clements Glacier, Glacier National Park, showed how readily ice can be deflected from a straight line of movement. Steps and trenches in thin-bedded argillite provided the irregularities, while striations on the walls and floors of these features demonstrated complex ice motion. Thus, a proportion of the ice crossing transverse trenches was deflected into and along the trenches, moving at 90° to the main ice flow (Fig. 6.2A), and in one or two cases, eddy-like currents had been set up in the ice (Fig. 6.2B), though the term 'eddy' is not suited to moving ice whose viscosity is many orders of magnitude greater than that of water. Such behaviour of the basal ice will be further considered in connection with plastically-moulded rock surfaces.

2 Friction cracks

Whereas glaciated fine-grained rocks often exhibit good striations and polished surfaces, rocks of medium-size grain are usually less clearly striated; on the other hand, the latter type of rock may have a much clearer development of friction cracks (N. Edelman, 1951). The term 'friction crack' was proposed by S. E. Harris (1943) to cover a variety of rock fractures induced by the passage of ice over bedrock. Some types

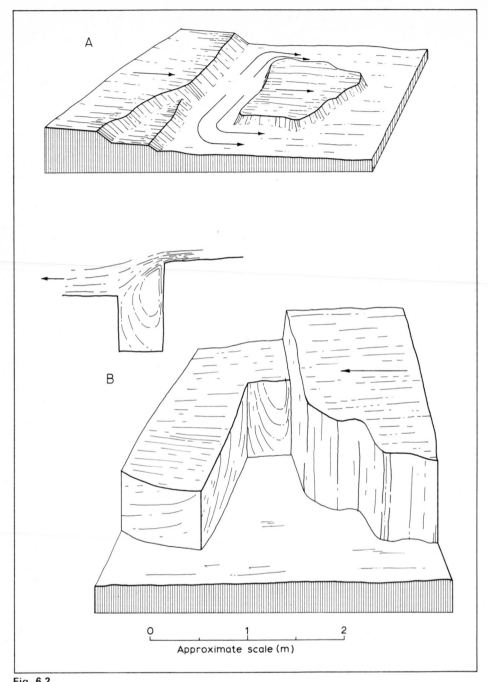

Fig. 6.2
A Striations recording deflected movement of basal ice crossing an irregular bedrock surface;
B Eddy-like currents in basal ice as recorded by striations (M. Demorest, *J. Geol.*, 1938, University of Chicago Press)

of fracture were noted by Chamberlin (1888, pp. 218–21)—gouges of crescentic shape, semi-vertical fractures with crescentic outcrops whose convexity opposed the ice motion, and chatter-marks, which Chamberlin considered to result from the sort of vibratory motion, well known to machinists, of a cutting tool being forced across a hard surface. The term 'chattermark' has also been used in a more general sense, synonymous with friction cracks (C. Laverdière *et al.*, 1968; Laverdière and C. Bernard, 1970) but this usage is not followed here. In the more restricted sense of repeated gouges or hollows along a groove, chattermarks were described by Agassiz in 1876. Crescentic gouges received detailed analysis from G. Ḱ. Gilbert in 1906, who studied numerous examples cut in granite in the Sierra Nevada. Varying in size from a few centimetres to as much as 2 m across, and occurring usually in sets of two to six, one behind the other, their form is shown in Fig. 6.3B. As the section shows, they are composed of two fractures, the gently dipping one clearly having been formed before the vertical one, and a chip of rock has been torn out between them. They are most commonly found on the upstream sides of projecting rock bosses.

Gilbert (1906a) argued that they were formed by a sizeable boulder in the basal ice transmitting pressure on to the bedrock surface, which at first deformed elastically and then ruptured along a conoid fracture (Fig. 6.3F). After fracture, the wedge of rock above the fracture tended to straighten and therefore lift upward slightly, when it was cracked off by the moving ice above. A puzzling feature is that the boulder responsible did not usually make direct contact with the bedrock, since the latter shows no sign of any major striation or grooving at the site where the crescentic gouge occurred. Gilbert suggested that perhaps a cushion of sand between the boulder and bedrock was able to transmit the pressure. The localization of gouges on the upstream faces of bedrock projections he attributed to increasing vertical compression of the ice as it passed over the projection, which in turn would increase the pressure on any boulder held in the lower layers of the ice.

F. H. Lahee (1912) was concerned with the second type of fracture noted by Chamberlin, which he termed 'crescentic fractures'. Fig. 6.3C shows the characteristic form of these, which he studied on quartzite roches moutonnées in New Hampshire. The fracture outcrops are concave forward and of hyperbolic form, and the axis bisecting the concavity is parallel to the striae. The near-vertical fracture is clean cut near the rock surface, often cutting the actual rock grains, but at depths below a few millimetres, the fractures tend to wander between the grains. Unlike the crescentic gouge, no rock has been removed.

Both crescentic gouges (Sichelbrüche) and crescentic fractures (Parabelrisse) were regarded by E. Ljungner (1930) as the products of conoids of fracture induced by oblique pressure on the bedrock, an adaptation of Gilbert's hypothesis. Ljungner added another form of rock fracture induced indirectly by ice pressure, that of the conchoidal fracture (Muschelbruch) where the fracture plane is concave upward (Fig. 6.3D).

Harris's review of friction cracks added yet another type, the lunate fracture (Fig. 6.3A), very similar to the crescentic gouge except for the fact that the horns of the crescent point forward with the ice motion. It thus became clear that crescent-shaped friction cracks could not be used for distinguishing between two possible directions of ice motion along the axis simply by recording the direction in which the horns

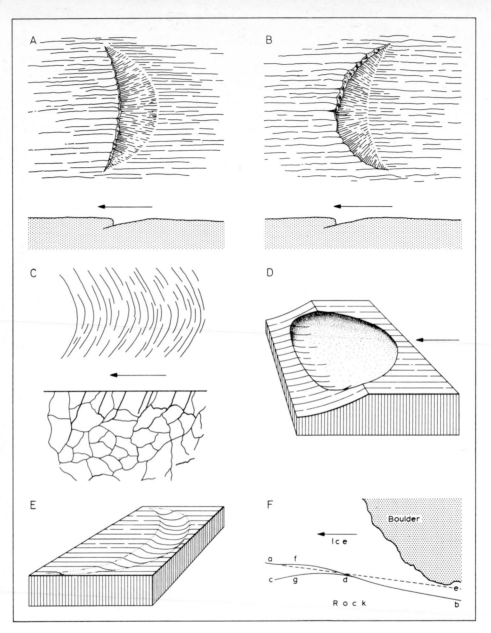

Fig. 6.3

Some small-scale features of glacial and fluvioglacial erosion. The horizontal arrow indicates the direction of ice motion:

A Lunate fracture (plan and section) (S. E. Harris, 1943); **B** Crescentic gouge (plan and section) (*ibid.*); **C** Crescentic fractures (plan and section) (*ibid.*); **D** Conchoidal fracture (E. Ljungner, 1930); **E** Sichelwannen (F. Hjulström, 1935); **F** Formation of a crescentic gouge (G. K. Gilbert, 1906)

a d e original rock surface
a f d b rock surface deformed by pressure transmitted from
        boulder in the ice
c g d conoid fracture
    f g secondary fracture formed when the wedge of rock
        f g d is broken off.

pointed, a fact amply confirmed by V. Okko's work in Finland (1950). However, Harris claimed that in all cases of friction cracks, direction of ice motion could be accurately determined by measuring the direction of dip of the 'principal fracture', which is always forward in the direction of ice motion. Not only may friction cracks have this advantage over striations for determining ice motion but, being larger, they are less prone to erasure by weathering, and they are also well developed on many rock types which are not suited to striation. It should be noted that Harris does not include chatter-marks, which have no principal fracture, under the general heading of friction cracks.

In experiments using a file or knife on glass, Harris succeeded in reproducing both crescentic gouges and lunate fractures. Further relevant experiments were described by P. MacClintock (1953): in scoring optical glass with a steel ball bearing, crescentic fractures formed with the horns pointing forward if the ball was forced forward without rolling, but if it was allowed to roll, the horns developed pointed backward. However, the form of the fracture in section did not support Ljungner's hypothesis that such fractures are conoids of percussion.

The idea that the dip of the principal fracture was a reliable indicator of ice move-ment was not supported by A. Dreimanis's work in Ontario (1953). In the case of crescentic fractures, which accounted for two-thirds of all the friction cracks examined, the majority of the fractures dipped not forward but backward against the ice flow. If Ljungner's theory of their association with conoids of percussion accepted, the back-ward dip could be produced by a high angle of impact. This would also be expected to produce crescentic gouges with steeply-dipping principal fractures, which in fact Dreimanis finds. The usual angle of dip of the latter he records as 50° to 70°, in contrast to the gentle or low dips said to be typical by Harris and Gilbert. Dreimanis confirmed Harris's claim that the principal fracture of crescentic gouges, regardless of how steeply it dipped, always dipped forward with the ice flow, but this contention has not sub-sequently been upheld. J. L. Andersen and J. L. Sollid (1971) have also noted reversed crescentic gouges in southern Norway in which the principal fracture apparently dips against the direction of ice flow. Much further work is needed, not only to improve our understanding of the relationships of these micro-forms with ice movement, but also the mechanics of the processes producing them.

3 Plastically-moulded surfaces

Glaciated rock surfaces that have suffered smoothing and scouring, as distinct from plucking, often exhibit complexly moulded forms. A recent study by J. Gjessing (1965) of the Portør area of coastal southern Norway provides a good illustration. The area consists of granite intruded by amphibolite dykes; it has obviously suffered considerable abrasion but there are few signs of plucking. It was overridden by south-eastward mov-ing ice in the last (post-Allerød) glacial advance. Numerous roches moutonnées occur, heavily striated, and with the striations curving in round the stoss sides. Sinuous grooves furrow the rock, and these also are followed by striae. Striae show deflections around obstructions as Demorest (1938) and Edelman (1949) described elsewhere, and can even be traced into and out of potholes. The problem is to account for the moulding

of a hard rock surface into such complex forms; undoubtedly the processes were connected with the last glaciation.

Plastically-moulded surfaces ('p-forms') have also been examined in detail in northern Norway by R. Dahl (1965). He classifies them as follows:

1 *Cavetto forms*, consisting of channels, oriented with the ice flow, cut on steep rock
   faces (sometimes as horizontal grooves along vertical walls). They may be up to half
   a metre deep, have sharp edges, and the upper part may overhang. Striae and crescentic gouges are found within them.
2 *Grooves* occur on open flat surfaces, cut in the hardest of rocks, and have rounded
   edges to distinguish them from cavettos.
3 *Sichelwannen* are the commonest small trough form (Fig. 6.3E). They may or may
   not be symmetrical; their axis always coincides with the ice flow (and their horns
   point forward with the ice); and they are seen on both flat and sloping surfaces.
   Scratches conformable with the ice flow are found in them, but these need not necessarily be striae.
4 *Curved and winding channels* are often generally reminiscent of fluvial activity. They
   may arise through the merging of Sichelwannen. Most striking are the deep almost
   tubular variety whose genesis, according to Dahl, 'seems to require a plastic medium
   whose movements gave rise to a blasting effect. Ice can hardly be involved here'
   (p. 127). N. Edelman (1972) has described details of their forms and relationships
   with jointing in south-west Finland.
5 *Bowls and potholes*. The former are commonest on granite and, curiously, were rarely
   found on horizontal surfaces. Potholes include both normal fluvial forms found
   on valley floors and others in anomalous situations, sometimes in rows down valley
   sides. Bowls and potholes will receive special consideration in a later section (p. 194).

Occasionally, p-forms have been seen in subglacial locations, for instance in cavities below the Glacier d'Argentière where R. Vivian and G. Bocquet (1973) describe
'solution cups, scallops and potholes' associated with striations and grooves. A problem
is how to record adequately the detail and complexity of the micro-relief. Photography and very large-scale mapping have been used, while Andersen and Sollid (1971)
have effectively employed plaster casts.

Gjessing considers that four media may be considered as possible sculptors of the
p-forms: basal ice containing rock debris, water-soaked ground moraine squeezed
between bedrock and the overlying ice, subglacial meltwater under pressure, and ice-
water compounds. The first of these has in the past often been accepted as adequate
(for example, Demorest, 1938), but there are some p-forms which are hardly compatible with such an origin by ice itself, in the light of recent studies of ice behaviour
under pressure. Observations are also available to show that basal ice cannot accommodate itself to the more abrupt bedrock irregularities but bridges over them (for example,
B. Kamb and E. LaChapelle, 1964). Nevertheless, scouring by basal ice has been recently invoked (Edelman, 1972) to explain meandering bedrock channels. Gjessing
himself favours the second medium, water-soaked ground moraine containing ice particles and moving in laminar flow according to differences in ice pressure. The pressure
differences would be generated by variations in ice thickness and bedrock irregularities,

and the viscous liquid would tend to move in the direction of ice surface slope and ice-flow. Eddy-like motion is said to have scoured out potholes, and the close connection between the latter and other p-forms is emphasized (Gjessing, 1967). There are many observations on the existence and consistency of water-saturated ground moraine; thus Okko (1955) at the southern outlet glaciers of Vatnajökull observed how the basal debris occurred as a 'porridge-like till', and that farther up the glaciers it sometimes appeared to be squeezed upward out of crevasses (pp. 33–4). G. Hoppe (1952) contended that his Veikki-type till ridges represented water-soaked debris squeezed into crevasses. Gjessing argues that the high proportion of clay material in the Ra moraine of southern Norway originated in the subglacial liquid sludge which had scoured the p-forms inland. The questionable point is, however, not the existence of such a substance but its capacity to model hard rock surfaces. We do not know what velocities of movement of the substance might be involved.

There is much evidence to favour the third medium—subglacial meltwater—but before examining this, the fourth medium, that of an ice-water mixture may be briefly mentioned. G. Johnsson (1956) considered the p-forms of southern Sweden to be eroded by rock flour moved by an ice-water mass, and compared the forms produced by normal fluvial or marine water currents. The viscosity of the ice-water paste would vary with pressure and temperature.

Subglacial meltwater as the sculpturing agent of nearly all p-forms is proposed by Ljungner (1930), E. Ebers (1961) and Dahl (1965). Ljungner regarded Sichelwannen, for instance, as purely fluvial forms induced by turbulence (pp. 303–47); potholes and giant kettles (Riesentöpfe) are even more suggestive of fluvial action. A very attractive explanation of Sichelwannen as cavitation forms was put forward by F. Hjulström in 1935. Cavitation erosion has been recognized since 1894, but not until 1930 was it intensively studied, mainly in the field of engineering. It is associated with water moving at very high velocities: at ordinary temperatures it appears above a limiting velocity of 14·3 m/s at sea-level pressure, but if turbulence is introduced, the limiting velocity may be reduced by 10 per cent or more. Metal, glass, and even quartz surfaces are attacked, as hollows within the water mass collapse violently. If the water contains particles, as all subglacial meltwater streams will, these impact on the channel sides at very high speeds, scouring and scratching them to give a sand-blasted effect. Hjulström showed miniature Sichelwannen forms on a metal plate attacked by cavitation, where a protruding metal bolt had set up turbulence. In the case of subglacial streams, fast flowing because of hydrostatic pressure, he suggested that a moraine block might act as an obstacle behind which cavitation occurred, locating the Sichelwannen.

Dahl enlarges on Hjulström's suggestions, and considers that p-forms, with the possible exception of the large cavettos which may be ice-cut grooves, are modelled by meltwater moving very rapidly in subglacial tunnels or englacial tubes. The high velocities may be produced by the sudden draining of englacial water-bodies through restricted tunnels. Sichelwannen are the initial cavitation forms which may be gradually modified by corrasion into potholes if the water flow at that point were prolonged. The action of cavitation can explain the curious fact that hard quartz veins in the p-forms are usually worn down as much as the surrounding rock, whereas on other glacially-eroded surfaces they stand up. The patchy occurrence of p-forms Dahl explains by suggesting that only in these places did the subglacial or englacial water contact the bedrock, and the orientation of grooves and the axes of Sichelwannen are said to be controlled by the fact that the meltwater was usually forced in the direction of ice-flow (compare esker tunnels, Chapter 16). Bedrock structure plays no significant part in their orientation, nor in the frequency of their occurrence.

4 Potholes

The association of potholes with p-forms has already been noted. As is well known, potholes may form in normal stream beds and are not necessarily connected with glaciation at all. However, some potholes occur in such situations that a fluvioglacial origin must be invoked. The literature contains numerous examples. In 1874, W. C. Brögger and H. M. Reusch communicated the results of their survey of 'giants' kettles' in the Oslo region; after removing infill, the largest pothole at Bäkhelagel proved to be 2·4 m in diameter and 10·4 m deep. The walls were smooth, and in some cases were cut in spiral form like 'the impression of a gigantic snail' (p. 756). Smoothly rounded stones were found in the lower layers of infill, one weighing 150 kg. The potholes could not possibly have been related to subaerial drainage because of their positions. Another celebrated instance is the Kailpot beside Ullswater noted by J. E. Marr in 1926. This is nearly one metre deep and of about the same diameter; it lies below steep crags at a height of only 1 m above lake level. In North America, gigantic potholes occur

Plate VIII
Potholes drilled by former glacial meltwater stream high above Devil's Lake, Wisconsin. The potholes now lie on a ridge top, entirely isolated from any present-day drainage. *(C.E.)*

at Taylor's Falls, Minnesota (up to 18 m deep and 3·6 m diameter), while on the steep quartzite bluffs above Devil's Lake in Wisconsin, small potholes are perched in an extraordinary position 120 m above the lake level (Plate VIII). Other examples of pot-holes in anomalous positions are found in the Gletschergarten of Lucerne (one is perched on top of a roche moutonnée), and one is noted by K. Faegri (1952) on top of a small island west of Bergen. Many of the larger potholes possess spiral grooves, and contractions or widenings in cross-section. H. Holtedahl (1967) describes how giant potholes (up to 20 m deep and 16 m across) occur in linear series near Fossli and the Flåm area (Fig. 6.4; and see p. 301), one pothole breaking into another to form miniature canyons.

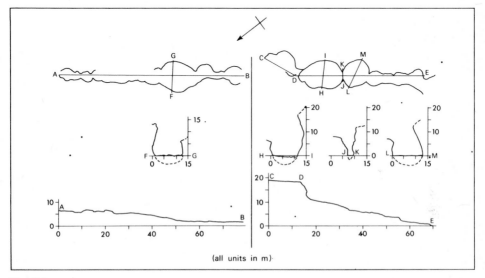

(all units in m)

Fig. 6.4
Sections through potholes and canyons in the Flåm valley, Norway (H. Holtedahl, *Geografiska Annlr*, *Stockholm*, 1967)

The hypothesis widely adopted until recently to account for such potholes was the glacier moulin hypothesis, set out by Brögger and Reusch (1874), and applied and amplified by later workers such as W. Upham (1900), G. K. Gilbert (1906b), M. B. Fuller (1925), J. E. Marr (1926), and K. Faegri (1952). In essence, it was held that streams of water pouring down crevasses or other openings in the ice (glacier moulins) would strike the bedrock and create a pothole at that point. If the jet of water struck the hole obliquely, it would spiral up out of it. Although some glacier meltwater streams do behave in this way, there are good reasons for thinking that this is not the correct explanation for most potholes. First, if the glacier was moving to provide the crevasses, how did the waterfall remain in one place long enough to cut a pothole? Secondly, there are clearly problems in explaining situations where potholes occur in great numbers. Thirdly, in all but very thin glaciers, it seems unlikely that the torrent could descend to bedrock in one unbroken fall. Fourthly, the potholes are quite different in detail from the plunge-pool holes normally found beneath waterfalls.

H. S. Alexander (1932) recognized three types of pothole. Plunge-pool holes, the first type, are broad, open, and have flaring walls; they are formed by waterfalls, are not usually narrow or deep, and lack spiral grooves. Secondly, gouge-holes are shallow depressions in stream beds caused by oblique impact of water currents maintaining a planar, not a spiral circulation. Thirdly, there are eddy-holes which may be very deep, possess sharp edges, and as confirmed by the spiral grooves on their walls, are produced by a vorticular circulation. They form on stream beds, and it is evident that nearly all potholes described in formerly glaciated regions are of this type—in other words they were the work not of moulin waterfalls but of fast-flowing subglacial or englacial streams.

In experiments with water jets entering a glass cylinder, Alexander showed by study of the movement of loose material that spiral motion does occur, that the energy available for bottom erosion falls off rapidly with increasing depth, and that the angle of the jet can be varied considerably (between 30° and 70°) without much change in the amount of vortex energy generated. However, the vortex energy falls off very rapidly to nil when the jet is vertical: thus spiral grooves in potholes are incompatible with moulin waterfalls. The infill of many potholes is probably of no great significance, since it could easily have accumulated after the hole was formed. The so-called grinding stones found by Brögger and Reusch and others may be merely water-worn pebbles subsequently washed into the hole, or even rounded morainic boulders. An important requirement for the formation of deep eddy-holes is that removal of sediment from the hole should at least equal the input, for otherwise the hole will be choked.

As noted in the previous section, potholes are included among p-forms by Gjessing and Dahl. Hjulström and Dahl both consider that cavitation may provide the initial locus for the eddy, and Dahl is convinced that the potholes result from high-energy subglacial or englacial streams striking the bedrock at an angle. R. Streiff-Becker (1951), C. G. Higgins (1957) and H. Holtedahl (1967) put forward similar ideas, and advocated that the moulin hypothesis should now be abandoned.

Potholes of supposed fluvioglacial origin have often been used to determine the limits and extent of former glaciation. Great caution is necessary, though, since potholes are also formed by normal subaerial streams, or even by the overflow waters of ice-dammed lakes, and there have been cases where weathering pits have been taken as potholes. For instance, M. B. Fuller (1925) claimed that there were over 2000 fluvioglacial potholes in part of the Front Range, Colorado, ranging from a few centimetres to 6 m (possibly even 15 m) deep. Although she noted that they occurred almost solely on flat-surfaced granite outcrops and often at joint intersections, and had a weathered appearance, no attempt to explain them as anything other than glacier moulin potholes was made. They were used to delimit the extent of ice in an early Pleistocene glaciation. F. E. Matthes (1930) was careful to distinguish in the Yosemite area between true potholes (as at the lower end of Tuolumne meadows) probably of subglacial origin, and the numerous basin-shaped cavities resulting from strongly localized granite weathering (as at Glacier Point), which he termed 'weather pits'. The latter, in contrast to potholes, had sharply overhanging rims, since enlargement by chemical action is proceeding at or below the level of water standing in them. W. D. Jones and L. O. Quam (1944) strongly refuted Fuller's attempts to use so-called potholes to determine glacial limits in the Rocky Mountains, Colorado; the potholes 'are adequately explained as a pitting in the granite of the sort found in many unglaciated areas' (p. 233). Yet there are cases where true potholes almost certainly do throw light on former ice extent. D. E. Hattin (1958) found numerous eddy-holes in different rock types, possessing smoothed and undercut walls, at over 1200 m on Mount Washington, New Hampshire: these do seem to demand a high-level ice cover in the last main glaciation here.

5 Plucked surfaces

Glacial erosion is thought to operate in two distinct ways: by abrasion of rock surfaces (giving them a smoothed or striated appearance) and by plucking or quarrying (also known as joint-block removal). Many consider that the second of these is by far the most important quantitatively; it also provides the tools essential for the first process. Matthes (1930) held strongly to this viewpoint, asserting that abrasive action alone was quite incompetent to account for the excavation of the deep valleys in the Yosemite region and that in tough massive rocks it achieved little but polishing (p. 89). But where the arrangement and spacing of joints was suitable, and where blocks and slabs of rock had already been loosened by chemical and frost-action, block removal by ice on a vast scale became possible, producing a characteristically plucked or quarried surface. Matthes pointed to the abundance and size of angular blocks in moraine, and in the form of erratics, as evidence of the efficacy of plucking. He considered that the process operated best in rocks with moderately coarse joint texture, with perhaps a mean joint spacing of 1 to 7 m, for larger blocks might be beyond the power of the ice to move while smaller material, though quarriable, would not provide such great bulk in total.

A. Cailleux (1952) also contends that glaciers moving over massive unfractured bedrock may be mainly agents of polishing and minor abrasion. On the other hand, deep ground freezing before the arrival of ice in a region (as M. Boyé (1950, 1968) has suggested in his so-called hypothesis of 'défonçage périglaciaire') may shatter the bedrock; then the advance of a temperate glacier, causing ground temperatures to rise to pressure melting-point or higher, would thaw out the bedrock and transport away the debris.

Evidence that certain rocks or strata are more susceptible than others to block-removal has been produced by many workers. Matthes himself demonstrated that the narrowing of the lower Yosemite valley into the Merced Gorge marked the point where well-jointed rocks terminate and the unquarriable El Capitan granite begins. W. O. Crosby (1928) argued that the massiveness of the granite flooring some Alaskan valleys inhibited their deepening by ice. Studying rock surfaces below the Altels and Rosenlaui glaciers, O. Flückiger (1934) found that ice had minimum adhesive and erosive effectiveness when moving down the bedding planes of steeply inclined formations. Another good example of the influence of rock structure on the ability of glaciers to erode effectively was given by I. B. Crosby (1945). Near Bingham in Maine, ice has been able to widen the Kennebec valley where this crosses a zone of hard quartz-mica schist, for the latter possesses good blocky jointing. But it has had little effect in an adjacent part of the valley where this is crossed by a belt of soft, but mainly unjointed, slate.

Similar conclusions were reached by J. H. Zumberge (1955) in Minnesota and Michigan. By quarrying, said to be the most important process of glacial erosion, ice was able to excavate rock basins in some highly jointed lava flows and jointed slate beds, but was unable to erode significantly in some massive sills. Till in the area is largely composed of fragments from the quarriable formations only.

In some cases, the sizes of angular boulders in moraine or till can be used to suggest the depths reached by glacial plucking (Chapter 9, p. 280). Demorest (1939) argued that the efficacy of plucking would in fact decrease when the ice had removed rock to depths where joints were fewer; thereafter, the supply of rock fragments to cause abrasion would also diminish and a hypothetical end-stage of erosion would be reached. The argument takes no account of the fact that new dilatation joints would probably develop as erosion proceeded (see p. 248).

The role of ice in encouraging dilatation (sheet) jointing will be considered more fully in Chapters 7 and 8, but there is also the question of vertical and steeply dipping joints: indeed, sheet jointing on its own is an incomplete explanation of the susceptibility of bedrock to quarrying. There is some evidence that moving ice may be able to open vertical and semi-vertical joints along pre-existing lines of weakness, and even a suggestion that some new vertical joints may be formed. F. W. Trainer (1973) measured joint patterns at twenty-one localities in California, Maine and New York State. Directions of ice flow were obtained from striations. He found that, on average in these areas, there are two sets of shear joints flanking the direction of ice advance and striking within 40° of it; two sets of extension joints flanking the direction of ice advance and striking within 10° of it; and two sets of release joints nearly perpendicular to the extension joints. Correlations with ice movement are unlikely to be coincidental; but the nature of any causal relationship is by no means clear, and many further investigations are needed.

The mechanism of glacial plucking is more complex than a tug-of-war between moving ice and unbroken rock, for the tensile strength of the latter is in all cases except unconsolidated rocks many times greater than that of ice. In most cases, pre-existing bedrock fractures, either tectonic or developed by pre-glacial frost shattering (Cailleux, 1952; Boyé, 1968; Birot, 1968), are an essential component, though there is the evidence outlined that glacial ice may reinforce these fractures by dilatation or by surface drag effects. Once the rock is broken, transporting mechanisms come into play to remove it. The degree of ice adhesion is a critical factor in this. Cold ice frozen to bedrock will in theory be the most effective agent of joint-block removal; temperate ice with a basal meltwater film will be ineffective. But in certain situations beneath temperate glaciers, freeze–thaw action may be operative, as in subglacial cavities (p. 139) or in bergschrunds (p. 237). Existing joints may be progressively filled with refrozen meltwater until blocks are effectively incorporated in the glacier, as W. D. Johnson (1904) observed. Boulders jammed against the rock bed by moving ice may also be able to gouge fragments from the bed (Chapter 8).

Plucking gives a characteristically shattered and broken appearance to a landscape (unless it is subsequently till-covered), particularly when viewed against the direction of ice flow, and the contrast between surfaces subjected mainly to abrasion and polishing and those affected chiefly by plucking (often adjacent to the former) is a most striking feature of glaciated terrain.

6 Roches moutonnées (stoss-and-lee topography)

Roches moutonnées, in which smoothed and plucked surfaces are juxtaposed, are 'the universal earmark of the invasion of an area by glacier ice' (O. Flückiger, 1934). The term was introduced by H. B. de Saussure in 1787 who likened roches moutonnées fields to the rippled appearance of wavy wigs styled *moutonnées* in his day. As R. F. Flint observes (1971, p. 97–8), the term has been both misapplied and mistranslated; he suggests it should be avoided, though there is already a mass of literature which uses the term. An alternative is stoss-and-lee topography. The smoothed stoss side is streamlined, with striae curving round it and coming together parallel again on the lee side (for example, Okko, 1955, p. 75), which is usually abruptly broken and shattered. This contrast can be used to indicate approximately the former directions of ice motion. Roches moutonnées usually occur in groups dominating considerable areas, as along the Norwegian coast or in the Åland Islands of the Baltic. The individual bosses vary greatly in size, and some writers such as O. D. von Engeln (1938) appear not to limit their size at all, thus extending the category of roches moutonnées to include many glacially streamlined landforms. W. V. Lewis (1947) showed that there is a continuous gradation through major roches moutonnées to valley steps.

Figure 6.5 shows an island off the west Swedish coast just north of Göteborg, last glaciated by ice moving from an east-north-easterly direction. As Rudberg's (1973)

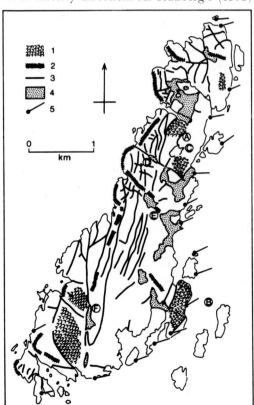

Fig. 6.5
Stoss and lee topography of the island of Härmanö off the coast of western Sweden (S. Rudberg, *Z. Geomorph.*, 1973)
1 Giant stoss sides; 2 Giant lee sides; 3 Micro-joint valleys; 4 Areas with fine-grained sediments, often excavated as rock basins; 5 Striae indicating ice movement, generally from the east-north-east.

mapping shows, a great part of the island (built of gneiss) consists of roches moutonnées of varying size. In some parts, the stoss (up-glacier) sides of the individual roches moutonnées are only separated from each other by very small lee faces a few deci-metres high, coalescing to form a larger-scale composite stoss side. Similarly, 'giant lee-sides' are composed of numerous minor plucked faces.

Among older theories of their origin, W. M. Davis (1909) regarded roches mouton-nées as the ice-moulded residuals of pre-glacial bedrock irregularities, implying that had glaciation lasted long enough, they would have been removed. Matthes (1930) explained them on the basis of variations in joint texture, the ice leaving as protrusions those parts of the bedrock which were more massively jointed. Many roches moutonnées fields do not in fact exhibit any such close relationship to joint patterns. O. Flückiger (1934), supported by von Engeln (1938), attempted to explain the streamlining of the stoss side as the product of moulding by giant stationary waves in the ice (which were not visible at the ice surface owing to ablation). The waves developed on the basis that where two fluids of different specific gravity move over each other, waves form at the contact plane, and in the same way, he supposed that waves would appear when ice flowed over bedrock or when clean upper ice flowed over sluggish dirty basal ice. The ice was said to press harder on the bedrock at the wave troughs and therefore erode more. Such stationary waves have rarely been observed, and as Demorest pointed out in 1939, the vital point ignored by Flückiger is the high viscosity of ice which compels it to move always in laminar flow. Flückiger and von Engeln held that roches moutonnées will be most prominently developed where ice action has been most vigorous and prolonged—they represent the end-stage of the glacial cycle. Demorest adopted a very different viewpoint, regarding them as 'pre-glacial structures which have successfully resisted the last abrasive action of the ice while the areas immediately surrounding them have been cut lower' (p. 600). In 1937, he described roches mouton-nées fields in west Greenland where structure (mainly the foliation of the gneiss) had been of paramount importance, giving rise even to 'reversed' forms with steep stoss and gentle lee slopes where the dip was away from the direction of ice advance.

The normal shattered and plucked lee sides of roches moutonnées probably represent freeze–thaw action beneath temperate ice. H. Carol (1947) succeeded in 1940–42 in penetrating to a subglacial cavity on the downstream face of a roche moutonnée be-neath the Ober Grindelwald glacier. Observing that the presence of the bedrock pro-trusion was locally increasing the pressure at the base of the glacier (which here was at pressure melting-point) and causing meltwater to form, he argued that in the cavity between ice and bedrock on the downstream face of the protrusion, reduced pressure would encourage the meltwater to refreeze and thus shatter the bedrock. The problems encountered by such freeze–thaw hypotheses are elaborated in Chapter 7. However, study of subglacial cavities under the Blue Glacier, Washington, by Kamb and LaCha-pelle (1964) and under the Glacier d'Argentière, France, by Vivian and Bocquet (1972) indicates that pressure-induced freeze–thaw does seem to occur. Regelation spicules were observed, weakly attached to the under-surface of the ice as it bridged these cavi-ties, suggesting alternations of freezing and thawing, probably induced by pressure changes. Fragments of rock were expelled from the basal ice as it reached the cavity, fell to the floor and were taken back into the glacier at the lower end. Whether the small pressure-induced freeze–thaw oscillations observed can be relevant to the large-scale shattering associated with the lee sides of roches moutonnées is a problem which has not yet been adequately answered. Another interesting question is whether roches moutonnées only form under temperate ice, which the hypothesis of subglacial freeze–thaw action demands. Many more studies of conditions in subglacial cavities are clearly needed.

7 Conclusion

Small-scale rock markings of formerly glaciated regions include such features as striations, polished surfaces, friction cracks, moulded and plucked surfaces, and potholes. Glacial striations and grooves are excavated by angular fragments pressed on to bedrock by larger boulders in the basal ice. In using such features for determining former directions of ice motion, care must be taken to exclude striations caused by non-glacial agents (such as snow slides) and to recognize that striations in one locality only indicate basal ice motion in that locality and need not be in accord with long-term regional ice movement. Crossing sets of striations in the majority of cases do not represent different glacial episodes but merely changes in ice-flow resulting from variations in ice thickness.

Friction cracks provide some of the best evidence of the power of moving ice to erode bedrock. Crescentic gouges and fractures represent elastic deformation and rupture of a bedrock surface by extreme pressure exerted by an ice-impelled boulder. The dip of the principal fracture of a crescentic gouge is in accord with the direction of ice motion.

Bedrock surfaces moulded rather than plucked by flowing ice have only very recently been systematically studied. The forms are complex, and include channels, grooves, bowls and Sichelwannen (small sickle-shaped troughs). Some may have been eroded by the abrasive effects of water-soaked subglacial till being squeezed into motion by differential ice pressure. The majority are probably cut by subglacial meltwater moving at high velocities under pressure.

Potholes in formerly glaciated regions were often the work of glacier meltwater streams. Some reach gigantic size—depths of 18 m are known—and may possess spiral-shaped grooves in their walls. The glacier moulin hypothesis must be abandoned except for the occasional plunge-pool holes which are different in being shallow and open. The majority of deep pot-holes are eddy-holes cut by fast-flowing subglacial streams or by englacial streams impinging on the subglacial bedrock surface. These forms, if correctly identified, can be used to determine the extents of a former ice cover.

Plucked surfaces represent the removal of joint blocks by ice. When the joint blocks are of a certain size (possibly 1–7 m), the ice may be able to remove very great quantities of bedrock. Another important factor may be bedrock shattering by a period of deep freezing prior to ice advance. Well-jointed rocks are much more susceptible to glacial quarrying than massive rocks, and even relatively soft formations if not well jointed may resist ice erosion by plucking.

The formation of roches moutonnées is far from being clearly understood. Probably the smoothed up-glacier slope represents glacial abrasion and polishing, while the shattered down-glacier side represents joint-block removal effected by freeze–thaw action in cavities beneath temperate ice. Meltwater passes from the high pressure up-glacier face of the obstacle to the low pressure down-glacier face where it refreezes.

8 References

AGASSIZ, L. (1838), 'On the polished and striated surfaces of the rocks which form the beds of glaciers in the Alps', *Proc. geol. Soc. Lond.* **3**, 321–2

(1876), *Geological sketches* (2nd series, Boston), 36–7

ALEXANDER, H. S. (1932), 'Pothole erosion', *J. Geol.* **40**, 305–37

ANDERSEN, J. L. and SOLLID, J. L. (1971), 'Glacial chronology and glacial geomorphology in the marginal zones of the glaciers Midtdalsbreen and Nigardsbreen, south Norway', *Norsk geogr. Tidsskr.* **25**, 1–38

BIROT, P. (1968), 'Les développements récents des théories de l'érosion glaciaire', *Annls Géogr.* **77**, 1–13

BOYÉ, M. (1950), *Glaciaire et périglaciaire de l'Ata Sund nord-oriental (Groenland)* (Paris; Expéditions polaires françaises, **1**)

(1968), 'Défense et illustration de l'hypothèse du "défonçage périglaciaire"', *Biul. peryglac.* **17**, 5–56

BRÖGGER, W. C. and REUSCH, H. M. (1874), 'Giant's kettles at Christiania', *Q. J. geol. Soc. Lond.* **30**, 750–71

CAILLEUX, A. (1952), 'Polissage et surcreusement glaciaires dans l'hypothèse de Boyé', *Revue Géomorph. dyn.* **3**, 247–57

CAROL, H. (1947), 'Formation of *Roches Moutonnées*', *J. Glaciol.* **1**, 57–9. See also LEWIS, W. V. 'Some comments on Dr H. Carol's article', ibid., 60–3

CHAMBERLIN, T. C. (1888), 'The rock scorings of the great ice invasions', *U.S. geol. Surv. 7th A. Rep.* (1888), 155–248

CLARK, T. H. (1937), 'Northward moving ice in southern Quebec', *Am. J. Sci.* **34**, 215–20

COSTIN, A. B., JENNINGS, J. N., BAUTOVICH, B. C. and WIMBUSH, D. J. (1973), 'Forces developed by snowpatch action, Mt Twynam, Snowy Mountains, Australia', *Arct. alp. Res.* **5**, 121–6

COSTIN, A. B., JENNINGS, J. N., BLACK, H. P. and THOM, B. G. (1964), 'Snow action on Mount Twynam, Snowy Mountains, Australia', *J. Glaciol.* **5**, 219–28

CROSBY, I. B. (1945), 'Glacial erosion and the buried Wyoming valley of Pennsylvania', *Bull. geol. Soc. Am.* **56**, 389–400

CROSBY, W. O. (1928), 'Certain aspects of glacial erosion', *Bull. geol. Soc. Am.* **39**, 1171–81

DAHL, R. (1965), 'Plastically sculptured detail forms on rock surfaces in northern Nordland, Norway', *Geogr. Annlr* **47**, 83–140

DAVIS, W. M. (1909), 'Glacial erosion in North Wales', *Q.J. geol. Soc. Lond.* **65**, 281–350

DEMOREST, M. (1937), 'Glaciation of the Upper Nugssuak Peninsula, West Greenland', *Z. Gletscherk.* **25**, 36–56

(1938), 'Ice flowage as revealed by glacial striae', *J. Geol.* **46**, 700–25

(1939), 'Glacial movement and erosion: a criticism', *Am. J. Sci.* **237**, 594–605

DIONNE, J-C. (1973), 'Distinction entre striés glacielles et striés glaciaires', *Rev. Géogr. Montréal* **27**, 185–90

DREIMANIS, A. (1953), 'Studies of friction cracks along shores of Cirrus Lake and Kasakokwog Lake, Ontario', *Am. J. Sci.* **251**, 769–83

DYSON, J. L. (1937), 'Snowslide striations', *J. Geol.* **45**, 549–57

EBERS, E. (1961), 'Die Gletscherschliffe und -rinnen' in *Der Gletscherschliffe von Fischbach am Inn* (ed. FEHN, H., *Landesk. Forsch.*, *München*, **40**)

EDELMAN, N. (1949), 'Some morphological details of the roches moutonnées in the archipelago of S.W. Finland', *Bull. Commn géol. Finl.* **144**, 129–37

(1951), 'Glacial abrasion and ice movement in the area of Rosala-Nötö, S.W. Finland', *Bull. Commn géol. Finl.* **154**, 157–69

(1972), 'Meandrande glacialrännor', *Terra* **84**, 104–7

ENGELN, O. D. VON (1938), 'Glacial geomorphology and glacier motion', *Am. J. Sci.* **35**, 426–40

FAEGRI, K. (1952), 'On the origin of potholes', *J. Glaciol.* **2**, 24–5

FLÜCKIGER, O. (1934), 'Glaziale Felsformen', *Petermanns Mitt.*, *Ergänz.* **218**, 55 pp.

FORBES, J. D. (1843), *Travels through the Alps of Savoy* (Edinburgh)

FULLER, M. B. (1925), 'The bearing of some remarkable potholes on the early Pleistocene glaciation of the Front Range, Colorado', *J. Geol.* **33**, 224–35

GILBERT, G. K. (1906a), 'Crescentic gouges on glaciated surfaces', *Bull. geol. Soc. Am.* **17**, 303–16

(1906b), 'Moulin work under glaciers', *Bull. geol. Soc. Am.* **17**, 317–20

GJESSING, J. (1965–6), 'On plastic scouring and subglacial erosion', *Norsk geogr. Tidsskr.* **20**, 1–37

(1967), 'Potholes in connection with plastic scouring forms', *Geogr. Annlr* **49**, 178–187

HARRIS, S. E. (1943), 'Friction cracks and the direction of glacial movement', *J. Geol.* **51**, 244–58

HATTIN, D. E. (1958), 'New evidence of high-level glacial drainage in the White Mountains, New Hampshire', *J. Glaciol.* **3**, 315–19

HIGGINS, C. G. (1957), 'Origin of potholes in glaciated regions', *J. Glaciol.* **3**, 11–12

HOLTEDAHL, H. (1967), 'Notes on the formation of fjords and fjord-valleys', *Geogr. Annlr* **49**, 188–203

HOPPE, G. (1952), 'Hummocky moraine regions with special reference to the interior of Norrbotten', *Geogr. Annlr* **34**, 1–72

HJULSTRÖM, F. (1935), *Studies of the morphological activity of rivers as illustrated by the River Fyris* (Uppsala)

JAHNS, R. H. (1943), 'Sheet structure in granites: its origin and use as a measure of glacial erosion in New England', *J. Geol.* **51**, 71–98

JOHNSON, W. D. (1904), 'The profile of maturity in Alpine glacial erosion', *J. Geol.* **12**, 569–78

JOHNSSON, G. (1956), 'Glacialmorfologiska studier i Södra Sverige', *Meddn Lunds geogr. Instn* **30**, 1–407 (with many illustrations and an English summary)

JONES, W. D. and QUAM, L. O. (1944), 'Glacial landforms in Rocky Mountain National Park, Colorado', *J. Geol.* **52**, 217–34

KAMB, B. and LACHAPELLE, E. (1964), 'Direct observation of the mechanism of glacier sliding over bedrock', *J. Glaciol.* **5**, 159–72

LAHEE, F. H. (1912), 'Crescentic fractures of glacial origin', *Am. J. Sci.* **33**, 41–4

LAMARCHE, R. Y. (1971), 'Northward-moving ice in the Thetford Mines area of southern Quebec', *Am. J. Sci.* **271**, 383–8

LAVERDIÈRE, C. and BERNARD, C. (1970), 'Bibliographie annotée sur les broutures glaciaires', *Rev. Géogr. Montréal* **24**, 79–89

LAVERDIÈRE, C., BERNARD, C. and DIONNE, J-C. (1968), 'Les types de broutures glaciaires (glacial chattermarks); observations effectuées au Québec', *Rev. Géogr. Montréal* **22**, 159–73

LJUNGNER, E. (1930), 'Spaltentektonik und Morphologie der schwedischen Skaggerrack-Küste', *Bull. geol. Instn Univ. Upsala* **21**, 1–478

MACCLINTOCK, P. (1953), 'Crescentic crack, crescentic gouge, friction crack, and glacier movement', *J. Geol.* **61**, 186

MARR, J. E. (1926), 'The Kailpot, Ullswater', *Geol. Mag.* **63**, 338–41

MATTHES, F. E. (1930), 'Geologic history of the Yosemite Valley', *U.S. geol. Surv. Prof. Pap.* **160**, 137 pp.

OKKO, V. (1950), 'Friction cracks in Finland', *Bull. Commn géol. Finl.* **150**, 45–50 (1955), 'Glacial drift in Iceland', *Bull. Commn géol. Finl.* **170**, 1–133

RUDBERG, S. (1973), 'Glacial erosion forms of medium size—a discussion based on four Swedish case studies', *Z. Geomorph.*, Suppl. **17**, 33–48

SHARPE, C. F. S. (1938), *Landslides and related phenomena* (New York)

SHELDON, P. (1926), 'Significant characteristics of glacial erosion as illustrated by an erosion channel', *J. Geol.* **34**, 257–65

SMITH, H. T. U. (1948), 'Giant glacial grooves in Northwest Canada', *Am. J. Sci.* **246**, 503–14

STREIFF-BECKER, R. (1951), 'Pot-holes and glacier mills', *J. Glaciol.* **1**, 488–90

TRAINER, F. W. (1973), 'Formation of joints in bedrock by moving glacial ice', *J. Res. U.S. geol. Surv.* **1**, 229–36

UPHAM, W. (1900), 'Giants' kettles eroded by moulin torrents', *Bull. geol. Soc. Am.* **12**, 25–44

VIRKKALA, K. (1951), 'Glacial geology of the Suomussalmi area, east Finland', *Bull. Commn géol. Finl.* **155**, 1–66 (1960), 'On the striations and glacier movements in the Tampere region, Southern Finland', *Bull. Commn géol. Finl.* **188**, 159–76

VIVIAN, R. and BOCQUET, G. (1973), 'Subglacial cavitation phenomena under the Glacier d'Argentière, Mont Blanc, France', *J. Glaciol.* **12**, 439–51

ZUMBERGE, J. H. (1955), 'Glacial erosion in tilted rock layers', *J. Geol.* **63**, 149–58

7

Cirques

I noticed in the valleys of Lanzo, and at heights between 2000 and 3000 metres, hollows in the form of amphitheatres or great cirques, *which have generally a shape like the interior of an armchair, or, more elongated, like a sofa.* (B. GASTALDI, 1873, in a letter to Sir Charles Lyell)

The glacial cirque has long been recognized as one of the most characteristic forms of glacial erosion. In its most easily recognized form, it consists of a rounded basin partially enclosed by steep cliffs and sometimes containing a small lake or cirque glacier; the cliffs at the back of the basin may rise to great heights and culminate in jagged peaks or arêtes. Glacial cirques are worldwide in their occurrence, but confined to areas of present or former glacierization. As a consequence of this broad distribution, they have received numerous local names, such as cwm in Wales, corrie or coire in Scotland, Kar in Austria and Germany, botn in Norway and nisch in Sweden. The term cirque is the oldest established of these in geomorphology, having been introduced as long ago as 1823 by Jean de Charpentier in a study of the Pyrenees, at a time when the glacial theory was only just being tentatively formulated.

It has long been accepted that most cirques in present-day or former glacierized areas are the result of erosive processes closely associated with the presence of ice in them. Indeed, the view that they are glacially sculptured features is now more than a century old. A. C. Ramsay in 1860 thought it 'probable that such rock basins were ground out by heavy loads of ice' (p. 104), and he was followed in 1873 by Gastaldi who, in reference to cirques, asserted that glaciers 'are well able to excavate for themselves deep beds in soft rocks and also in rocks relatively hard' (p. 397). Attempts to explain the development of cirques without appealing to processes of ice erosion encountered grave difficulties. T. G. Bonney (1871, 1873, 1877), A. J. Jukes Browne (1877), J. W. Gregory (1915) and others have suggested that they were the products of the work of running water, but failed to explain, for example, the presence of rock basins commonly present in the cirque floors or to consider those cirques which have developed at or near the summits of mountains (the 'crater cirques' of T. C. and R. T. Chamberlin (1911)). Detailed examination of cirques soon showed that their features were not compatible with theories of water erosion. As W. V. Lewis (1938) observes in regard to British cirques, the streams which now drain them are 'trickles by comparison with the size of the cirques which they occupy and as the streams dwindle towards their sources, the great amphitheatres in which they lie have been incised more and more deeply into the ancient upland surface' (p. 250).

Much more recently it has come to be recognized that the distinctive geometry of the cirque landform may not always be wholly or even partly glacial in origin. Other groups of processes are capable of producing forms similar to, and often difficult to distinguish from, cirques of glacial origin. Nivation processes may lead to the formation of nivation cirques (Embleton and King, 1975, Chapter 6); nivation, snow avalanching and running water combine to produce the funnel-shaped 'rasskars' described by H. W. Ahlmann in 1919. Several processes of mass movement may produce cirque-like features representing the scar from which the material has been derived, the mass of slipped or fallen rubble often superficially resembling morainic debris, though careful observation of the size, form and structure of the debris will indicate its non-glacial origin. Different interpretations of the same cirque-like feature may, however, still prevail. The Bizzle amphitheatre on the north side of Cheviot, for instance, has been regarded as a possible example of a landslip scar by R. Common (1965) though he notes it has the appearance and expected location of a glacial cirque. C. M. Clapperton (1970) is more convinced of its glacial origin, and interprets the debris on its floor as moraine, while accepting that avalanching and scree sliding from the cirque walls may have contributed. Some now favour the use of the term 'cirque' in a purely morphological sense, qualifying it genetically by adjectives such as glacial or nival where the origins are clear. This is preferable to the use of such ambiguous terms as 'pseudo-cirque' (O. W. Freeman, 1925) for non-glacial cirques. In this chapter, and normally elsewhere in this book, the term 'cirque' will not be qualified, the adjective 'glacial' being understood for the sake of brevity except where otherwise stated.

The close connection between cirques and areas of present or former glacierization, observations of cirques containing cirque glaciers, and the virtual impossibility of devising suitable non-glacial hypotheses in all cases, gradually forced universal acceptance of the view that most cirques are features of glacial erosion, though even now, the processes responsible are not fully understood. W. M. Davis (1909) described the cirques of North Wales as having been 'significantly enlarged and deepened by glacial erosion, with pronounced steepening of the side and head walls, partly by direct glacial action, partly by superglacial weathering', but admitted that 'no one knows just how glaciers can erode retrogressively so as to enlarge as well as to deepen their névé reservoirs'. In the discussion following his paper, Davis referred only briefly to Willard D. Johnson's earlier and very significant contribution to the problem of cirque formation (1904). Johnson's observations of conditions in a bergschrund led him to deduce that alternate freezing and thawing of water in rock joints was a major process in the erosion of cirque headwalls, a process also suggested by A. Helland, one of the pioneers of the glacial theory, as long ago as 1877, but Davis doubted its effectiveness in depth, where temperature variations might be small, and wondered whether pressure changes might aid the freeze-thaw process. As regards cirque floors and their rock basins, ignorance of process was even more striking; vague references to the scooping action of the cirque glacier characterize the literature of the early part of this century.

At the same time, there were those who recognized the connection between cirques and cirque-glaciers, yet allotted to the cirque glaciers a very minor role. J. W. Evans (1913) held that cirques were mainly 'the result of frost action facilitated by mountain streams fed by snowfields' and would go no further than admit that 'glaciers have in many cases taken a minor part in [cirque] development', wearing away the floor of the cirque (p. 298). E. J. Garwood (1910) contended that 'ice, on the whole, erodes less rapidly than other denuding agents, and that under certain conditions, it may act relatively as a protective agent' (p. 311). Discussing cirques, he attributed the shattering of the headwalls to frost action above the glacier surface. The glacier itself, occupying the lower parts of the cirque and protecting them from frost action, was held to be responsible only for overdeepening the upper end of the cirque basin.

Clearly, the problem of the origin of cirques could only be settled by intensive studies of those cirques now vacated by ice, by studies of the structure and behaviour of cirque glaciers, and by investigation of the conditions which exist at the contact between ice and bedrock. Another problem which has long attracted the attention of geomorphologists is that of the evolution of cirques and the sequence of landform development which accompanies the enlargement of cirques, up to the stages where the intersection of one cirque with another begins to result in virtual destruction of the pre-glacial relief.

1 Cirque morphology

The term cirque really embraces a whole family of landforms, of very varied appearance and dimensions. At the one extreme, there are small and shallow depressions a few tens of metres across, some containing firn or thin ice, which may represent the beginnings of cirque formation; at the other, there are the great cirques of Antarctica and the Himalayas whose widths may be measured in terms of kilometres and whose backwalls may attain heights of hundreds or even thousands of metres. G. Taylor (1926) quotes a width of 16 km and a backwall 3000 m high for the Walcott cirque on Mount Lister near McMurdo Sound, while the Western Cwm on Mount Everest has a width approaching 4 km at a maximum, and a headwall which, if the ice were removed, would probably amount to 2800 m in height. However, the largest cirques are not necessarily the most perfectly formed. F. E. Matthes (1900) claimed that several conditions had to be satisfied before maturely fashioned cirques would result: the cirques must not be so close that in their development they interfere with one another, and the rock type should be reasonably homogeneous in order that the cirque form should not be distorted by the influence of structure. Furthermore, the glacierization of the area must not be so extensive that few if any parts of the landscape remain unsubmerged by ice. The size to which a cirque will grow depends on many factors. Among them, competence of the rock to withstand failure will clearly be important in limiting the heights of the side and head walls; Cwm Cau in the Cader Idris massif (Wales) possesses rock walls up to 430 m high in the volcanic rocks but only 240 m in mudstone. Most of the largest and best developed cirques in Britain, for example, in the Cuillins of Skye or in Snowdonia, are associated with igneous or metamorphic rocks, and this is generally true for large cirques elsewhere in the world. In areas of sedimentary rock, cirques are often only poorly developed, as in southern and central Wales, or in the southern uplands of Scotland. The size of the mountain mass in which cirques are forming will also obviously affect the size to which any one cirque can grow, and the duration of glaciation will be another factor, for as the cirques enlarge, the intervening ridges will be gradually cut away and the cirques themselves will eventually lose their outlines and identities.

Agreement on the precise definition and principal characteristics of a glacial cirque is far from being reached. Since, as will be shown later, cirques can develop by glacial erosion from hollows or valley-heads of non-glacial origin, there is the problem of defining at what point such a hollow or valley-head begins to take on cirque-like features.

Plate IX
Cirque on Hafrafell, Iceland. Little ice or firn now remains beneath the present-day winter snow. Note the moraine ridges and the frost-shattered debris. (*C.E.*)

Furthermore, several elements combine to form a cirque, and in many cirques one or more of these elements may either be weakly developed or entirely missing. Thus it is not surprising that estimates of how many cirques there are in a given region vary widely—D. L. Linton (1959) numbered 473 cirques in Scotland while J. B. Sissons (1967) counted 347. The problem of enumerating cirques is exacerbated by the existence of nested cirques, coalescing cirques, cirques of complex and anomalous form owing to strong structural control, and so on.

W. V. Lewis (1938) considered that there were four elements characteristic of a cirque. First, there are the head and side walls, steep and usually shattered; secondly, there is the rock floor, showing evidence of smoothing and polishing, and often of rock basin form; thirdly, near the junction of headwall and cirque floor there is sometimes a projecting node of rock; and fourthly, there is the lip or threshold of the basin, convexly rounded and often shattered on the down-valley side. Figure 7.1 shows the long

Fig. 7.1
Cirque morphometry and definition of morphometric variables, **A** in section; **B** in plan (J. T. Andrews and R. E. Dugdale, *Quaternary Research*, 1971, Academic Press)
1 Height of mountain crest above reference datum
2 Height of cirque headwall crest above reference datum
4 Height of cirque threshold above reference datum
5 Maximum depth of cirque
6 Cirque length
7 Maximum breadth perpendicular to long axis
8 Ratio of 6 to 7
9 Ratio of 6 to 5
10 Azimuth of cirque long axis
12 Height of cirque headwall
13 Ratio of 12 to 6
14 Azimuth of regional slope into which cirque is cut.

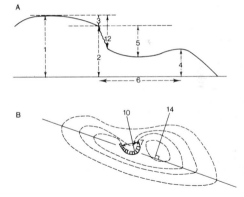

profile and appearance in plan of an idealized cirque. The numbers refer to various geometrical dimensions used by J. T. Andrews and R. E. Dugdale (1971) to describe a cirque. Some of these numbered dimensions are self-explanatory; others, including certain dimensionless ratios, are listed in the legend beside the Figure. Andrews and Dugdale used seventeen variables as the basis of an analysis of the form of 165 cirques in part of Baffin Island, from which an idealized 'median' cirque typical of this area was computed. Some of the most significant cirque form ratios are between length and width, and between length and height. For the former, Andrews and Dugdale find a range from 0·258 :1 to 2·001 :1 for 88 per cent of the population. The length–height ratio ranges from 0·629 :1 to 13·148 :1, but for 95 cirques the limits are 0·629 :1 and 4·8 :1. The median is 4·29 :1.

Length–height ratios for cirques were first examined by G. Manley (1959) in the English Lake District. He claimed that the ratio normally lies between 2·8 :1 and 3·2 :1 for well-developed cirques. Manley's definition of cirque 'height', however, differs from that of Andrews and Dugdale since he took it from the crest of the cirque threshold, not from the lowest point of the floor. Nevertheless, there is still probably a significant difference between the Baffin Island and the Lake District cirque populations in this respect—the latter are on the whole more deeply cut. In part of Labrador, Andrews (1965) found an average length–height ratio nearer 2·1 :1. More statistical analyses of cirque geometry will be valuable and undoubtedly help us to obtain a clearer idea of the range of variation of cirque form, which in turn must be explained. Andrews and Dugdale raise the interesting question whether the difference between the Baffin Island and Lake District cirques could reflect the difference in action between predominantly cold ice and temperate ice respectively.

Another relevant point is that Manley's length–height ratio, treated as a cotangent, yields a value for the angle of slope from the cirque lip to the top of the backwall. This angle approximates to the maximum slope of the surface of any glacier occupying the cirque, and since it seems likely that angles of the order of 12° are necessary before significant rotational slipping of the ice will occur (see p. 230), it becomes possible to gauge, albeit very roughly, the degree of expected erosional activity. The median value for East Baffin Island is 13°, compared with 18–20° for the Lake District.

I. S. Evans (1969) has proposed two other useful measures of cirque geometry. The first refers to the degree of closure of a cirque in plan, that is, whether it shows as merely a shallow indentation in the contours or whether it is a deep recess. The measure proposed is the range in azimuth possessed by the longest contour. 360° would represent a totally enclosed hollow (not a cirque!), 180° would be the case of a cirque whose sides are parallel. Values for two of the cirques shown on Fig. 7.2 are: Vesl-Skautbotn 166°, Blea Water 199°. The latter has the greater degree of closure. Evans' second measure is of the degree of closure in long profile—the difference between the steepest backwall slope and the minimum outward floor slope (if the cirque has a reversed floor, the two values are added). Values for Vesl-Skautbotn are 81°, for Blea Water 86°, the latter showing slightly greater depth of incision.

The actual form of cirque long profiles has been analysed by V. M. Haynes (1968). She points out that earlier attempts (e.g., by W. V. Lewis, 1960) to describe the mathematical form of the long profiles as approximating to arcs of circles were often unsatisfactory (though D. E. Sugden, 1969, finds a reasonable fit for some Cairngorm cirques), and suggests that logarithmic curves usually provide a better fit. The family of curves proposed is of the form $y=k(1-x)e^{-x}$, and such curves provide good fits for 54 out of a sample of 67 Scottish cirques. k values normally range between 0·5 (shallow cirques with almost level or even outward-sloping floors) and 2·0 (deeply incised cirques with basin floors). k values approaching 2 probably indicate well-developed rotational movement of the cirque glacier responsible, whereas for small k values rotational slipping is unlikely. Haynes finds that, in her sample, the main controls on k, which is a useful parameter in describing cirques, are probably structural (see p. 214).

It should be noted that all measures of cirque form are limited by the accuracy of the original data, normally taken from published maps. Contours are of greatly varying reliability, and there are problems of operator variance, for instance, in defining the central axis of a cirque and its azimuth.

Fig. 7.2A shows an example of a beautifully proportioned cirque in the Jotunheim of Norway, Vesl-Skautbotn. Sloping at approximately 60 degrees, the headwall has a maximum height of 350 m above the rock floor: the latter is concealed by the small cirque glacier but its elevation has been approximately determined (J. G. McCall, 1952). The ratio of height to length, as defined by Manley, is about 2·8 : 1. A small cirque glacier, the subject of detailed investigation, occupies part of the hollow and conceals the lower one-third of the backwall. The almost circular nature of the basin in plan is striking; the spurs enclosing it on each side approach one another and unite in a low moraine-covered rock bar which impounds the small lake. A substantial moraine within the cirque marks the present toe of the glacier and separates the glacier from the lake. Above the glacier, the rock walls are highly shattered, and the breaking-off of rock fragments by frost action is a process active today, adding material to the inner moraine. The rock floor is not visible, except where a small patch was exposed at the inner end of a tunnel driven through the glacier; here, the rock surface was mostly smoothed, but not highly polished (J. G. McCall, 1960). Vesl-Skautbotn is of the same order of magnitude as the cirque containing Blea Water (Fig. 7.2B) in Westmorland (average length : height ratio = 2·8). Analysis of the morphology is easier in the case of Blea Water where no glacier is present, and comparison with Vesl-Skautbotn suggests that little change in the cirque has occurred since the last ice vacated it. The backwall rises 240–335 m above the lake, which itself is up to 63 m deep; the water surface of the lake in turn lies over 30 m below the central part of the moraine-covered lip of the cirque (W. V. Lewis, 1960). The average angle of slope of the rock wall is between 40 and 50 degrees.

The shapes and proportions of many maturely developed cirques seem remarkably little influenced by geological structure, apart from the fact already established that the strength of the rock will limit the size and steepness of the cirque walls. Cirques are known to occur in almost all types of rock. But the outlines of cirques may cut across well-marked lithological boundaries with practically no change, as in Snowdonia (where Lewis, 1938, quotes the example of Cwm-y-Llan) and in the Cuillins (A. Harker, 1901). In the Jotunheim of Norway, M. H. Battey (1960) comments that in the cases of Vesl-Skautbotn and Veslgjuv-botn, 'the lack of variety amongst the principal rocks of the area disposes at once of the possibility that lithology has influenced the positions or forms of these particular cirques' (p. 6). On the other hand, there are exceptions to this. K. M. Strøm (1945) contrasts the Rondane cirques in sparagmite with the shorter steeper cirques of Moskenesøy cut in tougher plutonic rocks. H. R. Thompson (1950), studying cirques in north-west Sutherland, concluded that the stage of development of individual cirques depended to a large extent on the distribution of quartzite and gneiss. He also attached importance to the existence of shatter-planes in controlling the location of well-developed cirques in the gneiss. In the English Lake District, P. H. Temple (1965) concluded that the majority of well-developed cirques were located in structurally weak zones.

L. H. McCabe (1939) in Spitsbergen also observed that cirques were intimately related to rock structure. Both in the Campbell range and in the Stubendorff mountains, the floors of the cirques were determined by a platform of more resistant strata, and those of the Stubendorff mountains were clearly eroded along lines of weakness caused by faults. Structures such as foliation, joints, faults and shatter-planes seem to

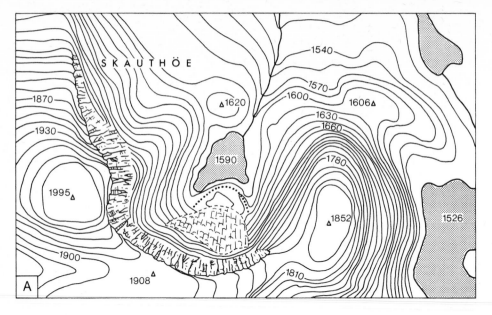

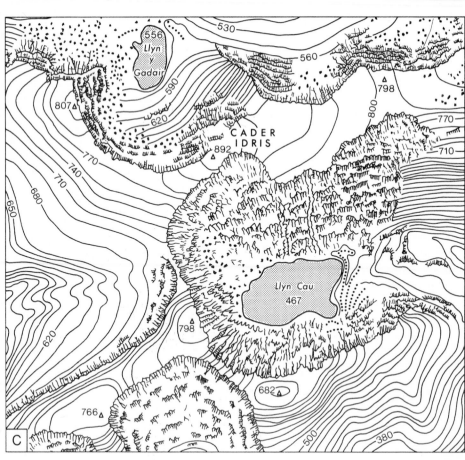

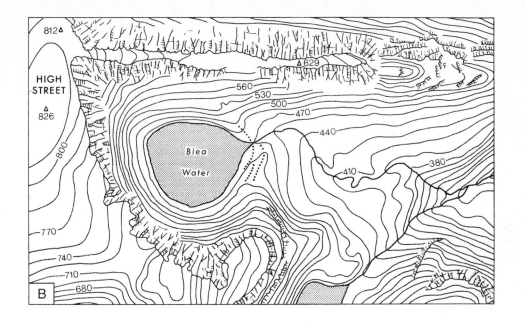

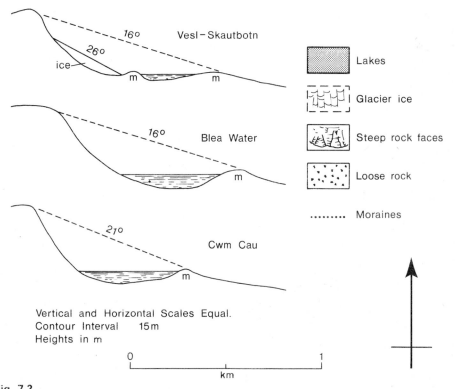

Fig. 7.2
A Vesl-Skautbotn, a cirque in the Jotunheim, Norway
B Cirque containing Blea Water, Lake District, England
C Cwm Cau and adjacent cirques, Merionethshire, Wales
The long profiles of each cirque are drawn without vertical exaggeration

Plate X
The arête of Lliwedd, seen from Snowdon summit, north Wales, with the rock basin containing
Llyn Llydaw on the left. (*C.E.*)

be relatively more important than lithology in cirque development. Battey (1960)
shows how thrust-planes outcropping in the walls of Veslgjuv-botn (Jotunheim) often
define the upper and lower edges of tabular masses of rock that have broken away
parallel to the free face and form the treads of shallow steps. He also discusses the important part played in cirque morphology by dilatation or pressure-release joints (Chapter
8). These, he suggests, strongly control the headward erosion of the cirque walls, which
retreat roughly parallel to themselves by the spalling off of sheets from the free face.
In forming thus approximately parallel to the ground surface, they help to maintain
the forms of glacial erosion, so that cirques grow in size without any essential change
of form.

The problem in evaluating the role of structure in cirque morphology is very often
one of trying to disentangle its possible influence from that of other factors. Haynes
(1968), in the case of some Scottish cirques, concluded that both structure and process
are completely combined. For instance, structure, affecting the pre-glacial form from
which cirque formation commences, may well decide whether rotational slip of the
glacier will occur initially or readily. On the other hand, once rotational movement
has begun, ice flow will play a selective part, exploiting certain structures such as joint
planes of suitable inclination and ignoring others. Figure 7.3 shows some examples of
how joint and bedding planes may influence the long profiles of cirques studied by
Haynes. She contends that the amount and direction of dip of these structures is a
major control on whether a basin is present in the floor or not (only 21 per cent of

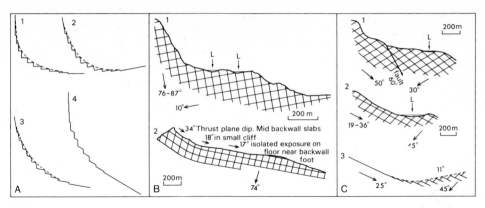

Fig. 7.3
A Idealized relationships between cirque long profiles and jointing
B Examples of effect of 'insloping' and 'outsloping' joints and other fracture planes on two cirque long profiles in Sutherland
C Cirque floors developed on opposed medium-angled joints. The first two are from actual examples (in Sutherland and the Cairngorms); the third is idealized (V. M. Haynes, *Geografiska Annlr*, *Stockholm*, 1968)

the cirques she sampled possess lakes), though other factors must not be forgotten—the upward component of flow at the toe of a rotating cirque glacier may be small in cases where rotation is taking place along *k*-curves rather than arcs of circles, and even the lack of any substantial moraine to obstruct the flow at the toe (*cf.* p. 229) may play a part.

The influence of pre-glacial relief on the location and morphology of cirques has been much discussed. Since in most cases one cannot establish with any precision the form of the ground surface before cirque erosion commenced, it is impossible to decide between theories which on the one hand attribute relatively little to glacial erosion and much to the influence of pre-glacial relief, and on the other hand, those which postulate radical transformation of the pre-glacial surface. It is generally agreed that though cirques will develop in a great variety of situations, the majority have originated in previously water-eroded features. W. M. Davis (1909) contended that pre-glacial valley heads favoured the formation of a local glacier and the excavation of a cirque. E. de Martonne (1910–11) also thought that pre-glacial valley heads or valley sides would provide suitable locations for the initial collection of snow and its transformation into moving ice; Strøm (1945) indicated that some cirques inherited pre-glacial valleys in the Rondane of Norway; and Harker (1901) made similar suggestions for cirques in Skye.

Clearly, there are many water-eroded forms, both major and minor, which have subsequently been taken over and resculptured by moving ice; the problem is in deciding how much influence the forms of water erosion have had on cirque development. In occasional cases where cirques are aligned along zones of rock weakness, for instance Cwm Cau (Cader Idris), or certain cirques in Sutherland (Thompson, 1950), it may be reasonably argued that pre-glacial river erosion was guided by the structural weaknesses and the heads of valleys thus created later became the sites of cirques. In the

Cader Idris massif, one can perhaps obtain some idea of the pre-glacial valley form by examining the head of the small valley which drains west from the summit ridge and contrasting its form with Cwm Cau on the other side of the ridge (Fig. 7.2C).

Although many cirques thus develop out of older water-eroded features, it is now accepted that many also arise through processes of nivation deepening any available hollows, however slight, which are able to collect snow. Lewis (1939) shows that sometimes these nivation hollows themselves are related to previously water-eroded features, but there are obviously numerous other locations in which banks or patches of snow may accumulate and persist. D. L. Linton (1963a) notes that mature cirques may develop in isolation on suitable hill-slopes even though these are almost unindented, and quotes the cirque at about 300 m on the north-east flank of Mount Eagle in County Kerry, the most westerly of British cirques. It is suggested that by various processes

Plate XI
View from Galdhöpiggen (2468 m), Norway. Note the series of arêtes and horns resulting from destruction of the preglacial relief by cirque enlargement and recession. A bergschrund parallels the cirque walls in the lower right part of the photograph. (*C.E.*)

(Embleton and King, 1975, Chapter 6) the snow patch will deepen its hollow, more snow will collect, and eventually ice will form. Once this begins to move, the transition to a small cirque glacier will have begun. In this way, cirques could arise in a great variety of topographical situations; in some cases, pre-glacial forms mightly strongly affect cirque development, while in others, the cirque might be a virtually independent feature.

A third major factor influencing cirque development, and one perhaps as important as either rock structure or pre-glacial relief, is the duration of glaciation. Standing on the summit of Galdhöpiggen (2453 m) in Norway (Plate XI), one can look down on an impressive series of snowfields contained by narrow and jagged arêtes, representing the walls of cirques half submerged by snow and ice. The ridge between Svellnosbreen and Storgjuvbreen is clearly being steadily consumed as the cirques containing these glaciers and standing back to each other, grow in size. Sections of the headwalls sur-rounding Tverråbreen and Svellnosbreen have already been partly demolished. It is difficult to resist the conclusion that the pre-glacial Galdhöpiggen massif is being slowly destroyed by glacial erosion, involving the recession of numerous cirques, and that the degree of destruction is roughly proportional to the duration of glaciation. Willard Johnson admitted as much in 1904 in respect of part of the Sierra Nevada. Viewing the highlands from Mount Lyell, he likened them to 'the irregular remnants of a sheet of dough ... after the biscuit tin has done its work', and he inferred that 'the suspension of glaciation had suspended as well a process which had threatened truncation of the range' (pp. 571–2). Ahlmann (1919) illustrated the evolution of cirques in the Lofoten Islands. 'On Moskenaesö there occur all stages in the develop-ment of a cirque. The first stage is represented by a shallow depression ...' The youthful stage is marked 'by a semi-circular cirque with steep, not particularly lofty sides and flattish, scoured, and slightly uneven floor' (pp. 229–30). The excavation of a rock basin indicates maturity, while further lengthening of the cirque owing to headwall recession will cause the formation of a round-headed trough, which according to Ahlmann represents old age.

W. H. Hobbs in 1926 put forward a scheme outlining hypothetical stages in the de-struction of uplands by cirque recession (Fig. 7.4A). The 'grooved upland' of youth still preserves much of the pre-glacial surface intact: the classic area is the Bighorn Range of Wyoming, but on a smaller scale, parts of North Wales east of Ffestiniog demonstrate this stage well, as do parts of the island of Arran. Hobbs's second stage—the 'early fretted upland'—is well displayed in the Snowdon massif of North Wales, where the smooth-topped ridge is being cut into by well-proportioned cirques. As the cirques enlarge and recede, so a series of arêtes and horns is created, typical of a 'mature fretted upland'. This stage can readily be matched in the Cuillins of Skye (Harker, 1904), in the Galdhöpiggen area of Jotunheim, in the Rondane (Strøm, 1945), in the Lofoten Islands (Ahlmann, 1919), or in the Ötztal Alps of Austria, for instance. Linton (1967) suggests a series of sub-stages in arête shortening as cirque enlargement and interosculation continue. Using the term 'grat', the German equivalent of arête, he recognizes 'interrupted grats' when cirque headwalls first begin to intersect, followed by stages of 'shortened grats' and finally 'vestigial grats'. If ice levels rise, or if the bounding arêtes are broken down so that ice overflows from one cirque to another,

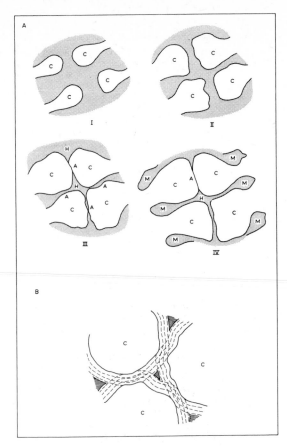

Fig. 7.4
A Stages in the enlargement of cirques (C) and their recession into an upland (W. H. Hobbs, *Earth Features and their Meaning*, Macmillan & Co., New York, 1926)
  I Grooved upland   II Early fretted upland
III Mature fretted upland   IV Monumented upland
A Arête   C Cirque   H Horn   M Monument
Preglacial surface shaded
B The development of dilatation joints parallel to cirque walls aids the destruction of arêtes, where the joints intersect at shallow angles, and the formation of pyramidal peaks (horns).

the resulting modified arêtes are classified as 'subdued grats'. Horns rise from the intersection of three principal arêtes (in the Alps, the Aletschhorn or the Dom, for instance) or of four (Gross Venediger, or the Dent Blanche); examples of more than four principal arêtes converging to a horn are fewer, but the Mönch and the Weissmeis in the Swiss Alps may be cited. It is clear that horns tend to survive long after the arêtes linking them have been destroyed; in some cases, secondary cirques may develop on one or more faces of the horn (for instance, Semeltind and Skardalstind in Jotunheim, Norway). Linton suggests (1963a) that the reason for late survival of horns in landscapes subjected to prolonged glacial erosion may be connected with the formation of dilatation joints. Arêtes might be expected to disintegrate more readily since they will develop two sets of intersecting dilatation joints (Fig. 7.4B) whereas the rock in the horns having been free for some time to expand outwards parallel to three or more free faces, may have reached a state of equilibrium. Thus, 'the assistance given to weathering and removal in the early stages by dilatation joints is considerably diminished when isolation has actually been achieved' (p. 21).

There is less agreement about landform evolution in the later stages of cirque enlargement and recession. Hobbs describes a stage of old age (his 'monumented upland') in which the breaking-down of arêtes creates cols which allow ice to spill over from one cirque into another. Beneath the ice, original cirque floors may coalesce into a surface of glacial denudation, surmounted only by horns. E. Richter (1906) attributed widespread planation (Reusch's 'Paleic Surface') of the Norwegian uplands entirely to this process, but no one would support this extreme view today. W. M. Davis (1900) argued that as the later stages of the glacial cycle were approached, the feeding grounds of the glaciers would become confluent, broken only by nunataks (representing the original horns); he speaks of an ice and snow shield covering a lowland of glacial denudation when old age had been reached. But as C. A. Cotton (1942) has pointed out, 'destruction of mountainous relief by glacial peneplanation is a theoretical abstraction in that no cases of such levelling down of mountains are known with certainty; if a glacial peneplain exists anywhere, it may be expected to exhibit a mammillated cirque-floor-like relief' (p. 195). Strøm (1949) observes that all stages of Hobbs's cycle of glacial erosion can be matched in the Jotunheim of Norway, and that in some areas, cirque glaciers 'have split up or wholly destroyed mountain massifs.... As (the cirque) backwalls cut ever farther into the mountain massif where they occur, a horizontal or somewhat rising platform is formed which is frequently difficult to distinguish from an older peneplain' (p. 21). He designates this pediment-like platform built out of cirque floors as a *fly*. E. Dahl (1946) claims that part of the modelling of the Norwegian strandflat took place in this way. In the Antarctic, Linton (1963b) describes the landforms characteristic of prolonged glacial erosion, where the ice, 'in the long time that has been available to it, has been able to gnaw away the greater part of substantial mountains' (p. 281), and where there is represented 'a degree of destruction ... not even hinted at in such classic areas of glacier study as Alaska or the Alps' (p. 282). The outlines of cirques have often disappeared, their bounding arêtes submerged by ice; ice-sculptured pyramids are the principal surviving masses standing above the ice-fields. One can only conjecture about the form of the land surface beneath the ice in such an area. But quite clearly, prolonged cirque erosion will result in completely radical transformation of the landforms.

It is important at this point to sound a cautionary note. As in many other branches of geomorphology, supposed sequences of landforms cited in support of an evolutionary model may be more imaginary than real. There is a real danger in substituting space for time (S. A. Schumm and R. W. Lichty, 1965). Nor is it possible to prove that one particular stage of cirque development claimed by Hobbs and others has necessarily taken a longer or shorter time to evolve than any other stage. Allowances must also be made for differences in structure and in process (for example, temperate *v.* cold ice). Altogether, there are so many unknown and unquantifiable variables that Hobbs's scheme, while attractive and superficially plausible, must be regarded as non-proven.

Cirque stairways, Kartreppen, or escaliers de cirques, are characteristic of some mountain areas. In Snowdonia, Wales, the cirque containing Llyn Glaslyn bears part-way up its headwall a less well-formed and smaller cirque directly beneath Snowdon summit, while on the lower side of Glaslyn there is a descent into the larger cirque holding Llyn Llydaw. The floors of the three basins, measured at their deepest points, lie successively at 808, 562, and 375 m above sea level. In the Cuillin hills of Skye, another example may be found in the valley descending southward from Sgurr nan Gillean into Harta Corrie. At its head, there is Lota Corrie; then a precipitous step down to a second cirque basin (floor level about 260 m); and a third sharp descent into the head of Harta Corrie. Ahlmann (1919) notes many examples in Norway, such as on the island of Moskenesøy; he suggests that, at certain stages of glacierization, ice flowed continuously from the highest cirque in a stairway to the lowest, broken only by ice-falls at the steps between the different cirques. It obviously becomes impossible in cases to distinguish between cirque stairways and glaciated valleys possessing a series of basins and steps. Indeed, Griffith Taylor (1914) has suggested that typical series of valley steps and basins may originate as isolated cirques in a stairway later modified and united by continuously flowing ice. There are many reasons why, in an area undergoing glaciation, cirques should develop at successively different levels. A rising snow-line might be the cause (see p. 223 for a discussion of the relationships between cirques and snow-lines), or intermittent uplift of the mountain mass (E. Fels, 1929). Cotton (1942) thinks that the cirques of cirque stairways might develop contemporaneously without such changes, lower cirque glaciers forming in lower niches out of avalanche ice descending from higher cirques.

The problem of the origin of cirque stairways is complicated by the multiple nature of Pleistocene glaciation; the cirques that we now see are mostly the products of not one but several glaciations of differing magnitude. However, contemporaneous occupation of cirques in a stairway by separate cirque glaciers has fairly been demonstrated for certain glacial episodes. B. Seddon (1957), for instance, shows that the cirques of Glaslyn and Llydaw on Snowdon, already described as members of a cirque stairway, contained separate cirque glaciers in the Late-glacial period, judging by their individual morainic accumulations on the lips of the cirques at about 610 and 425 m respectively. But it is also clear that these cirques were not formed solely by the Late-glacial ice, but merely reoccupied. Seddon deduces that the size of glacier in each cirque was roughly proportional to the amount of precipitation in their respective accumulation areas. In earlier glaciations of greater magnitude, ice from the higher cirques must have overflowed into the lower ones.

2 Cirque orientation and elevation

In the middle latitudes of the northern hemisphere, it is well known that the majority of well-developed cirques face in directions between north and east. Conversely, in the restricted land areas of middle latitudes in the southern hemisphere, a southerly or south-easterly aspect is dominant (see, for example, E. Derbyshire, 1964), although the number of cirques involved is relatively small. The factors determining cirque orientation are several. Structure is held to be locally of importance, and McCabe

(1939) considered its influence paramount in controlling cirque orientation during phases of oncoming glaciation in West Spitsbergen, but its influence as a primary factor is denied by others, such as P. H. Temple (1965). Protection from solar radiation is a more general control, for it determines to a large extent how much snow can accumulate in the cirque and survive through the summer. In the British Isles, most well-developed cirques are orientated away from the sun: Lewis (1938) noted that 29 out of 44 major British cirques face between north-west and east-north-east. In the Dingle peninsula, a marginal area in respect of Pleistocene glaciation, C. A. M. King and M. Gage (1961) draw attention to an even more pronounced concentration, for out of 34 cirques, all but three face in directions between north-west and east. More detailed quantitative analyses of orientation have been made by V. Schytt (1959) in a comprehensive survey of Swedish glaciers, and for smaller areas, by B. Seddon (1957) and P. H. Temple (1965), for example. Temple has investigated the orientation of 73 cirques in the west-central English Lake District (Table 7.1). He distinguishes carefully between the *location* of a cirque on a particular hill mass, and the *aspect* of a cirque. These two parameters are often but not necessarily similar. Seddon's findings for the aspect of cirques in Snowdonia are substantially similar: 75 per cent face between north and east. In part of Labrador, J. T. Andrews (1965) found that 72 per cent of cirques have this orientation.

Table 7.1   Position and aspect of cirques in the west-central English Lake District

Quadrant/Aspect	Per cent located in each quadrant	Per cent facing each direction
North-east	53·4	52·0
South-east	19·2	19·2
North-west	24·7	23·3
South-west	2·7	5·5

The mean aspect of cirques in a group or region can be determined by simple graphical or trigonometrical methods (Evans, 1969). In Fig. 7.5, each numbered leg is plotted with a length (L) proportional to the number of cirques possessing the azimuth (α) of that leg. Azimuths are normally grouped in intervals of, say, 5° or 10° because of the difficulty of precisely measuring the aspect of any individual cirque, and these groups must be ranked in order (as the numbers on Fig. 7.5 show). The vector mean angle is given by the thickened arrow and is the angle whose tangent equals

$$\frac{\sum L \sin \alpha}{\sum L \cos \alpha}.$$

The length of the resultant is

$$\sqrt{(\sum L \sin \alpha)^2 + (\sum L \cos \alpha)^2}.$$

In this way, Evans arrives at the mean aspect for 437 cirques in Scotland as 48°; for the 73 cirques in the west-central Lake District examined by Temple it is 40°. Thirty-

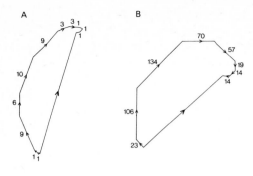

Fig. 7.5
Analysis of cirque aspect (I. S. Evans in *Water, Earth and Man* ed. R. J. Chorley, Methuen, 1969)
A For 45 cirques in British Columbia;
R (length of resultant vector)=32·7
L (strength of resultant vector)=72·6 per cent
θ (azimuth of resultant vector)=18°±14·4°
B For 437 cirques in Scotland;
R=221·4
L=50·7 per cent
θ=48·1°±6·4°
Note that different scales are employed for plotting **A** and **B**.

seven cirques in Jotunheimen have a mean aspect of 12°, much nearer north than the British averages and similar to a mean of 18° for 45 cirques in British Columbia. A southern hemisphere group, 50 cirques in New Guinea, has a mean aspect of 185°. Table 7.2 shows a more detailed breakdown of cirque aspect in northern Snowdonia, Wales (D. J. Unwin, 1973) where it is interesting to note that for no group in this mountain area does the vector magnitude fall below 40 per cent and in one group reaches the unusually high figure of over 90 per cent.

Table 7.2 Cirque orientation in northern Snowdonia, Wales (Unwin, 1973)

Mountain group	Vector mean, degrees	Vector magnitude, per cent
Hebog	56·5	40·4
Snowdon	41·2	49·3
Glyder	42·8	91·7
Carnedd	56·1	53·4
All cirques (84)	47·5	57·3

    The shade factor may be of most importance in middle latitudes, for in the tropical zone, the angle of the sun's rays is much greater and no clear distinction can be drawn between shady or sunny sides of the mountains. Shade is not the only climatic consideration, however; the strong easterly component in the orientation of British cirques, for instance, requires further explanation. F. Enquist (1917) in the Alps showed that the largest cirques and the largest glaciers lie on the lee side of the mountains with respect to the prevailing snow-bearing winds, and much more recently, R. W. Gloyne (1964) indicated how lee slopes steeper than about 10° will provide a shelter zone whose width is proportional to the actual slope angle. In Britain during glacial periods, the main snow-bearing winds were probably from directions between south and west; hollows most favourable for snow accumulation and therefore cirque glacier growth were those facing in the opposite direction, where drifting snow could most effectively collect. In the Front Range of the Rockies in Colorado, cirques are also largest and most numerous on the eastern leeward flanks, and the only surviving cirque glaciers occur on this side too, for during the winter season in which snowfall is now concentrated, winds blow dominantly from the west. In the Falkland Islands (latitudes 51°–52° south), most cirques and nivation hollows face between north-east and south-east, backing the dominant snow-bearing winds from between south-west and north-west. Although the north-easterly slopes receive more insolation than south-easterly ones, they carry 41 per cent of the cirques, an anomaly for the southern hemisphere that is largely explicable in terms of wind direction (Clapperton, 1971). Shade and wind direction are therefore the most powerful controls affecting snow accumulation, and, in turn, cirque orientation. There are notable exceptions, however, apart from those cases where structure is a major influence. In the southern half of the Sierra Nevada, as is well shown on the Mount Whitney quadrangle (U.S.G.S. 1:125,000), the greatest number of well-developed cirques lies on the west of the main divide, for the eastern slopes, although representing the lee side with respect to the dominant moisture-bearing winds, are relatively arid because of the rain-shadow effect; snow accumulation on the eastern slopes has therefore never been as great as on the windward side. In detail, however, the position and orientation of cirques on the windward side of these mountains is still closely influenced by the availability of shade.

In addition to climatic control of cirque orientation, relief undoubtedly plays a part. Certain slopes may be either too steep or too gentle for efficient collection of concentrated snow patches. Many cirques, as already noted, appear to have developed in preglacial valley-heads and D. E. Sugden (1969) notes that in the Cairngorms, Scotland, the dearth of such valley-heads facing the most suitable climatic aspect, north-east, is probably largely responsible for the unusual bimodal distribution of Cairngorm cirques, the majority of which face either east or west-north-west.

The elevation of a cirque is closely related to the elevation of the local snow-line, or firn-line. The relationship is a very significant one, since it can be used to estimate positions of the regional or climatic snow-line during the formation of the cirque. The firn-line on a cirque glacier usually occurs about three-fifths of the way between the snout and the upper limit of the ice; for instance, the present small cirque glacier in Vesl-Skautbotn (Jotunheim, Norway) has a firn-line at about 1650 m, a lower ice limit of 1580 m, and an upper limit of 1710 m (Fig. 7.2A). For individual cirques now vacated by ice, rough estimates can therefore be made of the height of the former firn-line (for example, G. Manley, 1959) and from these the climatic snow-line during cirque formation may be obtained. J. T. Andrews (1965) for instance, in the northern Nain-Okak section of coastal Labrador, puts the climatic snow-line at about 240 m above the local cirque firn-line at the time the cirques were forming, and deduces a rise in mean summer temperatures of 6°C since that time. A crude approximation can even be made using the elevations of the cirque floors, since the climatic snow-line will not usually lie at more than a few hundred metres above these. R. F. Flint (1971, p. 475) shows a map of generalized contours on a theoretical cirque-floor surface for the western United States which reflects in some measure the varying heights of the snow-line over this area, possibly related to the last glacial maximum. In Rocky Mountain National Park (Colorado), the average altitude of the former snow-line implied by the mean level of cirque floors is about 3000 m, compared with 4200 m for the present snow-line. M. M. Miller (1961), studying more than 200 abandoned cirques in the Alaska–Canada

Boundary Range, finds that their floor elevations are concentrated at five levels (110, 340, 530, 750, and 960 m), which he relates to former firn-lines in the Wisconsin glacia-tion. It is possible that the different levels reflect an intermittently rising snow-line in this period. Miller finds no evidence which might support the view that lithology, bedrock structure, or pre-glacial relief might have been responsible for the successive cirque levels.

Other workers have questioned such views; in some areas the altitudes of cirque floors appear to be too variable and almost random, and it is pointed out that the aspect of the cirque, and therefore the duration and effectiveness of ice erosion, may result even in neighbouring cirque floors having very different altitudes, even though their formation may have been synchronous. In any analysis of cirque-floor elevations as a means of fixing former snow-lines, therefore, it is well to use only well-developed cirques and to avoid cirques whose altitude is obviously affected by geology or pre-glacial relief. It should also be borne in mind that a cirque may be composite in the sense that it may have been reoccupied and deepened in successive glacial episodes.

Trend-surface analysis has recently been applied in studying the vertical distribution of cirques. Unwin (1973) uses linear and quadratic regression surfaces to show that in northern Snowdonia there is a mean increase in cirque-floor altitude from south-west to north-east of 13·3 m/km, confirming Seddon's (1957) findings from less sophis-ticated analysis. In Scotland, G. Robinson et al. (1971) based trend-surface analysis on 175 cirques and found that the best fit to the data was provided by a cubic surface rising to a maximum of about 800 m in the Cairngorms.

3 Structure and behaviour of cirque glaciers

An essential preliminary to understanding the origin of cirques is to consider the nature of cirque glaciers and their movement.

The banded appearance of glaciers was first discussed in some detail by J. D. Forbes (1843). Nearly a hundred years passed before further studies of glacier bands were undertaken. In 1939, G. R. Gibson and J. L. Dyson published the first detailed map of some of the ice bands of a cirque glacier, the Grinnell glacier of Glacier National Park, Montana, and showed how significant these bands might be in understanding the mode of accumulation and the nature of ice movement. They regarded the ice bands as annual accumulation layers, each representing some 15 to 20 m of winter snow deposited in the firn area, and becoming tilted and slightly deformed as they moved down the glacier. The outcrops of the bands at the glacier surface formed sweep-ing curves roughly parallel with the back wall; they dipped up-glacier at 12 degrees near the firn-line, and up to 45 degrees nearer the glacier front. They had no evidence as to the disposition of the bands within the glacier. In 1948, J. M. Grove (see J. M. Clark and W. V. Lewis, 1951, and J. M. Grove, 1960) commenced an investigation into the structure of the cirque glacier Vesl-Skautbreen, whose surface banding is un-usually simple and striking. It is, moreover, a conveniently small cirque glacier, and one which was simultaneously being studied in its other aspects by a team of research workers. Not only was Mrs Grove able to study the surface banding but to examine the internal behaviour of the banding along two tunnels excavated through the glacier.

Her findings on Vesl-Skautbreen, and also on nearby Veslgjuv-breen, are of great significance for all cirque glaciers, and will therefore be presented in some detail.

The Vesl-Skautbreen (Fig. 7.2A) consists of a series of ice layers, separated by discontinuities, some of which are emphasized by the occurrence of small quantities of organic or mineral debris. In the past, the discontinuities have been referred to as dirt bands, but this term is undesirable since they are not bands and not always dirty. The ice layers (Plate XII) each represent one year's increment of snow in the firn region (a few layers may be composite), and are therefore termed 'accumulation layers'. The discontinuities represent ablation in the summer periods, and are termed 'ablation surfaces'. Where they outcrop at the surface of the glacier, the ablation surfaces yield debris in varying amounts as the surface of the glacier wastes downward. This, together

Plate XII
Sedimentary layers (dirt bands) exposed in an ice cliff, Veslgjuv-breen, Norway. Melted-out debris is accumulating at its foot inside a winter snowbank. (*C.E.*)

with the fact that the next higher accumulation layer often tends to protrude or hang slightly above the ablation surface, led W. V. Lewis (1947,1949) to suggest that many ablation surfaces were planes of differential movement, with debris being brought up to the glacier surface by the slow overthrusting of one accumulation layer over the ablation surface beneath. That the ablation surfaces of Vesl-Skautbreen are not thrust planes was conclusively shown, particularly in a simple experiment by J. F. Nye (see J. G. McCall, 1960, p. 49); and it is now agreed that the overthrust appearance is simply the effect of differential ablation. This does not, of course, preclude the existence of thrust-planes on other glaciers, particularly near the termini of valley glaciers, but they are not to be equated with ablation surfaces. As well as the accumulation layers and ablation surfaces, Mrs Grove describes blue layers of ice within the accumulation layers, which she contends result from formation of surface crusts or ice layers by freezing of meltwater in the firn zone; blue veins cutting across the layers probably represent the freezing of meltwater seeping down cracks.

The density of the white ice in the accumulation layers is about 0·9; thus an annual accumulation layer 1 m thick is probably the equivalent of about 15 m of freshly-fallen snow in the firn zone. As this is more than the mean annual snowfall for this part of Jotunheimen, it is clear that drifting, avalanching and eddy deposition have caused a concentration of snow in this cirque, which is backed by a broad upland plateau to windward. The cirque of Veslgjuv-breen 8 km away shows in some years spectacular effects of eddy deposition of new snow under the backwall. The central tunnel through Vesl-Skautbreen showed that this glacier consisted of a minimum of 88 accumulation layers, as well as an uncertain number of layers compressed in the base of the glacier, and others in the firn zone above the tunnel. The bulk of the ice in this small glacier was therefore probably less than 100 years old. The tunnel showed too that most of the ice was extraordinarily clean, a finding which was borne out also in the case of Veslgjuv-breen. What debris there was in the glacier, apart from that in the basal layer next to the rock floor, was clearly not originating from the rock floor but from debris fallen or collected on the firn zone, especially as the result of rock falls from the backwall, and thereafter passing down into the glacier. One ablation surface in particular, outcropping a little above the central tunnel entrance, was, on the left-hand half of the glacier, a prolific yielder of debris which must have represented a series of large rock falls on the firn zone about eighty years ago.

The greatest amount of rock debris in the glacier was found in the basal layer or 'sole' next to bedrock. The ice here, seen at the inner end of the central tunnel, had a blackish appearance, containing material ranging from rock flour to large boulders; it was also sheared and slickensided indicating a tearing movement on the rock bed. Since the ice here was also bubble-free, it probably originated not from compaction of firn but from refreezing of meltwater (compare the regelation layer of B. Kamb and E. LaChapelle, 1964, discussed in Chapter 4), the meltwater seeping down the back of the glacier. Rock debris fallen into the gap between the upper part of the glacier and the headwall was almost certainly the source of most of the debris in the glacier sole.

The accumulation layers have been deformed by ice movement. In the firn zone, the latest accumulation layers dip down the glacier at about 30°, and as shown

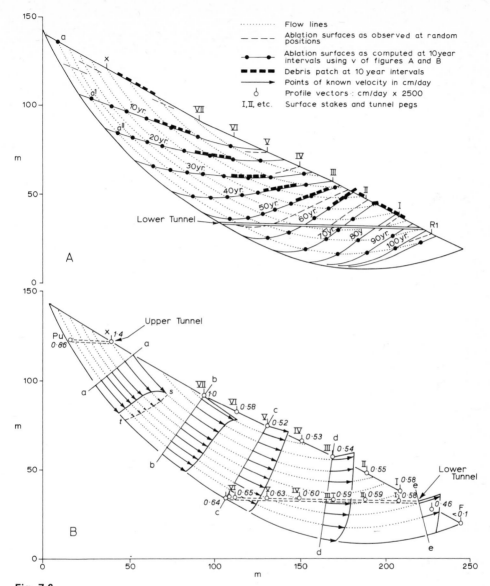

Fig. 7.6
A Longitudinal section of Vesl-Skautbreen showing flow lines and ablation surfaces.
B Velocity distribution in a longitudinal section of Vesl-Skautbreen (J. G. McCall, 'Norwegian Cirque Glaciers', Royal Geographical Society, *Research Memoir*, 4, 1960)

in Fig. 7.6A, the dips of the layers exposed below the firn-line are nearly horizontal at their outcrops, but steepen so as to dip up-glacier farther down the ice. At the entrance to the central tunnel, they dip up-glacier at 26 to 28°, but below this, the dip once again flattens out, becoming virtually nil near the glacier snout. The section provided by the central tunnel showed a synclinal arrangement of the ice layers; at its inner end, they curve so as to meet the rock bed tangentially. Thus, the deformation of the layers by glacier movement suggested in itself how that movement was taking place. Clearly, the layers were gradually being rotated as they passed down the glacier; at the base of the glacier, they were also being stretched and bent upward by basal drag.

More precision was given to such theories of ice motion in this glacier by J. G. McCall's careful theodolite surveys of the changing position of pegs both on the ice surface and in the tunnels. From these records, McCall was able to interpolate the velocity distribution and to plot velocity profiles (Fig. 7.6B). The five velocity profiles here shown indicate the shapes and positions of originally straight section lines (a-a to e-e) after undergoing seven years (=c. 2500 days) movement. The flow lines were based on observed velocities plus an assumed regular curve (approximating to the bedrock surface) for the lower limit of ice movement. The flow lines show downward movements in the firn zone, movement parallel with the surface at the firn-line (which lies between stakes V and VI), and pronounced upward movements in the lower tongue. The velocity vectors reveal the more rapid movement of the surface layers of the firn (profiles a-a and b-b), the retardation of the basal ice (profiles b-b through e-e), and a very slightly lower rate of movement for the surface ice than for the deeper ice in profiles c-c and d-d. The hypothesis which best explains the observed surface and internal movements of the ice, and the deformation of the accumulation layers, is that of rotational sliding. Suggested by W. V. Lewis in 1947 and elaborated in 1949, the hypothesis implies sliding of the glacier along a roughly arcuate bedrock floor and, simultaneously, rotation around an approximately horizontal axis. These points are consistent with the observed rotation of the accumulation layers, with the forward and downward motions recorded in the firn zone, and with slight upward movement near the toe of the glacier. Furthermore, the central tunnel, excavated originally with a slight outward slope to facilitate meltwater drainage, over a period of time gradually tilted to become horizontal, and later developed an inward slope; lowering of the tunnel floor near the entrance became necessary to prevent water accumulating in the inner parts of the tunnel. The motivating force behind such rotational movements appears to be quite simple; the main weight of winter snow accumulates on the upper part of the glacier beneath the headwall, whereas the lower part of the glacier in summer experiences most losses by ablation. Thus the overall surface of the glacier is effectively steepened, and rotation occurs to restore equilibrium when the forces are sufficiently large to overcome friction between the ice and bedrock.

An approximate figure for basal shear stress at the inner end of the lower tunnel can be obtained in the usual way by calculating the value of ρ gh sin α. Taking h as 50 m and α as 25°, this yields a value of 1·9 bar, somewhat higher than might be expected because of the relatively high angle of surface slope. J. Weertman (1971) argues that a correction factor of the order of 0·4 should be applied to the term ρ gh sin α in the case of a rigidly rotating cirque glacier, giving in the case of Vesl-Skautbreen what is probably a more realistic value of about 1 bar.

Although rotational sliding can explain much of the observed behaviour of Vesl-Skautbreen, it does not explain everything. That the accumulation layers are deformed indicates some degree of creep in the ice; the observed angles of dip of the accumulation layers in the central parts of the glacier surface are greater than would be expected if it were a question solely of the rotation of a rigid mass; and Fig. 7.6B shows that forward motion of the ice ceases almost entirely at point F against the moraine and that in sections c-c and d-d, the locus of maximum velocity occurs neither at the surface nor near the bed, but within the lower half of the glacier. It is apparent that some force system within the glacier is slightly retarding the surface ice movement; McCall argues that a very weak form of extrusion flow is superimposed on the rotational sliding because the surface zone is behaving as a sloping dam beneath and against which the interior ice is pushing. The surface zone is slightly more rigid since in the lower part of the glacier it is braced against the terminal moraine at its foot, and since it is colder in the winter and early summer than the deeper ice. The result is that the lower part of the glacier tends to bulge slightly; and this in turn is sufficient to allow ice movement along the bed virtually to cease at point F by the moraine.

These theories of ice movement have been developed in respect of Vesl-Skautbreen, but it is highly probable that they apply directly to most other simple cirque glaciers, and less directly to compound types of cirque glacier. Thus, observations on the Arapaho cirque glacier in the Front Range of the Rocky Mountains, the largest in Colorado, show that it too is moving primarily by rotational sliding (H. A. Waldrop, 1964); studies of the double cirque glacier of Veslgjuv-breen also reveal rotational movements, and some slight bulging of the ice surface in the zone of maximum dip of the accumulation layers (J. M. Grove, 1960). The significance of these observations and theories with respect to cirque morphology will be considered in the next section.

4 The problem of cirque erosion

Cirques possess a number of features which can only be explained in terms of glacial erosion or by processes closely associated with ice. The principal features are the rock basins in cirque floors, the rock lip or threshold at the cirque exit, the steep head and side walls, and the presence of smoothed and abraded rock surfaces mainly but not exclusively on the cirque floors, and of broken and shattered surfaces especially on the head and side walls. The processes responsible for these features are still matters of controversy, and their investigation is rendered difficult by the fact that direct observation of processes operating beneath a glacier is usually impossible. The main processes now generally considered responsible for cirque formation are those of abrasion, joint-block removal, and freeze–thaw action or frost riving. Nivation processes play an important part in the initial stages of cirque formation.

4.1   *Abrasion*

The nature of this process has already been considered in relation to glaciated rock surfaces in Chapter 6. Here, we are specifically concerned with the abraded surfaces of cirque floors and sidewalls, and in this connection, the subglacial rock surface exposed at the inner end of the central tunnel through Vesl-Skautbreen is very instructive. Unlike the rock surfaces which may be readily examined in ice-free cirques, its features obviously owed nothing to subaerial weathering. The surface remained damp and unfrozen throughout the year, suggesting that frost riving was not a relevant process in its erosion, at least at the present. Abrasion and joint-block removal were more plausible possibilities as sculpturing methods. With respect to the first of these, projecting parts of the rock surface were seen to be smoothed (but not polished) and striated. A plumb-bob suspended from the ice above was observed to move exactly in the direction of striation. Short-period shear-meter records indicated an even rate of sliding of the ice over the rock bed. McCall and others who examined the surface considered that a convincing case could be made for regarding abrasion as the chief erosional process operating here today, and noted that as much as 90 per cent of the motion of this glacier was contributed by basal sliding. On the other hand, the sliding velocity is comparatively small at present—only 2·3 m/year at the inner end of the lower tunnel—and it may be that Vesl-Skautbreen is today much less active than it was a few hundred years ago when undoubtedly its bulk, and consequently the forces of rotation, were larger.

Surfaces showing abrasion and striation are most commonly associated with cirque floors and thresholds, but are also to be found, as Lewis noted in 1938, in places on cirque headwalls. It is here that they are much more difficult to explain. W. Dort (1957) noted striated surfaces high on the headwalls of cirques in northern Idaho, and attempted to explain them as being carved in a period when the cirques were over-filled with ice—when even the divides above the cirque headwalls were submerged by moving ice. Support for such ideas has grown. Sugden (1969) contends that many Cairngorm cirques show signs of subsequent over-running by ice: for example, the upper cliffs of the east-facing cirques in Glen Derry appear to have been smoothed and moulded by ice sweeping across the Derry Cairngorm spur. Convincing evidence for ice-sheet overriding of cirques is collected by R. P. Goldthwait (1970) for the Presidential Range in New Hampshire. Here, the steep headwalls of cirques up to 300 m deep possess grooves, some 4 m long and 6 mm deep, plunging straight down them and in some cases obliquely to the cirque axis. But all such striations possess a common regional trend which is, furthermore, consistent with the carriage of erratics by an over-riding ice sheet.

Abrasion by moving ice is thought to be responsible for the basin form of cirque floors. The hypothesis of rotational slip is very relevant to this problem. W. V. Lewis in 1949 suggested that 'when a section of a glacier is moving in a rotational slip, the bedrock beneath must be moulded into a basin fitted to this rotational movement' (p. 156). Gibson and Dyson in 1939 thought that the rotational movements were the effect and not the cause of the rock basin, but Clark and Lewis (1951) were convinced that the tendency to rotate was inherent in certain glaciers and that this might, under favourable circumstances, lead to the scouring-out of basins in the rock floor beneath. The fact that the long profile of the rock floor in the Vesl-Skautbreen cirque is nearly arcuate has already been illustrated (Fig. 7.2), though the information on the ice-covered segment is meagre. Lewis (1960, pp. 97–8) shows how the long profile of the cirque holding Blea Water (Westmorland) also approximates closely to an arc of a circle (Fig. 7.2); on physiographical evidence, he restores the probable form of the ice surface of a small glacier in this cirque. With a surface slope of 13° or more, such a glacier would probably not be in equilibrium, but would tend to rotate about an approximately horizontal axis 600 m above the cirque floor, until its surface slope were reduced to 7 or 8°; 'there would be no difficulty in producing the required force not only to move the bottom debris uphill out of the corrie basin, but also to scour this basin actively' (p. 99). Such concepts of rotating cirque glaciers are undoubtedly oversimplified—more complex forms of ice motion are most certainly involved in larger cirque glaciers and Haynes' analysis of the detailed long-profile shapes of cirques already discussed on page 210 must be borne in mind—but such simple concepts form a useful working hypothesis which must be further tested.

Lewis considered that the relatively high velocities of the basal ice in a simple rotational slip were a great advantage in excavating the rock basin. In the profile c-c measured in a direction perpendicular to the rock floor at c (Fig. 7.6B), for instance, the velocity of the ice is slightly greater next to bedrock than it is at the glacier surface. In addition, the very rapid diminution of the rate of basal ice movement in Vesl-Skautbreen (resulting from the 'damming' effect described on p. 229), from 6·4 mm/day at the inner end of the central tunnel to less than 1 mm/day at the toe of the glacier (stake F), can be related to the reduced amount of corrasion on the cirque threshold (at the toe of the glacier) as compared with the Skautbreen basin floor which lies at least 12 m below its threshold. The rotational slip hypothesis provides a simple mechanism for the uplift of ice over a cirque threshold and for the evacuation of debris uphill from the basin floor.

4.2   *Joint-block removal*

The details of this process and its effects are also presented in a wider context in Chapter 6 and will also be referred to in Chapter 8. It should be recalled that it is not a question of ice simply 'plucking' rock from the bedrock surface, since the yield stress in shear for ice at 0°C is only about 1 kg/cm². There is also the question of the difference between cold and temperate ice. The latter will be normally separated from the rock by a film of basal meltwater and the strength of the attachment of the ice to an irregular rock floor will be zero or, at best, extremely small. Cold ice adhesion to rock is about an order of magnitude greater than the average yield stress of ice in shear; in this case the latter will be the limiting factor. It seems likely, then, that even in favourable circumstances attachment of ice to bedrock can do little more than remove material already loosened by frost-riving, for example, or well-jointed rock. At the inner end of the Vesl-Skautbreen central tunnel, the bedrock surface showed no signs of frost-riving or weathering. It was smoothed and striated in places as already described, but it also possessed surface angularities which led Lewis (1954, p. 419) to describe its appearance as 'freshly fractured when the all-pervading coating of rock flour was removed'. The basal ice of this temperate glacier did not adhere to the rock but could be pulled off easily and cleanly, so that McCall found it difficult to imagine how any further plucking of rock could take place, even though it might have occurred in the past.

Lewis in 1954 was the first to consider the possible role of dilatation joints in joint-block removal by ice (Chapter 8); with J. W. Glen, he argued that 'the effects of glacial quarrying on such [jointed] rocks might be far greater than a simple calculation based on the relative strengths of ice and rock would suggest' (1961, p. 1118). As already mentioned on p. 214, Battey showed the probable existence of dilatation joints roughly parallel to the cliff faces of the Vesl-Skautbreen and Veslgjuv-breen cirques of Jotun-heim, Norway, and considered their rôle a most important one in cirque development. Linton (see p. 218) contends that they play a vital part in the destruction of arêtes and divides, while also assisting in the survival of pyramidal peaks. It is conceivable, therefore, that dilatation joints developing in cirques greatly assist joint-block removal by ice from the headwalls and floor, and that the surface exposed in the Vesl-Skautbreen central tunnel was one formed in the recent past by massive joint-block separation along dilatation joints parallel with that surface, and later abraded and striated by continuing ice-flow over it.

Plate XIII
The cirque containing Veslgjuv-breen, Norway. The difference in altitude between the lake and the summit of Galdehöe is 400 m. Note the winter snow banked against the foot of the rock walls and the pattern of dirt bands in the glacier. (*C.E.*)

Another process that may play a part in preparing the bedrock surface for erosion, particularly block removal, by cirque glaciers, is that of pre-glacial frost action. L. Lliboutry (1953), for instance, has argued that the rock in the cirques is first broken up under periglacial cold conditions. Later, the ice cleans out the debris and smooths the rock floor.

4.3   *Freeze–thaw action*

Repeated freezing and thawing of water in rock crevices has often been suggested as the mechanism behind the shattering of cirque walls since the observations of Lorange on cirques in Norway quoted by A. Helland in 1877. In 1904, W. D. Johnson gave evidence that the most important concentration of freeze–thaw action at the back of a cirque glacier was associated with the bergschrund. He descended an unusually deep bergschrund in a glacier on the north of Mount Lyell in the Sierra Nevada, reaching the base of the cirque headwall at a depth of 45 m. For the last 6–10 m of his descent, he observed that the rock wall of the cirque was exposed, the ice forming the other wall of the bergschrund 2 or 3 m away. 'The rock face, though hard and undecayed, was much riven, its fracture planes outlining sharply angular masses in all stages of displacement and dislodgement. Several blocks were tipped forward and rested against the opposite wall of ice; others, quite removed across the gap, were incorporated in the glacier mass at its base. Icicles of great size, and stalagmitic masses were abundant; the fallen blocks in large part were ice-sheeted; and open seams in the cliff face held films of this clear ice' (p. 574). Meltwater was evidently penetrating any available crevices in the rock wall, and Johnson concluded that the change of air temperatures in the bergschrund across freezing-point, perhaps occurring diurnally in the summer, was the effective agent of rock disintegration.

Although workers such as G. K. Gilbert (1904), A. Penck (1905), and W. H. Hobbs (1910) welcomed this hypothesis of cirque wall shattering and erosion, it was not long before several difficulties became evident. I. Bowman (1920) and R. E. Priestley (1923) contended that bergschrunds were not always present in cirques still occupied by glaciers today and that many bergschrunds failed to reach bedrock. R. von Klebelsberg (1919) observed that meltwater was absent from certain bergschrunds penetrated by military tunnelling operations in the eastern Alps. Priestley and others found it difficult to apply the hypothesis, with its emphasis on activity within a bergschrund no more than 30–60 m deep, to cirques with headwalls many hundreds of metres in height. T. C. and R. T. Chamberlin (1911) thought that diurnal temperature changes might be more significant. But once an enclosed bergschrund is occupied by relatively dense cold air, how can warmer air ever displace it?

In 1938 and 1940, Lewis put forward certain modifications of W. D. Johnson's original bergschrund hypothesis, notably in suggesting that much of the meltwater appearing in the bergschrund was of external origin, from either summer rain or the melting of winter snows high above the bergschrund, and that such meltwater could seep down narrow cracks far below the limits of the deepest bergschrunds. Moreover, he rightly observed that the position of the bergschrund on a cirque headwall would vary slightly each season and to an even greater extent during the waxing and waning of a glacial period when the ice surface would rise and fall with changes in the glacier's budget. The observations of Lewis and others on cirque glaciers in humid maritime climates such as those of Iceland or Norway have shown that in normal years meltwater is plentiful. Indeed, the supply of meltwater is no problem on any temperate glacier in the summer season. In the case of polar glaciers, meltwater is more difficult to provide; Priestley (1923) and others have recorded melting taking place on rock outcrops in Antarctica when air temperatures were well below freezing-point, but the amount of melting can only be very slight in such conditions (see also Chapter 3).

A most serious problem encountered by the bergschrund hypothesis in its application to temperate glaciers is that of the freezing of the meltwater, and the removal of latent heat (80 cal/g at 0°C) in the process. Prior to 1950, it had too readily been assumed that temperatures in bergschrunds would fluctuate considerably above and especially below freezing-point, aiding freeze–thaw action. McCabe (1939) was one of the few who had made actual measurements of temperatures in bergschrunds; he recorded (in August) 0°C at a depth of 7·6 m in one case and 0·5°C at 1·8 m in another, both in Spitsbergen. These figures were not unexpected, for H. U. Sverdrup (1935) had shown that the temperature at depths of more than 10 m in the firn of Isachsen's Plateau, Spitsbergen, was 0°C even in winter. As noted in Chapter 3, this is typical of temperate glaciers which consist of ice whose temperatures at any depth are at pressure melting-point for that depth, apart from a surface layer of cold ice, up to 20 m or so in thickness, which is cooled by the atmosphere well below freezing-point in winter. There is no way by which meltwater can be refrozen simply by contact with ice at pressure melting-point; and the deeper one goes in a bergschrund, the less possibility will there be of any air of sub-zero temperatures penetrating from the outside atmosphere.

An ingenious attempt to overcome these problems was made by C. D. Holmes in 1944. His hypothesis did not depend on either meltwater or cold air penetrating deep behind a cirque glacier, but on local thawing and freezing at the base of isothermal ice induced by variations in ice pressure, for the pressure exerted by 30 m of ice, for instance, will lower the pressure melting-point fractionally (by 0·0192°C). Cirque glaciers vary seasonally in thickness and thus small changes of temperature will occur at the contact between isothermal ice and rock. A thin film of meltwater produced during winter when the snow–ice mass is thickest would penetrate any available rock crevices and refreeze when pressure was later reduced. The process, Holmes argues, need only be very slow—0·5 m of rock removed from the cirque floor every 200 years would lead to a cirque 1000 m deep during the probable period of Pleistocene ice activity. But are such slow and minute temperature changes at all effective in rock shattering? Evidence brought forward by W. R. B. Battle (1951, 1960) suggests not (see also H. R. Thompson and B. H. Bonnlander, 1956).

In an attempt to provide more data on conditions in bergschrunds, W. R. B. Battle carried out a number of temperature measurements in various bergschrunds in Greenland, Norway and Switzerland between 1948 and 1953 (Fig. 7.7, A and B). Using thermographs and freeze-recording units, he studied particularly the bergschrunds at the head of Tverråbreen (Jotunheim), at the Jungfraujoch, and at the head of Glacier 32, Pangnirtung Pass, Baffin Island. In the Tverråbreen bergschrund, air temperatures

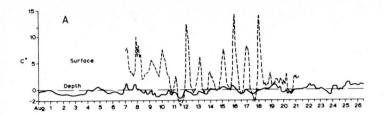

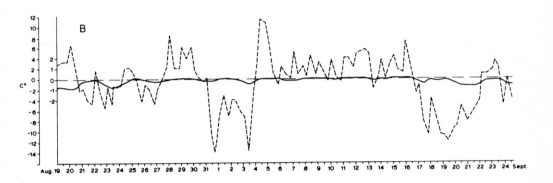

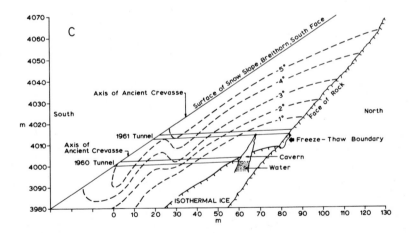

Fig. 7.7
A Temperature record from the Tverråbreen bergschrund, August 1951 (W. R. B. Battle, 'Norwegian Cirque Glaciers', Royal Geographical Society, *Research Memoir*, 4, 1960);
B Temperature record from the Jungfraujoch bergschrund, August-September 1951 (W. R. B. Battle, *op. cit.*);
C North-south cross-section of the glacier on the south face of the Breithorn, showing isotherms (°C), tunnels and water-filled cavern (J. E. Fisher, *J. Glaciol.*, 1963, by permission of the International Glaciological Society)

fluctuated only from 0°C to −1°C in August 1950, and from 0°C to −2°C in August 1951, even though this was a fairly open bergschrund. There was no sign of any daily temperature cycle. At the Jungfraujoch, the fluctuations were −2°C to −5°C in March 1951, and 0 to −1°C in August and September 1951. The rate of temperature change was in all cases very slow, apart from one drop of 1·2°C in 6 hours (Jungfraujoch, March 1951). Nor did the small fluctuations of air temperature inside the bergschrunds always accord with the much greater changes of external air temperatures: on 19 March 1951, at the Jungfraujoch, external air temperatures fell by 5°C at a time when bergschrund air temperatures rose by nearly 2°C. In the bergschrund of Glacier 32, air temperatures at a depth of 27 m ranged from +0·5°C to −3·7°C. The positive temperature recorded here was unusual but of short duration.

The freeze-recorders confirmed the thermograph evidence though they did not operate very satisfactorily; their failure to respond to changes of temperature which frequently only amounted to 1°C suggested that such small changes might not be very effective in rupturing rocks. Moreover, it became clear that under certain conditions water might be supercooled to −1°C without freezing. Battle conducted laboratory experiments on the freezing of water-saturated porous stones, which showed that only when temperatures fell below −5°C did marked deterioration set in. On the other hand, smaller falls of temperature sufficed to affect cracked non-porous rocks, provided the drop in temperature occurred at a rate of at least 0·1°C per minute. This allowed freezing of water to occur first at the top of any crack, producing a closed system in which pressure could build up (for a more detailed discussion, see Embleton and King, 1975, Chapter 1). But changes of 0·1°C per minute had never been recorded in bergschrunds—the maximum was probably not much more than 0·004°C per minute. Such experiments also serve to dispose of C. D. Holmes's hypothesis of subglacial erosion in which the postulated temperature fluctuations are fractional and probably very slow.

Reviewing Battle's work, it is clear that deep or closed bergschrunds belonging to temperate glaciers are not favourable environments for freeze–thaw action, and yet it is only in the deeper parts of true bergschrunds that the rock wall is exposed. But the deeper one penetrates a bergschrund behind a temperate glacier, the more closely will the temperatures approximate that of the isothermal ice. Only in the upper parts of such bergschrunds can refreezing of any incoming meltwater take place, since here the air temperature may be several degrees below zero. Within the bergschrund, there are no large-scale or rapid fluctuations of the air temperature which might favour alternate freeze–thaw action; on the contrary, in the bergschrunds examined, meltwater from outside appeared to freeze more or less permanently on to the rock wall wherever this was exposed, covering it with layers of ice and thereby effectively protecting the rock from all possibility of attack by frost-riving. This ice coating of the rock wall was noted by Battle in 1948 behind the Grif Gletscher in Greenland, in the Tverråbreen and Jungfraujoch bergschrunds 1950–51, and in Baffin Island in 1953. These facts led him eventually to abandon the bergschrund hypothesis in the form developed by Lewis or Johnson; by 1953, he seems to have adopted the view that freeze–thaw action was much more likely to be encountered in the upper parts of open bergschrunds, or generally in the zone where rock headwall, glacier surface and freely circulating air meet.

Investigations by McCall and others in the headwall gap or randkluft behind Vesl-Skautbreen certainly supported this view. Plenty of rain and meltwater penetrated this impressively large gap between the firn and the rockwall, and in winter would undoubtedly refreeze since air circulation was rapid. Even in summer, the air temperature was fractionally below freezing ($-0.6°C$) immediately after the tunnel into it through the firn was excavated. Rock fragments clearly prised off the headwall by freeze–thaw action were falling down the gap to provide debris for the basal sole of the glacier. But could not such action be equally effective on the exposed headwall well above the glacier or firn surface, provided meltwater was available from higher snow patches to wet the rock surface occasionally? In this way, a glacier consisting mainly of isothermal ice would play a protective role in regard to freeze–thaw action, for this can then only take place at or above the ice surface. But the glacier also has the very important function of a transporting agent, removing frost-shattered material and preventing the accumulation of protective talus against the rock wall. And as Battle observed (Thomson and Bonnlander, 1956, p. 768), as the glacier has thinned or thickened, so more or less of the headwall would be exposed to destructive action.

Some observations made in cavities and tunnels below temperate valley glaciers are relevant to this discussion. There is no doubt that in certain subglacial cavities, alternate melting and refreezing does occur. R. Vivian and G. Bocquet (1973), for instance, reported that in the network of subglacial cavities and interconnecting passages on the downstream side of the Lognan rock bar, 100 m below the surface of the Glacier d'Argentière, air currents indicated a more or less open system, in which small changes of temperature (amounting to some tenths of a degree C) appeared to be related to external changes. Subglacial streams rushing through the passages aided the air circulation. Refrozen meltwater observed probably related to winter freezing mainly; in summer, warm air and water dominated. In tunnels beneath the margin of Østerdalsisen, Norway, R. G. Bennett (1968) also noted penetration of meltwater in summer and autumn at times when temperatures in the tunnels were still sub-zero; and freshly broken rock fragments were observed adhering to the under-surface of the glacier in winter.

It seems, then, that the possibility of freeze–thaw action in subglacial locations should not be entirely dismissed, either in the case of temperate valley-glaciers or temperate cirque-glaciers. But clear evidence of it has proved remarkably difficult to establish. In the case of cold glaciers, a new lead was provided by J. E. Fisher's investigations (1955, 1963) on some high-level ice and firn areas in the Alps. Fisher claimed that 'it is to these areas of cold ice, bergschrunds or no bergschrunds, that one must look ... to understand processes of cirque erosion' (1963, p. 514). In 1954, he organized the excavation of a tunnel through cold ice on the north slope of Monte Rosa, at 4240 m. When the rock face was reached after 92 m, the temperature was $-13°C$; sub-zero temperatures probably characterize the ice-rock interface here down to an elevation of 3800 m, below which isothermal (temperate) ice may be expected. The actual junction between cold ice and isothermal ice at the rock face, marked by the 0°C isotherm, was almost encountered by a later tunnel (completed 1961) on the Breithorn, at a little over 4000 m (Fig. 7.7C). In neither case was there any gap (bergschrund or randkluft) between ice and rock. The only meltwater encountered was in a tunnel driven in 1960 on the Breithorn (a little below the level of the 1961 tunnel) which, because it struck a large water-filled cavern in the ice, never reached the rock face. The origin of the very considerable quantity of water here was a mystery. It was inconceivable that it could represent meltwater seeping down from above, for it was present in too large a quantity, would have had to penetrate cold firn (temperatures down to −6°C) in which no cracks were visible, and moreover the quantity of water in the cavern actually increased in winter. Fisher suggested that it represented water forced up by hydrostatic pressure from lower levels where it was derived from melting of isothermal ice. It may be suggested that if, in other situations, meltwater were similarly forced up so as to reach the rock wall behind cold ice, refreezing could occur and rock disintegration would thus be induced at that level, but this remains a speculation, and the circumstances on the Breithorn may have been exceptional. Fisher preferred to associate rock wall shattering with the 'boundary between cold ice, frozen to the bedrock, and wet isothermal ice. This boundary is invited by external conditions to oscillate slowly back and forth, with powerful shattering and removal of rock as a result' (1963, p. 518). Such oscillations, he thought, would be measured in terms of perhaps hundreds of years, the result of minor climatic fluctuations, a very different time-scale from the daily fluctuations postulated by W. D. Johnson and sought for by Battle. Fisher further suggests that this mechanism, the concentration of freeze–thaw action at the 0°C boundary in certain cold glaciers, may provide an explanation for the approximate uniformity in level of cirque floors in any given area, for the location of the 0°C horizon will be related to the prevailing mean atmospheric temperatures.

Thus cirque wall shattering, according to Fisher, will be associated with those types of cirque glaciers possessing cold ice overlying temperate ice. Those built entirely of cold ice or entirely of temperate ice will not normally provide favourable environments for freeze–thaw action at depth. In the varied climatic conditions of the Pleistocene, most cirque glaciers probably passed through phases when cold ice overlay isothermal ice.

5 Conclusion

The glacial erosion origin of 'glacial' cirques is firmly accepted today. The principal morphological features of a well-developed cirque include steep and shattered headwalls, a rock threshold sometimes capped with moraine, a rock basin of circular outline enclosed by the headwalls and the threshold, and a ratio of height to length of about 1 to 3. Many cirques show a relationship with rock structure, but some do not. The majority have developed by glacial action out of previously water-eroded features or from nivation hollows. Enlargement of cirques leads to the development of arêtes and horns, and in the later stages of this process, the greater part of original mountain masses may be destroyed to leave only residual pyramidal peaks rising above an irregular cirque-floor surface. Cirque elevation and orientation are closely controlled by meteorological factors.

Headwall shattering by freeze–thaw action is one of the chief factors in cirque development, perhaps even the most important one. The association of this process with the bergschrund, however, is not entirely justified; it is apparent now that the most favourable environment for freeze–thaw action may well be the exposed rock wall above the glacier or firn limits, or in the upper parts of any headwall gap or randkluft between the glacier and the rock wall. Alternatively, the process may be associated, as Fisher believes, with particular climatic conditions when the boundary between cold ice and temperate ice in a glacier migrates up or down part of the headwall, sapping it intensively within this zone. This might tend to undercut the headwall and thereby cause its retreat, a possibility noted by McCall at Vesl-Skautbreen. Headwall sapping by freeze–thaw action will serve to enlarge the cirque overall, whereas abrasion by rotationally moving ice will tend mainly to deepen it. The relative potency of the two is difficult to judge; Flint (1971) thinks it generally probable that the walls retreat far more rapidly than the floors are deepened and McCall (1960) concluded that the Vesl-Skautbreen cirque had been enlarged primarily in a horizontal or headward manner by sapping. Hence, cirque-floor elevations may not change radically during glacierization. Nevertheless, the rock basin form of cirque floors does indicate at least a minimum of deepening by abrasion, and the rotational slip hypothesis seems best adapted to explain this feature. The third process, joint-block removal, is one which may operate over the whole surface area of a cirque, but it will be particularly relevant to the erosion of the headwall where the existence of well-developed joints will be of the greatest help to freeze–thaw action. Finally, it seems unlikely that cirques are formed or enlarged under conditions of glacierization by cold ice, though, once formed, cirques may be occupied by cold glaciers as in parts of present-day Antarctica.

6 References

AHLMANN, H. W. (1919), 'Geomorphological studies in Norway', *Geogr. Annlr* **1**, 1–148 and 193–252

ANDREWS, J. T. (1965), 'The corries of the northern Nain-Okak section of Labrador', *Geogr. Bull.* **7**, 129–36

ANDREWS, J. T. and DUGDALE, R. E. (1971), 'Quaternary history of northern Cumberland Peninsula, Baffin Island, N.W.T.; Part V: Factors affecting corrie glacierization in Okoa Bay', *Quaternary Res.* **1**, 532–51

BATTEY, M. H. (1960), 'Geological factors in the development of Veslgjuv-botn and Vesl-Skautbotn', in 'Norwegian cirque glaciers' (ed. LEWIS, W. V.), *R. geogr. Soc. Res. Ser.* **4**, 5–10

BATTLE, W. R. B. (1960), 'Temperature observations in bergschrunds and their relationship to frost shattering', in 'Norwegian cirque glaciers' (ed. LEWIS, W. V.), *R. geogr. Soc. Res. Ser.* **4**, 83–95

BATTLE, W. R. B. and LEWIS, W. V. (1951), 'Temperature observations in bergschrunds and their relationship to cirque erosion', *J. Geol.* **59**, 537–45

BENNETT, R. G. (1968), 'Frost shatter and glacial erosion under the margins of Østerdalsisen, Svartisen', *Norsk geogr. Tidsskr.* **22**, 209–13

BONNEY, T. G. (1871), 'On a cirque in the syenite hills of Skye', *Geol. Mag.* **8**, 535–540

(1873), 'Lakes of the north-eastern Alps and their bearing on the glacier erosion theory', *Q. J. geol. Soc. Lond.* **29**, 382–95

(1877), 'On Mr Helland's theory of the formation of cirques', *Geol. Mag.* **14**, 273–7

BOWMAN, I. (1920), *The Andes of Southern Peru* (*Am. Geogr. Soc. Spec. Publ.* **1**), 295

CHAMBERLIN, T. C. and R. T. (1911), 'Certain phases of glacial erosion', *J. Geol.* **19**, 193–216

CHARPENTIER, J. DE (1823), *Essai sur la constitution géognostique des Pyrénées* (Paris)

CLAPPERTON, C. M. (1970), 'The evidence for a Cheviot ice cap', *Trans. Inst. Br. Geogr.* **50**, 115–27

(1971), 'Evidence of cirque glaciation in the Falkland Islands', *J. Glaciol.* **10**, 121–5

CLARK, J. M. and LEWIS, W. V. (1951), 'Rotational movement in cirque and valley glaciers', *J. Geol.* **59**, 546–66

COMMON, R. (1965), 'Slope failure as a factor in rates of erosion', *Rep. Br. geomorph. Res. Grp, Symposium at Bristol, 1965,* 3–5

COTTON, C. A. (1942), *Climatic accidents in landscape-making* (Christchurch, N.Z.)

DAHL, E. (1946), 'On the origin of the Strand Flat', *Norsk geogr. Tidsskr.* **11**, 159–71

DAVIS, W. M. (1900), 'Glacial erosion in France, Switzerland, and Norway', *Proc. Boston Soc. nat. Hist.* **29**, 273–322

(1909), 'Glacial erosion in North Wales', *Q.J. geol. Soc. Lond.* **65**, 281–350

DERBYSHIRE, E. (1964), 'Cirques, Australian landform example no. 2', *Aust. Geogr.* **9**, 178–9

DORT, W. (1957), 'Striated surfaces on the upper parts of cirque headwalls', *J. Geol.* **65**, 536–42

EMBLETON, C. and KING, C. A. M. (1975), *Periglacial geomorphology*

ENQUIST, F. (1917), 'Der Einfluss des Windes auf die Verteilung der Gletscher', *Bull. geol. Instn Univ. Upsala* **14**, 1–108

EVANS, I. S. (1969), 'The geomorphology and morphometry of glacial and nival areas' in *Water, Earth and Man* (ed. CHORLEY, R. J.), 369–80

EVANS, J. W. (1913), 'The wearing down of rocks', *Proc. Geol. Ass.* **24**, 241–300

FELS, E. (1929), 'Das Problem der Karbildung', *Petermanns Mitt., Ergänz.* **202**, 85 pp.

FISHER, J. E. (1953), 'Two tunnels in cold ice at 4000 m on the Breithorn', *J. Glaciol.* **2**, 513–20

(1955), 'Internal temperatures of a cold glacier and conclusions therefrom', *J. Glaciol.* **2**, 583–91

FLINT, R. F. (1971), *Glacial and Quaternary geology*

FORBES, J. D. (1843), *Travels through the Alps of Savoy* (Edinburgh)

FREEMAN, O. W. (1925), 'The origin of Swimming Woman Canyon, Big Snowy Mountains, Montana, an example of a pseudo-cirque formed by landslide slipping', *J. Geol.* **33**, 75–9

GASTALDI, B. (1873), 'On the effects of glacier erosion in Alpine valleys', *Q.J. geol. Soc. Lond.* **29**, 396–401

GIBSON, G. R. and DYSON, J. L. (1939), 'Grinnell Glacier, Glacier National Park, Montana', *Bull. geol. Soc. Am.* **50**, 681–96

GILBERT, G. K. (1904), 'Systematic asymmetry of crest lines in the High Sierra of California', *J. Geol.* **12**, 579–88

GLEN, J. W. and LEWIS, W. V. (1961), 'Measurements of side-slip at Austerdalsbreen, 1959', *J. Glaciol.* **3**, 1121

GLOYNE, R. W. (1964), 'Some characteristics of the natural wind and the modification by natural and artificial obstructions', *Scient. Hortic.* **17**, 7–19

GOLDTHWAIT, R. P. (1970), 'Mountain glaciers of the Presidential Range in New Hampshire', *Arct. alp. Res.* **2**, 85–102

GREGORY, J. W. (1915), 'The geology of the Glasgow district', *Proc. Geol. Ass.* **26**, 162

GROVE, J. M. (1960), 'The bands and layers of Vesl-Skautbreen', in 'Norwegian cirque glaciers' (ed. LEWIS, W. V.), *R. geogr. Soc. Res. Ser.* **4**, 11–23; 'A study of Veslgjuv-breen', ibid., 69–82

HARKER, A. (1901), 'Ice erosion in the Cuillin Hills, Skye', *Trans. R. Soc. Edinb.* **40**, 221–52

(1904), 'Tertiary igneous rocks of Skye', *Mem. geol. Surv. Gt Br.*

HAYNES, V. M. (1968), 'The influence of glacial erosion and rock structure on corries in Scotland', *Geogr. Annlr* **50**A, 221–34

HELLAND, A. (1877), 'On the ice-fjords of north Greenland, and on the formation of fjords, lakes, and cirques in Norway and Greenland', *Q.J. geol. Soc. Lond.* **33**, 142–76

HOBBS, W. H. (1910), 'The cycle of mountain glaciation', *Geogrl J.* **35**, 146–63 and 268–84

(1911), *Characteristics of existing glaciers*

(1926), *Earth features and their meaning*

HOLMES, C. D. (1944), 'Hypothesis of subglacial erosion', *J. Geol.* **52**, 184–90

JOHNSON, W. D. (1904), 'The profile of maturity in Alpine glacial erosion', *J. Geol.* **12**, 569–78

JUKES BROWN, A. J. (1877), 'The origin of cirques', *Geol. Mag.* **4**, 477–9

KAMB, B. and LACHAPELLE, E. (1964), 'Direct observation of the mechanism of glacier sliding over bedrock', *J. Glaciol.* **5**, 159–72

KING, C. A. M. and GAGE, M. (1961), 'Note on the extent of glaciation in part of west Kerry', *Ir. Geogr.* **4**, 202–8

KLEBELSBERG, R. VON (1919), 'Glazialgeologische Erfahrungen aus Gletscherstollen', *Z. Gletscherk.* **11**, 156–84

LEWIS, W. V. (1938), 'A melt-water hypothesis of cirque formation', *Geol. Mag.* **75**, 249–65

(1939), 'Snow patch erosion in Iceland', *Geogrl J.* **94**, 153–61

(1940), 'The function of meltwater in cirque formation', *Geogrl Rev.* **30**, 64–83

(1947), 'Valley steps and glacial valley erosion', *Trans. Inst. Br. Geogr.* **14**, 19–44

(1949), 'Glacial movement by rotational slipping', *Geogr. Annlr* **31**, 146–58

(1954), 'Pressure release and glacial erosion', *J. Glaciol.* **2**, 417–22

(1960), 'Norwegian cirque glaciers', *R. geogr. Soc. Res. Ser.* **4**, 104 pp.

LINTON, D. L. (1959), 'Morphological contrasts between eastern and western Scotland' in *Geographical essays in memory of Alan G. Ogilvie* (ed. MILLER, R and WATSON, J. W., Edinburgh), 16–45

(1963a), 'The forms of glacial erosion', *Trans. Inst. Br. Geogr.* **33**, 1–28

(1963b), 'Some contrasts in landscapes in British Antarctic Territory', *Geogrl J.* **129**, 274–82

LINTON, D. L. (1967), 'Divide elimination by glacial erosion' in *Arctic and alpine environments* (ed. WRIGHT, H. E. and OSBURN, W. H.), 241–8

LLIBOUTRY, L. (1953), 'Internal moraines and rock glaciers', *J. Glaciol.* **2**, 296

MANLEY, G. (1959), 'The late-glacial climate of North-west England', *Lpool Manchr geol. J.* **2**, 188–215

MARTONNE, E. DE (1910–11), 'L'érosion glaciaire et la formation des vallées Alpines', *Annls Géogr.* **19**, 289–317; **20**, 1–29

MATTHES, F. E. (1900), 'Glacial sculpture of the Bighorn Mountains, Wyoming', *U.S. geol. Surv. 21st A. Rep. (1899–1900)*, 167–90

  (1930), 'Geologic history of the Yosemite Valley', *U.S. geol. Surv. Prof. Pap.* **160**, 137 pp.

MCCABE, L. H. (1939), 'Nivation and corrie erosion in West Spitsbergen', *Geogrl J.* **94**, 447–65

MCCALL, J. G. (1960), 'The flow characteristics of a cirque glacier and their effect on glacial structure and cirque formation', in 'Norwegian cirque glaciers' (ed. LEWIS, W. V.), *R. geogr. Soc. Res. Ser.* **4**, 39–62

MILLER, M. M. (1961), 'A distribution study of abandoned cirques in the Alaska-Canada Boundary Range', *Proc. 1st int. Symp. Arctic Geol. (1960)*, **2**, 833–47

PENCK, A. (1905), 'Glacial features in the surface of the Alps', *J. Geol.* **13**, 1–19

PRIESTLEY, R. E. (1923), *Physiography (Robertson Bay and Terra Nova Regions)*, *Rep. Br. Antarct. Exped. (1910–13)*

RAMSAY, A. C. (1860), *The old glaciers of Switzerland and North Wales*

RICHTER, E. (1906), 'Geomorphologische Beobachtungen aus Norwegen', *Sber. wien. Akad. Math. Naturw.* **105**, 1

ROBINSON, G., PETERSON, J. A. and ANDERSON, P. M. (1971), 'Trend-surface analysis of corrie altitudes in Scotland', *Scott. geogr. Mag.* **87**, 142–6

SCHUMM, S. A. and LICHTY, R. W. (1965), 'Time, space and causality in geomorphology', *Am. J. Sci.* **263**, 110–19

SCHYTT, V. (1959), 'The glaciers of the Kebnekajse massif', *Geogr. Annlr* **41**, 213–27

SEDDON, B. (1957), 'Late-glacial cwm glaciers in Wales', *J. Glaciol.* **3**, 94–9

SISSONS, J. B. (1967), *The evolution of Scotland's scenery*

STRØM, K. M. (1945), 'Geomorphology of the Rondane area', *Norsk geol. Tidsskr.* **25**, 360–78

  (1949), 'The geomorphology of Norway', *Geogrl J.* **112**, 19–27

SUGDEN, D. E. (1969), 'The age and form of corries in the Cairngorms', *Scott. geogr. Mag.* **85**, 34–46

SVERDRUP, H. U. (1935), 'Scientific results of the Norwegian-Swedish Spitsbergen Expedition in 1934; part III', *Geogr. Annlr* **17**, 53–88

TAYLOR, G. (1914), 'Physiography and glacial geology of east Antarctica', *Geogrl J.* **44**, 365–82, 452–67, and 553–71

  (1926), 'Glaciation in the south-west Pacific', *Proc. 3rd Pan-Pacific Congr., Tokyo, 1924*

TEMPLE, P. H. (1965), 'Some aspects of cirque distribution in the west-central Lake District, northern England', *Geogr. Annlr* **47**, 185–93

THOMPSON, H. R. (1950), 'Some corries of north-west Sutherland', *Proc. Geol. Ass.* **61**, 145–55

THOMPSON, H. R. and BONNLANDER, B. H. (1956), 'Temperature measurements at a cirque bergschrund in Baffin Island: some results of W. R. B. BATTLE's work in 1953', *J. Glaciol.* **2**, 762–9

UNWIN, D. J. (1973), 'The distribution and orientation of corries in northern Snowdonia, Wales', *Trans. Inst. Br. Geogr.* **58**, 85–97

VIVIAN, R. and BOCQUET, G. (1973), 'Sub-glacial cavitation phenomena under the Glacier d'Argentière, Mont Blanc, France', *J. Glaciol.* **12**, 439–51

WALDROP, H. A. (1964), 'Arapaho glacier, a sixty-year record', *Univ. Colo. Stud. Ser. Geol.* **3**, 1–37

WEERTMAN, J. (1971), 'Shear stress at the base of a rigidly rotating cirque glacier', *J. Glaciol.* **10**, 31–7

8

Glaciated valleys

The length of time during which these glaciers have existed must have been so great, and there is so much evidence of their dimensions having been formerly greater than now, that a considerable portion of the erosion, or of the widening and deepening of the valley must be attributed to ice-action. But to what extent the valleys in which the Swiss glaciers move were excavated by rivers before the valleys were filled with ice we have no positive data at present for deciding. (SIR CHARLES LYELL, 1866)

The features characteristic of glaciated valleys are among the most easily recognized and the most spectacular results of glaciation. The method of formation of these features is, however, not nearly so obvious, and doubt still remains concerning their formation in many instances. The great glacial troughs, such as those of the Alps and Norwegian mountains, carved out to depths of hundreds or even thousands of metres, demonstrate clearly the immense erosive activity of which glaciers are capable under certain circumstances and which was recognized by J. Hutton in 1795 when he wrote: '...in the Alps of Switzerland and Savoy, there is another system of valleys, above that of the rivers, and connected with it. These are valleys of moving ice, instead of water. This icy valley is also found branching from a greater to a lesser, until at last it ends upon the summit of a mountain covered continually with snow. The motion of things in those icy valleys is commonly exceedingly slow, the operation however of protruding bodies, as well as that of fracture and attrition, is extremely powerful' (Vol. 2, p. 296).

In the Pleistocene, as today, glacial erosion was most intense when thick, well-nourished temperate glaciers were flowing rapidly down steep slopes towards a free outlet. These conditions were realized on the windward slopes of highland regions which received heavy snowfall and which sloped steeply down towards the sea. This situation was particularly characteristic of western Norway, north-west Britain, British Columbia, southern Chile and south-west New Zealand. Under these conditions, fast-flowing glaciers were generated, usually following valleys prepared for them by pre-glacial rivers and enlarging them into spectacular troughs.

The glaciers that flowed through these valleys originated in various ways. Some of them extended out from cirques as the snow-line fell with increasing cold; others origi-

Plate XIV
Valley at Turtagrø, Norway, showing glacial trough carved below a higher level bench, itself glacially scoured. The difference in level between the river and the highest peaks (Skagastölstind) is about 1500 m. *(C.E.)*

nated on upland plateau areas, forming the outlet glaciers of mountain ice caps. Vatna-jökull in Iceland and Jostedalsbreen in Norway provide examples of this type of ice mass from which many valley glaciers descended and still descend to lower levels. The present outlet glaciers are only very small remnants of their much larger Pleistocene predecessors, whose effects are now clearly revealed in the parts of the valleys from which the glaciers have retreated.

Glaciated valleys develop characteristic transverse and longitudinal profiles. Breached watersheds are another feature of interest associated with the action of large valley glaciers. In many areas, glacial activity has not been sufficiently prolonged for the landforms to develop beyond a youthful or early mature stage; in others, however, there is evidence of the more prolonged effects of valley glaciers on the landscape. Details of the processes operating to produce these forms will be considered.

1 The cross-profiles of glaciated valleys

One of the most striking characteristics of a well-developed glaciated valley is the U-shape of its cross-profile. The side walls are considerably over-steepened, approaching the vertical in places, as for example in the Austerdalsbreen valley in Norway (Plate XVI, p. 250) and the Lauterbrunnen valley in Switzerland. The valley floor, on the other hand, is flatter than in river valleys, a feature which is partly or wholly the result of deposition of valley train deposits. The extreme flatness of this type of fluvioglacial deposition is discussed in Chapter 18.

Because of the steepness of glacial trough walls and their frequently craggy nature, there have been few attempts at surveying accurate cross-profiles, and reliance on carto-graphic data as a substitute, or even photogrammetric measurements, is far from satis-factory. Published height data and map contours may be sparse and vary widely in reliability, and are too often of a low order of precision. Nor is it possible, using profiles constructed from maps, to know whether the profile is on bedrock or superficial de-posits—a crucial point in interpretation.

From work in the Lapporten valley, Sweden, H. Svensson (1959) claimed that the typical cross-section form was parabolic. The best-fit equation relating the height of a point above the valley floor (y) to its horizontal distance from the valley centre (x) was of the form $y = ax^b$. Values of the exponent b (2·045, 2·177) were close to 2 (which would define a perfect parabola), but the claim is based on too few observations. W. L. Graf (1970) argues that knowledge of b on its own, moreover, is not sufficient, and that a more complete description of the form of the valley is obtained if valley depth (D) and valley top width (W_t) are also taken into account, defining a form ratio, $F = D/W_t$. Using data from the Beartooth Mountains (Montana–Wyoming) based on 1:62,500 photogrammetric maps with 80-foot (24·3 m) contour intervals, the following values were computed:

b	1·463 to 1·839
a	0·0024 to 0·0134
F	0·242 to 0·445

F values increase approximately in proportion to b: deeper narrower valleys more closely approach a true parabola. The results confirm a tendency for cross-profiles to approach a parabolic form, but tend to veer from a normal parabola $(b=2)$ towards a semi-cubic parabola $(b=1·5)$. D. J. Drewry (1972) finds values of $b=1·629$ and $b=2·273$ for two sides of a subglacial valley in Antarctica.

The formation of the glacial parabolic or U-shaped cross-profile could be achieved by the widening of the valley from a pre-glacial V-shaped form without overdeepening. Alternatively the widening could be accompanied by overdeepening, and when the longitudinal profile is discussed, it will be apparent that the latter can often occur. The deepening of the cross-profile depends on the ability of the glacier to cut into the rock of its bed.

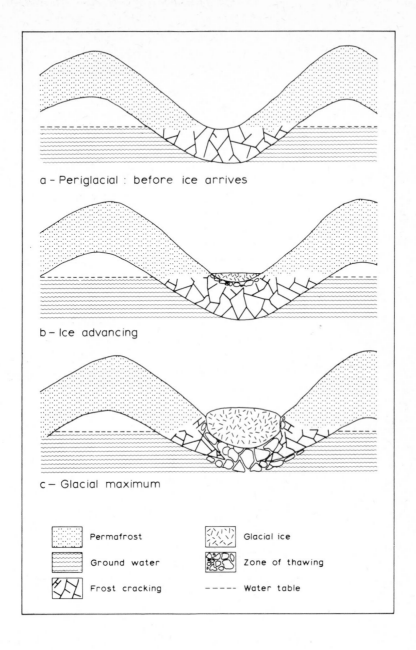

a – Periglacial : before ice arrives

b – Ice advancing

c – Glacial maximum

Permafrost		Glacial ice	
Ground water		Zone of thawing	
Frost cracking		Water table	

Fig. 8.1
Diagrams to show the preparation of the valley floor by permafrost and the effects of glacial erosion
(J. Tricart and A. Cailleux, *Le Modelé glaciaire et nival*, III, Société d'Edition d'Enseignement
Supérieur, Paris, 1962)

An interesting suggestion has been made by M. Boyé (1950, 1968) (see Chapter 6) to account for this aspect of glacial erosion. He considered that the preparation of the valley floor ('défonçage périglaciaire', 'défonçage par la gélivation') into which the advancing ice is penetrating is important. Permafrost could prepare the ground for effective glacial erosion, by breaking up the rocks in zones where there is plenty of moisture. Slope water would tend to drain to the bottom of the valleys, increasing the water content in this area. It may be suggested, therefore, that the valley bottoms are subjected to particularly intensive frost action and the rocks in this situation are broken up to a greater extent than those on the drier valley sides. The advancing glaciers take advantage of this; if of temperate type, thawing of the frost-broken rock will occur under the glacier, and a greater proportion of material will be removed from the valley bottom than from the sides. The ice thickness and hence the stress of the ice on the bottom is greatest in the valley floor, and this also helps to concentrate glacial erosion in this zone (Fig. 8.1).

Another possible mechanism is the action of pressure release beneath thick ice. W. V. Lewis (1954), D. L. Linton (1963) and others have pointed out that, as a glacier bites more deeply into its bed, rock of density 2·5 or more is replaced by ice possessing a much lower density (0·9). This reduction of density on the floor of the glacier might lead to the development of horizontal dilatation joints, especially if there are pre-existing stresses in the rock. Breaking-up of the rocks in this way would enable the glacier to carry away further material from its bed and increase the depth of the valley. Where glacial erosion has produced a very deep valley, such as that of the Lauterbrunnen or in parts of the Sognefjord system, it seems likely that much of the overdeepening must have been achieved by glacial plucking of previously loosened joint blocks. As was noted in Chapter 6, scouring of the bed would not be able to move such a large volume of material, although this process has no doubt also been active. The preparation of the rock bed to a state which enables the rock to be readily removed by the moving ice is therefore a process worthy of close examination, and one which, if effective, would help to account for the deep U-shape of glacial valleys.

Confirmation of the pressure release process has come from quarrying and mining operations, in which rock is known to expand following the removal of overburden pressure (G. W. Bain, 1931; R. H. Jahns, 1943). The dilatation joints normally develop parallel to the surface from which the load is removed, and thus on valley floors or flat surfaces they tend to be horizontal. There are fine examples of horizontal dilatation jointing in the Rapakivi granite of the Åland Islands in the Baltic, an area which at times was subjected to the pressure of nearly 3000 m of ice. Rapid thinning of the ice allowed sudden reduction of pressure on the rock, which in turn may have caused the development of the dilatation jointing. There are also many examples of dilatation jointing developing parallel with the sidewalls of glaciated valleys and also parallel with the walls of cirques as noted in the previous chapter. It is possible that the well-known sheet structures in the granite of Yosemite valley in California (F. E. Matthes, 1930) result from rapid excavation of the valley, first by river action and later by ice. As the sheeting develops parallel to the surface being eroded, it helps to perpetuate the same cross-profile during valley enlargement, so that glaciated valleys often show very similar cross-profiles regardless of scale. Just as in the case of cirques, the

Plate XV
Glen Sannox, Arran, Scotland, showing well developed parabolic cross-profile. Planes of dilatation joints in the granite bedrock lie sub-horizontally in the floor of the valley, curving up laterally to match the form of the valley sides. *(C.E.)*

detached shells of rock are concentric, but in the case of glacial troughs, they are concentric about a horizontal axis. Many glaciated valleys in Britain could be quoted to show dilatation jointing developed in conformity with a parabolic cross-profile, for instance, Glen Sannox in north Arran (Plate XV). Linton (1967) shows how the process may lead to the gradual elimination of divides between glacial troughs (p. 270).

P. Birot (1968) and others have drawn attention to horizontal as well as vertical stresses in the bedrock. Horizontal stresses may be relics of former tectonic or igneous events, and may be several times greater in magnitude than vertical stresses due to dilatation. As glaciers cut into bedrock under such stresses, rock bursting into the valley sides can take place, and up-arching of the valley floor may lead to further jointing and facilitate deeper excavation.

W. B. Harland (1957) has suggested a hypothesis to augment the possible effects of dilatation (and other joint-producing mechanisms) in breaking-up bedrock. Glacial retreat would not only unload the bedrock, but also expose it to subaerial freeze-thaw action in the presence of abundant meltwater. The two mechanisms combined might be very effective in loosening bedrock prior to its removal in a subsequent glacier advance. Harland's freeze–thaw hypothesis is, in fact, a variant of Boyé's hypothesis of ground preparation by deep freezing.

These preparatory mechanisms are exceedingly important. Without them, it is diffi-

Plate XVI
Austerdalsbreen, Norway, showing truncated spurs and hanging valleys. Thin morainic debris rests on glacially-scoured rock in the left foreground. (*C.E.*)

cult to see how glacial erosion could be so effective. Joint-block removal would be impossible without existing fractures in the rock; and freeze-thaw action likewise cannot operate in the absence of fractures. Some details of these processes were presented in the previous chapter (p. 231).

The development of the typical parabolic cross-profile also involves processes of abrasion and gouging by rock fragments carried in the ice moving past the floor and sidewalls. Abrasion processes do not depend on preparation of the bedrock, though this is helpful, but they can only occur effectively in the case of temperature glaciers. Cold glaciers, frozen to bedrock, will be capable of only limited abrasion (see p. 138). The processes of abrasion can be examined with reference to the slip of a glacier past its sidewall. Measurements of side-slip in relation to Austerdalsbreen have been made and analysed by J. W. Glen and W. V. Lewis (1961). At a number of points along the side of this glacier, it was possible to measure side-slip directly, as explained in Chapter 4. The rock is hard gneiss, which is well striated and shows evidence of glacial gouging. Measurements were made at several situations on the glacier (Fig. 8.2). One was near the base of the ice-fall on both sides of the glacier and other measurements were made midway between the ice-fall and the glacier snout. The measurements

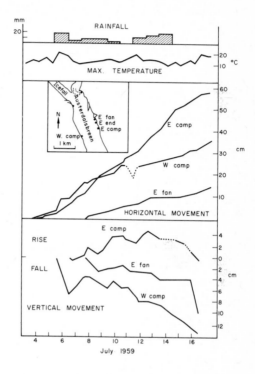

Fig. 8.2
Side-slip along the margins of Austerdals-breen, Norway (J. W. Glen and W. V. Lewis, *J. Glaciol.*, 1961, by permission of the International Glaciological Society)

near the base of the ice-fall showed that the side-slip here amounted to as much as 65 per cent of the maximum centreline velocity (marginal velocity 26 cm/day, compared with centreline velocities of 39·8 cm/day a little farther up-glacier and 19·4 cm/day a little down-glacier) (Glen, 1958). This relatively high side-slip suggests Block-schollen flow (Chapter 4), expected near the foot of a fast-flowing ice-fall. In the lower part of the glacier, the measurement sites were situated at either end of a line of stakes used to measure a transverse velocity profile. At these sites, side-slip amounted to about 20 per cent of the maximum velocity on the profile.

The observations provide important evidence on the way in which a temperate glacier erodes its sidewalls. It was noted that boulders occurred everywhere between the striated and plucked rock walls and the ice moving past them. There were also many boulders on the ice surface, some of which, disturbed by melting, fell into the sidewall gap, adding to the debris there. The movement of these boulders in the gap was clearly responsible for the striations so well displayed on the rockwall. There were also signs of quarrying activity where blocks had been riven off the down-glacier side of protrusions.

The process of joint-block removal from the valley floor and sides has been examined by J. G. McCall (1960). He notes that the yield stress of ice in shear, assuming perfect plasticity, is only about 0·06 per cent of that of granite. If a metre cube of granite is considered to be resting on the glacier bed or side, then the theoretical pressure of the ice against the cube consists of a compressive stress on its up-glacier face of a maximum of 20,000 kg. A total maximum shear stress of 30,000 kg would theoretically be exerted on the top and sides of the cube. If the opposing effect of ice on the down-glacier side is ignored, the maximum theoretical stress would therefore be 50,000 kg. If the value actually developed were only half this, and assuming the shear strength of granite to be 160 kg/cm², the force impelling the block would be adequate to shear off a projection in the bedrock having a bearing surface of 25000/160=156 cm². Rocks are generally stronger in compression than in tension; the compressive strength of granite, for instance, is about 1400 kg/cm². If the rock cube pressed only on one point on the bedrock, it could crush a protrusion whose area theoretically could be 25000/1400=18 cm². In the case of gouging, McCall shows that up to 16 cm² could be carried away, using the same assumptions. These relative values show that quarrying by the impact of rocks carried in a moving ice stream can be highly effective.

An actively eroding temperate glacier must carry a considerable number of boulders at the ice-rock interface, for it is these boulders that the glacier uses as tools to accomplish its erosion. The tools can be used both for grinding and prising. Their frequent occurrence at the side of glaciers, in the form of lateral moraine deposits, may help to account for the widening that produces the typical U-shaped glacial valley. Frost action on the exposed valley walls will also produce fragments to fall into the glacier sidewall gap. Erosion of this type cannot, however, take place unless the glacier is moving actively along its sidewalls.

The effectiveness of boulders in carrying out erosion increases to some extent with depth beneath the ice. Near the surface, the hydrostatic pressure is small and the ice can flow round the boulder instead of exerting its full pressure against it. At depths of about 22 m, on the other hand, the ice can exert a pressure of about 2 bar, equivalent to its yield stress in compression of about 2 kg/cm² at 0°C, as suggested by McCall (1960) on the basis of the perfect plasticity theory. On this basis, the ice could not exert any greater pressure on the boulder at depths greater than 22 m. The force exerted on the boulder would be limited by the yield strength of ice and would depend solely on the size of the boulder or group of boulders. If a power flow law is adopted, the force that can be exerted on the boulder will increase with depths greater than 22 m, but the question is one that has never been fully investigated. H. Rothlisberger (1968) also points out that erosion also depends on the boulders being in contact with the bed. Bottom melting of temperate glaciers plays a part in this process. Its rate is proportional to the basal shear stress and the sliding velocity. If $\tau_b=1$ bar and $V_b=30$ m/year, the sliding friction will melt about 1 cm of ice per year. The situation will usually be complicated, however, by the presence of other circulating subglacial water.

The actual velocity of the moving ice is a less important factor. This is because the force exerted by ice on the boulder only increases as the $\frac{1}{4}$ power of the ice velocity, according to Glen and Lewis (1961). An increase of ice velocity has relatively little effect on the ice pressure, although it will assist erosion to the extent that the number of boulders carried past a point will increase with the velocity. The distribution of velocity in the ice is usually more significant for glacial erosion. In particular, Block-schollen flow with its high rates of side-slip and basal slip may be very effective in shaping the glacier bed (W. Pillewizer, 1969). There is some evidence that many Pleistocene advancing glaciers showed Blockschollen movement even on low gradients.

Although there are many factors affecting the size which a glacial trough may attain, one of the primary controls is likely to be the amount of ice flowing through it; conse-quently, some relationship between trough capacity and catchment area (from which the trough derives its ice) is expectable. V. M. Haynes (1972) has attempted to demon-strate such a relationship in the case of some glacial troughs in the area of the Sukker-toppen ice cap in western Greenland. Troughs such as Sondre Strømfjord carrying glaciers from the main inland ice are larger than those fed by only locally derived ice. Relating cross-sectional area of trough (y) to catchment area (x), the equation

$$\log y = -0{\cdot}23 + 0{\cdot}3827 \log x$$

yields a correlation coefficient of $0{\cdot}71$. The coefficient value increases to a highly signifi-cant value of $0{\cdot}93$ when continentality, in terms of distance from the sea, is included as a second variable in the analysis. A problem in such analyses is that of quantifying the catchment area sufficiently accurately; and it is important also to allow for other factors affecting trough size, particularly rock structure.

It is sometimes possible for valleys in which glaciers have flowed to exhibit V-shaped sections in a generally U-shaped valley. The V-shaped sections are usually associated with reaches of greater gradient than the U-shaped portions, as in the case of the Ötztal, Austria (Plate XVII), the valley below Turtagrø, Norway (Lewis, 1947a), and the Aberglaslyn Pass, North Wales (C. Embleton, 1962). The Gorge du Guil in the French Alps, another example, has been examined by J. Tricart (1960). It lies above the Würm snow-line and is 15 km long. It crosses a variety of rock types, but it is particularly narrow where it cuts across massive limestones. Some morainic deposits have been found in the gorge, extending nearly to the bottom of the valley. The gorge predates the Würm glaciation as pockets of moraine occur within small basins along its length. These basins occur where softer rocks have been exploited by the ice. The morainic till has been partially water-sorted, the fine material has been washed out, and it has been consolidated by lime-impregnated waters. Tricart suggests that the gorge was cut by subglacial meltwater flowing under hydrostatic pressure. The glacier in this section of the valley would have been relatively thin, probably less than 300 m, and deep crevasses and moulins would have allowed meltwater to reach the base of the decaying glacier. The meltwater, heavily loaded with abrasive material, would rapidly scour a deep gorge, having a steep V-shaped cross-profile, beneath the ice. The stream that has occupied the valley after the ice retreated has only been able to erode the bottom of the gorge by about 5 m. The evidence does not support a hypothesis that the Gorge du Guil might have been cut by overflow waters from a lake in the main basin upstream of the gorge, for there is no sign of lacustrine deposits in this basin.

These examples illustrate the strong erosive capacity of subglacial streams flowing under thin, heavily crevassed ice on steep slopes (see also Chapter 12). It is their activity

Plate XVII
Part of the Ötztal, Austria, showing V-shape and interlocking spurs in the lower part of a major glacial trough. (*C.E.*)

that can account for the formation of V-shaped sections within U-shaped glaciated valleys, the V-shaped sections being cut while the whole valley is still occupied by ice. In the situations where glacial meltwater can be particularly active, as in the Guil gorge, the activity of the ice itself is often reduced as a result of thinning and crevassing. Where the slope is gentler and the ice is thicker, meltwater may be unable to reach the bed and is less likely to become concentrated into one stream. Conspicuous fluvio-glacial erosion beneath the glacier is unlikely in this situation. In this way, both U-shaped and V-shaped sections of a glaciated valley can form together, the longitudinal slope being the determining factor in the distribution of the two types of cross-section.

2 The longitudinal profiles of glaciated valleys

The essential difference between the long profile of a glaciated valley and that of a river is the irregularity so commonly characteristic of the former. Apart from the U-shaped cross-profile, the most striking features of a glaciated valley are the basins and steps that diversify its long profile. The basins frequently contain lakes or have been later infilled with alluvium while the steps often show marked signs of glacial scouring and plucking. The steps are rather similar, but on a much larger scale, to roches mouton-nées, as discussed by Lewis (1947b). Other features associated with glaciated valley long profiles are the hanging tributaries and the truncated spurs that occur along the valley sides. All these features indicate clearly the erosive capacity of the ice, but this action must clearly be selective to produce the observed effects.

In the early years of this century, there was considerable controversy between the protectionist school of thought and those who considered that the features of glaciated valleys were erosional features. The protectionist view was put forward forcibly by E. J. Garwood in 1910, but it has been generally abandoned in connection with the forma-tion of valley steps and hanging valleys. However, the concept of glacial protection is still valid under some conditions, as will be discussed in Chapter 9.

D. E. Sugden's (1968) work on the Cairngorms provides clear evidence that glacial erosion can be highly selective. In this area, some idea of the pre-glacial relief can be established from the pattern of sheet jointing in the granite, which parallels the smooth upland plateau surface into which deep glacial troughs are cut. Together with the presence of tors and rotted granite, the sheeting suggests that the upland surface is of pre-glacial age and enables its form to be reconstructed. Strikingly discordant with the gentle contours of the upland plateau are the steep-sided glacial troughs (Fig. 8.3). Three major troughs end abruptly in ice-smoothed cliffs 300 m high, but the trough heads are not cirques. On the contrary, the evidence of their form suggests that the troughs were cut by ice moving off the plateau—Linton's Icelandic type (see p. 269). Sugden assembles the evidence bearing on ice flow, including the pattern of deep-ened troughs, the carriage of erratics, the ice-moulded bedrock forms and breached watersheds, and concludes that glaciation was of the ice-cap type, the plateau being almost completely covered at the glacial maximum and the troughs representing the major routes taken by ice flowing off the plateau. This can only mean that, on the plateau and interfluves, ice played a relatively protective role, for here pre-glacial

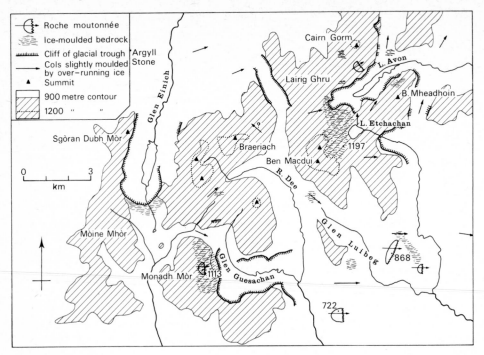

Fig. 8.3
Forms of glacial erosion in the Cairngorm Mountains, Scotland (D. E. Sugden, *Trans. Inst. Br. Geogr.*,
1968)

features such as the tors have survived with little modification; while the troughs attest
to the considerable erosive power of ice flowing faster down steeper gradients and select-
ing routeways favourable for its discharge. Erratics around the tors and the existence
of some ice-smoothed tors show that the plateau ice was not entirely immobile, but
the contrast with the behaviour of ice cutting the troughs is remarkable. Altogether,
there is clear evidence that, as is known to occur in Greenland and Antarctica today
(p. 111), ice can move in streams within an ice sheet, and be highly selective in its
erosive action.

    Before considering the actual characteristics of the landforms, the processes acting
subglacially, where the longitudinal profile is fashioned, will be discussed. The side-
slip observations can give direct information on the processes operating at the side of
the valley, but it is less easy to observe the processes operating at the base of a thick
glacier. Information must be obtained indirectly and theoretically, apart from the
limited data provided by bore-holes and tunnels.

    The extent to which a glacier moves over its bed (Chapter 4) is clearly relevant
to a study of the formation of rock bars and basins. The conditions under which it
can erode its bed require consideration, and in this connection, it is important again
to emphasize the difference between temperate and cold, polar glaciers. The former
are at their pressure melting-point and are able to slide over their beds by regelation
and creep. Cold glaciers are frozen to their rock bed and hence differential movement

occurs in the lowest layers of the ice rather than at the ice-rock interface. B. Kamb
and E. LaChapelle (1964) have studied the basal sliding of the temperate Blue Glacier,
Washington. A tunnel was excavated, entering the ice near the top of an ice-fall and
reaching bedrock where the ice thickness above the tunnel was 26 m. The movement
of the ice relative to the rock was recorded with dial micrometers. A stake, anchored
in the ice 10 cm above the bedrock, moved at a mean speed of 1·6 cm/day parallel
to the bed. Another stake 150 cm above the bedrock moved 12 per cent faster. Further
stakes spaced at 10 cm intervals showed that most of this differential movement took
place in the lowest 50 cm; in fact, 90 per cent of the 1·6 cm/day movement took place
as slip over the bed. The movement of the surface was almost identical with that at
150 cm above the bed. At the site, the bed had a slope of 22°, and it steepened
to 55° down-glacier from the end of the tunnel. The calculated basal shear stress
was 0·7 bar and the normal stress, resulting from the weight of the ice, was 1·6 bar.
The rate of movement at the base of the glacier was irregular, varying over periods
of the order of seconds, and by as much as 10 per cent from day to day. The existence
of a regelation layer up to 3 cm in thickness at the base of the ice has already been
described in Chapter 4. This layer contained considerably more debris than the overlying ice, but never more than 10 per cent by volume. Similar debris-laden layers have
been reported from other glacier tunnels (for example, Vesl-Skautbreen; McCall,
1960), and they indicate that the basal ice can act as an abrasive agent. The debris
tends to pile up on the up-glacier side of bedrock protuberances, but where the ice
flows over the down-glacier side, cavities may form. The processes whereby the ice
flows over and around obstacles have been discussed in Chapter 4.

In order to account for the irregular erosion at the base of a glacier, by which deep
basins and intervening bars may be formed, variations in glacier flow down the valley
must be considered. The theory of extending and compressing glacier flow (see p. 134)

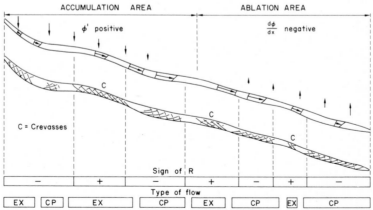

Fig. 8.4
Longitudinal section of an idealized valley glacier. The upper diagram shows the velocity distribution
calculated from a simple flow law assuming perfect plasticity. The lower diagram shows the slip-
line field and theoretical zones of crevassing. EX—extending flow; CP—compressing flow; R—radius
of bed curvature; ϕ—rate of discharge (J. F. Nye, *J. Glaciol.*, 1952, by permission of the International
Glaciological Society)

put forward by J. F. Nye (1952) may throw some light on this matter. Figure 8.4 shows the theoretical pattern of slip planes or potential shear surfaces associated with the two types of flow. Where the surface flow is accelerating or extending, a series of planes of weakness is likely to develop within the ice, curving down tangentially to the glacier bed. In the ablation zone and those parts of the glacier where the flow is decelerating (compressing), the planes of weakness will tend to develop curving upward tangentially from the bed in a down-glacier direction. These latter planes become pronounced near the snout of the glacier and also at the base of steep slopes. It is in these zones, therefore, that the ice may be expected to carry debris up from the glacier floor into the ice and hence it can act more effectively as an eroding agent. In this way, basins at the base of an ice-fall, once initiated, may become self-perpetuating.

The form of the bedrock surface over which the ice is flowing exerts an influence on the ice motion and is in turn itself modified by the ice movement. The former relationship has been shown to be true by Nye (1959) from observations made on Austerdalsbreen. Observations of the longitudinal strain rate were made in identical positions relative to the bed of the glacier in two successive years. The general strain rate in the area studied was strongly compressive as the site was at the base of the ice-fall. It was found that the strain rates were very similar in the two successive years at the same positions relative to the glacier bed. These rates, although in general compressive, showed local areas of extending flow, indicated by a tensile stress. Nye related these areas of extending flow to irregularities in the glacier bed. The results showed that the irregularities in the bed influence the glacier flow in a systematic way, and in turn, the different types of flow can influence the bed. In this way, the features of the glacier bed tend to become self-generating: once irregularities in the bed have appeared, they will tend to become accentuated.

Nye's theory of extending and compressing flow in relation to glacial erosion and the long profile forms of glaciated valleys was further elaborated by Nye and P. C. S. Martin in 1968. They argue that the potential slip-line field in the moving ice may play an important role in limiting the depths to which rock basins can be excavated. The basis of their discussion and the form of the slip-line field was described on p. 135 in the chapter on ice flow. Fig. 8.5 illustrates the argument. The bed is assumed to consist of three sloping plane sections. The cycloidal α lines theoretically generated for a thick glacier will have radii of curvature of $4h$ where h is the ice thickness. The slip-line AY in diagram (a) meets the bed tangentially and the ice below it (stippled) will in theory remain stationary. Consequently, no bed erosion can occur here so long as AZB remains unchanged. But as the bed around B is lowered by erosion, so AZB will be progressively affected by moving ice until it is modified to the form of AY. The bed cannot thereafter develop a concavity greater than the curvature of AY. On the other hand, there is no limit to the degree of convexity at C. The argument so far as been based on a plastic flow law; if a more realistic flow law is adopted, boundaries such as AY cease to be sharp but become zones of high shear strain rate, and the velocity of ice in the zone ABZ will not be zero but sluggish. However, the erosional implications are similar. Thus the form of the glacial valley floor depends, first, on the slip-line fields (which are controlled primarily by ice thickness and bed slope) and, secondly,

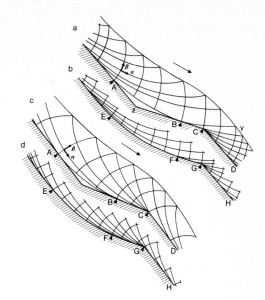

Fig. 8.5
Slip-line fields (J. F. Nye and P. C. S. Martin,
Int. Ass. scient. Hydrol., 1968)
a compressing flow, before erosion
b compressing flow, after erosion
c extending flow, before erosion
d extending flow, after erosion

on the relative rates of erosion of different segments of the floor, which may be related to rock resistance, valley constrictions, speed of flow and so on.

The irregularities of the longitudinal profile of the glacier show some correlation with variations in the cross-profile. The steeper portions of the long profile are often associated with narrower and often more V-shaped sections of the cross-profile, as has already been mentioned. This relationship agrees with Nye's theory of extending and compressing flow. In the steep reaches, where the flow would be extending, bottom movement would not be so conducive to erosion as in those places where the slope was less (or even reversed), giving rise to compressing flow.

The formation of rock steps, bars or riegels, is associated with the basins that usually occur above them. Many possible reasons for the position of rock bars have been suggested: among the most common are changes in lithology, changes in the pre-glacial gradient of the valley, constrictions in the valley walls, and the junction of tributary valleys. J. P. Bakker (1965) has drawn attention to some earlier views on the formation of glacial stairways and rock basins. He mentions the confluence of two glaciers, the variations of rock structure, and the importance of pre-glacial steps in the valleys as possible factors in the formation of glacial stairways. He lays most stress, however, on S. Passarge's views concerning Tertiary differential deep weathering. Evidence of differential deep weathering in immediate pre-glacial times in central Europe is found in the vegetation associated with waterfalls. These falls, of pre-glacial origin, are associated with rocks of greater resistance to weathering. In the flatter areas between the falls, deep weathering occurred, extending both across and down the valley. Glacial ice advancing over an area of very uneven depth of weathering exploits the weathered zones. By removing the weathered material, the ice can readily form rock basins and stairways. The riegels across the valleys represent the sites of the pre-glacial falls on the unweathered rocks. This process of basin formation does not require the ice to remove fresh and unweathered rock to any great extent. Glacial erosion is particularly effective where deep weathering had taken place in the Tertiary period, and this process may help to account for some of the intrusive troughs, described by D. L. Linton and discussed on p. 270.

Good examples of rock bars occur in the valley of Austerdalen, Norway (C. A. M. King, 1959), the uppermost still concealed by glacier ice. There are eleven rock bars in the valley between the point where it joins a tributary of Sognefjord and the snout of the present glacier, which lies just behind a rock bar. A high valley step underlies the ice-fall, which links Jostedalsbreen ice cap above to Austerdalsbreen below (Plate III, p. 135). The whole area is formed of massive gneiss so that the bars cannot be explained in terms of varying lithology. Four rock bars lie between the glacier snout and Veitestrondsvatn and five more are hidden beneath the lake waters. These submerged bars were revealed by echo-sounding traverses across and along the lake. Another rock bar dams up the lake, and the lowest bar dams the lake of Hafslovatn (Fig. 8.6).

The rock bars of Austerdalen illustrate several of the situations in which rock bars are frequently found. The uppermost exposed rock bar, near the present glacier snout, occurs where the glacier is forced round a sharp bend. It is noticeable that the bar is highest at the inside of the bend where the flow would be most restricted by a preglacial spur extending into the valley at this point. The spur would tend to dam up the ice which would thicken on the upstream side, and the ice would escape over the bar mainly at the outside of the bend. The valley wall above the constriction shows very marked glacial scouring on the outside of the bend, where the flow would be most rapid. The second rock bar, situated near Tungasaeter, forms a steep barrier across the valley. Meltwater, flowing subglacially in the initial stages, has cut a deep narrow cleft through this bar. The bar probably represents the hang of Austerdalen as a tributary to Langedalen just below the rock bar. Another small rock bar occurs just below the junction of the two valleys. This bar may have been caused by increased erosion under the thicker ice below the confluence of the two glaciers, the classical hypothesis put forward in 1905 by A. Penck. The next major bar is also similarly situated close to the confluence of Austerdalen and Snauedalen, which is an important tributary leading by a col into the adjacent valley of Fjaerland.

A further series of five rock bars is hidden beneath Veitestrondsvatn, smaller features than the major rock bar that dams up the whole lake. The basins that separate the bars in the lake have depths of about 60–100 m, and the bars occur where there are bends in the valley. The large rock bar at the end of the lake occurs where the valley is very constricted, before it opens out into the broad basin of Hafslovatn. The latter hangs 168 m above Sognefjord. Ice from the basin reached Sognefjord by two different routes, both of which have large rock bars in their upper reaches. There is a considerable fall of level between Veitestrondsvatn and Hafslovatn, a fall that suggests a thinning of the ice and less vigorous erosion at the point where the bar occurs. Erosion subsequent to ice retreat appears to have lowered the level of Veitestrondsvatn by about 60 m. Supporting evidence for this is found in exposed lake deposits at the head of the lake.

In assessing the factors responsible for the formation of the rock bars, the pre-glacial

Fig. 8.6
The rock bars in Austerdalen and Veitestrondsvatn, Norway (C. A. M. King, *Geogrl J.*, 1959, Royal Geographical Society)

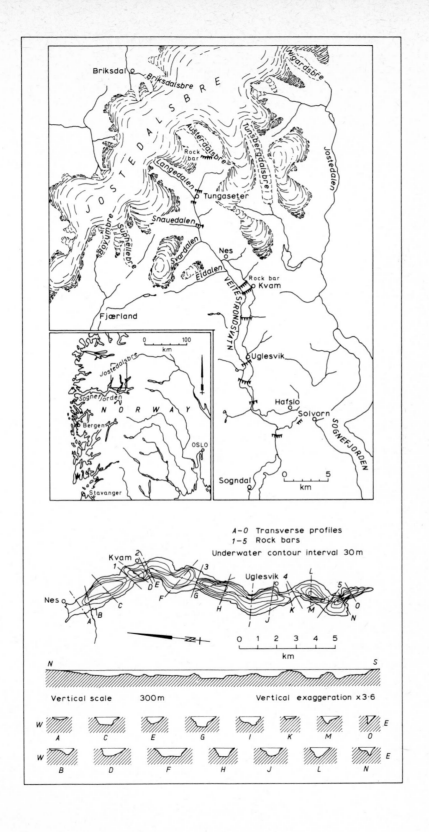

Briksdal
Briksdalsbre
Nigardsbre
JOSTEDALSBRE
Alsterdalsbre
Rock bar
Tunsbergdalsbre
Jostedalen
Langedalen
Tungaseter
Snauedalen
Bøyumbre
Supfiellebre
Svardalen
Nes
Eldalen
Rock bar
Kvam
Fjærland
VEITESTRONDSVATN

0   100
km

Jostedalsbre
Sognefjorden
N O R W A Y
Bergen
OSLO
Stavanger

Uglesvik

Hafslo
Solvorn
SOGNEFJORDEN

Sogndal

0   5
km

A–O Transverse profiles
1–5 Rock bars
Underwater contour interval 30 m

Kvam 2
1   3
D E
C   F
Nes
A B   G
Uglesvik 4
L
5
H   M   O
I J   K   N

0 1 2 3 4 5
km

N   S

Vertical scale   300 m   Vertical exaggeration ×3·6

W   E
A   C   E   G   I   K   M   O

W   E
B   D   F   H   J   L   N

form of the valley must be taken into account. On the basis of arguments already put forward it seems likely that any pre-glacial irregularity in the long profile will be accentuated rather than removed by glacial erosion. Austerdalen probably possessed well-developed knick points, associated with the recent elevation of the pre-glacial erosion surface that forms the conspicuous plateau on which the Jostedalsbreen ice cap rests. The major knick point of the series probably lay at the site of the present ice-fall. This huge step is about 760 m high and, according to Nye's calculations, the ice cover is probably only about 6 m thick at the top and 46 m thick near the base of the ice-fall.

The situations of the rock bars in this valley may be summarized by stating that they occur both above and below tributary junctions, at bends in the valley and near constrictions in the valley. At all these positions, there would tend to be changes in the velocity and thickness of the ice. These changes would lead to zones of extending and compressing flow. The different types of flow would in turn lead to areas of reduced or enhanced erosion respectively on the bed of the glacier. Once initiated, the steps and bars would tend to be self-perpetuating and increase in magnitude during glaciation. The pre-glacial stream pattern and irregular long profile would determine the position of the bars. In deglacial phases, subglacial meltwater streams would begin to cut V-shaped gorges through the bars (Plate XVIII). These gorges would be further deepened by post-glacial river action, resulting in the eventual draining of any lakes in the intervening rock basins not yet silted up.

Plate XVIII
The terminus of the Pasterze Gletscher, Austria. The meltwater stream, originally flowing subglacially, has carved a gorge through a rock bar. The lake is artificial, part of a hydro-electric scheme. (*C.E.*)

Lewis (1947b) has studied the formation of valley steps in a variety of areas, including North Wales and the Cuillin Hills of Skye. He points out that the flat treads of the steps usually show well-marked glacial scouring, while quarrying is more in evidence on the steep risers of the steps. The steps resemble giant roches moutonnées in form. Because of the similarity of the two features, Lewis suggests that they may have a common mode of origin, and contends that the processes operating under thin glaciers differ from those responsible for the formation of steps under thick ice. Throughout, he is concerned with temperate glaciers. When the ice is thin, he suggests that meltwater is able to penetrate to the bed of the glacier (see p. 148 for a discussion of the maximum depths of crevasses in active ice). If it were then refrozen in any cracks in the rocks, shattering of the rock would ensue. This process might operate best on the steep risers of the steps, where plucking is in evidence, and the process would tend to increase the height of the step, but it could not cause the overdeepening of the basin at some considerable distance from the step. When the ice is thick, Lewis suggests that changes in pressure as temperate ice moves over the step could result in alternate melting and refreezing.

Lewis comes to the conclusion that the most effective step formation probably occurs under thick glaciers (by implication of temperate type). This assumes that freeze–thaw is an important process in their formation, a process discussed in more detail in Chapter 7. It seems likely that freeze–thaw is not necessarily important under thick glaciers and may not even be significant under thin ice, since temperature changes are very slow beneath ice. The prising action described by Glen and Lewis (1961) is probably more important when the ice is moving at its base, armed with suitable tools in the form of boulders.

The basins separating the bars and steps form the complementary features and are the most important single characteristic of a glaciated valley. The basins indicate the ability of the basal glacier ice to flow uphill. It has already been suggested that there are limits to the degree of concavity that a valley floor may develop, and thus for any given size of basin there is a theoretical limit to the depth it may attain, related primarily to the thickness and slope of the glacier responsible. If the basins have not been later infilled with alluvium, and if the bar at the lower end has not been cut through by fluvial action, the hollows may contain ribbon lakes, which are characteristic of heavily glaciated mountain areas; the depths of the lakes may be augmented by the deposition of moraine. Lakes of this type can reach considerable depths. Loch Morar, for example, the deepest of British lakes, attains a depth of 315 m. As Linton (1957, 1959) has pointed out, most of the deep glacial lakes of Scotland occur in the west of the country, radiating from the area where precipitation is at a maximum at present. It is likely that these areas would receive the heaviest snowfall under glacial conditions. The glaciers of the western side of the Scottish Highlands would, as a result, have been well nourished, while they also flowed down steep slopes to deep water and were able to achieve the maximum amount of erosion. Where other conditions were also suitable, basin formation could take place effectively. Similar rock basins, whether filled with water to form lakes or with sediment to form valley flats, are characteristic of all mountain groups that have nourished vigorous valley glaciers. There are many good examples in the Lake District, western Ireland, North Wales and in many other parts of the world.

Not all long valley lakes, however, have necessarily been formed by glacial erosion, despite their appearance on casual inspection. E. Watson (1962) has drawn attention to a ribbon lake that is not of glacial origin in the Tal-y-Llyn valley in North Wales. This straight and deep glaciated valley links a number of pre-glacial drainage lines because the ice has exploited the Tal-y-Llyn fault. Within the trough are several basins, separated by barriers; one of the basins contains the Tal-y-Llyn lake. The barrier retaining the lake rises about 53 m above the valley floor and might at first sight appear to be moraine. However, there are outcrops of solid rock visible in the barrier. On closer examination, the rock in these outcrops can be seen to be contorted and shattered, and further, it is not *in situ*. If the valley side above the barrier is examined, it can be shown that the barrier represents a large landslide feature and that the basin is not primarily a glacially eroded landform. It can no longer be regarded as the southern-most rock basin in Wales. Care must, therefore, be taken to confirm that the rock bars do in fact consist of solid rock, as is normally the case.

Some of the largest ice-scoured hollows are those now occupied by the large lakes at the southern boundary of the Alps in northern Italy. The three largest of these lakes are Lake Garda, with a maximum depth of 346 m, Lake Maggiore, 362 m deep, and Lake Como, 410 m deep. These lakes are situated where the glaciers spread out into piedmont lobes and built very large arcuate moraines. The moraines increase the depth of the lakes by adding to the height of the rock bars beneath them. The upper portions of the lakes are narrow, but they spread out into broad expanses near their outlets. The origin of the piedmont lakes may be connected with the strongly compressive flow that must have taken place as the glaciers emerged from the mountains on to the plains. This type of flow has already been associated with possible zones of erosion. By this stage in their courses the glaciers would carry a large amount of debris; they would also be thick and hence the basal shear stress would be great. All these factors would, therefore, tend to concentrate erosion in these situations. Similar factors also applied on the northern side of the Alps, where other large glacial lakes occur. A rock basin occurs beneath the piedmont Malaspina glacier (Alaska) at present; conditions in this glacier are similar to those at the Alpine margin in many respects.

3 Glacial diffluence and transfluence: watershed breaching by ice

Where vigorous glaciers are flowing down valleys whose outlet is constricted or impeded, the ice level in the valley will rise. This situation may arise either if the outlet is blocked by other ice, or if the level of the ice in the valley is rising more rapidly than the ice can escape down the valley to lower the level. If the ice thickness continues to increase, the glacier surface will eventually rise above the lowest col available, and the surplus ice will then escape across the col into the neighbouring valley. With increasing glaciation, the level may rise to such an extent that all available cols are used by a series of diffluent ice streams. The term 'transfluence' may be used when this stage has been reached. These phenomena were clearly recognized by A. Penck (1905) who drew attention to the examples afforded by the Rhine Valley at Sargans, and also at Zell-am-See.

Glacial transfluence has also long been recognized in Scandinavia, where erratics from Sweden have been carried westward across the main watershed to the Atlantic coastlands of Norway. This resulted from the accumulation of ice over the Baltic and Scandinavia until the ice-shed lay to the east of the main mountain divide. The ice then flowed westward across the latter, carving breaches where the ice flow was concentrated in existing cols. Linton (1949) has demonstrated similar events in the Scottish Highlands where there was also a tendency for the ice thickness to be greatest east of the pre-glacial watershed, and where erratics were also transported from east to west across the backbone of the country.

Many examples of glacial diffluence are known in Britain. In North Wales, pre-glacial watersheds were overridden and breached by ice at the heads of the Nant Ffrancon valley, the Llanberis valley, the Gwynant valley, and the Nantlle valley (C. Embleton, 1962). In the Lake District, an area which was entirely glaciated by local ice moving down the major valleys, R. K. Gresswell (1952) has examined the results of glacial diffluence. In the Langdale valley, for instance, rock steps were initiated by glacial diffluence as well as by varying lithology. Ice moving south down Langdale into the Grasmere basin escaped by two routes to the south. The western stream crossed Red Bank and the eastern one flowed by way of Rydal Water to the Windermere Basin. This latter ice stream was joined by several others. However, it split again into two channels farther downstream. One of these passed south-westward into the Esthwaite Water valley, while the bulk of the ice moved south down the main trough. This ice gouged out the lake to a depth of 27 m below sea level and 66 m below its present surface. Some ice also escaped eastwards along the Gowan valley, where it was joined by ice flowing down the Kent valley. Thus a complex system of glacial diffluence and confluence was in existence at the stage when the glaciers had shrunk back into their valleys, towards the end of the last glaciation.

The most extensive evidence of glacial diffluence and the most spectacular instances of watershed breaching by ice are to be found, however, in Scotland. A recent study of Loch Lomond, the largest lake in the kingdom (Fig. 8.7), illustrates several aspects of the process (Linton and H. A. Moisley, 1960). The lake attains a length of 36 km and a maximum depth of 190 m. The upper part is long and narrow, but it opens out in the southern portion which lies in lower country. In pre-glacial times the lake area was drained by streams belonging to at least three different drainage basins according to Linton and Moisley. A watershed extended across the northern end of the lake. From this watershed, a stream drained northward to the Tay by way of Glen Dochart along a zone of weakness in the rocks. To the south of this northern watershed, a stream flowed eastward to the Teith through the valley in which Loch Katrine now lies. A major watershed ran east–west across the present lake from Ben Lomond eastward. South of this watershed, the area was drained by streams flowing south-eastward, possibly two separate streams draining to the Forth and the Vale of Leven. The positions of these pre-glacial watersheds can be identified by the greatly over-steepened valley sides of the breaches through the watersheds, which show considerable evidence of severe ice scouring and moulding.

The ice accumulated in the northern basin to such an extent that it overflowed and breached the surrounding watershed in about six places. The southern breach, just east of Ben Vorlich, was the deepest. This breach diverted much of the drainage of

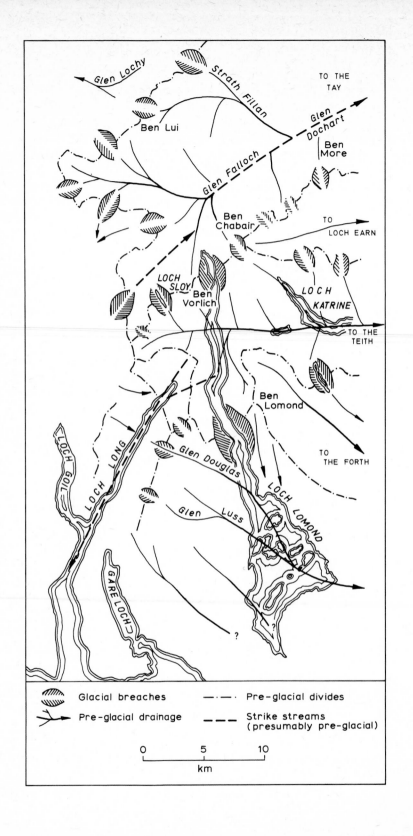

Glen Lochy

Strath Fillan

TO THE
TAY

Glen Dochart

Ben Lui

Glen Falloch

Ben
More

Ben
Chabair

TO
LOCH EARN

LOCH
SLOY

LOCH
KATRINE

Ben
Vorlich

TO THE
TEITH

Ben Lomond

Glen Douglas

TO
THE FORTH

LOCH GOIL

LOCH LONG

Glen   Luss

LOCH LOMOND

GARE LOCH

?

?

Glacial breaches	—·—·—	Pre-glacial divides
Pre-glacial drainage	———	Strike streams (presumably pre-glacial)

0   5   10

km

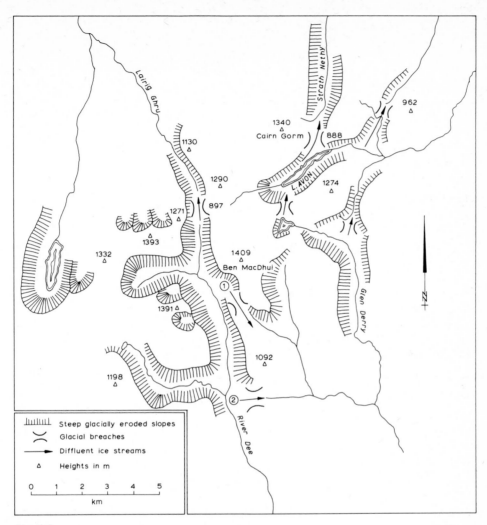

Fig. 8.8
Watershed breaching by ice in the Cairngorms, Scotland (after D. L. Linton, *Trans. Inst. Br. Geogr.*, 1949)

Glen Falloch south into Loch Lomond. The southern basin also became congested with ice, which overflowed across the watershed to the south and south-east. The southern diffluent channel, west of Ben Lomond, was lowered enough to divert the drainage of the upper part of the valley, including Loch Arklet, to the south into the Loch Lomond basin. In this way, the three separate sections of the present lake became linked in one major basin, draining south into the Clyde. The erosion achieved by these diffluent glaciers was such that a valley draining north at levels between 274 m and 458 m

Fig. 8.7
Map of the Loch Lomond area, showing glacial breaches (D. L. Linton and H. A. Moisley, *Scott. geogr. Mag.*, 1960, Royal Scottish Geographical Society)

was converted into a trough 150 to 180 m deep. This erosion was severely localized, however, and was restricted to the valleys and the cols through which the ice was concentrated by the relief control imposed upon it.

The effects of glacial diffluence depend on the relative power of the original and the diffluent glaciers. The diffluent glaciers that crossed the former cols and formed the continuous trough in which Loch Lomond now lies were able to lower their troughs to such an extent that the drainage was permanently diverted. Not all glacial diffluent channels, however, operate for long enough or carry sufficiently vigorous ice streams to achieve this effect. In these circumstances, the diffluent channel is left on the side of the main valley as an eroded col, which normally has a U-shaped form.

Various stages in the development of a diffluent glacier and its effects on the landscape can be seen in the Cairngorms. A good example is the col or trough, 244 m deep, leading out of the Loch Avon valley (Fig. 8.8). The glacier in the Avon valley became constricted by ice moving across its mouth, causing the level to rise until a col into Strath Nethy was crossed by the ice. The escaping ice then lowered the col and widened Strath Nethy into a glacial trough. The Nethy valley, which was only fed by ice passing through the diffluent col, was not lowered enough to cause any drainage diversion. An equally conspicuous example of glacial diffluence occurred in the neighbouring valley of Glen Dee to the west. The southern outlet of the Glen was also blocked by another powerful ice stream, causing the ice level to rise until ice escaped northward through a striking col into Lairig Ghru. The pre-glacial height of this col is estimated by Linton to be at least 1070 m and its present height is 840 m, indicating a removal of a minimum of 230 m of granite (see Chapter 11). The col now shows clearly the effects of glacial action and ice moulding. In neither of these cases were permanent drainage changes effected. There are, however, in this part of Scotland and particularly farther west many examples of glacial diffluence that has proceeded far enough to cause drainage diversions: one example is given in Chapter 12. Many sharp bends and otherwise unaccountable drainage patterns can be explained by glacial diffluence.

An interesting feature is often found in association with diffluent glacier cols. This is a small lake lying in the col, which has been lowered by the diffluent glacier. There are several examples of this in North Wales, for instance, Llyn Ogwen, Llyn Cwellyn and a small lake near the diffluent col that leads into the Nantlle valley. Many lakes of this type, usually of small dimensions, can also be found in the Scottish Highlands. The reasons for their formation in this watershed position are not apparent; however, they may indicate enhanced glacial erosion in association with the acceleration of flow at the position where the glacier starts to diverge. The freer outlet of the diffluent glacier would allow some speeding up of the flow as the ice moved over the col.

Glacier transfluence is associated with conditions in which the ice volume increases rapidly in relation to ice dispersal. All available cols are then used by the ice. These conditions can be seen at present in the glaciated parts of the Alaskan mountains and in Antarctica. Similar conditions must have occurred in the western part of the Highlands of Scotland, where precipitation was heavy and the ice thickness very great. Linton has shown how the main Grampian watershed has been breached in several places by transfluent glaciers. The amount of scouring involved in the production of the through passages reaches 366 m in Glen Tilt, and Loch Treig shows overdeepening to the extent of at least 460 and possibly 550 m (Chapter 11).

4 The classification of glacial troughs and their development

Four main categories of glacial trough have been identified by Linton (1963): Alpine troughs, the Icelandic type, the composite type, and the intrusive type.

1 The main essential of the Alpine type is that the accumulation area of the glacier should be surrounded and overlooked by higher ground. The source of the glacier is often in a cirque or series of cirques, and the glacier occupies a pre-glacial valley. The Haweswater and Ullswater valleys in the Lake District provide good examples of this type of trough, which is ideally a self-contained unit with no external source of ice and no ice escape by diffluence. These conditions are relatively rarely fulfilled.

2 The ice forming the second type of trough, the Icelandic type, originates on a mountain ice cap and flows down as a series of outlet glaciers, often through ice falls. The outlet glaciers from the Jostedalsbreen ice cap in Norway illustrate the situation. Austerdalsbreen, already referred to in this and earlier chapters, is one of these glaciers that has created a very deep and steep trough of this type. In Britain, there are examples of troughs of the second type in eastern Scotland, where ice accumulated on the high, pre-glacial erosion surfaces, such as the Grampian plateau. Similar flattish upland surfaces provided the gathering ground for the ice that formed the troughs in the Pennines. Bishopdale and upper Wharfedale provide examples of troughs that show marked glacial modification in the Pennines.

3 The third type is the composite type, and describes those systems that did not find the pre-existing valleys sufficient for the discharge of the ice. In the third type, parts of the troughs are eroded entirely by the ice, these parts linking modified portions of pre-glacial valleys. In the composite troughs, glacial diffluence and transfluence have been active and by this means, new valleys have been created and drainage changes initiated.

   Five categories of composite troughs are suggested by Linton. The first (a) consists of simple diffluent troughs, for example Strath Nethy, which has already been described. The second (b) consists of multiple diffluent troughs, such as those leading away from Loch Fyne. The third (c) consists of troughs formed by simple transfluence. These differ from the first category in that the new trough branches out from the valley head and not from its side. The pass through Beattock summit is a good example. Ice from the north passed south across the former col into Annandale, transferring the watershed 6·5 km to the north in the process. The Nant Ffrancon pass in Snowdonia is another example, where the former watershed has been completely worn away and the watershed displaced eastward. The fourth category, (d), is the multiple transfluence system of troughs where the ice spilled over a number of cols at the head of a valley. For example, ice spilled over many cols at the head of Glen Falloch, in the area already described in the discussion of the formation of Loch Lomond. Ice escaped through more than ten cols around the upper part of the Falloch valley, forming watershed breaches and ice-modified cols leading down to glacial troughs. The final group of composite troughs, *(e)*, consists of radiative dispersal systems. This applies to those troughs that display a radial pattern, as in the Lake District, southern Norway, the south-west Highlands of Scotland, and Fjordland in south-west New Zealand. In all these areas, the ice spread out from a centre along pre-existing valleys and also cut across watersheds wherever the ice pressure was sufficient to carry the ice over a col, which was then lowered by erosion.

4 The fourth type of trough is the intrusive or inverse trough. The ice in this type pushes up the valley against the direction of pre-glacial drainage. Such troughs are likely to occur in the lower hill areas where the local ice was not sufficiently powerful to keep out the more powerful ice from more vigorous centres. There are good examples of these troughs in the hill areas near the southern boundary fault of the Highlands of Scotland. Three well-marked troughs occur at the south-eastern side of the Kilpatrick, Campsie and Ochil group of hills. Two of these troughs, originally excavated below present sea level, are now filled with drift, but the third is an obvious feature and forms the sections of Clydesdale between Hamilton and Lanark. Upstream of Hamilton, the Clyde is flowing in a deep narrow valley, clearly modified by ice action. Erosion by ice has reduced the valley level to below 60 m at a position far inland. Both the drainage gradients and drainage pattern have also been modi-fied.

The effects of prolonged erosion by valley glaciers in a cold climate have been studied by Linton (1963, 1967) in such areas as Antarctica where rugged topography has been heavily submerged by ice for a long period. The marginal parts of Antarc-tica best illustrate these conditions. Under the severe climatic conditions that occur in this area, the volume of ice increases down-valley as tributaries merge, because ablation is largely by calving. The lack of melting prevents loss of ice on the scale associated with glaciers in more temperate climates. For this reason the cross-section of the valley must increase downstream to accommodate the increasing ice volume. As the valley cross-section becomes enlarged, adjacent valleys in a sub-parallel system of troughs will begin to intersect each other. If the valleys are parallel, the intervening divides will be lowered equally along their lengths (Fig. 8.9A), but if the valleys converge and join, enlargement of the valleys will progressively shorten the interven-ing divides (Fig. 8.9B). Linton recognizes stages of shortened, vestigial and subdued divides (compare p. 217) and observes that such an evolution is characteristic not so much of temperate mountain regions but of areas such as Spitsbergen and Antarc-tica where the snow-line is at or near sea level. Reindalen in Spitsbergen is an example of a trough which steadily increases in width downstream as successive tribu-tary valleys come in and their divides with the main valley are eliminated. At latitude 78°N its width is about 3 km; after a distance of 10 km it has increased to 6·4 km and after a further 12 km the width attains no less than 10 km. The divides between the major valleys in these situations have been shorn of all their smaller spurs and represent long and straight ice-moulded forms. In the Antarctic, these valleys still

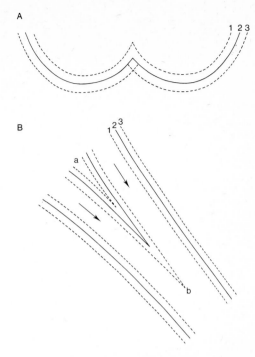

Fig. 8.9
Divide elimination by glacial erosion (after
D. L. Linton, *Arctic and alpine environments*
(ed. H. E. Wright and W. H. Osburn),
1967)
A Between adjacent glacial troughs, seen
in cross-section
B Between converging glacial troughs,
seen in plan

lie underneath the ice, but it is easy to imagine that this process is taking place actively
at present, owing to the volume of ice that fills the valleys. As erosion continues,
the divides between two neighbouring valleys become progressively narrower. They
are slowly consumed and shortened as erosion continues further. As the divide is
traced upward to the nunatak peaks, it becomes sharper, forming an arête where
it emerges above the ice. The exposed arête is attacked by frost shattering and eventually only pyramidal peaks marking junctions of former arêtes remain. It is by this
process of divide destruction that valley glaciers can most readily widen their valleys.
Once the divides have been removed, the ice can combine into one lowland ice sheet
and produce the ice-moulded surface characteristic of such an area (Chapter 9).

5 Conclusions

It is evident that ice can produce spectacular glacial features by vigorous erosion under
suitable conditions. Ice can, however, under other conditions remain almost static and
can be more protective than erosional. In order to erode effectively, the ice must exceed a
certain minimum thickness and it must flow reasonably fast as a concentrated stream with
high velocities of basal sliding. These conditions are best achieved in temperate valley
glaciers that are fed by well-nourished accumulation areas. There must also be a free
outlet for the ice to lower ground or out to sea so that its flow can be maintained.
    One of the most notable characteristics of glacial erosion is its irregular incidence
along the length of the glacial trough. The rock bars and intervening basins, which
are so common in glaciated valleys, are evidence of this irregularity of erosion. The

explanation for these zones of alternating excessive erosion or reduced erosion must be sought both in the rocks of the valley, in the form of the valley, in the response of the ice to this configuration, in additions of ice volume owing to incoming tributary glaciers and in losses of ice due to glacial diffluence. Zones of compressing flow, associated with the form of the valley, may well cause an increase in erosion, owing to the nature of the ice movement within the glacier. Ice-flow theory also suggests that there are limits to the degree of concavity that rock basins in the valley floor may develop.

It is likely that the ground is prepared in some way, possibly by deep freezing or deep weathering, before the ice advances over it. Another factor assisting erosion of bedrock by a valley glacier is pressure release. This causes the development of dilatation joints as ice of relatively low density replaces rock of relatively high density. The glacier can both scour and prise material from its bed, although both processes are only achieved effectively with the aid of rock fragments carried near the base or sides of the glacier. Movement past the rock walls at the sides and base of the glacier is essential for active glacial erosion. The presence of meltwater in temperate glaciers aids this basal movement.

6 References

BAIN, G. W. (1931), 'Spontaneous rock expansion', *J. Geol.* **39**, 715–35

BAKKER, J. P. (1965), 'A forgotten factor in the interpretation of glacial stairways', *Z. Geomorph.* NF **9**, 18–34

BIROT, P. (1968), 'Les développements récents des théories de l'érosion glaciaire', *Annls Géogr.* **77**, 1–13

BLACHE, J. (1960), 'Les résultats de l'érosion glaciaire', *Méditerranée* **1**, 5–31

BOYÉ, M. (1950), *Glaciaire et périglaciaire de l'Ata Sund nord-oriental (Groenland)* (Paris; Expéditions polaires françaises, **1**)

  (1968), 'Défense et illustration de l'hypothèse du "défonçage périglaciaire"', *Biul. peryglac.* **17**, 5–56

CAROL, H. (1947), 'The formation of *Roches Moutonnées*', *J. Glaciol.* **1**, 57–9

COTTON, C. A. (1941), 'The longitudinal profiles of glaciated valleys', *J. Geol.* **49**, 113–28

  (1942), *Climatic accidents in landscape-making* (Christchurch, N.Z.)

DREWRY, D. J. (1972), 'The contribution of radio echo sounding to the investigation of Cenozoic tectonics and glaciation in Antarctica', *Inst. Br. Geogr. Spec. Publ.* **4**, 43–57

ELLISTON, G. R. (1962), 'Seasonal changes of speed in temperate valley glaciers', *J. Glaciol.* **4**, 289

EMBLETON, C. (1962), *Snowdonia* (Geogr. Ass., 'British Landscapes through Maps')

GARWOOD, E. J. (1910), 'Features of Alpine scenery due to glacial protection', *Geogrl J.* **36**, 310–39

GLEN, J. W. (1958), 'Measurement of the slip of a glacier past its side wall', *J. Glaciol.* **3**, 188–93

GLEN, J. W. and LEWIS, W. V. (1961), 'Measurements of side-slip at Austerdalsbreen, 1959', *J. Glaciol.* **3**, 1109–22

GRAF, W. L. (1970), 'The geomorphology of the glacial valley cross-section', *Arct. alp. Res.* **2**, 303–12

GRESSWELL, R. K. (1952), 'The glacial geomorphology of the south-eastern part of the Lake District', *Lpool Manchr geol. J.* **1**, 57–70

HARLAND, W. B. (1957), 'Exfoliation joints and ice action', *J. Glaciol.* **3**, 8–10

HAYNES, V. M. (1972), 'The relationship between the drainage areas and sizes of outlet troughs of the Sukkertoppen ice cap, West Greenland', *Geogr. Annlr* **54**A, 66–75

HUTTON, J. (1795), *The theory of the Earth with proofs and illustrations*

JAHNS, R. H. (1943), 'Sheet structure in granites: its origin and use as a measure of glacial erosion in New England', *J. Geol.* **51**, 71–98

KAMB, B. and LACHAPELLE, E. (1964), 'Direct observation of the mechanism of glacier sliding over bedrock', *J. Glaciol.* **5**, 159–72

KING, C. A. M. (1959), 'Geomorphology in Austerdalen, Norway', *Geogrl J.* **125**, 357–69

LEWIS, W. V. (1947a), 'The cross sections of glaciated valleys', *J. Glaciol.* **1**, 37–8
(1947b), 'Valley steps and glacial valley erosion', *Trans. Inst. Br. Geogr.* **14**, 19–44
(1954), 'Pressure release and glacial erosion', *J. Glaciol.* **2**, 417–22

LINTON, D. L. (1949), 'Watershed breaching by ice in Scotland', *Trans. Inst. Br. Geogr.* **17**, 1–16
(1957), 'Radiating valleys in glaciated lands', *Tijdschr. K. ned. aardrijksk. Genoot.* **74**, 297–312
(1959), 'Morphological contrasts of Eastern and Western Scotland', in *Geographical Essays in Memory of Alan G. Ogilvie* (ed. MILLER, R. and WATSON, J. W.)
(1963), 'The forms of glacial erosion', *Trans. Inst. Br. Geogr.* **33**, 1–28
(1967), 'Divide elimination by glacial erosion' in *Arctic and alpine environments* (ed. WRIGHT, H. E. and OSBURN, W. H.), 241–8

LINTON, D. L. and MOISLEY, H. A. (1960), 'The origin of Loch Lomond', *Scott. geogr. Mag.* **76**, 26–37

MATTHES, F. E. (1930), 'Geologic history of the Yosemite Valley', *U.S. geol. Surv. Prof. Pap.* **160**, 137 pp.

McCALL, J. G. (1960), 'The flow characteristics of a cirque glacier and their effect on glacial structure and cirque formation', in 'Norwegian cirque glaciers' (ed. LEWIS, W. V.), *R. geogr. Soc. Res. Ser.* **4**, 39–62

MILLER, M. M. (1958), 'Phenomena associated with the deformation of a glacier borehole', *Un. géod. géophys. int., Symposium at Toronto, 1957*, **4**, 437–52

NYE, J. F. (1952), 'The mechanics of glacier flow', *J. Glaciol.* **2**, 82–93
(1959), 'The deformation of a glacier below an ice fall', *J. Glaciol.* **3**, 386–408

NYE, J. F. and MARTIN, P. C. S. (1968), 'Glacial erosion', *Int. Ass. scient. Hydrol., Gen. Assembly Bern (1967), Commn Snow Ice, Publ.* **79**, 78–86

PENCK, A. (1905), 'Glacial features in the surface of the Alps', *J. Geol.* **13**, 1–19

PILLEWIZER, W. (1969), 'Die Bewegung der Gletscher und ihre Wirkung auf den Untergrund', *Z. Geomorph.* Suppl. **8**, 1–10

PIPPAN, T. (1957), 'Anteil von Glazialerosion und Tektonik an der Beckenbildung am Beispiel des Salzachtales', *Z. Geomorph.* NF **1**, 71–100

ROTHLISBERGER, H. (1968), 'Erosive processes which are likely to accentuate or reduce the bottom relief of valley glaciers', *Int. Ass. scient. Hydrol., Gen. Assembly Bern (1967), Commn Snow Ice, Publ.* **79**, 87–97

SUGDEN, D. E. (1968), 'The selectivity of glacial erosion in the Cairngorm Mountains, Scotland', *Trans. Inst. Br. Geogr.* **45**, 79–92

SVENSSON, H. (1959), 'Is the cross-section of a glacial valley a parabola?' *J. Glaciol.* **3**, 362–3

TRICART, J. (1960), 'A subglacial gorge: La Gorge du Guil (Hautes-Alpes)', *J. Glaciol.* **3**, 646–51

TRICART, J. and CAILLEUX, A. (1962), *Le modelé glaciaire et nival* (Paris)

WATSON, E. (1962), 'The glacial morphology of the Tal-y-Llyn valley, Merionethshire', *Trans. Inst. Br. Geogr.* **30**, 15–31

9

Glacial erosion in areas
of low relief

A very remarkable action of existing glaciers is to chafe and polish the rocks over which they are pushed or dragged.... The fact is certain that ... there is continued contact between the supporting rock ... and the glacier itself. Its stupendous unwieldy mass is dragged over the rocky surface, it first denudes it of every blade of grass and every fragment of soil, and then proceeds to wear down the solid granite, or slate, or limestone, and to leave most undeniable proofs of its action upon these rocks. (J. D. FORBES, 1843)

The features most frequently associated with glacial erosion are those that have been described in the last two chapters. These are features associated with areas of considerable relative relief, where the ice is generally confined to the valleys; hence its erosive power is concentrated and can be exerted effectively. These forms will now be contrasted with those produced by ice action in areas of low relief. The essential difference between these two environments is the relative relief and not the absolute relief. Glacial erosion on an elevated plateau will not differ so significantly from that on a lowland plain as it will from glacial erosion in an area of steep relief. In the case of areas of high relative relief, valley and cirque glaciers will be the principal forms of ice accumulation unless, as in Antarctica, the amount of ice is so enormous that all relief of whatever form is buried. In areas of low relief, ice accumulation will always be of the ice-cap or ice-sheet type.

In the upland areas of low relief, the ice will be situated mainly in the ice accumulation zone. This may produce rather different results from the effects of ice erosion in a lowland of gentle relief, where the ice will usually be in the ablation zone, except in very cold climates. The lowland area is more likely to become an area of glacial deposition when the ice has retreated. Therefore the effects of glacial erosion in lowlands can often only be studied by examining the sub-drift forms.

The early controversy between the glacial erosion school of thought and the glacial protectionists was partly the result of a failure to appreciate that ice can behave both as an agent of erosion and an agent of protection according to the characteristics both of the area and of the ice concerned. The most important variables in this connection are the relief of the area and the nature of the glacier régime. Where the régime is active, in temperate areas with heavy accumulation and rapid ablation, the flow will be correspondingly fast and the ice will be able to effect considerable erosion where the relief features are suitable. On the other hand, where both accumulation and ablation are small and temperatures low, the ice will mostly move slowly and erosion, especially in areas of low relief, may well be negligible.

1 Régime and temperature: evidence from present-day ice caps and ice sheets

The temperature of the ice plays an important part and will determine whether the ice sheet belongs to the cold or temperate category. Cold ice masses have temperatures below the pressure melting-point and therefore are normally frozen to their beds. Meltwater is confined to the surface or surface layers and cannot penetrate to the bed of the ice sheet. Ice masses of this type will be considerably less effective as agents of erosion than those that can move over their beds on a thin film of water. Temperate ice masses have this characteristic, as they are at the pressure melting-point throughout their thickness. It should be remembered, however, that different parts of an ice sheet may have different characteristics. It has been suggested that bases of the Antarctic and the Greenland ice sheet may be at the pressure melting-point in their central parts. This is because of the greater pressure of the thicker ice and the accumulation of geothermal heat. Near their margins where the ice thickness decreases, the ice becomes colder as atmospheric cold can penetrate farther and the pressure is less. Thus in these parts, the ice is frozen to its bed and basal sliding becomes minimal or zero.

Basal ice velocity cannot be measured directly under an ice sheet so that indirect observations and theoretical relationships must be used to obtain information concerning ice movement and ice behaviour. C. Bull (1957) in a study of the Greenland ice sheet measured the surface slope of the ice and related this to the ice thickness and the shear stress at the ice-sheet bed. These types of measurements will eventually provide the quantitative information necessary to assess the way in which an ice sheet can erode its bed. There are, however, complications, for example those apparent in Greenland where the shear stresses were found to differ in the southern and northern parts of the ice sheet. The differences were thought to be caused by the base of the ice being at the pressure melting-point in the north, but frozen to the bed in the south.

S. Orvig (1953) has studied the variation of shear stress on the bed of the Barnes ice cap in Baffin Island. The thickness of the ice was measured by gravimetric survey on the southern lobe of the ice cap, where the ice reached a maximum thickness of 460 m. The slope of the ice surface was also measured and from these observations, the average shear stress was found to be about 0·4 bar, the value increasing in a north-easterly direction. This increase was supported by indications of more active flow in this direction. Towards the southern margins of the lobe, the ice was retreating. The low value of the shear stress of 0·4 bar in the southern part is less than that suggested by Nye for valley glaciers (Chapter 4). It would be associated with small strain rates and hence low ice velocities, especially as the ice in this area is thought to be colder than its pressure melting-point and hence frozen to its bed. Other investigations confirmed that the ice was nearly stationary (see p. 125). In this instance, therefore, there is good evidence that the ice cap was acting in a protective capacity. It is, however, possible that the shear stress values were somewhat too low as they were not measured along lines of maximum slope.

The higher shear stress values in the north-east were associated with faster flow, which may have been caused by the recent lowering of ice-dammed lakes, impounded along this margin of the ice cap. The lowering would remove a considerable amount of support from this part of the ice cap and induce more rapid flow in this direction. There may be a significant link between the presence and height of ice-dammed or other marginal lakes and the movement of the ice cap.

The Meighen ice cap is another example from Arctic Canada showing little sign of activity either now or in the recent past. Covering 85 km², its maximum thickness is about 120 m and it is frozen to its bed with basal temperatures of $-16°C$ (W. S. B. Paterson, 1969). The maximum measured longitudinal strain rate is only 0·00045 per year. Its edges are thin and gently sloping, and the ground beyond shows no obvious signs of either erosion or deposition. Recent slight recession has even uncovered patches of dead vegetation and patterned ground, confirming the view that erosional activity is virtually absent. Paterson concludes that 'These ice caps thus seem to protect the underlying land surface' (p. 341).

J. F. Nye (1959) has calculated the velocity distribution along a vertical line in the Greenland ice sheet; the results probably apply to much of the Antarctic ice sheet also and possibly to the large Pleistocene ice sheets of Northern Europe and America. In this study, he made the very important distinction, as far as geomorphological effect of the ice is concerned, between the velocity of movement on the bed and the differential movement within the ice. He comes to the conclusion that in the ice sheets of Greenland and Antarctica, the essential movement may be considered as almost entirely made up of rapid shearing in the lower layers. Some simplifying assumptions were, however, made in arriving at this conclusion. The reason for this result is partly the relatively higher temperature in the lowest layer and the dependence of the strain rate to a considerable degree on the temperature. The higher temperature causing this effect results from the accumulation of geothermal heat and heat generated by the shearing in the basal layers. Although the temperatures are higher near the base of the ice, they may well still be considerably below freezing-point. The lower layers will respond more effectively to the stresses within the ice sheet for these reasons. The bottom layer in which shearing is concentrated could be as much as 100 m thick or it could be much thinner than this value, and the actual base is frozen to bedrock in the marginal areas as noted above.

Lower shear stresses are not necessarily associated with lower velocities. If the temperature increases, the shear stress may be reduced while the velocity remains the same. A temperate ice sheet probably increases in basal velocity of flow towards its margin. The rates of surface flow also differ greatly from place to place around the margin. This variation has been clearly demonstrated around the Antarctic continent. Observations of ice-flow here have shown that some parts of the marginal regions have flow rates that are at least an order of magnitude greater than the flow rates along wide adjacent stretches. The higher rates of flow are characteristic of the outlet glaciers and vary from about 100 m/year to over 7 km/year, with an average of perhaps 0·5 km/ year (see p. 52). Between these fast-flowing streams of ice, and over large areas of the inland ice, velocities may be only 10–20 m/year. Under these circumstances, those parts of the underlying area in which the ice flow is concentrated must be undergoing considerably more rapid erosion than the neighbouring stretches where the ice is moving much more slowly.

In summarizing the factors that affect the way in which ice erodes the ground over which it is flowing, the importance of ice thickness and velocity should be stressed. Observations of ice velocities confirm that ice sheets in general flow considerably more slowly than active glaciers, and over wide areas their erosive effects may be correspondingly less. But under favourable conditions of ice streaming, as exemplified in Chapters 4 and 8, they may become locally very effective agents of erosion. Ice streaming will be particularly encouraged when the ice has a ready outlet so that its velocity can be maintained to the edge. These conditions apply particularly in cases where ice sheets descend into the sea. The well-nourished Pleistocene ice sheets, flowing westward over the north-west highlands of Scotland and calving into deep water in the Atlantic provide an example. These conditions may be contrasted with those that occurred on the east of Scotland, in the Irish Sea and on the eastern side of Scandinavia.

In eastern Scotland the ice moving down into the low ground came into contact with ice from Scandinavia, which impeded and reduced its flow. Anglesey also provides an example, where ice erosion does not appear to have significantly altered Tertiary and early Pleistocene planation surfaces (C. Embleton, 1964). This lack of effective erosion was related to the great distance of the Irish Sea from the ice sources and the blocking effects of the Welsh mountains and glaciers. However, where the Irish Sea ice was funnelled through gaps in the hills in the Dee and Mersey estuaries it may have been able to gouge out hollows in the bedrock (R. K. Gresswell, 1964). In eastern Scandinavia and the lowlands to the east, the ice moved down on to low ground and could not escape by calving, so it built up into a very slowly moving, extensive mass. In areas of this type, where the ice was not able to move fast, whether because of blockage or distance from its source, the ice sheet would not be expected to be an effective agent of erosion.

2 Landforms of glacial erosion in areas of low relief

2.1   Small- and medium-scale relief forms

One of the most characteristic forms of glacial erosion in areas of low relief is the 'knock and lochan' formation discussed by D. L. Linton (1963). It occurs both at low and high elevation and is particularly well displayed in the north-west highlands of Scotland. The ice in this area was moving fairly actively over an outcrop of Lewisian gneiss westward to the Atlantic. As a result of this glacial activity, the landscape now consists of many ice-moulded knobs, with intervening lochans that have been eroded along lines of structural weakness in the rocks. The features often show strong structural control: faults, weak dykes and other zones of weakness were exploited by the ice. Where the structural control is strong, as in the Lewisian gneiss, the direction of ice movement is not immediately obvious from the landscape lineations.

In many other areas, however, the ice imposes its direction of flow on the landforms it creates. Such features have been called glacial fluting, especially where the features are of small dimensions and considerably elongated. Good examples occur in the low-lands of central Canada, such as north-eastern Alberta where the features have formed on a large scale. The glacial flutings and lineations are eroded in the Pre-Cambrian granite-gneiss of the Canadian Shield. They occur in the area north of Lake Athabasca and are parallel to the glacial striations, but at an angle to the bedrock trends. They trend at about 60° east of north. In this area differential erosion is subdued compared with the fluting pattern. Dykes cutting across the flutings have been eroded to the same extent as the softer metasediments into which they were intruded.

In the Mackenzie District the glacial flutings resemble giant grooves (see also Chapter 6). The ridges are 3 to 8 m above the troughs, about 90 m apart, and up to 1·6 km long. It seems likely that they are elongated parallel to the ice movement. The fluting wave length has a preferred value of 90 to 120 m in a variety of rock types and a second smaller preferred spacing value of 180 to 214 m. It is, therefore, thought that the movement of the ice was in some way responsible for the fluting pattern (C. P. Gravenor and W. A. Meneley, 1958). A possible mechanism could be found if there were parallel bands of varying pressure zones in the ice. The material would be eroded from the high pressure zones to a greater extent than from the low pressure zones. The solid rock flutings merge into forms composed of drift, and both types may be erosional in character. The forms in both instances show that the ice was moving actively in the low relief areas in which they occur. This situation supports the view that ice sheets possibly erode more effectively on low-lying ground than on high ground in the accumulation area when the relief at both elevations is of low amplitude.

Somewhat larger streamlined forms also occur. These forms resemble drumlins in shape, but they differ from true drumlins in that they are composed of rock. The features have been termed 'rock drumlins'. Linton (1959) has described features of this type on the plateaux of well-jointed Lower Old Red Sandstone lavas of Lorne. Similar features are also formed a little farther south in the harder Dalradian schists and quartzites. The rounded streamlined hills are formed in the more resistant rocks within the series. Similar though larger features occur in south-west Iceland, where they are up to 8 km long. There are also examples in northern Iceland, where the alignment clearly reflects the direction of ice movement. The ice must have been about 500 m or more thick to have moved sufficiently rapidly to produce these ice-moulded forms in this area of relatively low relief.

Ice-moulded bedrock and drift features are well displayed in the Forth valley of central Scotland. M. J. Burke (1969) has analysed the orientation and vector strength of the forms and correlated the results with other evidence of the direction of ice flow. The results show great consistency in trend from south of west to north of east. The median direction is E 10°N, which agrees with evidence from striations. An inverse relationship exists between moulding and altitude, the moulding being stronger below 120 m. Ice moulding intensified the pre-existing east–west valleys, the ice responsible coming from the Highlands to the west in two streams down the Forth valley.

The hummocky glaciated surface characteristic of ice sheet erosion is especially well developed in areas of old rocks on the ancient shields, for example around the Baltic and on the Canadian shield. These are areas where relative relief rarely exceeds several tens of metres. They occur in Karelia in Finland, in the Norrland interfluve area of Sweden, central Finland, central Labrador, parts of Quebec, and north of the Great Lakes of America. Most of these areas are plateaux of about 100 to 600 m in elevation and relatively undissected by through-valleys. The relief is rather haphazard and the drainage confused. Any structural weaknesses have been clearly exploited and revealed by the ice action. The hills often have fairly steep slopes of about 10 to 15° and the intervening hollows show feeble drainage. The hollows are often peaty or contain standing water in innumerable small lakes. The exposed rocks often show very clear evidence of glacial modification in the form of polished and striated surfaces, some in the form of roches moutonnées. There are also in places small patches of blocky moraine, usually in the sheltered hollows in the lee of the bosses. These moraine patches are usually thin and discontinuous and represent ground moraine.

2.2   *Evidence of depths of erosion in areas of low relief*

The features eroded by actively moving ice sheets are generally asymmetrical, having steeper proximal slopes than distal slopes. The amount of erosion has been shown by R. H. Jahns (1943) to be considerably greater on the lee side. He based his views on a study of the sheet structure of the granites in New England. He contended that these structures were pre-glacial and were such that the sheets become thicker and flatter with increasing depth. It is thus possible to use the structure to arrive at an estimate of the depth of glacial erosion (Chapter 11). The increase of thickness of the sheets with depth is consistent. The size of the boulders derived by ice from the different sheets therefore gives a clue to the depth from which the boulders were derived. The results of the study suggest that the bosses have been lowered by about 3 to 4·5 m on the north or proximal sides and that the depth of erosion was rather greater on the east and west slopes and on the summits (Fig. 11.4). The lee or south slopes seem to have suffered the maximum erosion, which was achieved largely by plucking, rather than by abrasion as on the other faces. The plucking and quarrying that were active on the distal slopes appear to have removed a maximum of 30 m of rock. These figures indicate the minimum amount of erosion accomplished by the ice sheet in this area.

The largest boulders denote plucking to the greatest depth, and seem to have come from a depth of about 29 m where the sheets were about 4·5 to 6 m thick. The largest boulder was from a sheet 8·2 m thick, and has remained fresh and angular. Such large blocks must have been derived by quarrying from considerable depths on the lee of the bosses. These quarried faces are fairly common features of the area. The method of estimating glacial erosion gave consistent results in this area when the observations were confined to one particular granite outcrop that lay athwart the ice flow. The results suggest that erosion by a continental ice sheet is not so effective as that produced by powerful valley glaciers under favourable conditions, for such glaciers can erode hundreds or even thousands of metres vertically.

2.3 *Glacial erosion in shield areas*

There has been considerable controversy over the extent to which the great Pre-Cambrian shield areas of the world have been modified by ice-sheet erosion. On the one hand, there are those who argue that glacial erosion of the bedrock was minor or even negligible and that the main forms have survived unaltered, except in detail, from pre-glacial times. H. Niini (1968), for instance, argues that glacial denudation of the surface of Finland amounts to no more than a few metres on average, though the floors of some strongly fractured valleys may have been lowered 'some tens of metres' (p. 47). On the other hand, some have claimed that deep erosion by continental ice has occurred, and that the salient features of the present shields are intimately related to Pleistocene or older glaciations. We will examine these views in turn.

J. W. Ambrose (1964) has assembled a considerable body of evidence to support the view that glaciation only modified the details of the North American shield and was not responsible for any of its major morphological features. The valley systems, for instance, accord precisely in form and direction with structure in many areas; being completely independent of the directions of ice movement, they have not been formed by glacial action. But the strongest evidence comes from studies of the relationships between the present shield surface and the stratigraphical succession. Palaeozoic sedimentary strata lying on the Pre-Cambrian basement dip off the margins of the shield and it is deduced from their disposition and the occurrence of outliers that they once covered the Pre-Cambrian rocks completely. Ambrose argues (and is supported by R. F. Flint, 1971) that the stripping of the Palaeozoic rocks from most of the shield took place in pre-glacial times, exposing the pre-Palaeozoic unconformity. Today, the actual surface of the shield departs little from the inferred plane of the unconformity, with the exception of some glacially eroded rock basins, and outliers of Palaeozoic sediments preserve beneath them forms of pre-Palaeozoic relief that are exactly matched in the surrounding exhumed surface. Glacial modification of the stripped unconformity has been remarkably slight and local in most areas. Additional evidence from New England, where lowering of the surface amounting to only a few metres is thought to have occurred, is cited on page 321.

Against this view it is possible to argue that the very stripping of the softer Palaeozoic cover was accomplished mainly by ice-sheet erosion, which did not proceed to any great depth below the level of the unconformity in many areas because of the greater resistance of the underlying Pre-Cambrian rocks. W. A. White (1972) goes further than this and maintains that the central zone of the shield was deeply eroded by ice. Fig. 9.1 indicates his views diagrammatically. In the central zone, there is a broad basin in the shield surface now partly submerged by Hudson Bay. The vertical amplitude of this basin is of the order of 600 m; the depth of the floor below the reconstructed sub-Palaeozoic unconformity must be more than this—possibly of the order of 1000 m— to allow for the original domelike form of the latter. A considerable part of this figure, which White estimates at 400 m (compare the figure of 460 m derived on p. 173), is represented by glacio-isostatic rebound; the remainder is attributed by White to glacial erosion. Beyond the central zone, there is a surrounding belt which is claimed to be the sub-Palaeozoic stripped unconformity modified in varying degree by ice action.

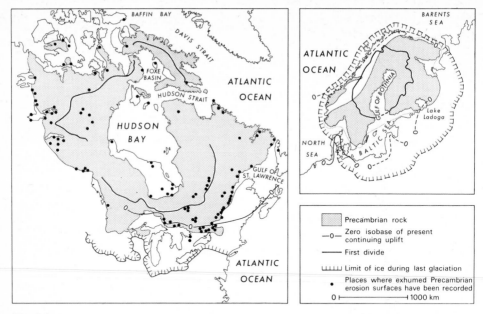

Fig. 9.1
Glaciated areas in North America and Fennoscandia, and concentric geomorphological zones of ice-sheet erosion according to W. A. White (*Bull. Geol. Soc. America*, 1972)

White argues that glacial erosion has been a major factor in the stripping process itself, removing more than 600 m of sedimentary rock.

White puts forward a general model of ice-sheet erosion in shield areas (Fig. 9.2). Zone I in the centre, a lowland that is often now partly submerged because of glacio-isostatic depression (e.g. Hudson Bay, Gulf of Bothnia), has suffered deep erosion amounting to hundreds of metres. Beyond the 'first divide' lies Zone II, an outward-sloping area with a progressively less damaged, exhumed, sub-Palaeozoic erosion surface. In Zone II are found many deep lake basins said to be accounted for by glacial erosion, such as Lake Superior (over 300 m deep), and also glacially eroded lowlands such as those of the St Lawrence and upper Mackenzie or, in the case of the Fennoscandian shield, the White Sea-Gulf of Finland–Baltic–Skaggerak depression. Beyond Zone II is the 'threshold of glacial erosion' followed by predominantly drift-covered terrain with little or no glacial erosion. The model is developed for the Laurentian and Fennoscandian shields, and it is said to be no coincidence that these were centrally located beneath Quaternary ice caps, but it is suggested that it could also apply to shields in tropical latitudes affected by Permo-Carboniferous glaciation.

These views are extreme and there is little incontrovertible evidence to support them. It seems improbable that any relics of glacial influence on the landforms of shield areas glaciated in Permo-Carboniferous times, some 250 million years ago, could have survived (the chance survivals of Dwyka tillite, etc., in South Africa are excepted). But there are more important arguments. First, lithology undoubtedly exerts a very strong influence on the relief of the shield areas, and it is just as plausible to argue that the major relief features were shaped by pre-Pleistocene non-glacial erosion as to suppose otherwise. Secondly, the central zones of ice sheets are unlikely to be zones of exceptionally intense glacial erosion: indeed, the reverse may be nearer the truth since velocities of ice flow are likely to be least here. Thirdly, White's model ignores the complex history of glaciation. As Flint (1971) notes, ice margins swept over the whole of the shield areas during ice-sheet growth, and there were important shifts in the centres of glacial outflow.

There is more general agreement over the glacial erosional origin of rock basins in shield areas ranging upwards in size to Lake Superior (see, for example, J. Tricart and A. Cailleux, 1962). Shield margin lakes resulting from glacial erosion were recognized by W. M. Davis in 1920 and termed 'roxen lakes' after Lake Roxen in Sweden. The Great Bear and Great Slave Lakes as well as many other smaller lakes occupy situations analogous to that of Lake Roxen at the margin of the Canadian shield. In Europe, the Gulf of Finland is elongated along the junction of the Baltic shield and the Russian platform with its cover of Palaeozoic rocks. Consisting of shales, mudstones and limestones, these weaker rocks have been vigorously attacked by the ice as is indicated by the large quantity of these materials in the till. It has been suggested that pre-glacial weathering prepared the way for differential glacial erosion, but in many areas the unweathered state of the rock fragments in the till makes this explanation seem unlikely.

2.4 *Glacial erosion of scarpland areas*

A distinctive type of lowland area is that where the pre-glacial relief consisted of upstanding scarps formed of more resistant rocks. In this instance, the advancing ice sheet might cause erosion which would reduce rather than enhance the pre-glacial

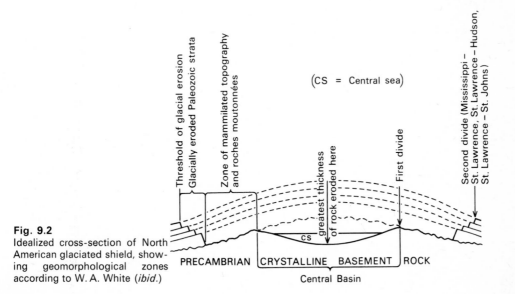

Fig. 9.2
Idealized cross-section of North American glaciated shield, showing geomorphological zones according to W. A. White (*ibid.*)

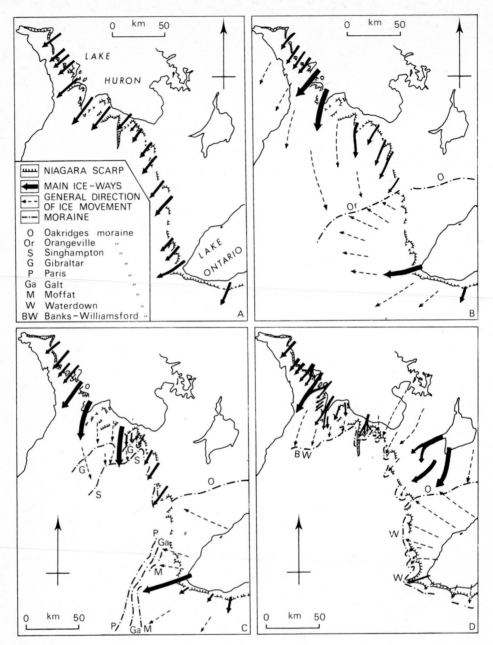

Fig. 9.3
Ice movements and moraines in the vicinity of the Niagaran escarpment, southern Ontario. Glacial stages are shown as follows:
A Early Wisconsin
B Late Wisconsin maximum
C Early Port Huron
D Late Port Huron
(A. Straw, *Bull. Geol. Soc. America*, 1968)

relief. An example of the reduction of relief is to be seen in the Finger Lakes region of western New York (K. M. Clayton, 1962). The hard rocks of this area form the Allegheny scarp, which is normally above 460 m high. It faces north towards the on-coming ice in the area south of the Lake Ontario basin, and it has been very severely reduced in elevation by glacial erosion. In a large embayment, eroded by the ice spread-ing out from the Lake Ontario basin, the height of the scarp has been reduced in some areas to less than 300 m. The ice sheet has also destroyed the pre-glacial watershed.

Farther west, ice moving up and over the Niagara escarpment has created some new re-entrant valleys according to A. Straw (1968). There were several ice advances in the Wisconsin, taking advantage of pre-existing cols and gaps in this escarpment and enlarging them, as in the Finger Lakes region. The Georgian Bay area illustrates the process well (Fig. 9.3); the Dundas valley (Fig. 9.4) is a major glacial trough. That the re-entrants are due to glacial erosion is supported by the pattern of moraines and drumlins in the area, and by the fact that their characteristics differ from those of normal stream-eroded valleys. Some may have been cut out *de novo* by eroding ice, on similar lines to the process of watershed breaching by ice described in Chapter 8.

Another example of ice-sheet erosion in a scarpland area is found in south-east Eng-land (Fig. 11.4). To the south-west of Hitchin, where active ice erosion did not occur, the Chalk forms the well marked scarp of the Chiltern Hills. This scarp has probably never been over-run by ice, at least in the Saale and Weichsel glaciations. Where the scarp is well developed, it has subsidiary benches formed of the Melbourne Rock in front of the main scarp and the Chalk Rock behind it. As the feature is traced north-eastward from the Hitchin gap, the scarp form becomes completely modified. This modification could be in part the result of the early Pleistocene marine transgression,

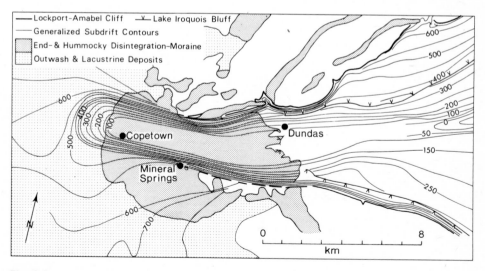

Fig. 9.4
Contours (in feet) on the sub-drift surface of the Dundas valley, southern Ontario (for location, see Fig. 9.3 where the Dundas valley is the large ice-way marked leading west from the western end of Lake Ontario (A. Straw, *Bull. Geol. Soc. America*, 1968)

but the disappearance of the Melbourne Rock bench from the scarp slope until north Norfolk is reached about 80 km to the north-east shows that this is not the only cause. The form of the main scarp is also considerably modified. It loses the steep scarp slope and becomes set back by many kilometres (see Chapter 11). Across much of Cambridgeshire and Norfolk, it forms a gentle slope, which is drift-covered in places and becomes a plateau-like surface, covering an extensive area. Farther north-east, the scarp disappears altogether and the streams draining north-west to the Fens rise 40 km east of the Gault, which outcrops west of the Chalk. The scarp does not reappear until Swaffham and then continues from this point to the Norfolk coast. Further evidence that this modification of the scarp was caused by erosive activity of the ice sheet is provided by the glacial deposits. The fabrics of the East Anglian tills indicate that ice moved from the overdeepened Fenland basin to the north-west of the Chalk and spread out over the ground to the south-east, while the large quantity of chalky material in the tills of East Anglia bears witness to the effectiveness of ice-sheet erosion of the upstanding chalk scarp.

Erosion may have been assisted in this area, and also in the Finger Lakes region, by an increase of ice thickness in the basins from which the ice was flowing over the surrounding areas of upstanding resistant rocks. Thus the smoothing and erosion of the upstanding parts of the pre-glacial landscape have been accompanied by extra deep scouring of the softer rocks that lay in the path of the ice before it reached the scarp features. Subsequent glacial deposition and the post-glacial infilling of the basins have partially masked the increase of relief related to the differential erosion of the softer rocks.

2.5 *Glacial erosion in soft-rock areas*

D. L. Linton (1962) demonstrated the range of features produced by ice in lowland areas composed generally of relatively 'soft' rocks, drawing examples from eastern Scotland. In the area of Old Red Sandstone, outcropping between Loch Lomond and the River Tay, distinctive ice-moulded forms have been fashioned. Smoothed valley sides, lacking even slight incisions by small streams now flowing down them, are attributed to ice erosion, as are tapered interfluves whose pre-glacial distal portions, where valleys converge, have been destroyed by the ice or buried by drift.

A. Straw (1958, 1963) has examined the effects of differential ice erosion in areas of Triassic and Jurassic sediments in the East Midlands. As in many such cases, the critical features have been largely concealed by subsequent deposition, but the subdrift contours of the Vale of Belvoir in the East Midlands show clearly the effect of scouring by the ice sheet moving south across the area from the lowlands of the Vale of York and Humber area. The ice was able to exploit the soft clay outcrops that run north–south to the west and east of Lincoln. In doing this, the ice sheet was able to destroy almost all trace of the presumed north bank of the proto-Trent, which in pre-glacial times is thought to have flowed across the Vale of Belvoir through the Ancaster Gap. The material eroded from the soft clay areas has been dumped on the rising ground to the south, where harder rocks outcrop. The drift forms the south Nottinghamshire Wolds, where the drift thickness exceeds 30 m.

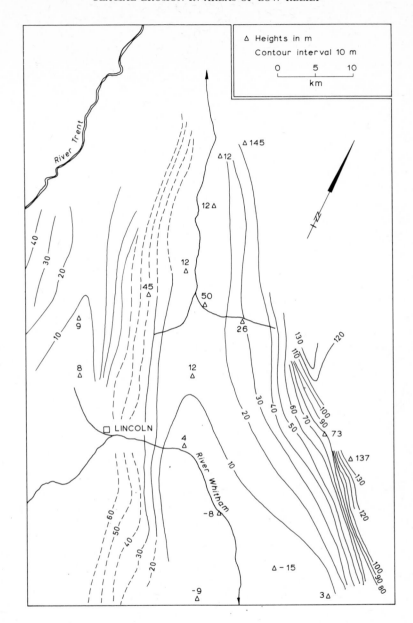

Fig. 9.5
The sub-drift contours of the Vale of Ancholme, Lincolnshire (A. Straw, *E. Midld Geogr.*, 1958)

Straw (1958) has also drawn attention to the sub-drift form of the Ancholme Valley, which occupies a clay strip in Lincolnshire between the outcrop of the Lincolnshire Limestone on the west and the Chalk of the Lincolnshire Wolds on the east. The sub-drift surface has been eroded to a little below sea level to the south of Lincoln (Fig. 9.5). To the north of the Witham, the sub-drift valley is very flat over a considerable distance at about present sea level. Had it been formed by fluvial action working to a lower base level, the valley would probably have been asymmetrical, reflecting the dip of the rocks. The valley is, in fact, symmetrical in form, suggesting that the southerly moving ice sheet was the formative agent. The ice removed the soft clays and carried them south to mingle with the drift of the area to the south, in which much Jurassic clay is incorporated with the chalk.

3 Glacial erosion and glacial protection

The examples just given from the south and east of England show that ice sheets, moving far from their mountain sources, can still exploit the soft rocks that lie across their paths. Under favourable conditions, they can also cause serious modification to harder rocks that are standing above the general elevation. It cannot be generally assumed that ice sheets, at the limit of their extent, must necessarily be incompetent as agents of erosion. They will be selective in their erosion, at times increasing the relief and at times decreasing it.

In their source regions, on the other hand, ice sheets may act in a protective role at times. The Lake District is normally thought of as an area deeply modified by glacial erosion. It is certainly true that the valleys of the area show clearly the influence typical of valley glacier erosion. In the interfluve areas and on the flattish hill tops, however, the evidence of glacial erosion is much less strong, and in fact there is evidence that in these areas the ice has been largely protective. The recognition by several geomorphologists of a series of pre-glacial planation surfaces in this area supports this view (R. B. McConnell, 1938). The same applies to neighbouring areas, including the Howgill Fells and the Southern Uplands of Scotland.

The inability of the ice to effect much erosion or to remove the previously formed planation surfaces of the flatter parts of these areas can probably be accounted for partly by their relatively low relief. The fact that the ice flow was concentrated in the valleys in such a way that the ice on the hill tops could not move effectively is also important. In the Howgill Fells, the inability of the ice to erode was partly caused by the blocking of the escape routes for local ice by other ice from neighbouring more active centres. This resulted in a reduction of velocity and the ice could no longer erode effectively. The same argument can be applied to the flatter summit areas of the Pennine Hills, on which there is little evidence of glacial erosion. Linton (1959) has drawn attention to the inability of ice to erode effectively on the high plateaux of the Cairngorm region of Scotland. He cites the presence of tors as evidence of the weakness of glacial erosion in these areas. He argues that the tors were formed by deep weathering in a warmer pre-glacial climate and were subsequently exhumed before the ice formed. The tors stand undestroyed on the upland surfaces. If the tors are in fact pre-glacial, it is clear that the ice on the plateau surfaces cannot have been an effective agent of erosion. Linton's views have been generally supported by D..E. Sugden (1968), as outlined in Chapter 8. Altogether, the evidence suggests that ice in its source area will act as a protective agent where it is resting on relatively gentle slopes, particularly where it has no ready access to lower ground and is receiving only a small amount of snow accumulation. Ice in these circumstances will move only slowly and apart from detailed modifications of the ground surface, such as the smoothing and stream-lining of some tors and bosses noted by Sugden, its role will be mainly a protective one.

Further evidence of the relative incompetence of ice sheets as erosional agents, in certain situations, is provided by observations made by J. T. Andrews (1965) in north-west Baffin Island. A magnetite dyke outcrops in the Striding Valley and sampling lines were set up at right angles to the dyke and parallel to the valley axis. All the occurrences of boulders of magnetite within a 5 m by 5 m square were counted at geo-metrically increasing distances of 10, 20, 40, 80, 160 m and so on, from the dyke. The results for observations in the direction of ice movement were plotted on double logarithmic paper. The relationship between the number of magnetite boulders and distance from the outcrop is a straight line on the graph. The rate at which the number drops off is extremely rapid, suggesting negligible transport after 4000 m; even by 400 m, the number of boulders has fallen to 4 per cent of the number in the first sample area. In the direction opposite to that of the ice flow, there were no magnetite boulders after a distance of 100 m.

The results of this experiment suggest that, on this upland plateau, there was only a minor amount of glacial movement and the ice was not capable of seriously modifying the bedrock over which it was moving. Nor was it able to transport material eroded from the bedrock for any great distance. The conditions of the ice in this area were such that it must have been frozen to its bed, at least in the marginal zone. The mean annual temperature is now about $-10°C$ at an elevation of 580 m in the vicinity of the magnetite dyke. The internal conditions of the ice when the boulders were deposited would suggest an internal movement of only 0·3 m/year. This value agrees with the observations made on the southern Barnes ice cap, where the basal shear stress is as low as 0·4 bar.

From a study of orientation patterns of surface boulders in such areas, Andrews has come to the conclusion that there may have been slight differential movement between the ice and the bedrock or till, even when the ice was frozen to its bed (see p. 138), although this movement was probably not at all vigorous. As the ice increased in thick-ness, however, the basal ice may have reached pressure melting-point and this would have facilitated movement between it and the bedrock or till, and also the development of preferred orientation in the underlying bouldery till deposits.

One reason why an ice sheet is probably a more effective agent of erosion on lowlands than in plateau areas has been mentioned by Tricart and Cailleux (1962). The zone of permafrost that develops in front of an advancing ice sheet has less time to form in the upland regions, because the ice covers the ground before the climate has been cold enough for deep permafrost to form. There is considerable evidence that deep permafrost had developed in the path of the ice sheets by the time they reached the lowlands in Europe and North America. The existence of deep permafrost on the lower ground is shown by the long-distance transport of fragile fossils in clays in Poland, where they have been carried for 300 km. This transport could only have taken place in areas where the ground moraine, incorporated in the ice, had been frozen before the ice sheet reached the area. The permafrost helps to break up the ground, as frost and ground-ice are very effective agents of mechanical weathering. The advancing ice sheet can then more effectively erode the prepared material (see Fig. 8.1). These processes operate most effectively in jointed rocks such as granite.

The relative effectiveness with which erosion occurs can be assessed by noting the proportions of the various strata crossed by the ice that are found in moraine. For example, the green rocks of Queyras in the French Alps, intrusive into schists lustrés, form upstanding peaks under normal erosion of non-glacial character, owing to their low rate of disintegration. This is because their joints are widely spaced and they therefore break into blocks that are too large for streams to transport. Under glacial erosion, the widely spaced joints, prepared by frost action, provide no obstacle to the removal of large blocks. Indeed, large blocks of this material constitute much of the morainic material, despite the fact that their outcrop occupies only between 2 and 5 per cent of the area. It seems probable that the ice itself cannot operate effectively on a low relief surface unless the rocks have been previously prepared by frost or other action. The character of the joints is of particular importance in this respect.

Another suggestion of Tricart and Cailleux concerns the effect of a general lowering of temperature of about 20°C in the rocks near the surface with the oncoming of glacial conditions. They suggest that such a general lowering of temperature will cause contraction of the rocks, and that this may initiate joints in previously sound rock. This would prepare the way for frost action and subsequent removal by the ice sheet when it reached the area.

Ice would first accumulate on the upland areas of low relief, when the level of the snow-line fell below the plateau surface. The snow would prevent the formation of permafrost and help to account for the relatively limited amount of glacial erosion where the relative relief was low. The uplands of Norway provide a good example of this (K. Strøm, 1948). The most conspicuous of the upland low relief surfaces of Norway is the broad plateau on part of which the Jostedalsbreen ice cap lies. The elevation of this surface during the Plio-Pleistocene period, combined with the falling temperature, may have initiated the formation of the great Scandinavian ice sheet. This upland plain, lying at an elevation of about 1800 m, has suffered relatively little glacial modification. It remains as a fairly level and undissected surface, showing only minor glacial moulding in areas where it is exposed away from the ice cap remnants. It is, however, deeply bitten into around its margins by the outlet glaciers that pour with high velocity through the ice falls around the edge of Jostedalsbreen.

The contrast between the high plateau surface (the 'fjell') and the serrated Jotunheim Mountains shows how the slow moving ice cap has protected the flat surfaces. The mountains of Jotunheim, on the other hand, stood out pre-glacially as mature hills above the plateau surface, while during the glacial period, their summits stood up above the ice as nunataks. They have been deeply sculptured by cirque and valley glaciers, and their summits have been severely riven by frost action to form jagged peaks. The upper hill slopes now consist of piles of huge boulders leading up to frost-shattered ridges.

In this part of Norway a great variety of glacial landscapes lies close together. There is the broad fjell, showing slight evidence of glacial action little modified by deep erosion. Above the plateaux, the isolated mountain groups show evidence of strong cirque formation and the peaks are characteristically frost-shattered nunataks. Cutting into the fjell are the deep fjords (Chapter 10), which demonstrate clearly the immensely powerful action of glaciers when they are steep, fast-flowing and thick. The difficulty of generalizing about ice erosion is apparent when all these forms are found in close proximity as in Norway and other mountain areas of this type. Greenland, Labrador, Scotland and parts of the south-west of the South Island of New Zealand provide other examples of similar assemblages of glacial forms.

4 Conclusions

In assessing the geomorphological role of glacial erosion in areas of low relief, whether they be at high or low elevation, each area must be considered individually. It is probably generally true, however, that ice sheets are more effective agents of erosion in lowland areas than in upland areas. The reasons for this may differ from one area to another. One reason is that the high-level areas of low relief are normally those that are relatively resistant to erosion, and for this reason are upstanding. Other significant reasons include the effect of permafrost on the low ground in front of an advancing ice sheet and the susceptibility of certain rocks to glacial erosion. Ice thicknesses also tend to be greater in the lowland areas than on the elevated plateaux. The examples that have been quoted from the Midlands of England and elsewhere illustrate these points. That ice sheets can erode effectively in some areas of low ground has been demonstrated by the differential erosion suffered by scarps, because of their upstanding position relative to the approaching ice.

One of the most characteristic landscapes of ice-sheet erosion in areas of relatively hard, homogeneous rocks with minor lines of weakness, is the hummocky knock and lochan relief, characteristic, for example, of parts of the north-west Highlands of Scotland and parts of western Connemara in Ireland as well as widespread areas of the Canadian and Baltic shields. In these regions, deposition is limited to thin spreads of basal moraine, deposited in the lee of the bosses.

A general relationship exists between the forms produced by the ice sheets and the general structural zones. The zone of fjell and fjord in western Scandinavia is succeeded eastward by a wide belt of hummocky boss country. This in turn gives way to the fringing roxen lakes and other water bodies along the margin of the shield. Farther from the ice centres, the glacial landforms become predominantly depositional. It is not clear how far this relationship depends on the pre-existing relief and structure and how far it depends on the zones of different action within the ice sheet. It is likely that both factors have combined to produce the observed effects. J. B. Bird (1959) has claimed that large ice sheets played a generally protective role over much of Canada and, in essence, such ideas have been supported more recently by Ambrose (1964) and Flint (1971). V. Tanner (1938) has given similar evidence for Finland. Extensive surfaces of planation have been preserved undissected and little modified in Swedish Lapland, where many pre-glacial features have survived. Much more radical are views such as those of White (1972) attributing deep bedrock erosion to thick continental ice sheets. In our present state of knowledge, it is not yet possible completely to reconcile such diverging viewpoints.

5 References

AMBROSE, J. W. (1964), 'Exhumed paleoplains of the Precambrian shield of North America', *Am. J. Sci.* **262**, 817–57

ANDREWS, J. T. (1965), unpub. Ph.D. Thesis, Univ. of Nottingham, Chapter 4

BIRD, J. B. (1959), 'Recent contributions to the physiography of Northern Canada', *Z. Geomorph.* NF **3**, 151–74

BULL, C. (1957), 'Observations in north Greenland relating to theories of the properties of ice', *J. Glaciol.* **3**, 67–71

BURKE, M. J. (1969), 'The Forth valley—an ice-moulded lowland', *Trans. Inst. Br. Geogr.* **48**, 51–9

CLAYTON, K. M. (1965), 'Glacial erosion in the Finger Lakes region (New York State, U.S.A.)', *Z. Geomorph.* NF **9**, 50–62

DAVIS, W. M. (1920), 'A Roxen lake in Canada', *Scott. geogr. Mag.* **41**, 65–74

EMBLETON, C. (1964), 'The deglaciation of Arfon and southern Anglesey, and the origin of the Menai Straits', *Proc. Geol. Ass.* **75**, 407–30 (see pp. 428–9)

FLINT, R. F. (1971), *Glacial and Quaternary geology*

GRAVENOR, C. P. and MENELEY, W. A. (1958), 'Glacial flutings in central and northern Alberta', *Am. J. Sci.* **256**, 715–28

GRESSWELL, R. K. (1964), 'The origin of the Mersey and Dee estuaries', *Geol. J.* **4**, 77–86

JAHNS, R. H. (1943), 'Sheet structure in granites: its origin and use as a measure of glacial erosion in New England', *J. Geol.* **60**, 71–98

LINTON, D. L. (1959), 'Morphological contrasts of Eastern and Western Scotland', in *Geographical Essays in Memory of Alan G. Ogilvie* (ed. MILLER, R. and WATSON, J. W.)

(1962), 'Glacial erosion on soft-rock outcrops in central Scotland', *Biul. peryglac.* **11**, 247–57

(1963), 'The forms of glacial erosion', *Trans. Inst. Br. Geogr.* **33**, 1–28

McCONNELL, R. B. (1938), 'Residual erosion surfaces in mountain ranges', *Proc. Yorks. geol. Soc.* **24**, 76–98

MELLOR, M. M. (1959), 'Ice flow in Antarctica', *J. Glaciol.* **3**, 377–84

NIINI, H. (1968), 'A study of rock fracturing in valleys of Precambrian bedrock', *Fennia* **97** (6), 60 pp.

NYE, J. F. (1959), 'The motion of ice sheets and glaciers', *J. Glaciol.* **3**, 493–507

ORVIG, S. (1953), 'On the variation of the shear stress on the bed of an ice-cap', *J. Glaciol.* **2**, 242–7

PATERSON, W. S. B. (1969), 'The Meighen ice cap, Arctic Canada: accumulation ablation and flow', *J. Glaciol.* **8**, 341–52

SHARP, R. P. (1954), 'Glacier flow: a review', *Bull. geol. Soc. Am.* **65**, 821–38

SMITH, H. T. U. (1948), 'Giant glacial grooves in Northwest Canada', *Am. J. Sci.* **246**, 503–14

STRAW, A. (1958), 'The glacial sequence in Lincolnshire', *E. Midld Geogr.* **2**, 29–40

(1963), 'The Quaternary evolution of the Lower and Middle Trent', *E. Midld Geogr.* **3**, 171–89

GLACIAL EROSION IN AREAS OF LOW RELIEF 293

(1968), 'Late Pleistocene glacial erosion along the Niagara escarpment of southern Ontario', *Bull. geol. Soc. Am.* **79**, 889–910

STRØM, K. M. (1948), 'The geomorphology of Norway', *Geogrl J.* **112**, 19–27

SUGDEN, D. E. (1968), 'The selectivity of glacial erosion in the Cairngorm Mountains, Scotland', *Trans. Inst. Br. Geogr.* **45**, 79–92

TANNER, V. (1938), 'Die Oberflächengestaltung Finnlands', *Bidr. Känn. Finl. Nat. Folk* (Helsingfors), **86**

TRICART, J. and CAILLEUX, A. (1962), *Le modelé glaciaire et nival* (Paris)

WHITE, W. A. (1972), 'Deep erosion by continental ice sheets', *Bull. geol. Soc. Am.* **83**, 1037–56

10

Glaciation in coastal environments

The mountains in Kangerdlugssuak Fjord, though a little lower than those in the sound by Uperni-vik Island, descend precipitously to the water at angles of eighty degrees from heights of full 1000 metres. The glacier at the head terminates in a steep wall advancing into the fjord.... Greenland is intersected by many large fjords which ... pierce deeply into the country. In front of the fjords near the open sea there is a 'Skärgård' of larger and smaller islands. (AMUND HELLAND, 1877)

There are certain coastlines in the world where the effects of glaciation are dominant, and where a distinctive environment results from the juxtaposition of glacially sculptured landforms and the sea. Among the features encountered in such coastal regions are fjords, representing glacial troughs now partly occupied by the sea, the fjärds and föhrdes of glaciated lowland coasts, and the partially submerged platforms known as strandflats.

1 Fjords

Fjords are best developed on the coasts of British Columbia, southern Alaska, southern Chile, eastern Canada (especially east Baffin Island), Greenland, Norway,* Iceland and Spitsbergen, the south-west of South Island (New Zealand), and finally Antarctica where to a considerable extent they are still occupied or buried by ice. They are more common on west-facing than east-facing coasts, they occur mostly in higher latitudes (more than 45° from the Equator), and they are usually backed by rugged and dissected highlands. Fjords have the features associated with glacial troughs, possessing a cross-profile of catenary form. As the lower part of the trough is submerged, only echo-sounding can provide adequate detail of the trough floor, which normally appears broad and relatively flat, though this may be the result of sediment infilling. The steepness of the trough sides, suggesting intensity of glacial down-cutting, is one of the most impressive features of many fjords, for example the Naeröyfjord in Norway or Milford Sound in New Zealand. Again in common with many glacial troughs, a cross-profile

---
* It should be noted that the term 'fjord' in the Scandinavian languages is used in a much broader sense, including lakes and coastal inlets of other types.

of composite form is often evident, showing successive stages in the deepening of the trough by both ice and water erosion. In the case of the Sognefjord, the largest of the Norwegian fjord systems, remnants of former high-level valley floors can be clearly seen above the steep sides of the main central trough—the bench at 400–500 m on the east of the entrance to Aurlandsfjord is a good example—providing evidence for the reconstruction of former fluvial valley systems draining west (see pp. 301–2).

Sounding has shown that fjords nearly always possess submerged rock basins, separated by rock bars which are termed thresholds when they lie at or near the sea entrance. Some claim that the threshold, with the associated rock basin behind, is the principal diagnostic feature of a fjord, most clearly distinguishing it from the ria of a submerged non-glaciated coast. The Sognefjord possesses one of the deepest known rock basins in the world. The main fjord head begins at Årdal, and the floor sinks evenly and gently to a depth of 640 m at the junction of Lustrafjord, 940 m at the entry of Aurlandsfjord, over 1000 m off Balestrand, and finally touches 1308 m between Höyangsfjord and Vadheimsfjord. Even this is not rock bottom: the flatness of the fjord floor in a transverse sense here (Fig. 10.1C) almost certainly indicates some sedimentary

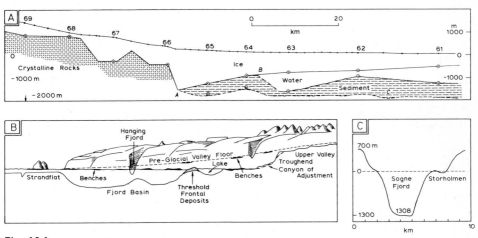

Fig. 10.1
A Section along Skelton Inlet, Antarctica, based on gravimetric and seismic observations (A. P. Crary, *Bull. geol. Soc. Am.*, 1966)
B Diagrammatic section along a Norwegian fjord, showing assemblage of characteristic features (J. Gjessing, *Norsk geogr. Tidsskr.*, 1966)
C Cross-profile of the deepest fjord basin in Norway, middle Sognefjord (J. Gjessing, *op. cit.*)

infilling. Depths in excess of 1200 m continue to be recorded as far west as the entry of Böfjord, beyond which the Sogne swings to the south-west and shallows rapidly to less than 200 m. The great Sognefjord has an entrance less than 200 m deep and only 3 km wide. The threshold is marked by the occurrence of numerous rocky islets or skerries; the threshold is clearly a rock bar and not morainic accumulation. Thus behind it lies an enclosed rock basin, 1100 m in depth. It is not clear why the Sognefjord should attain such great depths compared with those of neighbouring fjords.

The deepest fjord in northern Norway is Tysfjord, 727 m, enclosed by a bar little more than 200 m deep which in turn is followed seawards by a further rock basin descending to −660 m. Other deep fjords in the world are listed in Table 10.1. In the case of the fjord system of north-east Greenland, one of the most remarkable and spectacular of any, the fjords are all enclosed by rocky thresholds rising to within 400 m of the sea surface and often much less—Vega Sound is practically blocked by rocky islands with only shallow straits between. In northern Ellesmere Island, in the

Table 10.1　Maximum known depths, in metres, below present sea level, of some fjords

Vanderford, Vincennes Bay	Antarctica	2287
Skelton Inlet	Antarctica	1933
Sognefjord	Norway	1308
Hardangerfjord	Norway	870
Upernavik Icefjord	Greenland	1055
North-West Fjord, Scoresby Sound	Greenland	1450
Chatham Strait	Alaska	883
Nansen Sound, Ellesmere Island	Canada	920
Messier Channel	Chile	1288
Baker Fjord	Chile	1244

Canadian Arctic, the bathymetry of the Nansen Sound–Tanquary Fjord system is described by G. Hattersley-Smith (1969) who notes how the deepest basins are often located at the junctions of tributary fjords. The floor of Nansen Sound descends to −920 m, seawards of which it rises to within 500 m of sea level. Other examples of deep submerged glacial troughs, reaching depths of 900 m, characterize the east coast of Baffin Island (O. H. Løken and D. A. Hodgson, 1971). The deepest fjords are usually also the longest (Sognefjord threshold to Årdal, 160 km), probably reflecting the fact that the largest glaciers with many tributaries were capable of the greatest amount of erosion.

　　Fjords display diverse relationships with rock structure and occur in a great many different types of rock, from Pre-Cambrian to Tertiary in age and including both sedimentary and igneous formations, but it is the older crystalline rocks that are the most competent to form the deeply cut and steep-sided troughs. The rectilinear pattern of fjord coastlines has for long been held as evidence that the fjords developed along two or more intersecting lines of fracture. R. I. Murchison urged this connection in 1870, and J. W. Gregory (1913) adopted the extreme and now generally abandoned viewpoint that fjord systems were largely of tectonic origin, the result of fracturing under tension; subsequent fluvial and glacial action was held responsible for only slight modification and evacuation of fault-shattered debris. In 1927, he applied this view in detail to the fjords of the Hebrides. Inadequate geological mapping still makes it difficult to disprove the tectonic theory in many areas. H. W. Ahlmann (1941) studying north-east Greenland found that the influence of faulting was difficult to assess, but pointed out that geological mapping to date had failed to reveal any major longitudinal faults, and that in any case 'tectonic phenomena...as orientating factors of the erosion agencies, [must] be carefully differentiated from the results of such agencies' (p. 171).

B. A. O. Randall (1961), examining the Lyngenfjord area of northern Norway, concluded that there was a very close relationship between fracture pattern and valley direction. A vector diagram of valleys and fjords in South Lyngen totalling 386 km in length revealed primary concentrations on bearings of 10–20° and 290°, with several secondary maxima. The most important directions of fracture (based on analysis of 95 fractures) were 10–20° and 40–50°. The latter corresponds to a secondary concentration of valley and fjord directions. Randall concluded that valley direction intimately depends on the local fracture pattern, and suggested that the actual sites of valleys (and fjords) are lines of fracture, thus supporting Gregory's contention. This view has been strongly challenged by R. Nicholson (1963). Nicholson's mapping in the Glomfjord district of northern Nordland shows that normal faults or veins are few in number; thrusts are present, but appear to be no more significant in terms of ease of erosion than ordinary bedding contacts. Fjords such as Glomfjord or Melfjord cut across contacts between metamorphics and the granite-gneiss, and no fractures or lines of fracture were visible on fjord walls, though these were exposed up to heights of 1000 m. Both Randall and Nicholson exclude joints from consideration, for, as indicated in Chapter 8, there is good evidence that joints of dilatation type may result from the excavation of deep glacial valleys and may therefore be of secondary origin.

Most workers would now agree that fractures are not necessary for fjord formation. All elements of rock structure, including strike direction and foliation, may affect fjord direction, and erosion merely selects the most important weaknesses in any region.

10.1 *Erosional development of fjords*

Since most fjords pass inland into normal glacial troughs, and both possess common characteristics, there is now no dispute that glaciation is responsible for the main features of fjord physiography. The extent of penetration of the sea today is simply determined by the depth of glacial excavation of the trough floor, for glacial erosion, unlike fluvial erosion, is not controlled by a base-level, though sea level does control the point at which the ice in a fjord begins to float. The magnitude of the rock basins now inundated by the sea (1100 m in the Sogne) attests the power of glacial erosion in these localities, for no other mechanism is available for the extraction and uphill removal of rock debris from these basins assuming that the basins are of erosional origin and that they are totally enclosed by solid rock. The fjord coasts all lie in areas conducive to the development in the Pleistocene or at present of thick and rapidly moving glaciers, descending steeply from highlands well nourished by snow. As well as gouging out deep basins, the trunk glaciers also cut down more rapidly than tributary glaciers, thus producing many examples of hanging valleys, and frequently the size of the glaciers was such as to enable them to spill over previous divides and create new routes for the discharge of ice. Thus the Nordenskjold glacier in north-east Greenland once moved through Mystery Valley towards Is Fjord, and in the same area, there is a good example of a partially demolished watershed between Franz Joseph Fjord and the head of Dusén Fjord.

Attempts have been made to disprove the existence of deep enclosed basins in fjords by claiming that the thresholds are merely moraines or the results of landslips, for example. Some thresholds may indeed not be solid rock, as at the mouth of Yakutat Bay, Alaska, but the dredging up of occasional erratics or loose boulders from a threshold by no means disproves that the feature consists of solid rock beneath superficial debris. Many thresholds are undoubtedly solid—witness the hard Devonian conglomerate of the Sognefjord threshold, or the rocky islets which often top the threshold. The thresholds of some fjords are much too enormous to be regarded as moraines and their low angle of back-slope into the fjord basin also militates against this view. It is unfortunate, however, that echo-sounding does not distinguish between solid rock and morainic thresholds.

Nor are the basins simply the result of regional back-tilting. The large amount of tilting necessary would be clearly revealed in the attitudes of other land forms. On the other hand, a degree of back-tilting is involved in many fjords as a result of incomplete isostatic recovery since the last glaciation, or in the case of fjords bordering Greenland and Antarctica which are still depressed by the weight of present ice. Some fjord basins will in time become somewhat shallower, but the amount involved is nowhere near that needed to account for the basins.

In short, the basins must be regarded basically as the result of glacial erosion. The question then arises why the maximum deepening by ice should normally occur in the central parts of the fjords and why a threshold should exist at the fjord mouth. It is not often a question of differential rock resistance; the deepest parts of the northeast Greenland fjords described by Ahlmann (1941) lie in crystalline and sedimentary rocks of pre-Devonian age, whereas the shallower outer parts of these fjords lie in a belt of younger and generally less resistant rocks. R. Koechlin (1947) contended that the threshold marked the approximate termination of the fjord glacier rapidly melting in contact with salt water. W. A. Don Munday (1947) thought that thresholds were the result of decreased erosion at valley mouths where the ice was able to spread out and its velocity thus became much reduced. Clearly also glacial erosion will cease at the point (the grounding line) where the lower glacier, reduced in thickness, begins to float and loses contact with the ground. Another explanation of the threshold was given by A. Cailleux (1952) who with M. Boyé emphasized the part played by deep freezing of the ground prior to the advance of ice in a region, as discussed in previous chapters. It can be argued that the depth of freezing and rock shattering would decrease rapidly to nil near the sea coast and that therefore in approaching the coast, a glacier would be able to erode less deeply, giving rise to a threshold. But the depth of rock basins behind thresholds seems much too great for this to be the sole explanation; the glaciers gouging out rock basins must have cut deeply below even the maximum depth of previous frost shatter. The same disadvantage attaches to J. P. Bakker's hypothesis (1965) of deep pre-glacial weathering to prepare the basins for glacial erosion (see Chapter 8): weathering to depths of many hundreds of metres is unlikely. Nevertheless, J. Gjessing (1967) contends that many Norwegian fjords are located in zones where deep weathering facilitated later water and ice erosion.

A recent study by A. P. Crary (1966) of an Antarctic ice-covered fjord, Skelton Inlet, has thrown some light on the development of fjord basins. Seismic and gravimetric observations enabled the probable profiles of the bedrock and ice under-surface, and the thickness of sediment infilling, to be deduced (Fig. 10.1A). At the southern end of the section illustrated, the fjord glacier merges with the Ross Ice Shelf. Point A represents the grounding line of the glacier, marked by a slight but sudden fall in ice surface elevation; from A to B the ice seems to be just touching sediments (morainic and marine, probably), as though a recent increase in ice thickness has occurred. The ice in the section A–B, alternatively, may be separated from the sediment by a water layer too thin to be detected by geophysical means. From B seawards, the floating ice gradually thins. The amount of seaward thinning can be deduced from the formula

$$11{\cdot}17 \quad \frac{\delta t}{\delta x} = \frac{\tau_s}{\gamma}$$

where δt is the change in ice thickness along the fjord over a distance x, τ_s is the shear stress (bar) on the sidewalls, and γ is the half-width of the fjord (km). Uncertainty of the values τ_s reduces the usefulness of this relationship, but some estimates, ranging from 0·8 to 1·2 bar, are made by Crary from studies of crevasses and ice movement, using Glen's power flow law. The floating ice thins down-valley as the sidewall shear stress decreases or as the valley widens. If, for any reason, sea level relative to the land falls slightly, or if the ice thickens owing to increased accumulation, then part of the floating ice will become grounded and bedrock erosion will be possible (C. Swithinbank and J. H. Zumberge, 1965). Since the inner thicker part of the floating ice will ground first, it will tend to erode a basin that shallows seaward.

It should be stressed that fjords are not the result of glacial troughs later being subjected to submergence. It is quite clear that the floors of fjords were eroded to their present levels when the land stood at approximately its present height with respect to sea level. The post-glacial rise of sea level, of the order of 100 m, is minute in comparison with most fjord depths, and is largely irrelevant. G. K. Gilbert in 1910 correctly maintained that glaciers entering the sea would continue to displace sea water and exert normal pressure on their beds until flotation actually occurred. For ice to erode the floor of the deep Sognefjord basin, it would have to be about 1600 m thick, assuming sea level to have been 100 m lower than at present, though evidence from erratics and striated surfaces at over 500 m above sea level on the islands in the mouth of the fjord indicate that, at times, the ice in the deep basin behind must have been at least 1800 m thick and probably substantially more—3000 m is mentioned by H. Holtedahl (1967).

There is abundant evidence, some of which has already been indicated, that the fjord glaciers were largely guided in their paths by well-defined valley systems. In north-east Greenland, J. H. Bretz (1935) thought that the ice had taken advantage of stream-eroded valleys cut into the plateau margins. Ahlmann (1941) pointed in this area also to the pattern of fjords, which strongly suggests original fluvial action. In some places elbows of capture and reversed drainage can still be picked out. Late-Tertiary and early Quaternary uplift caused rapid incision of the streams, producing well-developed valleys for the glaciers subsequently to occupy. In western Scotland, S. Ting (1937) described the existence of a pre-glacial valley system draining south and west, south of Skye and Lewis, and generally north farther north. As J. A. Steers describes (1952), large parts of these valleys were glaciated and subsequently became fjords or sea-lochs. It is in western Scotland that the deepest British rock basins are found, such as Loch Morar (depth 308 m), and in the Inner Sound, east of Skye (324 m) (A. H. W. Robinson, 1949).

Although most workers attribute the greater part of fjord erosion to ice, some maintain that the ice merely modified valleys already of substantial depth and carved by other agents. A novel suggestion, not yet fully evaluated, is that of J. H. Winslow (1966, 1968), who discusses the undoubted problems encountered by the simple glacial hypothesis. Among these problems are the need to explain longitudinal fjords running across the general direction of ice movement from land to sea, the explanation of the deep submarine extensions of many fjords on the continental shelf (see p. 303), and the apparent existence of well-developed fjords in locations where ice accumulation must have been limited. A good example of this last problem lies in the Faröe Islands, where many of the fjords are paired head-to-head with only low ground separating them (for instance, at Klaksvik). To account for these and other anomalies, Winslow suggests that fjords represent glacially modified raised submarine canyons. The fjords are said to have acquired many of their distinctive characteristics as submarine canyons before glaciation, and were eroded initially by submarine processes, possibly turbidity currents, before uplift and glaciation. The apparent lack of submarine canyons continuing off-shore from certain fjords may be explained by subsequent infilling with sediment, and some seismic evidence to support this idea has been obtained for the Marlborough Sounds in New Zealand.

Winslow's views have been disputed by J. M. Soons (1968) who considers that the hypothesis applied to most fjords in New Zealand creates more difficulties than it resolves. Clearly in this controversy a great deal hinges on how much glacial modification of pre-glacial features (whether fluvial forms or submarine canyons) is said to have taken place, and on whether particular examples may fairly be described as 'fjord-like' or not. There is no doubt that both submarine canyons and fjords separately exist, and there is no *a priori* reason why glacially modified submarine canyons should not also exist. The difficulty lies in recognizing and distinguishing them in practice.

In Norway, J. Gjessing (1956, 1966) has attempted to separate features of fluvial and glacial erosion in connection with fjord and fjord valley development (Fig. 10.1B). He accepts the glacial origin of the fjord basins and intervening rock bars sometimes capped with morainic deposits. Another feature of the fjord valleys, and often a very impressive one, is the abrupt trough end. Many fjord valley heads in Norway share this characteristic: examples are the head of Isterdalen at the Trollstigveien, the southern end of Flåmsdal terminating in a 400 m cliff leading up to Myrdal and the Hardangervidda plateau, and the Måbödal gorge which in one direction passes down to Eidfjord and in the other heads in the giant Vöringfoss where the river leaps 150 m from the Hardangervidda at Fossli. The steep ends of the valleys heading in the Jostedalsbreen plateau are genetically similar, though in many cases they are still submerged by glacier ice, as in the ice-falls of Austerdalsbreen or of Brixdalsbreen. Many previous workers regarded these trough ends as pre-glacial (or possibly interglacial) knick points of water erosion, retreating up-valley after a period of Tertiary and Pleistocene uplift.

Gjessing argues, however, that the trough ends are of Pleistocene glacial origin, that the pre-glacial valley profile had no trough end (Fig. 10.1B), and that this feature has developed successively in glacial and inter-glacial episodes. The trough ends often possess a striated and ice-worn appearance, and are clearly at variance with the work of the present streams which are cutting 'canyons of adjustment' into them. The cutting of these notches into the trough ends may have been partly the work of subglacial streams (as in the case of gorges through valley steps discussed in Chapter 8) and partly the result of waterfall cutting when the melting ice-margin lay near the brink of the falls during deglaciation. H. Holtedahl (1967) considers that subglacial stream erosion was of major importance in the formation of the canyons at the heads of the Måbö, Flåm and other fjord valleys and may have played some part in the deepening of the fjord troughs themselves. Giant potholes (Fig. 6.4) up to 20 m deep and 16 m in

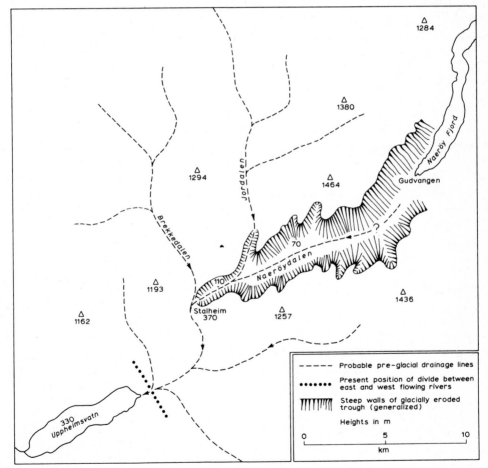

Fig. 10.2
The region around Naeröydalen, Vestland, Norway. A glacially eroded trough, partly submerged to form Naeröyfjord, ends abruptly at Stalheim; its excavation has disrupted the pre-glacial drainage and displaced the watershed between east and west flowing rivers by at least 20 km

diameter in and around the canyons provide striking evidence of the power of sub-glacial stream erosion.

A small area around Stalheim above the head of Naeröyfjord illustrates some of these relationships (Fig. 10.2). Pre-glacial drainage was to the west through the valley of Uppheimsvatn; Brekkedalen and Jordalen at one time joined near Stalheim and escaped by this route, whose valley floor is still visible at 370 m around Stalheim. The older view involved diversion of this drainage by river capture into Naeröydal, and the subsequent modification of the latter by glacial action. The alternative view, urged by O. Holtedahl (1960, p. 520) and others, is that the Naeröydal trough and fjord beyond are basically the result of glacial erosion. The 200 m step or trough end at Stalheim was eroded by rapidly flowing ice escaping to the north-west and the Sogne-fjord, and the step was notched to a small extent by waterfall erosion. The pre-glacial watershed was displaced by 10–16 km. Basins of glacial erosion lie in Naeröyfjord, for instance in the section (up to 60 m deep) between Gudvangen and the islands by Naeröy village, and in the fjord valley above Gudvangen where they are infilled by alluvium.

2 Skjaergård and Strandflat

Extending seawards from many fjord coasts there is often to be found a zone of low rock islands (or skjaergård) and partially submerged platforms of erosion (strandflats). The features are undoubtedly best developed off the Norwegian coast, where the strandflat and skjaergård appear around Stavanger and continue (with breaks) to the North Cape region, reaching a width of 50 km off Ranfjord (latitude 66°). Other coasts possessing similar off-shore features are those of western and southern Iceland (there are typical skjaergård coasts off Breiðafjörður and Faxaflói, but in some other parts the strandflat has been covered by the sandar), Spitsbergen (strandflat up to 10 km wide), and western Greenland (up to 30 km wide at Gothaab).

The Norwegian strandflat is essentially a horizontal cut into the land mass (K. Strøm, 1948), though there are probably several different closely spaced levels involved. Platforms at 8 to 18 m and 30 to 40 m have been claimed for instance, as have submarine platforms at shallow depths, and it also seems likely that isostatic tilt-ing has occurred. Smöla and Fröya are good examples of large strandflat islands each measuring 200–300 km², with most of their surfaces below 20 m in height, and sur-rounded by countless small islets and reefs. Some platforms seem to be virtually horizon-tal, or possess seaward slopes as small as 1 in 1000; cases of slope towards the mainland probably represent incomplete isostatic recovery. The strandflat often ends abruptly at its inner margin, though it can be traced a short way up some fjords as narrow benches (for example, Sognefjord). It cuts indiscriminately across geological structure, and also across some fjord rock basins where these extend seaward of the mainland. The skjaergård islets merely represent the post-glacial accident of partial submergence of certain strandflat levels. These islets should be distinguished genetically from the skerries of south-eastern Norway and Sweden, for instance, where the irregular glaci-ated rock surface *slopes* gently outward beneath the sea.

Both strandflat and skjaergård are intimately connected with glaciation. The rounded knobs of hard crystalline rock forming the islets bear witness to severe glacial erosion, and the strandflat levels themselves may be largely of Pleistocene age. B. G. Andersen (1965) summarizes prevailing opinion that the strandflat is a composite feature, formed by the combined action of waves, sub-aerial denudation, frost-wedging in the surf-zone and glacial erosion. The part played by marine action, however, is not easy to assess. Although the bevelling of the rock structures and the abrupt inner margin strongly suggest wave action, the platforms are very broad, cut in very resistant rocks, and possess little or no seaward slope. Simple wave erosion over such a width and gradient is very unlikely, if not impossible. Speculation about the evolution of the Norwegian strandflat is of long standing and it is impossible here to present all the different suggestions. Nansen's work published in 1922 provides a survey of previous views, and presents his own hypothesis, which includes a pre-glacial period of denudation to a peneplain, followed by wave-cutting at various sea levels in the Pleistocene. In cold periods, wave-cutting was facilitated by sea-spray freezing in rock crevices and thus loosening material for waves to transport. A novel suggestion was that of E. Dahl (1947) who claimed that much of the strandflat surface was once produced by the coalescing cirque and valley floors of coastal glaciers, the walls of the cirques and valleys having been demolished to subdued forms and the whole then somewhat smoothed by ice-sheet erosion. Outlines of former cirques can perhaps be detected on parts of the strandflat. H. Holtedahl (1959), working on the strandflat in the Møre-Romsdal area, also concluded that erosion by coastal glaciers played an important part in the modelling of the strandflat. Strøm (1948) emphasized marine abrasion, showing how the strandflat was broad and well-developed only on wave-exposed coasts (contrast the east- and west-facing coasts of Moskenesøy, for example). He regarded the several wave-cut levels as having been modified by ice erosion in successive glaciations, the last glaciation having reached only as far as the inner parts of the strandflat. The width of the strandflat in many places, however, demands some earlier period of preparation and lowering of the land surface near to present sea level before Pleistocene wave-cutting commenced, and in this respect pre-glacial peneplanation (as Ahlmann strongly invoked in 1919) has much to commend itself, unless one accepts the view that much of the strandflat is a cirque-floor surface.

An interesting feature of the strandflat and the continental shelf beyond is the series of deep submarine troughs and basins that often continue the lines of fjords far out from the mainland. The features are beautifully displayed on the maps prepared by O. Holtedahl (1940): Storfjord, for instance (Holtedahl, Sheet 5), has extensions reaching out 80 km from the large coastal islands. These troughs may be of glacial origin, but it is far from clear how or why ice extending this distance from the mainland on an almost flat surface, presumably as a piedmont ice sheet, should be able to erode selectively, or even to erode at all. J. H. Winslow (1966) has suggested that the submarine troughs are not glacial features but fall in the category of submarine canyons. Other possibilities are that they are tectonic features, or that they may represent in part the work of sub-glacial stream erosion referred to in the next section of this chapter.

3 Glaciation of lowland coasts

While such coasts lack the spectacular scenery associated with glacial erosion in mountain-backed coastlands, they possess some distinctive forms worthy of investigation and
analysis. A good example of such a coast is that of Maine, USA, north and north-
east of Portland. Numerous narrow inlets and islands produce a highly irregular coastline, yet neither the peninsulas in their distal portions nor the islands rise much more
than 30–50 m above the sea. Glacial erosion has played an important part in determining the present form of the inlets. Most of them are developed on weaker belts of Palaeozoic sedimentary and metasedimentary rocks trending approximately north–south and
coincident with the direction of ice outflow from the interior of New England, suggesting differential glacial erosion. The submerged floors of the inlets are typically irregular,
as can be readily appreciated from the submarine contours of the USGS topographical
maps (see, for instance, the Boothbay Quadrangle, 15 minutes series), and shallow
rock basins and rock bars as well as morainic accumulations are present. Where the
ice-flow was impeded, as at Mount Desert Island (a resistant granite mass), notches
and troughs indicating more intensive glacial erosion appear, and Somes Sound, unlike
other coastal inlets, has a close resemblance to a true fjord, with one wall rising 300 m
above the bottom of the sound.

　　The coasts of southern Sweden possess a range of similar features. The local name
'fjärd' is applied to the inlets here, which E. Werth (1909) linked with the system of
ribbon lakes and valleys inland whose pattern is one radiating from the interior to
the coasts and accordant with ice flow. He also suggested that subglacial streams followed these routeways, further eroding them, in their endeavour to reach the edge
of the ice and regions of lesser ice pressure. The ice, the main agent which sculptured
the fjärds, was clearly guided in most cases by pre-glacial valley systems, as in the case
of fjords; and again, it was able to erode without respect to base-level. The present
state of submergence of the fjärds is therefore an accident, contingent on the amount
of glacial erosion, the amount of sea-level change since glaciation, and also the degree
of post-glacial isostatic recovery.

　　It is important at this point to note that the terms fjord (Danish and Norwegian),
fjärd (Swedish), Föhrde (German) and firth (Scottish) are linguistically all variants
of the same word (K. Hansen, 1971). All refer simply to a marine inlet whose length
is greater than its width. Geomorphologically, the term fjord has traditionally come
to be applied to the partially submerged glacial troughs of mountain-backed coastlands.
There is some justification for using the word fjärd, as Werth proposed in 1909, to
denote the glacially eroded inlets of lowland coasts, differentiated from non-glacial
ria forms by the presence of shallow rock basins in their floors and other features of
glaciated lowlands. There are, however, practical difficulties of distinguishing lowland
coastal inlets resulting from glacial erosion, from superficially similar inlets resulting
from the submergence of sub-glacial stream-eroded valleys, which must next be considered.

　　In Denmark, the term 'tunnel valley' (*tunneldal*) is commonly used for the large
'fjords' in eastern Jylland, such as Vejle Fjord, Flensburg Fjord and Limfjord, and
their continuations as sub-aerial valleys inland. The shapes of these inlets are very vari-

able but usually they narrow inland from wide mouths. Their banks are low, composed of glacial deposits, and their depths are usually less than 20 m. The sub-glacial hypothesis of their origin dates back to the beginning of the century (to 1884 in north Germany) and is discussed more fully in Chapter 12. Essentially they were thought to have been cut by sub-glacial meltwater escaping from under the Baltic Ice towards ice-free ground: the longest can be traced from the Kattegat to the ice limit in Jylland at the eastern border of the great outwash plains (S. Hansen, 1965). They end suddenly at the ice limit, several often joining together at this point, rising abruptly, as much as 100 m, to the surface of an outwash fan.

Similar features, for which a sub-glacial stream origin has been proposed, have been described in Britain and the North Sea. A. W. Woodland (1970) maps a score of possible examples in East Anglia, some now infilled with other deposits, such as the Upper Cam-Stort through-valley, the Hitchin-Stevenage gap and the Little Ouse-Waveney depression. Some of the buried channels in these gaps are over 100 m deep and descend below sea level. Woodland suggests that wide tunnel valleys were probably formed by lateral migration of a much smaller ice tunnel and that the small eskers often encountered on their floors represent the infilling of the last ice tunnel. R. V. Dingle (1971) describes steep-sided linear erosion features at depths of up to 250 m from the north-west of the Dogger Bank in the North Sea. They are infilled with sediments but are revealed by seismic surveys. Similar V-shaped deeps, with closed basins along their floors, are described farther north in the North Sea by D. Flinn (1973). Again, an origin by sub-glacial stream erosion is proposed, at times when sea level was 100 m or so lower than today and ice from Scandinavia covered the area.

The sub-glacial hypothesis for tunnel valleys has recently been questioned. P. Woldstedt (1961) has returned to a glacial erosion hypothesis for the Danish–north German 'fjords', tunnel valleys and their ribbon lakes, though some modification by meltwater is allowed for. K. Hansen (1971) states his view that 'the large fjord valleys [of Denmark and north Germany] must be presumed to have been excavated by glacier tongues' (p. 303); as the last ice penetrated into the dissected country left from the Saale glaciation, ice tongues pushed up the valleys with greater erosive power than that of the ice covering the interfluves. Hansen does not deny the existence of tunnels and meltwater streams in the ice but thinks they were more often englacial, debouching at the margin to form the well-known alluvial cones and only occasionally reaching bedrock where the ice was thin.

4 Conclusion

Although there is still some disagreement on the amount of erosion to be attributed to the ice itself, fjords (using the term in its traditional geomorphological sense) represent partially submerged glacial troughs. The degree of submergence is determined by the amount of glacial deepening, and only to a minor extent by the postglacial rise in sea level. Fjords contain impressive examples of rock basins enclosed by thresholds at their entrances. The thresholds can in some cases be proved to consist of solid rock but further investigation is necessary. The basins, occasionally exceeding 1000 m in depth, are explained by various hypotheses, involving either direct glacial erosion or glacial erosion following a period of preparation of the ground by deep freezing or weathering. Fjord glaciers will erode bedrock as far down-valley as the point of flotation. Beyond this, they thin gradually as sidewall shear stresses decrease or as the valley widens. Crary (1966) associates the reversed slope from the basin to the threshold with the thinning of the floating ice down-valley. His hypothesis can also explain multiple basin formation. The inner ends of fjord valleys, sometimes at the termination of the sea inlet but more often farther inland, are frequently represented by huge steps or trough ends. These have been interpreted by Gjessing (1966) as glacial rock steps, into which subglacial meltwater and subsequently normal rivers have cut waterfall notches.

Certain features of fjords are not glacial but inherited from a previous valley system. Many fjords bear a close relationship with lines of structural weakness, though Gregory's extreme view (1913) of the tectonic origin of fjords cannot now be entertained. Surviving features of pre-glacial river-cut valleys have been described, and one recent suggestion (Winslow, 1966) regards fjords as glacially modified raised submarine canyons.

The strandflat with its accompanying skjaergård is closely linked with fjords and the glaciation of coastal areas. Its origin is uncertain; it may represent a pre-glacial peneplain subsequently trimmed by wave-action at various closely spaced levels, and modified to a greater extent by ice erosion. Fjärds are inlets on a lowland coast modified by glaciation; other coastal inlets, the so-called fjords of Denmark and the Föhrden of northern Germany, may be the submerged portions of tunnel valleys cut by subglacial stream erosion, though some maintain that, even here, glacial erosion may have played a part.

5 References

AHLMANN, H. W. (1919, 'Geomorphological studies in Norway', *Geogr. Annlr* **1**, 3–148 and 193–252

(1941), 'The main morphological features of north-east Greenland', *Geogr. Annlr* **23**, 148–82

ANDERSEN, B. G. (1965), 'The Quaternary of Norway' in *The Quaternary* (ed. RANKAMA, K.), **1**

BAKKER, J. P. (1965), 'A forgotten factor in the interpretation of glacial stairways', *Z. Geomorph.* NF **9**, 18–34

BRETZ, J. H. (1935), 'Physiographic studies in East Greenland', *Am. geogr. Soc. Spec. Publ.* **18**

CAILLEUX, A. (1952), 'Polissage et surcreusement glaciaires dans l'hypothèse de Boyé', *Revue Géomorph. dyn.* **3**, 247–57

CAMERON, R. L. (1965), 'The Vanderford submarine valley, Vincennes Bay, Antarctica' in *Geology and Palaeontology of the Antarctic* (ed. HADLEY, J. B.), *Am. geophys. Un., Washington*, 211–16

CRARY, A. P. (1966), 'Mechanism for fjord formation indicated by studies of an ice-covered inlet', *Bull. geol. Soc. Am.* **77**, 911–29

DAHL, E. (1947), 'On the origin of the strand flat', *Norsk geogr. Tidsskr.* **11**, 159–71

DINGLE, R. V. (1971), 'Buried tunnel valleys off the Northumberland coast, western North Sea', *Geologie Mijnb.* **50**, 679–86

FLINN, D. (1973), 'The topography of the sea floor around Orkney and Shetland and in the northern North Sea', *J. geol. Soc. Lond.* **129**, 39–59

GILBERT, G. K. (1910), *Harriman Alaska Expedition* **3** ('Glaciers')

GJESSING, J. (1956), 'Om iserosjon, fjorddal- og dalendedannelse', *Norsk geogr. Tidsskr.* **15**, 243–69

— (1966), 'Some effects of ice erosion on the development of Norwegian valleys and fjords', *Norsk geogr. Tidsskr.* **20**, 273–99

— (1967), 'Norway's Paleic surface', *Norsk geogr. Tidsskr.* **21**, 73–4

GREGORY, J. W. (1913), *The nature and origin of fjords*

— (1927), 'The fjords of the Hebrides', *Geogrl J.* **69**, 193–216

HANSEN, K. (1971), 'Tunnel valleys in Denmark and northern Germany', *Meddr. dansk geol. Foren.* **20**, 295–306

HANSEN, S. (1965), 'The Quaternary of Denmark' in *The Quaternary* (ed. RANKAMA, K.), **1**

HATTERSLEY-SMITH, G. (1969), 'Glacial features of Tanquary Fjord and adjoining areas of northern Ellesmere Island, N.W.T.', *J. Glaciol.* **8**, 23–50

HOLTEDAHL, H. (1959), 'Den norske strandflate, med saerlig henblikk på dens utvikling i kystområdene på Möre', *Norsk geogr. Tidsskr.* **16**, 285–303

— (1967), 'Notes on the formation of fjords and fjord-valleys', *Geogr. Annlr* **49**, 188–203

HOLTEDAHL, O. (1940), *Dybdekart over de Norske kystfarvann med tilgrensende havstrøk* (Norske Videnskaps-Akad., Oslo)

— (1960), ed., *Geology of Norway* (*Norges geol. Unders.* **208**)

KOECHLIN, R. (1947), 'The formation of fjords', *J. Glaciol.* **1**, 66–8

LØKEN, O. H. and HODGSON, D. A. (1971), 'On the submarine geomorphology along the east coast of Baffin Island', *Can. J. Earth Sci.* **8**, 185–95

MUNDAY, W. A. DON (1948), 'Formation of Norwegian fjords', *J. Glaciol.* **1**, 202

NANSEN, F. (1922), *The strandflat and isostasy* (Oslo)

NICHOLSON, R. (1963), 'A note of the relation of rock fracture and fjord direction', *Geogr. Annlr* **45**, 303–4

RANDALL, B. A. O. (1961), 'On the relationship of valley and fjord directions to the fracture pattern of Lyngen, Troms, N. Norway', *Geogr. Annlr* **43**, 336–8

ROBINSON, A. H. W. (1949), 'Deep clefts in the Inner Sound of Raasay', *Scott. geogr. Mag.* **65**, 20–5

SOONS, J. M. (1968), 'Raised submarine canyons: a discussion of some New Zealand examples', *Ann. Ass. Am. Geogr.* **58**, 606–13

STEERS, J. A. (1952), 'The coastline of Scotland', *Geogrl J.* **118**, 180–90

STRØM, K. M. (1948), 'The geomorphology of Norway', *Geogrl J.* **112**, 19–27

SWITHINBANK, C. and ZUMBERGE, J. H. (1965), 'The ice shelves' in *Antarctica* (ed. HATHERTON, T.), 199–220

THOMPSON, H. R. (1953), 'Geology and geomorphology in southern Nordaustlandet, Spitsbergen', *Proc. Geol. Ass.* **64**, 293–312 (strandflat, p. 298)

TING, S. (1937), 'The coastal configuration of western Scotland', *Geogr. Annlr* **19**, 62–83

WERTH, E. (1909), 'Fjorde, Fjärde und Föhrden', *Z. Gletscherk.* **3**, 346–58

WINSLOW, J. H. (1966), 'Raised submarine canyons: an exploratory hypothesis', *Ann. Ass. Am. Geogr.* **56**, 634–72

(1968), reply following paper by J. M. Soons, *Ann. Ass. Am. Geogr.* **58**, 614–34

WOLDSTEDT, P. (1961), *Das Eiszeitalter* (Stuttgart), **1**

WOODLAND, A. W. (1970), 'The buried tunnel valleys of East Anglia', *Proc. Yorks. geol. Soc.* **37**, 521–78

11

Quantitative aspects of glacial erosion

Then why is it that the rivers that flow from glaciers are so muddy? I venture to say ... that glaciers, by erosion, seriously affect their beds.... The mud of rivers is chiefly derived from this incessant ice waste, and that is why it is so unearthly, so clean, fresh, and impalpable.

(A. C. RAMSAY, 1864)

There has long been controversy over the ability of glaciers and ice sheets to erode their rock beds. Although the hypothesis of glacial protection is today only retained in certain circumstances (see Chapter 9), there are still considerable differences of opinion as to how much glacial erosion has occurred in particular localities or regions. Various methods have been employed in attempts to quantify glacial erosion; none has yet provided either accurate or wholly reliable figures, but some have given useful approximations.

1. Erosion of artificial markers by glacial ice

The classic experiment of this type dates back to 1846, when F. Simony cut a cross 3 mm deep on a rock in front of the Dachstein glacier. The glacier advanced in 1856, scraping the rock and, after retreat, left it bare and polished with no sign of the original mark. Others have occasionally succeeded in obtaining similar short-period measures of glacial erosion: A. de Quervain in 1920 reported bedrock erosion by ice of 0·5 to 1·5 mm in six months; O. Lütschg gave evidence of larger amounts in 1926. Instead of carving marks of limited depth, de Quervain used small boreholes drilled out and filled with a measured depth of wax. But as a method of measuring glacial erosion, it is unreliable and limited in its application. Much depends on being able to predict a glacier advance over a particular spot, so that although many markers have been set up, few results have been obtained, especially as the last hundred years have witnessed a general recession of glaciers. The few results obtained include an unknown quantity of rock weathering before the ice advanced, as R. von Klebelsberg (1943) stressed; they are not representative since the terminal ice is not the main agent of glacial erosion, and extrapolation from necessarily short-period measurements is dangerous.

Some of the problems associated with the method may be overcome by studying rates of abrasion in subglacial tunnels, though we still have no means of knowing how representative short-term measurements are. G. S. Boulton and D. Dent (1972) report on preliminary results of observations beneath the edge of Breiðamerkurjökull where a marble slab fixed to bedrock in 1970 has given an abrasion rate of 1 to 3 mm/year.

2. Measurements of sediment transport by glacier meltwater streams

Where meltwater streams emerge, usually subglacially, at the terminus of a temperate glacier, they appear milky white in colour owing to the high quantity of fine sediment (rock flour) in suspension, the colour sometimes giving the name to the river (for example, Hvitá = White River, in Iceland). R. S. Tarr (1908) described how the Kwik river in southern Alaska emerged from a subglacial tunnel: 'So heavily burdened is it that a pail of water dipped from the surface has an inch of mud and sand on its bottom in a minute or two' (p. 90). The streams are the main agents evacuating solid debris from the glacier basins behind, and the silt content of the streams in general reflects the amount of erosion in progress in these basins. However, not all the debris produced by glacial erosion is carried away by the meltwater streams; in particular, the larger boulders will be piled up in moraines, but estimates of the quantity of rock involved in morainic accumulation can be made from sample studies, and the quantity of debris accumulated here in any given period of time is usually surprisingly small compared with the loads of the meltwater streams. For the latter, we have to assess not only the suspended load but also material in solution, and the bed load. Problems, however, arise. Measurements of stream discharge and load should ideally be kept regularly and frequently for a prolonged period to even out short-term variations (G. Østrem et al., 1967, pp. 277–9), and all outlets of the glacier basin must be covered. Even if this be done, it must be recognized that not all the material in the stream need be the product of glacial erosion—some can be produced by subaerial erosion above the ice surface (a consideration which also applies to the morainic accumulations), and some may be the result of subglacial stream erosion.

One of the early applications of the method was by H. F. Reid (1892) to the Muir glacier, Alaska, where the silt removed by the stream was equivalent to an overall annual lowering of the subglacial rock surface by 19 mm. The figures employed, however, were very uncertain. J. B. Rekstad (1911–12) used an interesting set of circumstances to measure erosion for Engabreen, an outlet glacier from Svartisen in Norway. Owing to an advance of Engabreen in 1904, the meltwater stream was displaced into a small existing lake whose depths had previously been sounded. By the summer of 1910, the lake was completely infilled, the stream now meandering across a gravel plain. The annual quantity of sediment filling in the lake was computed at 200,000 m³. Assuming that the lake acted as a complete sediment trap, and taking the area of Engabreen as 36 km², we arrive at a figure of 5·5 mm/year of erosion (Rekstad's original figure of 11 mm/year allows for additional material piled up as moraine or swept downstream from the lake). It is interesting to compare such figures with more precise modern estimates given in Table 11.1.

S. Thorarinsson in 1939 gave more reliable estimates of subglacial erosion for Hoffellsjökull. Eighteen samples were taken from the main Hoffellsjökull river under normal conditions, and also two from the Skeiðará jökulhlaup of 1938. The samples were taken in various locations—large turbulent meltwater streams provide extremely difficult conditions for representative sampling—at distances up to 4000 m from the ice margin. Silt content varied closely with discharge. The average suspended load of the main Hoffellsjökull river was 1337 mg/l, and six samples gave 271 mg/l for the average quantity of material in solution. Comparable figures for a nearby non-glacier fed river were 766 and 52 mg/l respectively. Using a figure of 3000 mm for the total annual precipitation less evaporation, Thorarinsson showed that 2800 m³ of sediment were being removed annually from each square kilometre of Hoffellsjökull, which, assuming a specific gravity for the rock debris of 1·5, is equivalent to 4200 tons or a layer 2·8 mm thick. But this makes no allowance for the bed load of the meltwater rivers, which could not be measured. Possibly, he thought, the last figure should be doubled (=5·6 mm) to allow for this. This represents a rate of denudation five times greater than in an adjacent non-glacierized area where denudation was calculated on the same basis. At the other extreme, the denudation represented by the Skeiðará jökulhlaup of 1938 (see Chapter 19), though occurring over the short period of the glacier burst only, amounted to 15 or 20 times that of the normal rate for Hoffellsjökull.

In the French Alps, the rate of denudation achieved by the small St Sorlin glacier has been investigated over several years (J. Corbel, 1962). The glacier covers 3·5 km² and its meltwater stream carries an average of 1500 mg/l. Employing a useful simple formula to give the amount of erosion in cubic metres per square kilometre per year ($4E.T/100$, where E=depth of meltwater stream in decimetres and T=weight of solids in mg per litre), Corbel arrives at a figure of 1200. To this must be added the quantity of debris being piled up by the ice as moraine, which Corbel estimates very roughly at 1000 m³/km², giving a total of 2200 m³/km². The corresponding figure for Hoffellsjökull according to Thorarinsson is more than 5600 m³/km², but the difference is hardly surprising considering the probable difference in the level of activity of the two glaciers. Indeed, the range of variation among glaciers is so large that doubt has been cast on the usefulness and validity of the method. Corbel himself notes the following estimates of glacial erosion (per km²/year): Muir glacier, Alaska—5000 m³; Heilstugubreen, Norway—1400 m³; Hidden glacier, Alaska—30,000 m³ (Corbel, 1959). O. Liestøl (1967) derives a figure of about 100 m³ for Storbreen in Norway, though the figure is based on uncertain and short-period estimates of sediment transport and discharge, and neglects bed load. W. M. Borland (1961) uses a method of extrapolation upstream from gauging sites on the Susitna river of Alaska, which gives a figure equivalent to about 30,000 m³ at the glacier front, though this estimate must include an element of fluvial erosion.

Two constantly recurring problems in the method are, first, how to obtain any accurate information on bed load and, secondly, how to obtain meaningful averages for discharge and sediment transport in these notoriously variable streams. The only satisfactory solution to the second is to set up a programme of regular measurement extending over several years, such as the joint project of the Norwegian Water Resources and Electricity Board and the Department of Physical Geography at Stockholm University, begun in 1967, to study sediment transport of several Norwegian glaciers

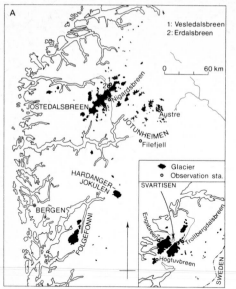
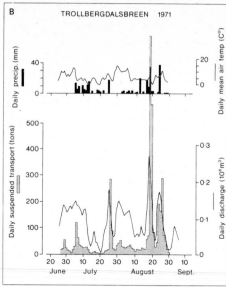

Fig. 11.1
A Location map of Norwegian glaciers and meltwater stream gauging stations.
B Record of precipitation, air temperature, meltwater discharge and suspended sediment load for Trollbergdalsbreen, June–September 1971 (H. C. Olsen and T. Ziegler, *Norwegian Water Resources and Electricity Board, Oslo,* 1973)

(H. C. Olsen and T. Ziegler, 1973). Seven are now being monitored (Fig. 11.1A), the aim being to measure discharge and suspended sediment five times a day throughout the main runoff season (approximately June to September) each year. Fig. 11.1B is an example of the 1971 record for Trollbergdalsbreen. The meltwater stream here shows vividly how enormous the variations in suspended sediment load may be: from virtually nil on certain days to 12,000 mg/l on 19 August. The problem of satisfactorily estimating bed load has not yet been overcome. Some data for Nigardsbreen have been obtained from measuring accumulation at the delta where the stream enters Nigards-vatn. Østrem (1971) reports on attempts to measure bed load in this stream directly, using a steel net with mesh size 2×6 cm. Over a period of 5 weeks, the net collected 400 tonnes, whereas suspended sediment for the same period amounted to 1270 tonnes, giving a proportion of 31·5 per cent for bed load as against suspended load. The steel mesh, however, allowed gravel-sized material to pass through and the true proportion may be nearer 50 per cent. Table 11.1 gives the estimated total transport for six glacier meltwater streams, in which bed load is arbitrarily taken as 30 per cent of suspended load. No allowance is made for material deposited by the glaciers as moraine. The most active glacier according to these figures, Trollbergdalsbreen, is lowering its basin at a minimum rate of 1·4 mm/year (=1·4 m/1000 years or approximately 10 m in post-glacial times, though the dangers of extrapolation have already been stressed). The final column in Table 11.1 shows comparative data for Filefjell, a non-glacial river basin south of Jotunheim. Measurements here showed that suspended sediment levels never exceeded 20 mg/l and, apart from three flood events, there was hardly any

Table 11.1 Estimated total transport, suspended sediment and bed load, for six Norwegian glacier streams and one non-glacial stream (m³/km²/year)

	Nigardsbreen	Vesledals-breen	Engabreen	Erdals-breen	Austre Memurubre	Trollberg-dalsbreen	Filefjell (non-glacial)
Area (km²):	40	4	39	11	9	2	154
Year							
1967		120		700	320		
1968	70	60		450	250		
1969	270	110	116	950	360		
1970	330	70	220	1180	500	1130	
1971	360	50	280	600	230	1670	
Average	257	82	205	776	332	1400	about 1–2*

*Estimate; based on suspended sediment sampling 21 May–14 June.

measurable transport in the basin at all during the period of flow. This surely must reflect the lack of glacial abrasion to provide material for stream transport; here, only slow weathering prepares debris for the stream.

Although the figures of erosion among glaciers based on meltwater stream loads vary widely, they suggest that the mean rate for erosion by large active glaciers probably lies in the range 1000 to 5000 m³/km²/year, which should be compared with what we know of rates of fluvial erosion. The relatively insignificant rates achieved by fluvial erosion and transport in the small Filefjell basin have already been mentioned. At the other end of the scale, the figure for the Mississippi basin is 50 m³, for the Colorado above Grand Canyon, 230 m³, and for the Hwang Ho, 1000 m³. It must be emphasized again that all these figures have a considerable margin of error, but taken together they do suggest that glacial erosion is several times more potent than fluvial erosion.

3 Reconstructions of the pre-glacial surface

If it were possible accurately to restore the form of the land surface before glacial erosion commenced, we would have measures of the amount of erosion during the period of glaciation at any one locality, and of the variations in depth of erosion from place to place. But in nearly every case, no two workers are agreed on the form of the pre-glacial surface; moreover, the possible effects of glacial erosion are inextricably entangled with those of fluvioglacial erosion, fluvial erosion in interglacial and post-glacial periods, and the work of other agents of denudation. Since the evidence on which pre-glacial surfaces are reconstructed and the circumstances of glaciation are so varied, it is only possible to discuss the method by examples.

Reconstructions of pre-glacial surfaces have often been attempted in connection with the excavation of glaciated valleys. W. M. Davis in 1909 wrote that 'the depth of glacial erosion in a main valley is roughly indicated by the discordant altitude of the hanging lateral valleys ... but ... allowance must be made for the glacial erosion of the lateral valleys' (pp. 340–1). He might also have added that glacial erosion is not the only possible process in overdeepening valleys and creating hanging tributaries. But he assumed that ice erosion was the cause of this overdeepening, and quoted figures of

Plate XIX
Yosemite Valley, California. El Capitan on the left rises 1200 m above the valley floor which is itself underlain by up to 300 m of sediment. (*C.E.*)

60–180 m as a minimum measure of the deepening of the main Snowdonian valleys in Wales by ice.

In Yosemite valley (Plate XIX), F. E. Matthes (1930) attributed a greater proportion of main valley deepening to pre-glacial rejuvenation, recognizing several stages of river down-cutting before ice erosion finally modified the valley to its present spectacular form (Fig. 11.2). Extrapolating sections of tributary long profiles unmodified by glacial erosion, he deduced that Yosemite valley had been deepened in three stages by pre-glacial river down-cutting, the last being termed the Canyon stage and represented, for instance, by the floor of Bridalveil creek above Bridalveil Falls. At this stage, Yosemite canyon had a floor level 180–370 m above its present alluvial floor. The latter is now known to be underlain by up to 600 m of sedimentary fill (B. Gutenberg *et al.*, 1956), so that glacial erosion of the bedrock floor of Yosemite valley may amount to as much as 800–1000 m in places. Part of this figure, however, may represent deepening by interglacial river action, to estimate which is a constantly recurring problem.

A third example of pre-glacial valley-floor reconstruction and its use in estimating the depth of glacial erosion is taken from part of the Mur valley in Austria. H. Spreitzer (1963) contrasts the north-bank and south-bank tributary valleys of the Mur east of Tamsweg, establishing that ice flowed down the former to join the trunk glacier of the Mur, but flowed *up* the southern tributaries to escape in part into the Drau basin beyond. In these southern tributaries, glacial erosion is thought to have been relatively

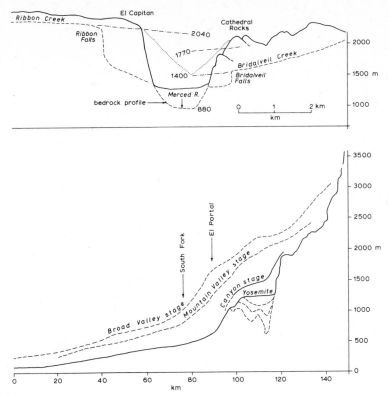

Fig. 11.2
Transverse and longitudinal profiles of Yosemite valley, California (C. Wahrhaftig, Guidebook for Field Conference I, *Congr. int. Ass. Quatern. Res.,* 1965)

weak since the ice was moving slowly uphill, and the pre-glacial valley-floors are there-fore well preserved, only notched by river-cut gorges where the main Mur valley is approached. But in the case of the northern tributaries, ice erosion was much more intense, and comparing these valleys with their southern counterparts, Spreitzer claims that the former have been deepened by ice in the Pleistocene by about 150 m. Once again, however, it is very difficult to assess how much deepening should be attributed to interglacial periods.

Minimum estimates of glacial valley erosion are, of course, given by rock basins, which must be wholly the work of ice; on this basis, F. A. Kerr (1936), for example, argued that the fjord valleys of northern British Columbia had been ice-deepened by at least 600 m. Similarly, Sognefjord (Chapter 10) bears witness to ice erosion of at least 1100 m in depth. These are minimum figures for ice erosion, and they cannot be contested. Likewise, the giant grooves described by H. T. U. Smith (1948) in north-west Canada are certainly the work of ice, and it is here calculated that a minimum of 1 km³ of rock has been removed to form the grooves over an area of 130 km², equal to a layer nearly 8 m thick if spread out evenly.

Just as attempts have been made to reconstruct pre-glacial valley-floors, so have there been attempts to restore the forms of pre-glacial watersheds which were sub-sequently breached by ice. The work of D. L. Linton (1949) in Scotland provides some interesting figures. Table 11.2 shows the amount of glacial erosion involved in creating breaches through certain watersheds.

Table 11.2 The depths of ice erosion represented by some glacially breached watersheds in Scotland

Location	a) Pre-glacial col level (m)	b) Present floor level (m)	c) Depth of ice erosion implied (m)
The Saddle (Cairngorms)	900	815	85
Head of Glen Derry (Cairngorms)	900	750	150
Lairig Ghru (Cairngorms)	1070	840	230
Glen Eagles (Ochil Hills)	340–370	270	70–100
Loch Treig (Grampians)	670–700	150	520–550
Geldie-Feshie col (Grampians)	670	385	285

The pre-glacial col levels are obtained by extrapolating supposed pre-glacial slope remnants on shoulders above the ice-carved troughs. The method is essentially that of W. M. Davis who in 1909 described the destruction of the col at the head of the Nantlle valley in North Wales and claimed that the divide here had been lowered by ice erosion. Figures corresponding to those in the table above would be:

$$a) \ 460–530 \qquad b) \ 230 \qquad c) \ 230–300$$

It is not known whether the cols were significantly deepened by non-glacial processes in interglacial (or even pre-glacial) periods, but this appears unlikely, and there is no doubt from other evidence that ice did move across these watersheds and through these breaches. The depths of glacial erosion are impressive when the nature of the rock involved is considered—the granite of Lairig Ghru, the metamorphic rocks around Loch Treig, or the slates and igneous rocks at the head of Nantlle.

So far, the reconstructions of the pre-glacial landscape and amounts of erosion have been in terms of vertical differences in height. But it is also possible, though less fre-quently, to attempt to restore the pre-glacial positions of certain landscape features in a horizontal sense. Examples will make the reasoning clear. The Driftless Area of Wisconsin is bordered roughly on the south-west in Iowa by an escarpment of the Nia-gara Formation whose regional dip is here to the south-west (Fig. 11.3). Although parts of this escarpment were overridden by older glaciations, the escarpment was never crossed by ice in the Wisconsin period. It possesses a highly irregular outline with numerous outliers of the Niagara Limestone many kilometres from the main escarpment, as near Platteville or at Blue Mounds. This highly crenulated escarpment may be contrasted with the smoothly outlined escarpment, also of the Niagara Forma-tion, in eastern Wisconsin and Illinois which was overridden by ice in the last glaciation, the ice moving generally southward and parallel to it. L. Martin (1916) argued that the smoothness and simple outlines of the Niagaran escarpment in the east were the results of glacial erosion removing any existing irregularities; to do this would, by com-parison with the form of the escarpment in the Driftless Area, involve the trimming back of parts of the eastern escarpment by as much as 8 or 16 km.

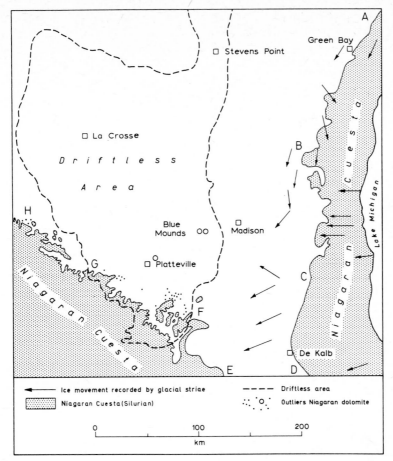

Fig. 11.3
The Driftless Area of Wisconsin, Minnesota, Iowa and Illinois, and the Niagaran Cuesta (L. Martin, 1916)
    A-B simple escarpment form, no outliers, ice moving parallel with escarpment
C-D, E-F escarpment deeply buried in drift, ice moving across the Niagaran outcrop
    F-G highly irregular unglaciated escarpment, with numerous outliers
    G-H irregular escarpment affected by pre-Wisconsin ice only. Irregular form may have developed since the Illinoian glaciation

Such simple conclusions, however, must not be unreservedly accepted. Although it seems likely that section F–G (Fig. 11.3) of the escarpment in Wisconsin was unglaci-ated, at least in the last glaciation, a crenulate form does not necessarily prove that this was so. On the contrary, in another part of the escarpment in southern Ontario, A. Straw (1968) has argued that the section showing deepest indentations is that which was most severely eroded by over-running ice (p. 285 and Fig. 9.3), and that the portion with a relatively smooth outline in plan has been relatively undisplaced from its pre-glacial position. There are clearly unresolved differences of opinion, and the nature and amount of glacial erosion along the escarpment are far from being agreed.

An area in England where there are analogous relationships between glaciation and a major escarpment, but where the evidence on glacial erosion is somewhat clearer, is that of the Chalk escarpment from Oxfordshire and Buckinghamshire (unglaciated) to East Anglia (glaciated). Fig. 11.4 shows the limits of the ice in the Saale (Wolstonian) glaciation: the area was not subsequently affected by ice.

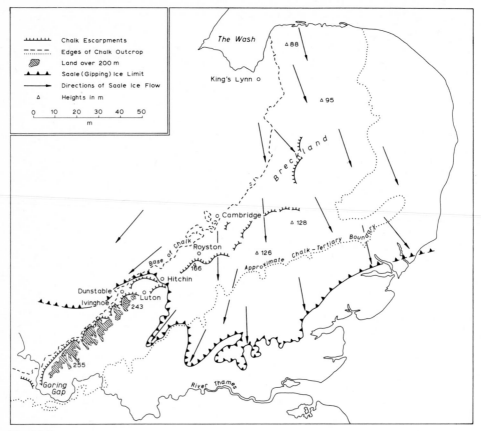

Fig. 11.4
The Saale (Gipping) glaciation of East Anglia, showing relation between directions of ice flow (as determined by R. G. West and J. J. Donner, *Q. J. geol. Soc. Lond.*, 1956) and the Chalk escarpment

In the Chilterns, from Goring Gap to Ivinghoe Beacon, the main scarp stands within 1·5 km of the basal outcrop of the Chalk, except in the vicinity of the principal wind-gaps. But north-east of Dunstable and Hitchin, the main escarpment is found progressively farther from the basal outcrop—at Royston, it is more than 8 km distant, and in the Breckland area up to 12–14 km. The escarpment also changes markedly in form, from a prominent and sharply defined edge south-west of Dunstable, to a very gentle and subdued form, much obscured by drift, in Cambridgeshire and East Anglia, as explained in Chapter 9. The difference is probably the result of glacial erosion of the Chalk escarpment in these latter areas—in other words, ice has displaced the line of the escarpment by up to 14 km.

In limestone areas, the relative abundance of caves as between glaciated and non-glaciated sections has been used to determine depths of ice erosion. The limestones of eastern (glaciated) Wisconsin are conspicuously lacking in caves or sink-holes, even allowing for some obliteration of these features by surface drift. On the other hand, the same formations in the driftless portions of Iowa and Minnesota possess abundant cavern systems down to the level of ground water. Martin (1916) therefore concluded that ice in eastern Wisconsin has stripped off up to 60 m of weathered and cavernous limestone. As with so many other attempts to measure glacial erosion already described, this is a hypothesis, not a proven fact.

There have also been attempts to reconstruct the pre-glacial surface using evidence from the smaller-scale features of glacial erosion. S. E. Hollingworth (1931) compared the smoothed and craggy slopes of stoss-and-lee topography in north-west England, and suggested that while the smoothed areas may not have been appreciably lowered by ice erosion, up to 30 m of rock may have been removed from the craggy uplands. More precise calculations based on similar landforms were made by R. H. Jahns (1943) in New England. An undulating area of granite bedrock in part of Massachusetts displays fine sheet structures of pre-glacial age whose attitudes can be used to reconstruct the form of the pre-glacial surface with considerable certainty. At the close of the Tertiary, the granite hills of this area were smooth domes, whose surface configuration

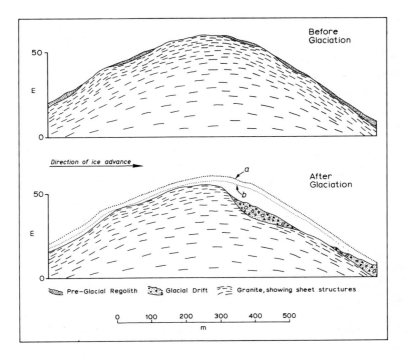

Fig. 11.5
Sheet structures (dilatation joints) in a granite hill, Massachusetts, and their use in reconstructing the pre-glacial form of the hill (R. H. Jahns, *J. Geol.*, 1943, University of Chicago Press)

was controlled by the sheet planes (dilatation joints). During glaciation, rock was plucked from the lee sides, and the minimum amount removed is given by the difference between the present land surface and the surface produced by extrapolating the highest sheet plane from the stoss side (Fig. 11.5). Figures of rock removal thus obtained for 8 hills are: 12, 8, 9, 12, 33, 9, 15 and 8–12 m. The absolute pre-glacial altitudes of the hills are unknown, so that these are minimum figures, but Jahns contends that at least 3–5 m of rock and regolith were stripped from the stoss sides, and that there-fore the figures given above should be at least increased by this amount. Confirma-tion of the depths of erosion is provided by the fact that one huge glacially transported boulder must have been plucked from a granite sheet 8 m thick, and that sheets of such thickness are not encountered until a depth of nearly 30 m below the surface is reached (Chapter 9).

An unknown quantity in all methods based on reconstructions of pre-glacial or pre-glaciation forms is the time factor. In none of the examples just given is it possible to say with any confidence how long a time interval was involved. One can only guess at the relative ages of the different episodes of down-cutting in Yosemite valley, for instance, while the description 'pre-glacial' applied to the col levels used by Linton (Table 11.2) in estimating depths of trough cutting cannot be supported by any chrono-logical evidence—the duration of the Pleistocene glaciation in Scotland is quite un-known. There is controversy, too, about the glacial chronology of East Anglia (Fig. 11.4), so that we do not know to which episode or episodes of the glaciation the partial demolition of the Chalk escarpment should be ascribed.

J. T. Andrews (1972) assumes a figure of 1 million years for the time during which cirques in east Baffin Island and Colorado were eroded which, combined with data on cirque volumes, gives estimates of rates of erosion varying from 0·06–0·2 mm/year (Baffin) to 0·6–1·3 mm/year (Colorado). From such figures he suggested that erosion by temperate glaciers might be an order of magnitude greater than in the case of polar or sub-polar glaciers, but the time-scales are so uncertain that the difference may be more apparent than real.

4 Computations of the volume of glacial drift

The value of this method lies essentially in its application to regions rather than indivi-dual localities. If, for a given region, we can determine the volume of glacial deposits and also determine with some precision the area from which they were eroded origin-ally, we have theoretically a measure of glacial erosion in that source area. But there are numerous difficulties. The time factor is once again largely unknown except in the vaguest possible way. Calculating volumes of drift in any region is by no means easy unless very detailed information on the form and position of the sub-drift surface is available. Some of the drift may have been removed by subsequent denudation and leaching, and some may have been carried out of the region beyond the coast and will now be invisible (Davis, 1909, p. 305). It is nearly always impossible accurately to identify the limits of the source region of the drift, and we cannot by any means be sure that the bulk of the drift has in fact been quarried by ice and not by other agents of erosion or a combination of these. Yet, regarding the last point, there is some evidence that most drift deposits are not simply re-worked alluvium or pre-glacial soils; many workers have found that drift consists predominantly of freshly quarried material. P. G. H. Boswell (1916), for instance, analysing the North Sea Drift in East Anglia showed that 'the large size and freshness of the mineral grains, together with the abundance of more easily decomposable ones, point to the fact, borne out by the evidence of the contained boulders, that the ice did not merely scour off the weathered material from the earth's surface, but that fresh rocks were also broken down' (p. 94).

As a method of estimating glacial erosion, it has long attracted attention. A. M. Hansen in 1894 thought that the rock basins in Scandinavia could be refilled and replenished many times over with the Scandinavian detritus now lying in such immense quantities on the North European plain, and later workers boldly produced figures to indicate that the quantity of drift on this plain was equal to an overall lowering of 150–300 m for Scandinavia. In a similar way, R. F. Flint (1957) calculated that overall glacial erosion of the Laurentian Shield, deduced from the amount of drift lying on and around it, was of the order of 10 m. Previously in 1930, he used the thinness of the drift and its strikingly local composition over the state of Connecticut as evidence that glacial erosion was here of very limited amount. This view was supported in New Hampshire by J. W. Goldthwait and F. C. Kruger (1938). Analysing the till stones, they showed that most of the coarse material in the New Hampshire drift lies within a few kilometres of its original source. The average thickness of the drift is 3–5 m, and there is evidence that part of this (possibly 1 m) consists of pre-glacial weathered material. In the north and east, some drift may have been carried out to sea, but there is no such way of disposing of drift in tthe south and west, where the ice was moving towards Connecticut and Rhode Island. The conclusion is that total glacial erosion in New Hampshire cannot have averaged more than about 3 m.

In eastern England K. M. Clayton (1965) has measured the volume of the Chalky Boulder Clay (Saale [Wolstonian] and/or Elster [Anglian] glaciation), as 300 km^3, or possibly 400 km^3 if allowance is made for later leaching and denudation of the original mass. The deposit is essentially a local one in eastern England, containing little far-travelled material, and its source must have been the Chalk outcrop of Cambridgeshire and East Anglia since the directions of ice flow are known. The quantity is, as far as it is possible to judge, commensurate with the amount of erosion of the Chalk cuesta and the displacement of its scarp several kilometres from the basal Chalk outcrop, as mentioned earlier.

These examples concern ice-sheet erosion in regions of pre-glacial low relief. Volumes of drift and corresponding depths of erosion may be much greater in other environments: the enormously deep (over 300 m) drift-filled hollows at the southern ends of some of the Finger Lakes in New York State (H. L. Fairchild, 1934) may well be related to the intensive glacial erosion which the Finger Lake and adjacent valleys have suffered. In the majority of highly dissected regions, however, it is neither easy to determine the sub-drift surface, to ascertain precisely the source region of the drift, nor to ensure that all the drift produced by glacial erosion has been included in the balance sheet.

Studies of moraine content in present-day ice sheets and glaciers has yielded no very positive contribution to the question of rates of erosion. It is virtually impossible to test if the data obtained in chance exposures, tunnels or boreholes are representative, or if they are significantly different between one locality and another. Some observations on the Antarctic ice have tended to suggest that rates of present erosion are small. S. A. Evteev (1964) claims that the debris content of the ice in Queen Mary Land, East Antarctica, ranges from 0·11 to 13·8 per cent. At one site, detailed measurement of a 40 m section showed negligible amounts of debris in the upper 20 m, but the debris content increased to 11·84 per cent towards the base. An average rate of erosion of only 0·05 mm/year was deduced, but the result should be regarded with caution. It is not known how much of the debris is local nor, as already mentioned, whether the site does indeed represent 'average' conditions for the area.

D. A. Warnke (1970) argues that, since modern Antarctic glaciers and icebergs seem to carry little or no debris, the Antarctic ice at present is playing a relatively protective role. The general absence of surface moraines (save for certain localities such as the 'dry valleys'), the lack of detritus on or around the ice except for eolian or fine-grained glacifluvial material, and the overall cleanness of the ice in the Byrd Station borehole (where only the last 5 m or so contained any significant rock content and there was no thick ground moraine—see pp. 49–50) all point to this conclusion. This obviously seems at variance with Linton's (1963) claim for great bulk removal by Antarctic glacial erosion. A possible solution is that glacial erosion in Antarctica is now much less intense than it was at some former time or times. Warnke concludes that 'the history of Antarctic glacial erosion is ... confused and unsatisfactory in detail. Nevertheless, there is agreement ... that Antarctic glacial erosion reached a peak long before the Pleistocene and has declined since then' (p. 287). Perhaps the period of maximum erosional efficiency was gradually phased out when easily erodible or weathered material was no longer available, or when some approximate state of geomorphological equilibrium was established, or, more fascinating in its implications, when ice temperatures fell below pressure melting-point so that basal sliding and erosion ceased.

5 Conclusion

All methods of estimating glacial erosion are therefore characterized by doubts and uncertainties. But the evidence accumulated so far certainly suggests that active temperate glaciers may be extremely potent agents of erosion, working at many times the speed of rivers of equivalent dimensions in terms of water discharge, while ice sheets, cold or temperate, moving sluggishly on surfaces with little regional slope may be comparatively powerless even given the whole span of the Pleistocene.

Attempts to measure amounts of glacial erosion in a given locality or region have included

1 the use of artificial marks on rock surfaces abraded by advancing ice;
2 measurements of sediment transport by temperate glacier meltwater streams and of the area of the respective glacier basins;
3 reconstructions of pre-glacial or interglacial land surfaces; and
4 estimates of the volume of glacial drift in a given region and its comparison with the area of the source region of that drift.

The first two methods apply to present-day glacierized regions, the last two to regions of Pleistocene glaciation. The least suspect and most meaningful results are those given by the second method. Further studies of sediment transport by glacier streams are urgently required. These should include collection of data over periods of at least several years, preferably a decade or more, and should be undertaken in as great a variety of glacier environments as possible.

6 References

ANDREWS, J. T., 1972), 'Glacier power, mass balances, velocities and erosion potential', *Z. Geomorph.* Suppl. **13**, 1–17

BORLAND, W. M. (1961), 'Sediment transport of glacier-fed streams in Alaska', *J. geophys. Res.* **66**, 3347–50

BOSWELL P. G. H. (1916), 'The petrology of the North Sea drift and upper glacial brick-earths in East Anglia', *Proc. Geol. Ass.* **27**, 79–98

BOULTON, G. S. and DENT, D. (1972), *Ice* **40**, 2

CLAYTON, K. M. (1965), 'Glacial erosion and deposition in eastern England', *Rep. 7th Conf. int. Ass. quatern. Res. (Colorado, 1965)*, Abstracts, 69

CORBEL, J. (1959), 'Vitesse de l'érosion', *Z. Geomorph.* NF **3**, 1–28
    (1962), *Neiges et glaciers* (Paris)

DAVIS, W. M. (1909), 'Glacial erosion in North Wales', *Q. J. geol. Soc. Lond.* **65**, 281–350

EVTEEV, S. A. (1964), 'Determination of the amount of morainic material carried by glaciers of the East Antarctic coast', *Soviet antarct. Exped. Inf. Bull.* **2**, 7–9

FAIRCHILD, H. L. (1934), 'Seneca valley physiographic and glacial history', *Bull. geol. Soc. Am.* **45**, 1073–110

FLINT, R. F. (1930), 'The glacial geology of Connecticut,' *Bull. Conn. State geol. nat. Hist. Surv.* **47**, 73
    (1957), *Glacial and Pleistocene geology* (New York)

GOLDTHWAIT, J. W. and KRUGER, F. C. (1938), 'Weathered rock in and under the drift in New Hampshire', *Bull. Geol. Soc. Am.* **49**, 1183–98

GUTENBERG, B., BUWALDA, J. P. and SHARP, R. P. (1956), 'Seismic explorations on the floor of Yosemite Valley, California', *Bull. geol. Soc. Am.* **67**, 1051–78

HANSEN, A. M. (1894), 'The glacial succession in Norway', *J. Geol.* **2**, 123–44

HOLLINGWORTH, S. E. (1931), 'The glaciation of western Edenside and adjoining areas, and the drumlins of Edenside and the Solway basin', *Q. J. geol. Soc. Lond.* **87**, 281–359

JAHNS, R. H. (1943), 'Sheet structure in granites: its origin and use as a measure of glacial erosion in New England', *J. Geol.* **51**, 71–98

KERR, F. A. (1936), 'Quaternary glaciation in the Coast Range, northern British Columbia and Alaska', *J. Geol.* **44**, 681–700

KLEBELSBERG, R. VON (1943), 'Die Alpengletscher in den letzten Dreißig Jahren (1911–41)', *Petermanns geogr. Mitt.* **89**, 23–32

LIESTØL, O. (1967), 'Storbreen glacier in Jotunheimen, Norway', *Norsk Polar-Inst. Skr.* **141**, 63 pp.

LINTON, D. L. (1949), 'Watershed breaching by ice in Scotland', *Trans. Inst. Br. Geogr.* **17**, 1–16

— (1963), 'The forms of glacial erosion', *Trans. Inst. Br. Geogr.* **33**, 1–28

LÜTSCHG, O. (1926), 'Beobachtungen über das Verhalten des vorstossenden Allalingletschers im Wallis', *Z. Gletscherk.* **14**, 257–65

MARTIN, L. (1916), 'The physical geography of Wisconsin', *Bull. Wisc. geol. nat. Hist. Surv.* **36**, 1–549

MATTHES, F. E. (1930), 'Geologic history of the Yosemite Valley', *U.S. geol. Surv. Prof. Pap.* **160**, 137 pp.

OLSEN, H. C. and ZIEGLER, T. (1973), 'Materialtransportundersökelser i Norske Bre-Elver 1971', *Vassdragsdirektoratet hydrologisk avdeling, Rep.* 4/73 (Oslo), 91 pp. (with English summary and foreword by G. ØSTREM)

ØSTREM, G. (1971), 'Sediment transport studies in glacier streams', *Ice* **35**, 8–9

ØSTREM, G., BRIDGE, C. W. and RANNIE, W. F. (1967), 'Glacio-hydrology, discharge and sediment transport in the Decade Glacier area, Baffin Island, N.W.T.', *Geogr. Annlr* **49**, 268–82

QUERVAIN, A. DE. (1919), 'Über Wirkungen eines vorstossenden Gletschers', *Vjschr. naurf. Ges. Zurich* **64**, 336–49

REID, H. F. (1892), 'Studies of Muir Glacier, Alaska', *Natn geogr. Mag.* **4**, 19–84

REKSTAD, J. B. (1911–12), 'Die Ausfüllung eines Sees vor dem Engabrae, dem grössten Ausläufer des Svartisen, als Mass der Gletschererosion', *Z. Gletscherk.* **6**, 212–14

SIMONY, F. (1871), 'Die Gletscher des Dachsteingebirges', *Sber. Akad. wien. math. naturw. kl.* **63**, 501–36

SMITH, H. T. U. (1948), 'Giant glacial grooves in Northwest Canada', *Am. J. Sci.* **246**, 503–14

SPREITZER, H. (1963), 'Grössenwerte des Ausmasses der glazialen Tiefenerosion (vornehmlich am Beispiel des oberen steirischen Murgebietes)', *Mitt. naturw. Ver. Steierm.* **93**, 112–19

STRAW, A. (1968), 'Late Pleistocene glacial erosion along the Niagara escarpment of southern Ontario', *Bull. geol. Soc. Am.* **79**, 889–910

TARR, R. S. (1908), 'Some phenomena of the glacier margins in the Yakutat Bay region, Alaska', *Z. Gletscherk.* **3**, 81–110

THORARINSSON, S. (1939), 'Hoffellsjökull, its movements and drainage', *Geogr. Annlr* **21**, 189–215

WARNKE, D. A. (1970), 'Glacial erosion, ice rafting, and glacial-marine sediments: Antarctica and the Southern Ocean', *Am. J. Sci.* **269**, 276–94

12

Glacial meltwater and meltwater channels; glacial diversion of drainage

Where engaged in gorge cutting, these meltwater streams work with great rapidity, for the volume is great, the sediment load heavy, and therefore, with sufficient grade to prevent deposit, they are active agents of erosion. (R. S. TARR, on the ice-marginal streams in the Yakutat Bay region, Alaska, 1908)

Meltwater may be found on all temperate glacier surfaces below the firn limit, but above this limit is relatively rarely encountered because of the great permeability of the firn and snow. It is not usually present on the surfaces of cold (high polar) glaciers, except temporarily and locally. Near to rock exposures, reflected heat in summer may cause some melting even when air temperatures remain well below zero, but any such surface meltwater on cold glaciers will soon refreeze without travelling far. Sub-polar glaciers in summer will be characterized by much more surface melting, so that surface and marginal streams will appear, but because of the low temperatures of the ice (except towards the base), meltwater will not be able to penetrate to any depth. In contrast, meltwater streams on temperate glaciers may attain considerably greater size but do not maintain surface or marginal courses for any great distance before losing themselves down crevasses in the ice, or down cylindrical melt-holes known as glacier mills, moulins or Mühle, to feed extensive englacial and subglacial drainage systems. On the surface, abandoned moulins and channels mark former supraglacial stream courses, for as melting proceeds, streams will be swallowed by the ice at progressively higher positions in their courses.

There are thus very important differences between the meltwater drainage systems of temperate, sub-polar and cold glaciers. Drainage on cold glaciers is ephemeral and restricted to small areas of the surface, and will not be further considered in this chapter. Most of what follows will concern temperate glaciers, the only ones able to support extensive and deeply penetrating systems of englacial and subglacial drainage, but sub-polar glaciers will be referred to where appropriate.

1 Some general characteristics of glacial drainage

The drainage networks of supraglacial streams are complex, consisting of both annual and perennial elements (K. J. Ewing, 1972). Perennial channels are deeper (some attain depths of 10 m or more) and are re-used from year to year each melt season. They run generally parallel to the direction of ice flow and are often regularly spaced. Details of the network are strongly influenced by structures in the ice such as fractures and foliation, by differential surface ablation and by the surface slope. The channels possess features similar to those of normal stream systems though the fact that the channel is forming in a soluble material introduces certain differences. A. D. Knighton (1972) has studied the forms of supraglacial stream meanders on Østerdalsisen, Norway. On flat or gently sloping surfaces the streams may be broad and sluggish, but on more steeply sloping glaciers, the flow may attain high velocities, for the ice channels offer little frictional resistance. The streams invariably carry little or no load. This is because the high velocities on the steeper slopes tend to keep the channels clean and, except near the margins, the debris content of the surface layers of glaciers is comparatively small. In marginal zones with greater dirt concentrations, there are usually also many more crevasses and thus supraglacial streams are rare in any case. Away from moraines, most of what sediment there is in the supraglacial streams is wind-blown.

The annual elements of supraglacial systems are ephemeral, rarely exceeding 0·5 m in depth because of their short life. They begin to develop at the end of the winter,

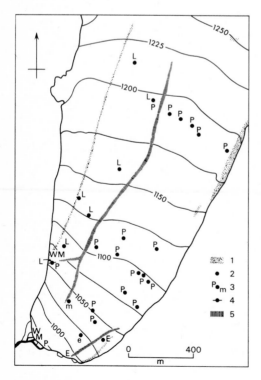

Fig. 12.1
Mikkaglaciären tongue, northern Sweden, showing moulins and outlets of internal drainage.
1 Supraglacial moraine (lateral and medial)
2 & 3 Moulins; the reference letters show connections with drainage outlets
4 Open crack in glacier
5 Deduced boundaries between zones of internal drainage (T. Stenborg, *Geografiska Annlr, Stockholm.* 1969)
Letters indicate connections proved between moulins and outlet streams.

as soon as the snow has become sufficiently soaked and before it has melted off the ice. Thereafter, feeding the perennial channels, they have a constant struggle to deepen their channels as fast as the overall ablation rate.

Most supraglacial water on temperate glaciers sooner or later finds its way into the englacial and subglacial systems; this is by far the greatest source of supply of the latter. In the percolation process, R. L. Shreve (1972) draws a useful analogy between glacier ice and limestone: temperate ice is slightly permeable over limited distances along inter-connected grain boundaries, but its main permeability is due to fractures, moulins and tunnels. Wherever ice temperatures and pressures permit, englacial streams flowing in such shafts and tubes will eventually reach the base of the ice to become subglacial streams. The internal hydrology of glaciers is just as complex as that in karst regions and similar techniques (salt-injection, dyes, etc.) are used in attempting to trace water movement. T. Stenborg (1969) has investigated the internal drainage of Mikkagla-ciären and Storglaciären in Sweden (Fig. 12.1). From evidence of connection or non-connection between twenty-six moulins and the outlets of several subglacial streams, he is able to show the existence of a major separation in drainage approximately down the middle of Mikkaglaciären, which is also the boundary between south-east- and south-west-striking crevasses. The latter are very important in thus diverting supra-glacial drainage away from the centre line, and this primary separation is shown by Stenborg to be maintained at depth in this example, giving two main semi-lateral systems. The average drainage rate is 0·7 m/s.

Such studies of internal drainage are rare and altogether there is still relatively little known about the behaviour of meltwater inside and beneath temperate glaciers. Apart from the valuable observations beneath the Glacier d'Argentière in France (R. Vivian and J. Zumstein, 1973) and a few other similar records, mostly in connection with hydro-electric power schemes or mining operations, there is a dearth of information on such topics as water pressure, the forms of channel in which the water flows, how much (if any) moves in sheetflow at the glacier bed, and how the channels are affected by ice flow and heat from the flowing water. As J. F. Nye and M. F. Meier (1973) point out, such information would not only be of academic interest but of great practical utility in connection with the development of water resources. Glaciers, often found in areas of high precipitation, represent valuable natural stores of water, provided their discharge can be controlled and the hazards of floods and glacier bursts over-come.

Englacial drainage systems can only exist in temperate glaciers and draw their supply almost entirely from supraglacial melting. Minor quantities may be added by internal melting due to the heat and friction of the supraglacially derived water or due to heat conducted down the pressure-melting gradient from above. If the englacial conduits are to continue to exist in a steady state, there must be an approximate balance between processes of ice deformation tending to close the tubes on the one hand, and melting of ice by frictional heat and any significantly positive temperatures in the running water on the other hand (H. Röthlisberger, 1972). Shreve (1972) argues that the larger englacial passages will grow more quickly than the smaller ones if there is enough sur-plus heat, for the meltwater in large tubes can carry more heat relative to the area of enclosing ice walls. Thus larger passages would gradually take over from smaller ones; the network would become arborescent, rather like a sub-aerial river network but in three dimensions.

In certain situations of impeded water flow, englacial water will accumulate within a glacier, filling all the passages to capacity and, if ice pressures permit, extending to the base of the glacier. Fig. 4.19 (p. 144) shows an example where a rock bar is obstructing the discharge of englacial and subglacial water. The borehole AR203 encountered a body of water held up in this way; the upper level of this water body, the englacial water table, must stand at 2186 m, the lowest point of the rock bar, or else slope down to this point if there is sufficient meltwater to build up a head. Englacial water tables have been postulated by geomorphologists in attempting to explain certain features of erosion and deposition. For example, subglacial chutes (p. 346) sometimes end abruptly at their lower limits, which may be related to the descending meltwater having reached a water table at this level; and conversely, some subglacial depositional forms rise to a common upper level which, together with other evidence, may mark a controlling water table in the ice (p. 489).

If it reaches the base of a glacier, englacial water flow becomes subglacial. Sub-glacial water can be present only at the bases of temperate or sub-polar glaciers. In the former case, it will be derived largely from supraglacial sources through the englacial system, but added to it will be small quantities resulting from the heat of friction of the sliding ice and from geothermal heat. At average sliding velocities (non-surging glaciers), J. Weertman (1972) suggests representative values of the annual thickness of ice melted would be up to 3 cm and 0·5 cm respectively. Thus, contrary to some early views (e.g., R. G. Carruthers, 1939), basal melting is relatively insignificant compared with surface melting. H. Hess (1935) calculated that the average basal melting for Alpine glaciers is probably less than 5·3 cm/year, whereas loss of ice by surface melting often amounts to several metres a year. Nevertheless, in situations where, because of ice thickness and pressure, supraglacial meltwater cannot reach the base of a temperate glacier, the small quantities of subglacial meltwater produced by sliding friction and geothermal heat are very important in lubricating the glacier bed, as described in Chapter 4.

In the case of those sub-polar glaciers which only attain pressure melting-point at or near their beds, englacial drainage systems will be absent, and therefore only small quantities of subglacial water, derived at the bed of the glacier, will be present. Basal meltwater is more abundant beneath those parts of the Antarctic and Greenland ice where a considerable thickness of the basal ice is at pressure melting-point (pp. 89–90).

Weertman (1972) considers three possible forms of subglacial water flow. First, there may be tunnels formed in the basal ice with a flat floor of bedrock ('Röthlisberger channels', Fig. 12·2A). In these, it is assumed that the tunnel exists solely because heat is available to melt enough ice to prevent the channel closing by plastic deformation, and that the subglacial stream (perhaps because its position is continually shifting) is unable to erode the bedrock. Such R channels are known to exist, for instance beneath the Glacier d'Argentière. Weertman shows that they must be largely fed from supraglacial/englacial sources. If the tunnel axis is aligned in the direction of maximum shear stress in the moving ice, they can only collect basal water from other sources for distances of the order of 100–200 times the tunnel radius. With shear stresses acting

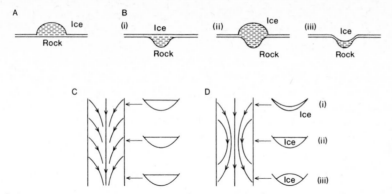

Fig. 12.2
A Cross-section of a Röthlisberger channel
B Cross-sections of Nye channels when these channels lie parallel to the direction of ice motion:
(*i*) completely filled with water, (*ii*) more water than can flow through the channel—a Röthlisberger channel exists above the Nye channel, (*iii*) partially filled channel
C Flow paths of water in sheet flow under a valley glacier in a channel of uniform cross-section and level ice surface
D Flow paths of subglacial sheet flow for a valley glacier whose surface in cross-profile is (*i*) concave, (*ii*) level, (*iii*) convex
(J. Weertman, *Reviews of Geophysics and Space Physics*, 1972)

transversely to such tunnels, Weertman finds that basal water would actually be driven away from the channel.

Secondly, there may be channels formed partly or wholly in bedrock ('Nye channels', Fig. 12.2B). Which of the three possibilities shown in Fig. 12.2B exist will depend on the balance between water pressure and ice pressure in the channel. The second of the three variants depicted will only be possible if the direction of ice flow is parallel to the channel. Nye channels will be more stable than R channels and potentially represent a very important means of evacuation of basal meltwater as well as carrying water from supraglacial/englacial sources. If they run even slightly transverse to the ice flow, there will be a high-pressure trap on the down-glacier side that will tend to force meltwater into them. On the other hand, the very existence of Nye channels is conditional on the rate of subglacial stream erosion equalling or exceeding the rate of glacial erosion.

The third possible way in which meltwater can exist at the base of a glacier is as a basal meltwater film, moving by sheetflow. Nye (1973) has calculated approximate values of the thickness (*t*) of this water layer (see also p. 143), assuming a bed of small sinusoidal waves (wavelength λ) and a viscous flow law for the ice:

$$t = k\lambda^{1/3}$$

For $\lambda = 5$ cm, $t = 0.5\,\mu$m, and for $\lambda = 5$ m, $t = 2.5\,\mu$m. Nye argues that in spite of the artificiality of the assumptions, these values are of the right order of magnitude, but are hardly compatible with Weertman's (1972) notion that the basal water film is an important means of transferring water along the glacier bed. Weertman also suggests that the thickness of the water film will vary with pressure differences beneath the glacier, being squeezed out to negligible thicknesses in high pressure zones at bedrock obstacles, for instance. But because of the concave shape of a glacial valley in cross-section, basal meltwater will also, in theory, be driven towards the centre, where a thicker water layer could help to explain faster sliding velocities of the central zone of the glacier. Fig. 12.2C demonstrates such an idealized situation. Fig. 12.2D shows how different theoretical forms of glacier cross-section might influence basal water flow. One of these (Fig. 12.2D, (iii)) is approximately matched by Stenborg's (1969) findings on Mikkagläciaren.

There is no reason why the three modes of subglacial water flow should not co-exist. R or Nye channels might serve as the main conduits for originally supraglacial water; the basal water layer could be trapped by oblique Nye channels, or if the transverse pressure gradient owing to valley shape was sufficient (e.g. if the valley was strongly V-shaped), the basal water layer might be driven into an R channel. Weertman considers it possible that an R channel runs down the centre of a glacier over a major part of its length collecting englacial/supraglacial water. The basal water layer might tend to flow as a thicker layer near the valley centre and, lower down the glacier, might move into the R channel.

Nye's (1973) views differ; because R channels are unstable and subject to closure by ice flow, channels eroded in bedrock (Nye channels) are thought to be the most likely means of moving the bulk of the drainage water. He contends that the basal water layer does not play any significant part in bulk transfer of meltwater since, first, too much meltwater in the basal layer will stop the regelation process of sliding and secondly, the water involved in the regelation process does not, of course, flow continuously down-glacier but only from high-pressure to low-pressure zones around obstacles. Therefore he suggests that the drainage water is likely to move in a system separate from the basal water layer. Against this, Weertman notes how variations in sliding velocity do seem to be related to variations in meltwater supply (Chapter 4), which would suggest some connection between the basal water layer and the drainage water.

There is, therefore, continuing controversy about the modes of subglacial water flow which will only be resolved with further field data. But about the geomorphological activity of subglacial meltwater there is more general agreement. Large subglacial streams seen issuing with great force from caves at the snouts of glaciers (Plate XXII) have long been regarded as potent agents of erosion and transport. Since such streams often attain high velocities and discharges, and are armed with rock particles washed out from basal moraine, they may be capable of active down-cutting even in hard bedrock, as shown by the gorges through many rock bars, such as the Lutschine gorge at Grindelwald or the Gorner gorge above Zermatt (see also Plate XVIII). It is obviously impossible to witness or measure such erosion going on under the ice, but the evidence from proglacial and ice-marginal meltwater streams is relevant. O. D. von Engeln (1911), for instance, described how marginal drainage of the Hidden Glacier in Alaska succeeded in cutting, in shales and sandstone, a gorge 1·6 km long and up to 3 m deep in less than 3 years. R. Vivian (1970) reported that, after the bursting of the Gornersee, Switzerland, in 1967, the subglacial torrent escaping from the ice moved more debris in 3 days than it moved in average conditions over a year, and up to 10 cm of erosion in granite bedrock was measured. Subglacial streams are not only, however, potential agents of erosion. Evidence of their massive transporting capacity is seen in the infillings of some Pleistocene subglacial gorges and buried channels (for example, south of Newport in Essex) and, most familiar of all, in the depositional forms of eskers and kames (Chapter 16). In contrast, supraglacial and englacial meltwater streams will leave relatively few traces of their former existence once the ice has vanished, unless their downward melting happened to have brought them into contact with bedrock.

There is some uncertainty over the question whether subglacial meltwater may play a corrosive as well as an erosive role. Certainly the geomorphological evidence favours the latter as the overwhelmingly dominant form of bedrock attack. Vivian and Zumstein (1973) suggest there may be a seasonal alternation of activity beneath temperate glaciers. In the warm season, discharge in the subglacial tunnels is high; the water fills them, travels with great speed and pressure, is highly laden with debris and mechanical erosion is dominant. Discharge for a subglacial torrent beneath the Glacier d'Argentière reaches 11 m³/s at this time of year. In winter, however, discharge falls to less than 1·5 m³/s and, they consider, tends to move more by sheetflow, washing the bedrock surface. Suspended sediment concentrations are less but dissolved load is significantly higher; pH values of 7·5 compare with 6·5 in the melt season. Possibly in winter chemical erosion becomes more important, attacking minerals such as feldspars and preparing the rock for mechanical attack in the next melt season.

The remainder of this chapter will concentrate on the results of erosion by meltwater (depositional forms will be dealt with in Part III), on the significance and use of these erosional forms in interpreting glacial and deglacial history, and on alterations of the pre-glacial drainage brought about by glaciation.

2 Glacial Lake spillways

Meltwater lakes may be impounded by ice blocking existing drainage outlets or by drift barriers (see Chapter 19). Although meltwater lakes are by no means rare today, they are more restricted in their occurrence than in the Pleistocene glacial periods when the extent of ice was greater. Those held up by Pleistocene glacial deposits have in many cases been drained by rapid down-cutting of the outlet streams where this flows across the barrier. Ice-dammed lakes around present-day temperate glaciers and ice sheets are not numerous since most of such glaciers on which meltwater is plentiful lie on high ground where drainage from the ice is not impeded. Conn and Bieler Lakes on the north-east of the Barnes ice cap, Baffin Island, and other examples associated with cold ice are discussed in Chapter 19. In the Pleistocene, however, ice-dammed lakes were in some areas numerous and extensive. Such lakes drained by a variety of routes—over the ice surface, into fissures in the ice, subglacially (a common means of escape), or over bedrock or drift cols to escape into an ice-free area. In the latter case, the over-spilling lake waters cut channels in the cols and carved deeper valleys beyond.

Until relatively recently, most channels recognized as being the result of meltwater erosion in the British Isles were interpreted as spillways of ice-dammed or drift-dammed lakes. This hypothesis has its origins more than a century ago; in 1863, for instance, T. F. Jamieson described the evidence for ice-dammed lakes in Glen Roy which drained through and scoured out cols at several different heights. The first detailed discussion of glacial lake spillways in Britain, however, dates from 1902 when P. F. Kendall claimed that most of the meltwater channels in the Cleveland Hills were cut by over-flowing waters from a series of glacially impounded lakes around the margins of the Hills. Indeed, assuming that the channels were lake spillways, he used them as a chief line of evidence for the former existence of ice-dammed lakes in the area, and regarded them as 'perhaps the most impressive memorials of the Ice-Age that our country con-tains' (p. 481). Four types of channel were recognized: 'direct overflows' such as New-ton Dale in the Cleveland Hills, where the water flowed away from the ice across a former watershed; 'severed spurs' where a minor watershed is breached; 'marginal over-flows' formed along the ice edge; and 'in-and-out channels' cut in a hillside by melt-water flowing round a projecting lobe of ice. In most cases, Kendall assumed that lakes of various sizes provided the principal sources of meltwater to cut the channels.

Several features were held to be characteristic of these 'overflow valleys'. They were said to be entirely independent of the natural drainage and seldom received tributaries with accordant floor level. Their transverse profiles exhibit steep sides and broad flat floors, while their long profiles usually steepen rapidly downstream. Sometimes meanders of considerable amplitude are present, and the lower end of the valley may 'open out inconsequently upon a steep hillside' (Kendall, 1902, p. 485) below which a mass of gravel may have accumulated.

Two channels which display many of these features and which were claimed to be typical of Kendall's direct overflow type by A. R. Dwerryhouse (1902), were later sur-veyed in detail by R. F. Peel (1949) (Fig. 12.3A and B). All the typical features listed by Kendall may be clearly seen; there is, however, an additional feature of great signi-ficance, namely the anomalous long profiles of the bedrock floor which in one case rises to a point 12 m higher than the supposed intake of the channel, a feature which led J. B. Sissons (1958a) to the view that these and other channels in southern North-umberland were not in fact glacial lake spillways (see p. 348).

Kendall's views have been presented in some detail, because they had a profound and to some extent undesirable influence on many subsequent British studies of melt-water channels. Many workers assumed that all meltwater channels were spillways, and postulated ice-dammed or drift-dammed lakes simply from the existence of the channels. Dwerryhouse's work already mentioned presented such an interpretation; so did F. W. Harmer (1907), who for instance regarded Goring Gap of the Thames crossing the Chilterns as cut by overflow from a supposed Lake Oxford. Ironbridge Gorge, intaking at 94 m and now carrying the diverted River Severn, was regarded by L. J. Wills (1924) as the overflow route of a supposed Lake Lapworth. The Ferryhill Gorge across the Magnesian Limestone cuesta in County Durham was said by A. Rai-strick and K. B. Blackburn (1932) to have drained a lake standing at 120 m in the Wear valley to the north. And in eastern England, several workers including Kendall postulated a Lake Humber and a Lake Ouse, barricaded by ice in the Vale of York and by Scandinavian ice along the east coast, and overflowing in the later stages by the remarkable through-valley transecting Norfolk and linking the Little Ouse and Waveney drainage near Diss at a level of about 24 m above sea level.

There is no doubt that some meltwater channels in Britain are true overflows and that some glacial lakes did exist, though, as discussed in Chapter 19, the evidence for

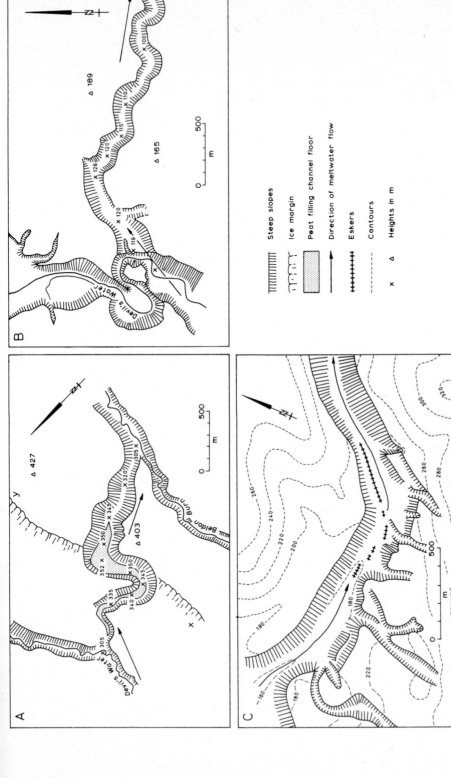

Fig. 12.3

A The Beldon Cleugh, Northumberland. X-Y marks an approximate suggested position of the ice margin when melt-water last flowed (southward) through this channel

B The East Dipton Channel, Northumberland (A & B after R. F. Peel, 1949; J. B. Sissons, 1961)

C The Deuchrie channel, southern Scotland (J. B. Sissons, *Scott. geogr. Mag.*, 1961, Royal Scottish Geographical Society). Spot heights and contour heights for all three diagrams are in metres

Steep slopes

Ice margin

Peat filling channel floor

Direction of meltwater flow

Eskers

Contours

× △ Heights in m

such lakes must be independent and not consist solely, as so often in the past, of the supposed overflow channels themselves. In the basin of the Warwickshire Avon, F. W. Shotton (1953) has given detailed evidence (Chapter 13) for the existence of a 'Lake Harrison' during the advance of the Wolstonian (Saale) ice, impounded between the Jurassic escarpment and three ice lobes (to the north and north-east, and to the west in the Severn-Avon valleys), and stretching from Leicester and Birmingham to More-ton-in-the-Marsh. Lake clays attaining a maximum altitude of 123 m near Moreton, and an erosion terrace at about 120 m mapped by G. H. Dury (1951) from Moreton towards Daventry, suggest a maximum lake level at about these heights. Three gaps in the Jurassic escarpment are considered by Shotton as possible overflow routes: the Watford Gap near Daventry at 137–140 m, the Fenny Compton gap (123 m), and the Dassett gap (125 m), the latter two converging to join at Banbury in the Cherwell val-ley. Clearly all three cannot have functioned simultaneously, and it seems unlikely that the higher level Watford Gap was ever used since there is no means of blocking off the other two lower level gaps at the same time. However, the levels of the Fenny Compton and Dassett gaps are so close that simultaneous use of these by overflowing lake waters is a possibility, and the marked deepening of the Cherwell valley which they enter may reflect erosion by the overflow.

In South Wales, extensive systems of glacial lakes were postulated by J. K. Charles-worth (1929) to account for the presence of numerous meltwater channels, mostly inter-preted as lake spillways. D. Q. Bowen (1964-66) and K. J. Gregory (Bowen and Gre-gory, 1965) have shown that most of the channels were not, in fact, eroded as overflows and that in nearly every case the hypothetical lakes of Charlesworth never existed. In the lower Teifi valley, however, O. T. Jones (1965) has presented definite evidence for an ice-dammed lake. Near Cardigan, bedded outwash shows meltwater moving up the valley from ice blocking it lower down; later, a lake came into existence in which laminated clays up to 5 m thick were laid down on top of the outwash and in which, at Lampeter, deltas were also constructed. Jones describes the related overflow channels of this lake.

The erosive action of glacial lake overflow waters is abundantly documented, some-times on a huge scale, in North America. One example will be taken from the Interior Lowlands, and one from the western Cordillera.

The largest of the late Pleistocene ice-margin lakes in North America was Lake Agas-siz, covering 280,000 km² (more than the combined areas of the present Great Lakes) at its maximum in the late Wisconsin glaciation and occupying the Red River basin and an area to the north. Present-day remnants include Lakes Winnipeg and Mani-toba. Warren Upham (1896) showed that for a considerable time Lake Agassiz over-flowed southward across the present continental divide into the valley of the Minnesota River, past Granite Falls to reach the Mississippi. The intake of this overflow channel (Upham's 'River Warren') is now marked by shallow Lake Traverse at 299 m (Fig. 12.4). The channel was not deepened below this owing to resistant granite outcrops downstream, and the lake level was held at this height for some time. J. A. Elson (1965) has recently mapped the associated Campbell shoreline on the west of Lake Agassiz, rising from 299 m at Lake Traverse to more than 430 m farther north in Canada as a result of later isostatic tilting. Higher strandlines show that overflow began at about

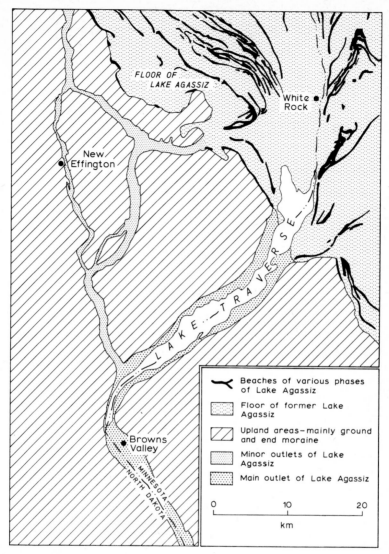

Fig. 12.4
Outlets of glacial Lake Agassiz (W. D. Thornbury, *Regional Geomorphology of the United States*, John Wiley & Sons, New York, 1965)

322 m, and the channel was deepened in stages (sixteen strandlines lie between the highest and the Campbell stage) as the overflow waters encountered resistant layers in Palaeozoic sedimentary rocks. The channel was abandoned when deglaciation permitted the lake waters to escape at lower levels eastward into Lake Superior. Elson shows that the story is in fact more complex than this since Lake Agassiz existed at more than one stage of ice advance in the late Wisconsin.

In the western Cordillera, spectacular erosive effects of meltwaters released by the

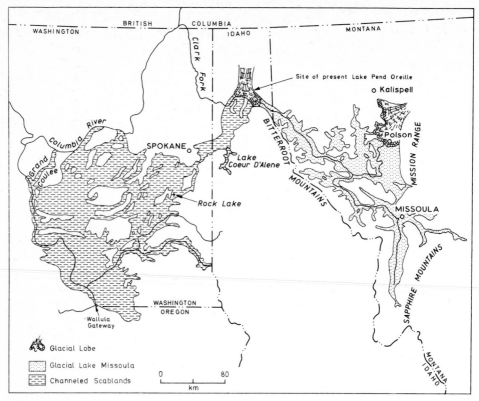

Fig. 12.5
Map of the Channeled Scablands and Pleistocene Lake Missoula in Washington, Idaho, and Montana
(W. D. Thornbury, *op. cit.*, 1965)

bursting of late Pleistocene lakes have been long known, though there has been great
controversy over the location of the lakes, their sizes, and the magnitude and duration
of the floods involved. The plateau of Columbia River basalts in the south-eastern
quarter of Washington State has been scoured, channelled, and largely swept bare of
sedimentary deposits by immense floods of glacial meltwater over an area of 5000 km².
This is the area known as the Channeled Scablands (Fig. 12.5). Numerous channels
or 'coulees' none of which now contains any permanent drainage owing to the dry
climate, cross the Scablands. The most famous is the Grand Coulee, 80 km long, up
to 8 km wide and 300 m deep at its head, but there are others almost equally immense.
Some of the coulees are braided and they possess normal features of fluvial erosion
such as dry waterfalls (up to 120 m high and 5 km wide), abandoned high-level
meanders, and interchannel islands with cappings of river-deposited gravel. Addition-
ally, there are unexpected and highly unusual features such as giant current ripples
and gravel bars noted by J. T. Pardee in 1942, which attain heights of 30 m and wavelengths of 80 m, and, most disturbing of all, rock basins (with or without lakes). One
such basin is that containing Rock Lake, 11 km long and 80 m deep, while other basins
up to 30 m deep lie on divide crests.

Theories to account for this remarkable assemblage of features in the Scablands have been many. I. C. Russell (1898) and Morris Leighton (1919) both generally suggest the work of glacial meltwaters; but it was J. H. Bretz who in 1923 first excited speculation by invoking a brief but gigantic flood caused by sudden bursting of a glacial Lake Spokane. I. S. Allison (1933) argued for a more extended flood caused by an ice dam in the Columbia Gorge, while R. F. Flint (1938) favoured normal erosion by meltwaters and a period of valley filling followed by partial re-excavation to account for the anomalous gravel deposits. W. H. Hobbs's theory (1943) of glacial erosion is not now supported.

The balance of opinion currently regards Bretz's theory of a sudden flood as most probable, but places the source of water in a glacial Lake Missoula (Fig. 12.5). The latter was first documented by J. T. Pardee (1910). It resulted from a Cordilleran ice-lobe blocking the Clark Fork of the Columbia River at Lake Pend Oreille (Idaho) and forming a lake over 7500 km², which rose to 1340 m in the Bull Lake glaciation and 1280 m in the Pinedale. The lake stood 600 m deep against the ice which was itself 1100 m thick. According to work by J. H. Bretz, H. T. U. Smith and G. E. Neff (1956), the ice dam burst seven times in the Wisconsin period alone, releasing catastrophic floods across the Scablands. The coulees were used and re-used by these floods which may well have occurred also in pre-Wisconsin glaciations. When Lake Missoula burst at its maximum level, Pardee showed that up to 40 km³ of water were discharged per hour, and probably 1500 km³ in a few days. The water flooding through the Wallula Gap is thought to have been no less than 250 m deep (Bretz, 1969). The Missoula flood also entered glacial Lake Columbia farther downstream, raising its level by 150 m and emptying through the Grand Coulee.

Repeated floods of such magnitude are capable of explaining all the features of the Channeled Scablands. Probably cavitation erosion associated with enormous velocities of escaping water was responsible for excavating the rock basins. Another area of scabland also attributed to a catastrophic flood of meltwater is the Snake River plain, Idaho, affected by sudden overflow of Pleistocene Lake Bonneville across the Red Rock pass, about 30,000 years BP. H. E. Malde (1968) estimates peak discharge at 1·4 km³/hour which, although far less than the Missoula flood, is nevertheless comparable in order of magnitude to present-day glacier bursts in Iceland, such as the Katlahlaup of 1918 (see p. 541).

3 Proglacial drainage

The term connotes drainage away from the ice, and therefore includes the examples of spillways in the last section. However, proglacial drainage need not depend on ice-border lakes, and the majority of proglacial drainage systems simply act as sluiceways for meltwater directly from the wasting ice. Volumes of meltwater often reach very high levels and fluctuate considerably. Heavy loads of debris give rise to extensive outwash deposits farther downstream (see Chapter 18). The courses of proglacial rivers are usually determined by pre-glacial stream channels.

In Britain, the middle Thames acted as a proglacial drainage system in the Saale glaciation, carrying meltwater from Oxfordshire through Goring Gap into the London

Basin. Large quantities of meltwater and gravel also entered the system down the Colne, Lea and Brent valleys at various stages. The Thames valley is now characterized by its thick infill and the series of gravel terraces produced during successive incisions of the river at different phases of the Pleistocene. The present river is a much shrunken remnant, occupying only a small part of its former floodplain, and its sluggish flow contrasts forcibly with the coarse gravels underlying the floodplain and in the terraces.

Central North America possessed the most gigantic system of proglacial drainage in the world, focusing on the Mississippi-Missouri-Ohio Rivers and collecting meltwater at times from nearly 3200 km of ice margin. Space does not permit a complete description of the system. One of the smaller members of the proglacial drainage net was the Illinois River. With its tributaries the Fox, Des Plaines and Chicago Rivers, it collected water from successive ice borders in Illinois. During the Valders phase, when an ice lobe occupied the northern two-thirds of Lake Michigan, the Illinois River was fed by huge discharge from the Calumet stand of glacial Lake Chicago (Bretz, 1959). As a proglacial drainage route, the Illinois River probably functioned longer and more continuously than any other outlet for glacial drainage in the Great Lakes area. In its lower section below Hennepin, its valley is remarkably wide and straight— projecting spurs are rare; it has a gentle gradient and very thick gravel infill. It is in this section following an earlier course of the Mississippi (Dubuque-Clinton-Hennepin-Illinois River), but the present characteristics of its valley result in no small measure from its function as a major proglacial sluiceway.

4 Marginal and sub-marginal meltwater channels

The surface of glacier ice, where it abuts against a hillslope, is often gently convex, owing to increased melting at the ice margin. As noted by I. C. Russell in 1893, this is a result of radiation of heat from the adjacent exposed ground, which itself is a better absorber of insolation than the ice. Thus a longitudinal hollow, one side of which is formed by the ice, may mark the ice margin and meltwater may collect in it (Plate XXI). The quantity of meltwater is not usually great since its source is purely local, though there are cases where drainage from other more distant sources is diverted along the ice margin (as along the north side of the Tasman glacier, New Zealand). The direction of movement of the meltwater will be controlled by the longitudinal gradient of the ice margin. If there is no gradient, the meltwater will accumulate as a marginal lake which may drain by way of a spillway or over, into, or beneath the ice. If the ice margin slopes longitudinally, meltwater will drain along the margin so long as there are no fissures in the ice, and will carve a marginal channel, partly in ice, partly in bedrock. In the early stages, only a narrow shelf may be cut in the hillside, but if the ice margin remains relatively stable in position, a true rock-cut channel may be eroded which deepens downstream. As the ice will normally possess many fissures and will rarely make a watertight seal with the hill slope, the meltwater will not adopt a marginal course for any great distance, rarely more than 1 or 2 kilometres, beyond which the water will disappear into or beneath the ice. Marginal channels are therefore characteristically short, comparatively straight, and run along a hillside making only a small angle with the contours. They are commonest on long gently-sloping hillsides,

Plate XX
Margin of the Barnes ice cap, Baffin Island, showing marginal drainage and moraine. *(C.A.M.K.)*

Plate XXI
Margin of Breiðamerkurjökull, Iceland, showing convexity of ice surface approaching bedrock slope, and marginal meltwater channel disappearing in the distance in a cave under the ice edge. *(C.E.)*

especially where the slope is veneered in drift, for then the meltwater can quickly erode a channel.

Marginal drainage of present-day glaciers in Alaska and Greenland has been vividly described in the classic reports of R. S. Tarr (1897, 1908, 1909) and O. D. Von Engeln (1911). They make it clear that simple marginal drainage is not as common as has sometimes been supposed. There is first of all the case where a marginal stream is diverted away from the ice margin by rock spurs. Tarr (1909) illustrates this on the eastern margin of Hayden glacier (Alaska): 'Below Floral Pass, a marginal valley is developed ... but at one point the ice presses against a low spur of the mountain, forcing the marginal drainage into a narrow rock gorge for an eighth of a mile. The stream then emerges into a broader valley, with one ice wall, and within a quarter of a mile disappears in an ice tunnel.... Below the ice tunnel for more than a mile there is no visible stream in the marginal valley whose bottom is hummocky, moraine-covered ice into which drainage disappears' (p. 83). Later, he relates how drainage from the Orange and Variegated glaciers 'flows in a marginal channel with ice for one wall and a gravel cliff for the other; then, where the ice crowds up against a granite spur, it enters a rock gorge more than half a mile long and 200 feet deep [in which] the glacial torrent rushes with great velocity and in a succession of falls' (p. 142).

Both Tarr and Von Engeln emphasize that under certain conditions there is no true marginal drainage, but instead a system of sub-marginal drainage. This is described as flowing 'under the ice but near its lateral edge' (Von Engeln 1911, p. 110), and is claimed to be more normal than marginal drainage when the ice is stagnant, or when the ice mass is active but stable in position. Sub-marginal streams may flow parallel with (but underneath) the ice margin, or obliquely beneath the ice; their exact courses are possibly influenced by structures (such as crevasse patterns) in the ice. There may be many sub-marginal streams beneath each margin of a glacier, often forming an anastomosing system leading down to truly subglacial or englacial drainage systems: their channels are of course only revealed after ice wasting uncovers them. Renewed advance of a glacier possessing mainly sub-marginal and subglacial drainage has been seen to result in the blocking or constriction of this drainage, and the reappearance of truly marginal streams: this was noted by Von Engeln (1911) following sudden advances of the Atrevida, Hidden, Variegated and Galiano glaciers in Alaska. Large marginal streams are nearly always associated with rapidly advancing valley glaciers.

Marginal and sub-marginal channels may be associated with both temperate and sub-polar types of glacier, though few studies of channels in sub-polar environments have been made. Fig. 12.6 shows the situation of marginal and sub-marginal streams in an area of Axel Heiberg Island, Canada, where englacial temperatures are negative to a considerable depth and there is permafrost in the bedrock at the glacier margins (H. Maag, 1969). Meltwater appears only in summer, on the glacier surface and along the margins, and during each summer, the lateral streams gradually work their way farther under the ice edge to sub-marginal positions. In some cases they also erode and melt ledges in the frozen bedrock, producing broader channels. In successive melt seasons, the streams tend to find and re-use old channels where these have not been completely blocked by ice collapse.

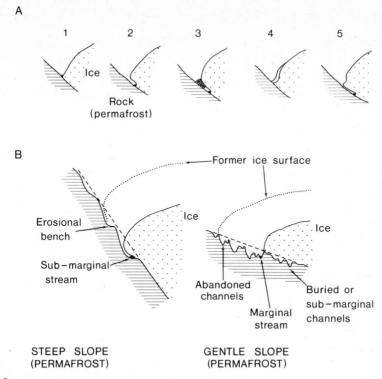

Fig. 12.6
A Development of marginal and sub-marginal channels beside a cold, sub-polar glacier
1 Beginning of melt season: marginal drainage
2 Marginal stream undercuts ice edge
3 Collapse of ice margin; sub-marginal drainage. End of melt season
4 Beginning of next melt season; snow covers previous year's marginal depression; previous year's ice cave survives
5 Drainage finds and re-uses former sub-marginal channel
B Marginal and sub-marginal channels beside a cold glacier, showing effect of steepness of slope
(H. Maag, *Axel Heiberg Island Research Report, McGill University, Montreal*, 1969)

In interpreting meltwater channels after deglaciation, no criteria enable sub-marginal and marginal channels to be distinguished infallibly, as J. D. Ives and R. P. Kirby (1964) have stressed. In every case, a sub-marginal origin must be considered to be at least as likely as a marginal origin. In studies of meltwater channels in Sweden, C. M. Mannerfelt (1945) claimed that the sub-marginal type was more common, but J. Gjessing (1960) in central Norway found it was 'not possible to separate with certainty the strictly marginal parts of the channel systems from the sub-marginal and subglacial parts' (p. 462).

Another complication in the lateral drainage of ice masses has been noted by V. Schytt (1956) along the northern border of the Moltke glacier in Greenland. Here there was not just one line of drainage marking the ice margin but several other parallel drainage channels within the ice. The latter appeared to be forming by differential ablation between the outcrops of dirt-laden shear planes which thus controlled

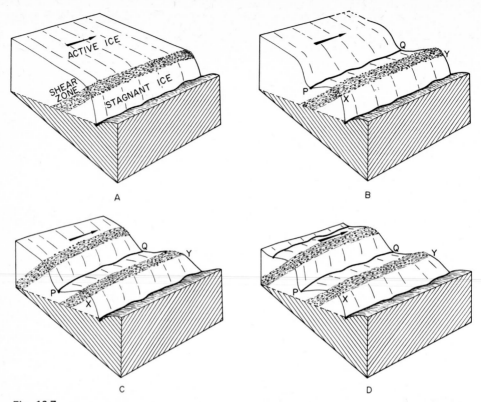

Fig. 12.7
Stages in the development of marginal drainage channels (V. Schytt, *Geogr. Annlr*, 1956)
A Stagnation of the glacier margin owing to thinning; active ice begins to shear past the dead ice.
B Debris melted out from shear zone protects ice from ablation and causes ridge X-Y to form. A new depression P-Q is melted out on the glacier side of this ridge.
C Further thinning of the glacier causes development of new shear zone. Stream P-Q continues to deepen its ice channel.
D A second ridge and another melt depression begin to form by differential ablation. Stream P-Q will shortly melt down to bedrock and become superimposed on it.

the position and spacing of the channels. The shear planes formed between the active ice of the glacier and its stagnating margins (Fig. 12.7). The innermost channel, just forming, was relatively shallow and cut entirely in ice; the next nearer the ice margin was deeper and possibly had melted down through the ice to bedrock; while the ice edge was marked by a truly marginal channel with one ice wall. Away from the ice margin, a series of older and now largely abandoned drainage channels, parallel to the ice margin, could be seen. Thus simultaneous formation of several lateral channels is possible.

In spite of the great difficulty of distinguishing truly marginal channels, many workers have used presumed marginal channels in reconstructing conditions of deglaciation. It has been claimed that marginal channels can provide estimates of ice margin gradient (which obviously cannot be steeper than the mean longitudinal slope of the channel floor); they may throw light on the annual melting rhythm of the ice;

and, as noted already, large marginal channels suggest active advancing ice during their formation.

Estimates of ice margin gradients have been used to reconstruct the form of the ice cover and also to correlate one marginal channel with another. As far back as 1908, J. L. Rich estimated the ice margin gradient in part of the Finger Lakes area, from the slopes of the Johnson Hollow meltwater channels, as 1 in 80. In Scandinavia, Mannerfelt (1945) calculated ice surface slopes in many areas. In Arådalen (eastern Norway, latit. 63°N.) he computed ice slopes between 1 in 33 and 1 in 100 depending on the part of the ice tongue selected; on Högfjället (Sweden, latit. 62°N.) the mean slope was 1 in 30. Mannerfelt (1949) stressed that care should be exercised in this procedure, for marginal drainage naturally tends to undermine the ice border, and thus erode channels which slope more steeply than the ice surface; even more important is not to confuse marginal with sub-marginal channels which may have slopes very much steeper than the ice margin. To avoid this pitfall, J. B. Sissons (1958b) in the Eddleston valley of southern Scotland used only the longest and most gently graded channels of the area to calculate the former ice slope (1 in 50, north-east) and thereby to establish ice margin positions when certain channels were in use. In this area, marginal and sub-marginal channels are numerous but mainly small, less than 5 m deep and 30 m wide; the smallest ones mapped were narrow benches about 10 m wide, cut by water flowing in channels one side of which was ice and the other bedrock or drift.

In north-west Wales, marginal channels were formed along the border of Irish Sea ice (C. Embleton, 1964b), and are distinguished from sub-marginal channels by their gentle slopes, some having gradients of only 1 in 300, such as that near Tregarth, south of Bangor. This example is about 2 km long, flanked on its northern side by drift mounds, and from an intake at 114 m slopes gently south-west to Rhyd-y-Groes where there is a small outwash flat at 110 m. In the same region, a good example of probable sub-marginal drainage occurs between Llanrug and Caernarvon, where anastomosing channels, each from 3 to 15 m deep, are cut in drift and possess gradients (1 in 25 to 1 in 45) that cannot possibly reflect a marginal origin. A series of channels near Conway with gradients between 1 in 20 and 1 in 50 was interpreted as marginal by Embleton in 1961, but most members of the group are now thought to be sub-marginal and sub-glacial.

Mannerfelt (1945, 1949) has claimed that the spacing of marginal channels in some sequences may have been controlled by the amount of annual ablation which the ice surface suffered, each shallow channel being used in one spring and summer only and being abandoned in late summer when subglacial drainage became dominant. In northern Sweden, he found the most complete series of lateral drainage channels and terraces on the west slope of Gråsidan, where no less than 78 consecutive marginal channels were measured, with an average vertical spacing of 3·5 to 4 m. Figures for other areas ranged from 3 to 5 m. Sissons (1958b) in southern Scotland deduced comparable figures of 3 to 3·6 m which he also suggested represented annual thinning of the ice. However, caution is again necessary. It has already been explained how, according to Schytt, several marginal channels can actively co-exist along or just within an ice border. Moreover, some workers have encountered vertical spacings much too large to be the result of annual ablation (for example, the figure of over 20 m given by C. Holdar, 1957), and both Mannerfelt (1960) and Sissons (1961a) have expressed doubts about the validity of earlier claims.

5 Subglacial channels

Although active subglacial streams cannot often be observed directly, there is no doubt as to their existence beneath many glaciers and the importance of their functions as outlined in the first part of this chapter. I. C. Russell (1893) deduced that 'the drainage of the Malaspina glacier is essentially englacial and subglacial' (p. 238) from observations of meltwater descending moulins and crevasses, and later escaping from the ice margin: Fountain stream 'comes to the surface through a rudely circular opening, nearly 100 feet in diameter, surrounded in part by ice. Owing to the pressure to which the waters are subjected, they boil up violently, and are thrown into the air to the height of 12 to 15 feet, and send jets of spray several feet higher. The waters are brown with sediment, and rush seaward with great rapidity, forming a roaring stream fully 200 feet broad' (ibid.). When the subglacial tunnels are abandoned by their streams, they can sometimes be located and explored, as for instance in the case of a small network of tunnels beneath the Martin River glacier in Alaska investigated by L. Clayton (1964). There is also clear evidence of subglacial (or sometimes englacial) drainage provided by the manner in which many glacial lakes periodically drain: the well-known Märjelen See has drained regularly into or beneath the Aletsch Gletscher (see Chapter 19), and G. de Boer in 1949 described the subglacial draining of a small lake beside Leirbreen in Norway. On a vastly larger scale, the Icelandic jökulhlaups represent great volumes of meltwater forcing a way beneath the ice. The work of subglacial streams is equally evident. On the one hand, their depositional activity is reflected in esker systems (Chapter 16), while their erosional activity has given rise to channels of unusual form, often cut in anomalous positions.

The maximum thickness of ice beneath which subglacial streams can exist will depend on several factors. First, the quantity and temperature of the meltwater are critical in determining the rate at which heat is conveyed to melt the ice and to enlarge and extend the waterways. Secondly, the water pressure will act to offset the tendency of the ice to close up any passages by plastic deformation and collapse. Thirdly, the degree of activity of the ice must be taken into account; stagnant, highly fissured ice, for instance, will favour very deep penetration by meltwater. Using Glen's flow law for ice, Fig. 12.8 (R. Haefeli, 1970) shows, on a graph plotting time against ice overburden pressure (or depth), the half-life (T) of an empty circular shaft. At a depth of 200 m where the pressure is about $18 \, kg/cm^2$, it will take theoretically 9–10 days for the diameter of the shaft to be reduced by one-half. At depths exceeding about 400 m, closure by collapse is likely to occur. If the shaft is filled with water, however, T will be increased: its value will then be determined by the difference between the hydrostatic pressure of the water and the overburden pressure of the ice. It can be seen, therefore, that englacial or subglacial channels in a large temperate glacier will only be able to survive the cold season (say 100 days) if they remain filled with water. Haefeli argues that it is likely that, in many such glaciers where circumstances do not favour retention

Plate XXII
The Snout of Kjendalsbreen, Norway, showing the emergence of a subglacial stream. (*C.E.*)

of water, the internal drainage will be reconstituted each year in the melt season. This helps to explain why ice-marginal lakes tend to reform each year, reaching a maximum level in summer followed by a sudden burst englacially or subglacially (Chapter 19).

Geomorphological evidence relating to depth of meltwater penetration shows some

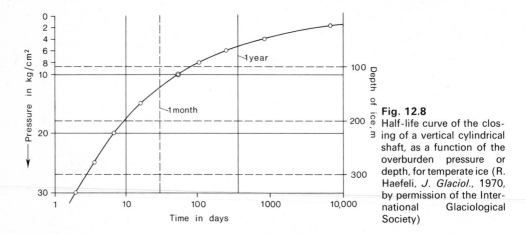

Fig. 12.8 Half-life curve of the closing of a vertical cylindrical shaft, as a function of the overburden pressure or depth, for temperate ice (R. Haefeli, *J. Glaciol.*, 1970, by permission of the International Glaciological Society)

agreement with Haefeli's suggestions. There is abundant evidence of meltwater, either now or in the past, penetrating to depths of 100–150 m, fewer cases of it penetrating to 200–300 m, and examples of penetration beyond 300 m are rare. Sissons (1963) states that 'many channels and some eskers ... prove that meltwaters penetrated 300 feet beneath glacier ice' (p. 110); Embleton (1964a) found evidence in the Wheeler valley (North Wales) for the existence of subglacial streams beneath about 120 m of ice. W. H. Mathews (1964) showed that meltwater in the South Leduc glacier (British Columbia) penetrated at least 110 m below the glacier surface, the location being about 2 km from the glacier snout and at the glacier's lateral margin. Meltwater under pressure was encountered by the borehole AR203 (Fig. 4.19) beneath the Glacier d'Argentière, 150 m below the glacier surface (Vivian and Zumstein, 1973). Glacier moulins have been explored to depths of 60 m (for example, G. Dewart, 1966) and occasionally a little deeper. J. Gjessing (1960) in northern Gudbrandsdalen, Norway, claims that, during deglaciation when the ice had become completely stagnant, esker and associated deposits were laid down by subglacial streams more than 250 m below the surface of the ice and that 'the first drainage that followed the valley bottom must have flowed more than 350 m below the ice surface' (p. 469).

The simplest form of subglacial channel is that termed the subglacial chute by Mannerfelt (1945, 1949). These were cut by meltwaters plunging directly down a hillslope beneath the ice. Often, they were fed by marginal or sub-marginal channels, whose meltwaters after flowing along a hillside suddenly abandoned that course for one more directly related to the bedrock slope. Sometimes there is evidence that the meltwater once continued farther along the slope before the chute was opened, and it seems probable that the opening and location of many chutes is dependent on the development of crevasses in the ice. In other cases, pre-glacial stream valleys have been used and enlarged as chutes, as shown by the way in which some marginal or sub-marginal channels terminate at such a stream valley (Sissons, 1961a, p. 20). The chutes can normally be distinguished from post-glacial gullies by their relations to other meltwater channels, their abrupt beginnings and endings (sometimes part-way down a slope), the absence of alluvial fans from their lower ends, and the fact that they are often now dry. Mannerfelt considered that chutes might open and operate mainly in the later part of each melting season, and Haefeli (1970) supports this idea. Subglacial chutes have only been recognized as such in Britain since about 1957, but since then many workers (for example, Sissons 1958b, Price 1960, Embleton 1964, Bowen and Gregory 1965) have mapped and described them. Their height range is usually no more than 100 m, supporting the theory of crevasse control (E. Derbyshire, 1961).

More complex forms resulting from subglacial drainage have long been known to exist in parts of northern and central Europe covered by the last ice sheet. In north

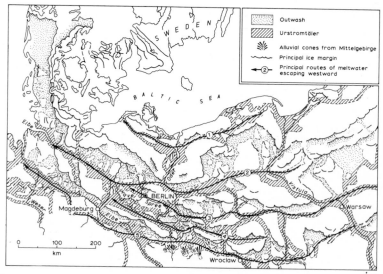

Fig. 12.9
Morainic limits, outwash and Urstromtäler of north Germany, Denmark and Poland (P. Woldstedt, *Norddeutschland und angrenzende Gebiete im Eiszeitalter*, K. F. Koehler Verlag, Stuttgart, 1955)

Germany and Denmark, there are the valley systems known as the Rinnentäler (channel valleys) and Tunneldale (tunnel valleys) respectively. They represent subglacial stream courses carrying meltwater westward and southward to the ice margin during the last glaciation (Fig. 12.9). The tunnel valleys of Denmark are up to 75 km long (A. Schou, 1949), relatively narrow and up to 100 m deep, with flat floors and sharply defined sides in transverse profile. The larger ones are so wide that it seems unlikely that they could have formed as single tunnels. Probably they represent a system of anastomosing tunnels, with ice masses falling from the roof causing the streams to shift about from one position to another. Some of the tunnel valleys can be traced from the inlets ('fjords') of the Baltic coast (p. 304) continuously to the sandy outwash plains of western Jylland. In long profile, their floors are characteristically irregular, for the subglacial streams, being under pressure, were able to force their way uphill in places and excavate depressions. As a result, ribbon lakes are frequently found occupying parts of the valley floors, such as Hald Sö and Lake Tjele near Viborg. Streams may follow parts of the tunnel valleys, but are usually diminutive in size compared with the valleys which the swollen glacial torrents carved. The system of tunnel valleys continues southward into Schleswig and north Germany, where their subglacial origin has been particularly studied over many years by P. Woldstedt (1926 *et seq.*). The valleys (Rinnentäler) tend to be orientated in the direction of regional ice slope, and therefore generally perpendicular to the ice margins, for the subglacial waters were forced out under the ice towards regions of lesser ice pressure. At the ice margins where hydrostatic pressure and therefore the load-bearing capacity of the streams were suddenly reduced, low outwash cones were built up at the points where the torrents emerged from the ice. As the outwash cones accumulated, so the terminal base-level of the streams was slowly raised, and as a result, the streams began to flow englacially and finally even supraglacially over collapsed ice filling the Rinnentäler (E. M. Todtmann, 1936). As in Denmark, the floors of the Rinnentäler are uneven in long profile, and Woldstedt considers that the majority of the north German lakes lie in such depressions (Rinnenseen).

More recently, Woldstedt (1961) and K. Hansen (1971) have considered that glacial erosion may also have played a part in excavating the larger tunnel valleys. Hansen does not deny the existence of tunnels and streams in the ice but now thinks they were mainly englacial and only in places reached bedrock where the ice was thin.

In Britain, modern studies of subglacial drainage systems were initiated by J. B. Sissons (1958a), whose attention was first drawn by R. F. Peel to two meltwater channels in southern Northumberland (Fig. 12.3a and b, and see p. 332) whose floors possess up-and-down long profiles. By mapping probable ice margins in the area from marginal channels, and by studying the relationships of these two channels with others in the area, Sissons showed that previous explanations of the humped long profiles—such as reversals of meltwater flow, post-glacial erosion of one end of the channel, or erosion of a channel across a col previously formed by a stream flowing in the reverse direction—were unsatisfactory, and that only a hypothesis of erosion by subglacial water flowing under hydrostatic pressure could adequately account for the features. Fig. 12.3c shows another example of a channel in East Lothian, the Deuchrie channel, where the subglacial hypothesis is applied by Sissons. It possesses a clear up-and-down long profile, is of unusually large capacity compared with nearby marginal and sub-marginal channels, and small but definitely subglacial eskers lie on its floor.

Two other examples of meltwater drainage systems in southern Scotland investigated by Sissons provide clear evidence that they were initiated subglacially. The first lies on the north side of the Tinto Hills (Sissons, 1961b) where an anastomosing system of relatively shallow channels receives at its upper end a series of eskers. These in turn emerge from the lower end of another system of meltwater channels, the largest of which crosses the Hills in a col and is of humped form. The intimate relationship between channels and eskers, and the anomalous long profile of the col channel point conclusively to a subglacial origin. The second example is in the Carlops area, a little south of Edinburgh. Here, Sissons (1963) has mapped and re-interpreted a complex braided system of channels which, he argues, developed first subglacially; later, as the ice cover wasted, the meltwater became concentrated in open ice-walled channels. The evidence supporting a subglacial origin is fourfold:

1 Several channels are so situated that ice must have been present on both sides of them when they were being cut
2 At least one channel possesses a humped long profile
3 One side of a large channel is cut into by gullies which can only be meltwater features and themselves formed under ice as subglacial chutes
4 The presence of eskers, some with a thin cover of ablation moraine, in the channel system.

The braiding of the channels around rocky knolls and hills points to contemporaneous formation of these sub-parallel channels. The initiation of the whole system is thought to have occurred by superimposition of englacial streams on an area of varied relief during the last period of ice downwasting. R. J. Price (1963) developed this hypothesis as an alternative to that which explains up-and-down channels as simply the product of meltwaters flowing under hydrostatic pressure. He suggests that, if an englacial stream by melting down gradually encountered parts of the underlying bedrock relief, it would not only become superimposed on these but would also probably develop an irregular long profile, for its rate of channel down-cutting would be much faster in ice than in bedrock. Two advantages of this hypothesis are that it accounts for the non-existence of channels in some areas (where the meltwater was flowing englacially), and that it avoids the necessity for postulating meltwater flowing subglacially beneath great thicknesses of ice.

S. A. Schumm and R. G. Shepherd (1973) have pointed out that normal fluvial erosion is capable of producing shallow enclosed basins in stream long profiles, representing scour holes (see also p. 337). Their argument that up-and-down long profiles of some meltwater channels do not, therefore, demand a subglacial origin takes no account of all the other evidence pointing to subglacial or englacial meltwater flow in such cases, nor that the humped long profiles of some channels bear no relation to scour holes.

Following Sissons's lead, many other workers in Britain have found strong evidence of the work of subglacial meltwaters in erosion and deposition. Subglacial drainage systems have been described from the Cairngorms (D. E. Sugden, 1970), the Tweed basin (Price, 1960 and 1963), the Cheviots (Derbyshire, 1961; C. M. Clapperton, 1968, 1971), Eskdale in Yorkshire (Gregory, 1965), North Wales (Embleton 1961, 1964a and 1964b), South Wales (Bowen and Gregory, 1965), and elsewhere. Clapperton (1968, 1971) and Sugden (1970) in particular support the view that superimposition of englacial streams is the most satisfactory way of explaining how the eroding meltwaters became subglacially emplaced in bedrock channels.

In Europe and North America, there is also a considerable modern literature on the subglacial drainage systems of the last deglaciation which there is no room to summarize here, beyond mention of such accounts as those of Gjessing (1960) for east-central Norway and Sissons (1960b) for part of New York State. Gjessing's account is based on detailed ground survey of fluvioglacial features over 1700 km².

6 Col channels

Meltwater channels are sometimes cut into existing cols; they may be of great size and raise special problems. As a particular type of meltwater channel, they were recognized by Mannerfelt (1945, 1960) who named them Sadelskäror or col gullies. A classic example occurs south-west of Drommen peak in northern Sweden (latitude 63°). Into the col separating Drommen from another peak to the south-west has been carved the spectacular Dromskäran, a canyon which begins abruptly with a 40-metre high cliff. The floor of the canyon at the foot of the cliff is eroded into a shallow rock basin, beyond which the canyon floor drops rapidly through a vertical interval of 110 m, with sidewalls about 50 m high. Mannerfelt visualizes the formation of the col channel as initiated by a subglacial stream forced through the col by hydrostatic pressure. Later, during downwasting of the ice, the hills and the col between them were gradually uncovered, and from the ice on the south-east of the col, meltwater began to pour across the col to regain the ice on the far side which, owing to regional ice slope, was at a slightly lower level. As downwasting proceeded, the col was more deeply eroded; a water-fall with plunge-hole at its foot developed and receded headward towards the crest of the col. A shallow ice margin lake fed the waterfall since strand-lines appear leading into the col, and a delta accumulated at the lower end of the canyon where another ice lake existed. Dromskäran is thus interpreted as a subglacial channel later functioning as a normal glacial lake spillway, though the dimensions of the lake were small and the water was essentially provided by the melting ice around.

A subglacial origin for col channels has been suggested by many other writers, though examples of such channels linked with lake strandlines as in the case of Dromskäran appear to be rare. The great size of some appears to demand more meltwater than might be locally available, some of them possess humped or irregular long profiles, eskers sometimes lead into the heads of the col channels suggesting that ice and not a lake lay behind them, and some col channels possess either double intakes or double outlets which are not easy to explain on a sub-aerial hypothesis. Derbyshire, studying col channels in the Cheviot Hills and in Labrador, concluded that the majority were cut by subglacial waters flowing under pressure, of such volume that they were not derived from their own immediate vicinity (Sissons, 1961a), and flowing in the direction of lesser ice thickness and therefore less pressure. Gjessing (1960) comments in regard to northern Atnedalen and Gudbrandsdalen (central Norway) that in accordance with the regional ice slope to the north and north-west, 'all subglacial water in the area between the culmination zone (i.e., the ice divide) and the present watershed (north and north-west of the culmination zone) was forced to ascend to a pass at the present watershed on its way to the valleys on the other side' (p. 443). Thus subglacial col channels, as in the case of the larger subglacial drainage systems generally, can be used to indicate approximately the slope of the ice surface and variations in regional ice thickness. The Sarn Galed and Sarn Adda col channels in Flintshire (Embleton, 1964a) provide good examples; both possess humped profiles, and were eroded by subglacial meltwaters being forced generally southward across cols in the direction in which the Irish Sea ice of this area was thinning. Similarly, the two Northumbrian spillways described by Peel (1949) and reinterpreted by Sissons (1958a) are essentially col channels carrying meltwater generally eastward across cols towards areas of thinner ice cover.

An interesting characteristic of col channels is the way in which they frequently become deeper as the crest of the col is approached. Moreover, any eskers that lead into them always stop short of the actual col (compare the way in which eskers are often broken where they cross humps or ridges). Clapperton (1968) suggests that this is because the meltwater became focused on the col floor, the side-slopes of the col forcing the meltwater to migrate down the ice-rock contact to the floor, so that the subglacial river became largest and fixed in one position at the col. Shreve (1972) also shows that the eroding and transporting capacity of the subglacial stream will be largest where it crosses a hump in its floor.

7 Open ice-walled channels

Present-day meltwater streams can sometimes be observed flowing in open ice-walled channels. Russell (1893) described the courses of the Osar, Kame and Kwik rivers on the eastern margin of the Malaspina glacier: 'Each of these issues from a tunnel and flows for some distance between walls of ice. The Kame ... issues as a swift brown flood partially choked with broken ice, from the mouth of a tunnel and flows for half a mile in an open cut between precipitous walls of dirty ice 80 to 100 feet high' (p. 239). B. Roberts (1933) noted similar behaviour of the Kreppa emerging on the north of Vatnajökull in Iceland, while R. P. Sharp (1947) described how the Wolf Creek (Yukon) emerged from an englacial tunnel to flow for 3 km between walls of ice. These ice-walled channels clearly originated by collapse of ice-tunnel roofs, and this is probably the commonest mode of origin, but there is also the possibility to be borne in mind that downcutting by supraglacial streams to bedrock can produce ice-walled trenches. In either case, the ice walls will guide the course of the stream and give rise to meltwater channels, after the ice has vanished, in anomalous positions. C. P. Gravenor and W. O. Kupsch (1959) have identified channels in east-central Alberta which they claim are of this type. Some have been filled with till, others are broad open troughs with sand and gravel on their floors. Some feed into esker systems, which might suggest a subglacial origin, but while the smaller ice-walled channels may have been true tunnels, the width of the larger trenches (more than 1 km) and the paucity of till in some of them, suggest that they were open to the sky. They developed in a pattern of parallel and intersecting elements during disintegration of stagnant ice over the area. In southern Illinois, Morris Leighton (1959) contends that the tendency for streams to be arranged in trellis pattern over Illinoian drift reflects the pattern of open crevasse meltwater channels developed in the stagnating Illinoian ice. Sissons (1960b) has put forward the view that some of the large channels in the Syracuse area of New York State may have operated as ice-walled channels after thinning of the ice and tunnel roof collapse. Limited ice thickness and the huge volumes of Great Lakes waters using them make it difficult to regard them simply as subglacial, while their locations and directions are not explicable in terms of marginal drainage. Some, moreover, such as the lower Butternut valley, contain outwash terraces which are unlikely to have formed in tunnels. In the lower Oneida valley, dead ice deposits show that a tongue of ice persisted here and was cut across by meltwater from the west giving rise to a transverse ice-walled channel, later infilled by sand and gravel. Ice-walled channels are also envisaged in east central Norway by Gjessing (1960) who describes how the subglacial rivers gradually melted 'the ceilings of the chambers and tunnels, so that drainage ascended up through the ice' (p. 471).

8 Superimposed channels

The possibility that meltwater channels with humped long profiles might be explained by superimposition of englacial streams has already been noted. But this is only one particular case of a process whereby any stream flowing on or in ice will lower its bed rapidly by solution and will eventually encounter bedrock. If the rate of ice channel lowering is not too rapid, the rock not too resistant, and the ice in firm contact with bedrock, then the stream will be superimposed on the rock. The process has been hinted at by many workers in the past. W. O. Field (unpublished work quoted by Flint, 1971, p. 231) observed how a meltwater stream from the Casement Glacier, Alaska, had been superimposed across resistant bedrock in 1935, and when next seen in 1941, had succeeded in cutting a gorge 8 m deep and 15 m wide. R. S. Tarr (1908) in the Yakutat Bay region referred to streams leaving the ice margin and cutting a gorge across a rock spur (p. 100); Von Engeln (1911) in the same area also commented on this phenomenon and stated that the stream would be permanently entrenched in such a course provided its rate of downcutting were equal to or greater than the rate of ice wasting (p. 128). The application of the process to formerly glacierized areas was not common until recently; J. W. Goldthwait's reference (1938) to superimposition of meltwater streams in New Hampshire was an exception. In Britain, R. Common (1957), Embleton (1961), Price (1960, 1963), and Sissons (1961a, 1963) have utilized the hypothesis to explain certain channels. Price has made the most complete analysis of the hypothesis in respect of meltwater channels in the Tweed basin. Channels resulting from superimposition may have long profiles of three forms:

1 Continuously and gently sloping
2 Gently sloping at first, becoming markedly steeper down-channel
3 Up-and-down character.

The last has been discussed already and demands superimposition of englacial rather than supraglacial streams. The second is characteristic of a stream superimposed on a rock outcrop and plunging beneath the ice once it has crossed the obstruction, while the first is typical of a channel cut by a supraglacial stream, whose downstream gradient accorded with the ice surface slope and which by downmelting has happened to encounter bedrock in part of its course. Supraglacial streams are usually of clear sparkling water and their channels in the ice seldom contain much debris, as noted by Russell (1893) on the Malaspina glacier, but when they reach the basal ice layers, a great deal more debris will become available for bedrock erosion. The meandering habit of supraglacial streams is well known, and the hypothesis of their superimposition may explain the occurrence of meandering channels on slopes where the form of the ground would not permit meandering streams to develop marginally or extra-marginally (Sissons, 1961a, p. 29).

9 Summary: the general significance of meltwater channels

Studies of meltwater channels throw much light on the conditions pertaining to episodes of deglaciation, provided they are correctly interpreted. In the past the majority of channels were regarded as glacial lake spillways or as marginal channels, but it is now clear that other types of channel, especially sub-marginal and subglacial types, are equally if not more important, and in those areas where the meltwater was not able to escape proglacially, were responsible for carrying most of the drainage during ice wasting. It is particularly important not to interpret sub-marginal channels as marginal, for gross errors in reconstructions of ice margin positions and gradients can thereby arise. If truly marginal channels can be identified, however, they can be of great use in such reconstructions, bearing in mind that under certain conditions the marginal zone of an ice mass may possess more than one line of lateral drainage; and in a few instances, closely spaced sequences of minor channels may even reveal the annual wasting rhythm of the ice. But marginal channels are not the only indicators of the form of the ice. Subglacial river systems are partly directed by variations in ice pressure and therefore tend to seek out regions of thinner ice; from this, it may be possible to deduce the regional ice surface slopes. On the other hand, structure is thought to play a part in influencing the form and alignment of meltwater channels in several areas (W. Barr, 1969; J. D. Ives and R. P. Kirby, 1964). Shreve (1972) points out that steep ice-surface gradients will most strongly affect the direction of flow of englacial and subglacial streams. As gradients lessen, bedrock configuration will more and more exert its influence on subglacial water flow, but englacial streams will continue to be directed by differential ice pressure.

The vertical range of regional channel systems will throw further light on the extent of the ice and its internal condition. The highest meltwater channels in a region may be used to indicate tentatively the minimum heights of the firn line during their formation while the lower limit of channel cutting may indicate the level of an englacial water table or the level below which crevasses and fissures at that time were too tightly closed to permit meltwater to circulate (Price, 1973).

Meltwater channels can also suggest the condition of the ice during deglaciation. Large subglacial channel systems indicate temperate ice at the time of their formation. Sub-polar glaciers will possess mainly marginal and sub-marginal systems. In addition, large marginal rivers are most typical of active and advancing glaciers, while subglacial drainage is more characteristic of stagnant ice or the thinner zones of active glaciers. Open ice-walled channels develop in the final stages of ice disintegration. The depths to which meltwater can penetrate ice depend on the quantity, temperature and pressure of the meltwater and the degree of activity of the ice. Channels can only be maintained beneath great thicknesses of ice if the corresponding hydrostatic pressure is also great enough to resist channel closure by plastic deformation. The greatest depths, possibly in excess of 400 m, would be attained by warm meltwater penetrating stagnant ice, coupled with a high englacial water table.

Subglacial drainage systems have been extensively studied in recent years, and their behaviour is now better known. They are responsible for impressive and unusual features of both erosion and deposition. They consist of both local drainage chutes, and the more extensive and complex systems fed by a wider melting area and controlled by ice-pressure variations. Subglacial channels may be distinguished by humped long profiles, by their anomalous positions in the landscape (which may rule out most other possibilities), and by their relation to esker systems. Humped long profiles are thought to have been cut by subglacial water being forced up or down hill by hydrostatic pressure, or to have resulted from superimposition of englacial streams. As ice melt proceeds, tunnel roofs collapse and open ice-walled channels replace the former sub-glacial drainage. It should be emphasized that many meltwater channels are of compo-site origin, not only having been re-used and re-excavated in successive glaciations but also having developed in different positions in relation to the ice. Thus channel systems may be initiated by subglacial drainage, pass into a phase of open ice-walled channels, and finally be occupied by proglacial streams.

10 Glacial diversion of drainage

Glaciation may interfere in several ways with existing river systems, both directly and indirectly. The simplest case is that of ice damming a former valley, but caution should be exercised in applying this hypothesis too readily since glacier ice, as shown above, is not often the impenetrable barrier to water that it might seem. Obstruction by moraine is common, as is complete plugging by drift and obliteration of former valleys. Glacial erosion can also radically derange the pre-glacial drainage by lowering certain valley floors or cols, thereby permitting new routes to be adopted by rivers post-glacially. Finally, river systems may be affected by glacio-isostatic movements conse-quent on the loading or unloading of the earth's crust during glaciation and deglacia-tion. Circumstances of glacial diversion vary greatly from one area to another, so that few generalizations are profitable; but generally drainage diversion by ice or drift obstruction produces the most spectacular rearrangements of river systems in regions of low relief. The processes will be discussed by reference to contrasting examples, both large and small in scale.

10.1   *The Baraboo-Devil's Lake area, Wisconsin* (Fig. 12.10)

This small area on the margin of the so-called Driftless zone in southern Wisconsin provides a classic model of drainage diversion. In the Cary phase of the Wisconsin glaciation, the Wisconsin River was ponded by ice at Wisconsin Dells, the resultant lake overflowing north and west by the Black River to the Mississippi. Below the Dells, the pre-glacial course of the Wisconsin River, now in part plugged by drift and recon-structed from well records in the plains sections (W. E. Powers, 1946) took it through the north rim of the Baraboo syncline (in a gap now utilized in the reverse direction by the Baraboo River), and through the South Range in a gap not now followed by any river but occupied by Devil's Lake. The latter is impounded both on the north and on the south-east by moraines of the Cary ice. With the melting of the Cary ice, the Wisconsin River was unable to resume its former course because of these morainic barriers and drift plugging, and now makes a detour over drift-covered country round the east of the Baraboo Ranges. At Devil's Lake, the former valley of the Wisconsin River is filled with at least 86 m of morainic and fluvioglacial material.

10.2   *The North European Plain* (Fig. 12.9, p. 347)

Rising in the central European uplands are the great rivers Weser, Elbe, Oder and Vistula which drain to the Baltic or North Seas. Their pre-glacial courses across the lowland of North Germany and Poland are not known, having been completely obliterated and altered by successive ice sheets and their deposits. The outstanding valley

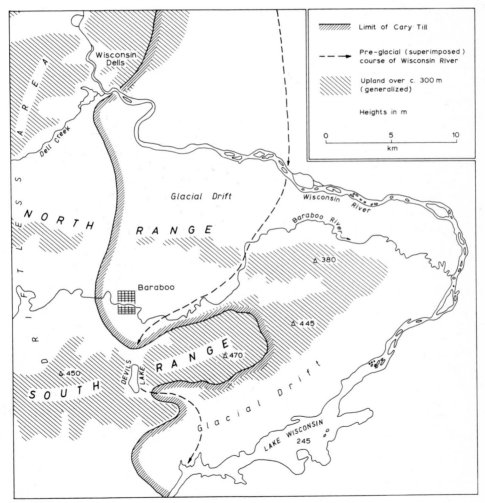

Fig. 12.10
Map of the area around Baraboo and Devil's Lake, Wisconsin, showing present and pre-glacial courses of the Baraboo river

systems of the present-day lowland are the Urstromtäler (Germany) or Pradoliny (Poland), great meltwater channels formed peripherally to the Baltic ice at various stages, and now carrying lengthy segments of the four great rivers just mentioned, together with many lesser streams. The Urstomtäler, cut largely into the sandy Geest outwash, are margined by moraines on their northern sides and are extremely capacious—that south of the Brandenburg moraine is thought to have once carried a river forty times the volume of the present Elbe, and possibly larger, the water coming both from melting ice to the north and normal land drainage to the south. Of the five main Urstromtäler (numbered on Fig. 12.9), the southernmost passing Wroclaw and Magdeburg and possibly continued in the lower Weser, is generally considered to date from the Warthe glaciation, while the remainder operated at various stages of Baltic ice retreat in the Weichsel. All the principal Weichselian Urstromtäler, with the exception of that lying north of the Baltic Ridge, converge on the lower Elbe which was in consequence in use for the greater part of the Weichsel glaciation. The Vistula, Oder, Spree and Havel were all forced to follow courses directed westward by the Urstromtäler in part, and now present curious zig-zag patterns as a result. None of the Urstromtäler, however, is occupied by continuous drainage, for the floor gradients, although very gentle, do not show a steady fall. Interruptions are due to formation of low outwash cones (such as where the Rinnentäler (p. 348) break into them on their northern sides), oscillations in the position of the ice margin, and deposits of windblown loess.

A major problem concerns the origin of the Durchbruchtäler, the breaches in the moraines and the cross-connections between the different Urstromtäler, such as the segments where the Oder successively breaks through the Brandenburg, Frankfurt and Pomeranian moraines to the north. Woldstedt's explanation of these channels as sub-glacial Rinnentäler flowing south has already been mentioned, and at present this seems the most likely hypothesis. Many other problems remain concerning the detailed evolution of the Urstromtäler; although the general scheme presented here, based on Wold-stedt's work, is largely accepted, the complexity of the landforms on a large scale is indicated by the fact that W. Behrmann (1950) finds evidence for seven Weichsel retreat stages in the Berlin area alone. Nevertheless, the broad principles of drainage diversion by barriers of both ice and moraine are clearly established.

There is no comparable situation farther east in the USSR, for here the main pre-glacial drainage lines of the Dniester, Dnieper, Don and Volga rivers were directed away from the later ice-covered areas, but in the northern Great Plains of the USA, interesting comparisons can be drawn with the pre-glacial west–east valleys draining towards the ice in the Dakotas and Nebraska. The lower parts of these were blocked and largely obliterated by successive glaciations from the Kansan onwards, and the present Missouri River to which they are now tributary in these states is essentially marginal to the limits of glaciation, as in the north German Urstromtäler, though far less is known about the details of its glacial history.

10.3   *The Teays* (Fig. 12.11)

This now-vanished river system of Ohio, Indiana and Illinois provides an unrivalled illustration of drainage diversion resulting from drift filling of former valleys (L. Horberg, 1945, 1956). Prior to the Kansan glaciation, as far as is known, the Teays flowed from the Appalachians past Fort Wayne (Indiana) into Illinois where it joined the ancestral Mississippi (Horberg, 1950). The whole of its course from Chillicothe (Ohio) to the Illinois River at Beardstown is now buried by drift, a hundred metres or more thick in Indiana as W. D. Thornbury and H. L. Deane (1955) and others have shown. Oil, gas and water well records, together with geophysical surveys, have located it precisely. No important streams follow the course now and it has little topographical expression: the one small exception to this is in Indiana where the Wabash River has by chance encountered the Teays valley in two places, and in each widens markedly as it crosses the soft infill (Thornbury, 1958). The infilling of the Teays valley was not, however, a one-stage process; it was begun in the Kansan (or earlier?), parts were subsequently re-excavated and deepened in the Illinoian, and later reburied beneath Illinoian and Wisconsin drifts.

The present Ohio has now replaced the Teays as the principal drainage line of the region; it developed in part to receive meltwater as a marginal stream to the ice sheets terminating on its northern side. In detail, the evolution of the Ohio is much more complex, and it is thought that the present river is made up of the following segments:

1 The lower Ohio, of pre-glacial origin and of small size compared with the Teays

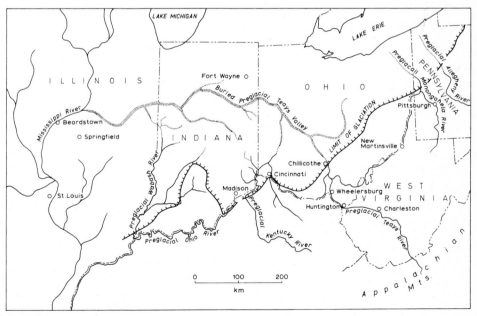

Fig. 12.11
Map of the pre-glacial Teays river system. Parts of this system were later integrated to form the present Ohio river system. The Mississippi above St Louis is shown in its pre-glacial course

farther north; there is controversy as to whether it headed near Madison (Indiana) or south-east of Cincinnati.

2 The short Wheelersburg (Ohio)–Huntington (West Virginia) section where the Ohio still follows the Teays closely.

3 The section from Huntington to New Martinsville represents a former tributary of the Monongahela, reversed in direction of flow by ice blocking.

4 The headwaters consist of the Monongahela and Allegheny, which pre-glacially were tributary to the Ontario-St Lawrence system.

10.4   *Diversion by glacial erosion*

The erosional activity of diffluent and transfluent glaciers has been referred to in Chapter 8. After dissolution of the ice, the drainage may be permanently deranged if valley floor levels have been suitably altered. This possibility has long been noted—

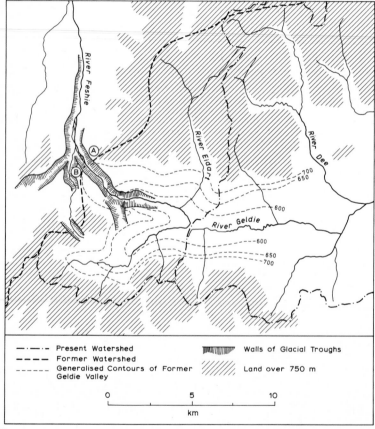

Fig. 12.12
The diversion of the upper Geldie to become the headstream of the Feshie, resulting from glacial breaching of the former watershed at A-B (D. L. Linton, *Scott. geogr. Mag.*, 1949, Royal Scottish Geographical Society)

A. Penck (1905) thought it the reason for the curious course taken by the Rhine at Sargans—but the first detailed application of the hypothesis to a number of instances of anomalous drainage was by D. L. Linton (1949a, b). His account of the diversion of the headstreams of the River Geldie in the Grampian Highlands of Scotland provides an excellent illustration (Fig. 12.12). Mapping of former valley floor remnants shows that the Geldie once possessed headstreams rising close to the former watershed marked A–B, and converging where the present divide between the Geldie and the Feshie is situated. Linton convincingly refutes the possibility that the Feshie has captured the Geldie, particularly on the grounds that the geological structure does not favour the breaching of the former watershed A–B by headward erosion of the Feshie. On the other hand, the features and position of the breach in the watershed are entirely in accord with a hypothesis of glacial erosion in which ice accumulated in the headwater basin of the Geldie, was impeded in its normal outflow to the east, and began in part to cross the former watershed A–B at its lowest point. This diffluent stream of ice, entering the Feshie basin on the far side of the watershed, eventually effected the cutting of a breach in the watershed to a total depth of 285 m, thus permanently diverting the upper reaches of the Geldie.

10.5   *Diversion by glacio-isostatic changes of level*

There is now abundant evidence testifying to the sensitivity of the earth's crust to changes in load (Chapter 5). The tilting of segments of the crust, induced by differential ice loading and unloading, has various important physiographical effects, such as the disturbance of certain delicately balanced river and lake systems. The Great Lakes of North America represent such a system and are located in an area where glacio-isostatic tilting has been particularly marked owing to the advances and retreats of the massive Laurentide ice sheet. The history of the Great Lakes is extremely complex (J. L. Hough, 1958) and in this section it is only proposed to outline certain relevant aspects of their post-glacial evolution. In an early post-glacial phase, Lakes Superior, Michigan, and Huron stood at the same level, submerging Sault Ste Marie to a depth of 15 m, and overflowing by way of the North Bay outlet and the Ottawa River valley to the St Lawrence. Lakes Erie and Ontario drained to the St Lawrence as they do now, but were separate from the other three lakes. Slow isostatic recovery in the period 9500–4000 BP raised the land in the north with respect to a hinge-line running from near Duluth towards Toronto. The North Bay outlet was abandoned, and in early Nipissing times, Lake Huron spilled over at the St Clair River into Lake Erie; and Lake Michigan, because of the raising of the water level, overflowed at the site of Chicago into the Des Plaines River. Downcutting of the Chicago River outlet was small owing to a bedrock bar, but the St Clair outlet in till was more easily lowered in stages until eventually, about 3000 years ago, the Chicago outlet was abandoned. The fall in water level exposed the Sault Ste Marie bar between Superior and Huron, so that these two lakes now differ in level by 7 m. Thus isostatic tilting resulted in the drying-up of the North Bay-Ottawa outlet and the transference of drainage to the St Clair and Chicago outlets; differential erosion rates later caused the abandonment of the latter. But the situation remains critical, for isostatic uplift is still in progress;

moreover, the Chicago outlet has been artificially resurrected, though carefully controlled.

Alteration of drainage by glacio-isostatic tilting is not of course confined to the Great Lakes area, though the effects are here on a grand scale and well documented. Similar changes in lake outlets and drainage directions are known to have occurred in the area of the Baltic ice and especially in Finland, where isostatic recovery is also still in progress, raising the area near the head of the Gulf of Bothnia 60 to 70 cm a century more than in southern Finland.

11 References

ALLISON, I. S. (1933), 'New version of the Spokane flood', *Bull. geol. Soc. Am.* **44**, 675–722

BARR, W. (1969), 'Structurally controlled fluvioglacial erosion features near Schefferville, Quebec', *Cah. Géogr. Québ.* **13**, 295–320

BEHRMANN, W. (1950), 'Die Umgebung Berlins nach morphologischen Formengruppen betrachtet', *Erde* **1**, 93–122

BOER, G. DE (1949), 'Ice-margin features, Leirbreen, Norway', *J. Glaciol.* **1**, 332–6

BOWEN, D. Q. (1964–66), 'On the supposed ice-dammed lakes of South Wales', *Trans. Cardiff Nat. Soc.* **93**, 4–17

BOWEN, D. Q. and GREGORY, K. J. (1965), 'A glacial drainage system near Fishguard, Pembrokeshire', *Proc. Geol. Ass.* **76**, 275–81

BRETZ, J. H. (1923), 'Glacial drainage on the Columbia Plateau', *Bull. geol. Soc. Am.* **34**, 573–608

  (1932), 'The Grand Coulee', *Am. geogr. Soc. Spec. Publ.* **15**

  (1959), 'The double Calumet stage of Lake Chicago', *J. Geol.* **67**, 675–84

  (1969), 'The Lake Missoula floods and the channeled scabland', *J. Geol.* **77**, 505–43

BRETZ, J. H., SMITH, H. T. U. and NEFF, G. E. (1956), 'Channeled scabland of Washington: new data and interpretations', *Bull. geol. Soc. Am.* **67**, 957–1049

CARRUTHERS, R. G. (1939), 'On northern glacial drifts', *Q. J. geol. Soc. Lond.* **95**, 299–333

CHARLESWORTH, J. K. (1929), 'The South Wales end-moraine', *Q. J. geol. Soc. Lond.* **85**, 335–55

CLAPPERTON, C. M. (1968), 'Channels formed by the superimposition of glacial meltwater streams, with special reference to the east Cheviot Hills, north-east England', *Geogr. Annlr* **50**A, 207–20

  (1971), 'The pattern of deglaciation in part of north Northumberland', *Trans. Inst. Br. Geogr.* **53**, 67–78

CLAYTON, L. (1964), 'Karst topography on stagnant glaciers', *J. Glaciol.* **5**, 107–12

COMMON, R. (1957), 'Variations in the Cheviot meltwater channels', *Geogr. Stud.* **4**, 90–103

DERBYSHIRE, E. (1961), 'Subglacial col gullies and the deglaciation of the north-east Cheviots', *Trans. Inst. Br. Geogr.* **29**, 31–46

DEWART, G. (1966), 'Moulins on Kaskawulsh glacier, Yukon Territory', *J. Glaciol.* **6**, 320–1

DURY, G. H. (1951), 'A 400-foot bench in south-eastern Warwickshire', *Proc. Geol. Ass.* **62**, 167–73

DWERRYHOUSE, A. R. (1902), 'The glaciation of Teesdale, Weardale, and the Tyne valley, and their tributary valleys', *Q. J. geol. Soc. Lond.* **58**, 572–608

ELSON, J. A. (1965), 'Western strandlines of glacial Lake Agassiz', *Rep. 7th Conf. int. Ass. quatern. Res. (Colorado, 1965)*, Abstracts, 126

EMBLETON, C. (1961), 'The geomorphology of the Vale of Conway, with particular reference to its deglaciation', *Trans. Inst. Br. Geogr.* **29**, 47–70

    (1964a), 'Subglacial drainage and supposed ice-dammed lakes in North-East Wales', *Proc. Geol. Ass.* **75**, 31–8

    (1946b), 'The deglaciation of Arfon and southern Anglesey, and the origin of the Menai Straits', *Proc. Geol. Ass.* **75**, 407–30

EWING, K. J. (1972), 'Supraglacial streams of the Kaskawulsh glacier', *Icefield Ranges Res. Proj. scient. Results* **3**, 153–62

FLINT, R. F. (1938), 'Origin of the Cheney-Palouse scabland tract, Washington', *Bull. geol. Soc. Am.* **49**, 461–523

    (1971), *Glacial and Quaternary geology*

GJESSING, J. (1960), 'Isavsmeltningstidens drenering, dens forløp og Formdannende virkning i Nordre Atnedalen' (The drainage of the deglaciation period, its trends and morphogenetic activity in northern Atnedalen), *Ad Novas* **3**, 492 pp.

GOLDTHWAIT, J. W. (1938), 'The uncovering of New Hampshire by the last ice-sheet', *Am. J. Sci.* **36**, 345–72

GRAVENOR, C. P. and KUPSCH, W. O. (1959), 'Ice disintegration features in western Canada', *J. Geol.* **67**, 48–64

GREGORY, K. J. (1965), 'Proglacial Lake Eskdale after sixty years', *Trans. Inst. Br. Geogr.* **36**, 140–62

HAEFELI, R. (1970), 'Changes in the behaviour of the Unteraargletscher in the last 125 years', *J. Glaciol.* **9**, 195–212

HANSEN, K. (1971), 'Tunnel valleys in Denmark and northern Germany', *Meddr. dansk geol. Foren.* **20**, 295–306

HARMER, F. W. (1907), 'On the origin of certain cañon-like valleys associated with lake-like areas of depression', *Q. J. geol. Soc. Lond.* **63**, 470–514

HESS, H. (1935), 'Die Bewegung im Innern des Gletschers', *Z. Gletscherk.* **23**, 1–35

HOBBS, W. H. (1943), 'Discovery in eastern Washington of a new lobe of the continental Pleistocene glacier', *Science* **98**, 227–30

HOLDAR, C. (1957), 'Deglaciations förloppet i Torne Träsk-området efter senaste nedisningsperioden med vissa tillbakablickar och regionala jämförelser', *Geol. För. Stockh. Förh.* **79**, 291–528 (with English summary)

HORBERG, L. (1945), 'A major buried valley in east-central Illinois and its regional relationships', *J. Geol.* **53**, 349–59

    (1950), 'Bedrock topography of Illinois', *Bull. Ill. St. geol. Surv.* **73**

    (1956), 'Pleistocene deposits along the Mississippi valley in central-western Illinois', *Rep. Invest. Ill. St. geol. Surv.* **192**

HOUGH, J. L. (1958), *Geology of the Great Lakes* (Univ. Illinois Press)

IVES, J. D. and KIRBY, R. P. (1964), 'Fluvioglacial erosion near Knob Lake, central Quebec-Labrador, Canada: discussion', *Bull. geol. Soc. Am.* **75**, 917–22

JAMIESON, T. F. (1863), 'On the parallel roads of Glen Roy, and their place in the history of the glacial period', *Q. J. geol. Soc. Lond.* **19**, 235–59

JONES, O. T. (1965), 'The glacial and post-glacial history of the lower Teifi valley', *Q. J. geol. Soc. Lond.* **121**, 247–81

KENDALL, P. F. (1902), 'A system of glacier lakes in the Cleveland Hills', *Q. J. geol. Soc. Lond.* **58**, 471–571

KNIGHTON, A. D. (1972), 'Meandering habit of supraglacial streams', *Bull. geol. Soc. Am.* **83**, 201–4

LEIGHTON, M. M. (1919), 'The road-building sand and gravel of Washington', *Bull. Wash. geol. Surv.* **22**, 307 pp.

(1959), 'Stagnancy of the Illinoian glacial lobe east of the Illinois and Mississippi rivers', *J. Geol.* **67**, 337–44

LINTON, D. L. (1949a), 'Watershed breaching by ice in Scotland', *Trans. Inst. Br. Geogr.* **15**, 1–16

(1949b), 'Some Scottish river captures re-examined', *Scott. geogr. Mag.* **65**, 123–31

MAAG, H. (1969), 'Ice-dammed lakes and marginal glacial drainage on Axel Heiberg Island', *Axel Heiberg Island Res. Rep., McGill Univ. Montreal*, 147 pp.

MALDE, H. E. (1968), 'The catastrophic Late Pleistocene Bonneville flood in the Snake River plain, Idaho', *U.S. geol. Surv. Prof. Pap.* **569**, 52 pp.

MANNERFELT, C. M. (1945), 'Några glacialmorfologiska formelement', *Geogr. Annlr.* **27**, 1–239 (with English summary)

(1949), 'Marginal drainage channels as indicators of the gradients of Quaternary ice-caps', *Geogr. Annlr* **31**, 194–9

(1960), 'Oviksfjällen: a key glaciomorphological region', *Ymer* **80**, 102–13. See also *Guidebook to Excursion E.SW.7, 19th int. geogr. Congr., Stockholm 1960*, 29–40

MATHEWS, W. H. (1964), 'Water pressure under a glacier', *J. Glaciol.* **5**, 235–40

NYE, J. F. (1973), 'Water at the bed of a glacier', *Symposium on the hydrology of glaciers* (Cambridge, 1969), *Int. Ass. scient. Hydrol., Publ.* **95**, 189–94

NYE, J. F. and MEIER, M. F. (1973), 'Preface', *Symposium on the hydrology of glaciers* (Cambridge, 1969), *Int. Ass. scient. Hydrol., Publ.* **95**, 5–6

PARDEE, J. T. (1910), 'The glacial Lake Missoula', *J. Geol.* **8**, 376–86

(1942), 'Unusual currents in glacial Lake Missoula, Montana', *Bull. geol. Soc. Am.* **53**, 1569–99

PEEL, R. F. (1949), 'A study of two Northumbrian spillways', *Trans. Inst. Br. Geogr.* **15**, 75–89

PENCK, A. (1905), 'Glacial features in the surface of the Alps', *J. Geol.* **13**, 1–19

POWERS, W. E. (1946 [1947]), 'The Dells and Devil's Lake region, Wisconsin', *Chicago Nat.* **9**, 74–86

PRICE, R. J. (1960), 'Glacial meltwater channels in the upper Tweed drainage basin', *Geogrl J.* **126**, 483–9

(1963), 'A glacial meltwater drainage system in Peebleshire, Scotland', *Scott. geogr. Mag.* **79**, 133–41

PRICE, R. J. (1973), *Glacial and fluvioglacial landforms*

RAISTRICK, A. and BLACKBURN, K. B. (1932), 'The Late-glacial and Post-glacial periods in the North Pennines', *Trans. nth. Nat. Un.* **1**, 16–36 and 79–103

RICH, J. L. (1908), 'Marginal drainage features in the Finger Lakes region', *J. Geol.* **16**, 527–48

ROBERTS, B. (1933), 'The Cambridge expedition to Vatnajökull, 1932', *Geogrl J.* **81**, 289–313

RÖTHLISBERGER, H. (1972), 'Water pressure in intra- and subglacial channels', *J. Glaciol.* **11**, 177–203

RUSSELL, I. C. (1893), 'Malaspina Glacier', *J. Geol.* **1**, 217–45
    (1898), 'The great terrace of the Columbia and other topographic features in the neighbourhood of Lake Chelan, Washington', *Am. Geol.* **22**, 362–9

SCHOU, A. (1949), 'The landscapes' in *Atlas of Denmark* (ed. N. NIELSEN) (Copenhagen)

SCHUMM, S. A. and SHEPHERD, R. G. (1973), 'Valley floor morphology: evidence of subglacial erosion?' *Area* **5**, 5–9

SCHYTT, V. (1956), 'Lateral drainage channels along the northern side of the Moltke Glacier, North-west Greenland', *Geogr. Annlr* **38**, 64–77

SHARP, R. P. (1947), 'The Wolf Creek glaciers, St Elias Range, Yukon Territory', *Geogrl Rev.* **37**, 26–52

SHOTTON, F. W. (1953), 'The Pleistocene deposits of the area between Coventry, Rugby and Leamington, and their bearing on the topographic development of the Midlands', *Phil. Trans. R. Soc.* **237**B, 209–60

SHREVE, R. L. (1972), 'Movement of water in glaciers', *J. Glaciol.* **11**, 205–14

SISSONS, J. B. (1958a), 'Sub-glacial stream erosion in southern Northumberland', *Scott. geogr. Mag.* **74**, 163–74
    (1958b), 'Supposed ice-dammed lakes in Britain, with particular reference to the Eddleston valley, southern Scotland', *Geogr. Annlr* **40**, 159–87
    (1960a), 'Some aspects of glacial drainage channels in Britain. Part I', *Scott. geogr. Mag.* **76**, 131–46
    (1960b), 'Subglacial, marginal and other glacial drainage in the Syracuse-Oneida area, New York', *Bull. geol. Soc. Am.* **71**, 1575–88
    (1961a), 'Some aspects of glacial drainage channels in Britain. Part II', *Scott. geogr. Mag.* **77**, 15–36
    (1961b), 'A subglacial drainage system by the Tinto Hills, Lanarkshire', *Trans. Edinb. geol. Soc.* **18**, 175–93
    (1963), 'The glacial drainage system around Carlops, Peebleshire', *Trans. Inst. Br. Geogr.* **32**, 95–111

STENBORG, T. (1969), 'Studies of the internal drainage of glaciers', *Geogr. Annlr* **51** A, 13–41

SUGDEN, D. E. (1970), 'Landforms of deglaciation in the Cairngorm Mountains, Scotland', *Trans. Inst. Br. Geogr.* **51**, 201–19

TARR, R. S. (1897), 'The margin of the Cornell Glacier', *Am. Geol.* **20**, 139–56
    (1908 [1909]), 'Some phenomena of the glacier margins in the Yakutat Bay region, Alaska', *Z. Gletscherk.* **3**, 81–110

TARR, R. S. (1909), 'The Yakutat Bay region, Alaska: physiography and glacial geology', *U.S. geol. Surv. Prof. Pap.* **64**

THORNBURY, W. D. (1958), 'The geomorphic history of the upper Wabash valley', *Am. J. Sci.* **256**, 449–69

THORNBURY, W. D. and DEANE, H. L. (1955), 'The geology of Miami county, Indiana', *J. Geol.* **48**, 449–75

TODTMANN, E. M. (1936), 'Einige Ergebnisse von glazialgeologischen Untersuchungen am Südrand des Vatna-Yökull auf Island (1931–34)', *Z. dt. geol. Ges.* **88**, 77–87

UPHAM, W. (1896), 'The glacial Lake Agassiz', *U.S. geol. Surv. Monogr.* **25**

VIVIAN, R. (1970), 'Hydrologie et érosion sous-glaciaires', *Rev. Géogr. alp.* **58**, 241–64

VIVIAN, R. and ZUMSTEIN, J. (1973), 'Hydrologie sous-glaciaire au glacier d'Argentière (Mont-Blanc, France)', *Symposium on the hydrology of glaciers* (Cambridge, 1969), *Int. Ass. scient. Hydrol., Publ.* **95**, 53–64

VON ENGELN, O. D. (1911 [1912]), 'Phenomena associated with glacier drainage and wastage, with especial reference to observations in the Yakutat Bay region, Alaska', *Z. Gletscherk.* **6**, 104–50

WEERTMAN, J. (1972), 'General theory of water flow at the base of a glacier or ice sheet', *Rev. Geophys. Space Phys.* **10**, 287–333

WILLS, L. J. (1924), 'The development of the Severn valley in the neighbourhood of Ironbridge and Bridgnorth', *Q. J. geol. Soc. Lond.* **80**, 274–314

WOLDSTEDT, P. (1920), 'Die Durchbrüche von Schtschara und Bug durch den westrussischen Landrücken', *Z. Ges. Erdk. Berl.* (1920), 215–25

(1926), 'Probleme der Seenbildung in Norddeutschland', *Z. Ges. Erdk. Berl.* (1926), 103–24

(1935), 'Bemerkungen zu meiner geologisch-morphologischen Übersichtskarte des norddeutschen Vereisungsgebietes', *Z. Ges. Erdk. Berl.* (1935), 282–95

(1950), *Norddeutschland und angrenzende Gebiete im Eiszeitalter* (Stuttgart)

(1961), *Das Eiszeitalter* (Stuttgart), **1**

Part III

Glacial and fluvioglacial deposition

13

Glacial and fluvioglacial deposition: general considerations

Say when, and whence, and how, huge Mister Boulder,
And by what wond'rous force hast thou been rolled here?
Has some strong torrent driven thee from afar,
Or hast thou ridden on an icy car?

(P. DUNCAN, C. 1840)

Glacier ice and its associated meltwater deposit a wide range of materials in many different forms. Some general points concerning glacial deposits are considered in this chapter and the following chapters in Part III deal with specific forms.

In any study of glacial drift, it is essential to examine the processes at work on the edge of and beneath present-day ice masses, but it should be borne in mind that modern conditions do not necessarily closely resemble those that operated around the Pleistocene ice sheets of Europe and North America. The present-day Greenland ice sheet, for example, is not forming moraine on its eastern margin, and whereas most of the Antarctic ice sheet today ends in the sea, most of the borders of the northern hemisphere Pleistocene ice sheets lay on land. The present-day ice masses are on the whole situated on old hard rocks from which any weathered material has long since been removed. They cannot, therefore, produce drift in quantities comparable with that produced by the earlier Pleistocene ice sheets which frequently advanced over deeply weathered areas. Thus the clayey till is often more bulky in the deposits of the Pleistocene ice sheets than it is in those of contemporary ice sheets. Nevertheless, a close study of present-day glaciers and ice sheets can provide valuable evidence on the processes of glacial deposition, and currently developing forms are described wherever possible in the subsequent chapters of this section.

One of the problems in studying glacial and fluvioglacial deposits is the confusion over terminology. J. K. Charlesworth (1957), for example, considers that the terms 'esker' and 'kame' are more or less synonymous, and he uses the term 'esker' to include features which some authors call kames. Thus, under the heading of 'eskers', he also discusses 'kame-terraces', 'kame-moraines' and 'osar'. In this book, eskers and kames are differentiated. Another term that causes problems is 'moraine', which has been used for a very wide range of features. Some of these are discussed in Chapter 15, but some of them, such as delta-moraines, are in reality fluvioglacial forms and these are considered in Chapter 16.

1 Classification

Glacial and fluvioglacial deposits have for long been referred to collectively as glacial drift, a term introduced by C. Lyell in 1840. The term dates from the time when it was widely believed that erratics and other glacial deposits were associated with major inundations. Lyell himself in 1841 supported the iceberg theory. It is useful to classify glacial drift into groups of sediments that have important properties in common. Further subdivision within the classification can then be based on the relationship of the deposit to the ice. The basic subdivision of drift is into glacial and fluvioglacial deposits. These groups can normally be differentiated by their sorting, for whereas water-laid fluvioglacial deposits are stratified, glacial deposits, laid down directly by the ice, lack this property. Further subdivision is based on the relationship of the deposit to the ice forming it (Table 13.1). The basis of such a classification was first suggested

Table 13.1   Classification of glacial and fluvioglacial deposits

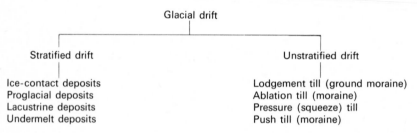

by T. C. Chamberlin in 1894. Ice-contact deposits are differentiated from proglacial deposits, which are laid down in front of the ice by meltwater. Lacustrine deposits, laid down in still water, differ significantly from the deposits laid down by rapidly flowing melt-streams. Undermelt deposits contain some characteristics of both lacustrine and ice-contact deposits and can best be recognized by their relationship to unstratified deposits under which they normally occur.

The terms 'till' or 'boulder clay' are often used to denote the unstratified deposits. A. Geikie used the term 'till' in 1863. The term had been used in Scotland previously to denote a coarse stony soil which forms on the bouldery drift of the north of Britain. It is a better term than boulder clay, as only occasionally does till consist of boulders and clay. The main distinction traditionally made in unstratified drift is between lodgement till and ablation till. These have been distinguished since at least 1877. The former is deposited under the ice while the latter is let down from the ice surface on to the ground as the ice wastes away. The distinction between lodgement and pressure till is that between a deposit laid down under relatively dry conditions, often by a plastering-on process, and one laid down either under water or with material in a water-logged state. Lodgement till is sometimes referred to as ground moraine, for example by L. Agassiz in 1837, and ablation till as ablation moraine. The use of the term 'moraine' is particularly confusing. It has been used in glaciological literature from very early days, H. B. de Saussure having employed it as early as 1799, and Agassiz having introduced it to a wider public. It was originally applied to topographical features and its use by Scandinavian workers as a synonym for till has led to confusion.

Another important distinction that can be made is between deposits laid down by active ice and those accumulating from stagnant ice. The importance of this distinction has been appreciated since the beginning of this century, and the interest of stagnant ice conditions is such that a separate chapter has been devoted to the features formed in this way.

2 Occurrence

The relative importance of these different types of glacial and fluvioglacial deposit will vary with a number of factors. Some of the most important include the nature of the land surface covered by ice, the nature of the climate, which partly determines the glacier régime, and the changes in the climate which cause advance and retreat of ice. The land surface over which the ice moves affects not only the nature of the ice flow but also the size of the ice mass. In addition, the rocks over which the ice moves determine to a major extent the character of the drift deposited by the ice. Normally, in mountain areas, the ice is near its source and movement may be expected to be fast if the slopes are steep and the snow supply plentiful. In the regions bordering the mountains, the ice often attains its greatest thickness and can still move actively. Ice in these areas may be expected to remain active even during frontal retreat. Thus in these mountain and marginal areas, the deposits are more likely to be those deposited by active rather than stagnant ice even during retreat. Examples of rapidly retreating glaciers which are still moving forward actively at their snouts have been cited in Norway. Observations on Austerdalsbreen (Chapter 4) have shown that the glacier is moving actively over its bed right up to its rapidly retreating snout. It is producing thin moraine that shows slight fluting, which is evidence of movement at the base of the glacier.

A situation that tends to restrict the amount of lacustrine stratified deposition is that in which the glacier margin is retreating upslope. This allows meltwater to drain away freely without being ponded in front of the glacier. Any stratified deposits that do occur will consist of proglacial outwash, apart from subglacial forms of deposition. The well-developed valley trains that fill the floors of Alpine valleys both in the European Alps and the Southern Alps of New Zealand are examples of this type of deposit. The easily eroded rock type and heavy precipitation in the latter area also help to increase the amount of material available for deposition. In mountain areas, depositional features associated with active ice are more likely to form. These features include drumlins and various types of moraine.

When climatic conditions allowed the development of very large ice sheets, such as those that covered much of north-west Europe and North America during the major ice advances of the Pleistocene, ice moved very far from its mountain sources down on to the low ground. It was natural, therefore, that when conditions ameliorated and general recession set in, these outer parts of the ice sheet would be the first to be separated from their accumulation areas and to become stagnant. In fact it appears that in north America, the ice margin at first retreated while the ice remained active, but that an acceleration of climatic amelioration coupled with topographical control led to widespread stagnation. R. F. Flint (1929) has shown that the presence of well-marked terminal moraines and clay-rich lodgement till in the outer parts of the glaciated area of north America indicate active frontal retreat in this area, which is 300 to 500 km wide (see p. 496, Chapter 17). The inner parts, north of the Great Lakes border moraines in north America and the Brandenburg moraines in north Europe, on the other hand, show mainly evidence of deposits laid down in connection with decaying ice, in which eskers, kames and fluvioglacial deposits abound.

The ice near the marginal zones of a large ice sheet will probably never have been so vigorous as that nearer the source. This is indicated by the inability of ice masses far from their source to override even fairly low hills in their path in some instances. Where hill masses extend above a stagnant ice mass, as in the Cleveland Hills of eastern England during the last glaciation, the deposits will tend to include many ice-contact meltwater features, deposited by the drainage from the hills escaping into and beneath the ice that is still filling the low ground around the hills. Stagnant ice lobes would be expected to be fairly common on low ground far from the ice source. The former presence of such dead ice masses can often be deduced to explain phenomena that are difficult to explain in other ways.

The formation of the trench of the Trent between Nottingham and Newark is a good example. This trench, which is about 3 km wide, is cut obliquely across the strike of the strata. It diverts the river from what appears to be an easier route, which it followed pre-glacially, across the Vale of Belvoir eastward to the Ancaster Gap and the Wash. It has been suggested that the route along the trench and through the Lincoln gap was initiated when the former route was blocked by a lobe of dead ice lying in the low ground of the Vale of Belvoir. Only an isolated lobe of dead ice could reasonably account for this diversion of the river to the north. The trench contains some fluvioglacial gravels, now part of the Hilton Terrace, to support this hypothesis. The Vale of Belvoir may have been previously deepened by glacial scouring of its soft clays when the ice was more vigorous.

When an ice sheet becomes stagnant and decays *in situ*, it is clear that a great deal of meltwater must be produced, although sometimes a relatively large loss may also take place by evaporation. Stagnant ice conditions are favourable to the preservation of fluvioglacial forms and deposits. Such conditions, in which lobes of stagnant ice develop at the edge of retreating ice sheets, occur at times when the ice source lies on a plateau surface. As long as the snow-line lies below the plateau surface, the ice mass will be reasonably well nourished, but as soon as the snow-line rises above the ice on the plateau, then the supply is suddenly cut off and the whole ice mass may become virtually dead at once. In these conditions marginal decay will be extensive. Such conditions appear to have occurred over part of the Prairie Provinces in Canada (C. P. Gravenor, 1955), where dead ice features are widespread.

Only a small change in climate leading to a slight raising of the snow-line can bring about the situation just described. Other climatic controls are also important to the nature of the glacial deposits. It is important to differentiate between those deposits that are associated with cold glaciers and those that occur in association with temperate ice in which meltwater can move freely. Examples of features formed along the margin of cold glaciers will be described in Chapter 15. Stratified subglacial meltwater deposits can only be associated with temperate ice masses in which it is possible for water to circulate beneath the ice (Chapter 12). For this reason many ice-contact stratified features must be associated with temperate glaciers. Nevertheless, cold-based ice sheets can and often do have meltwater flowing along their margins and held up in ice-dammed lakes alongside them, because, being frozen to their beds, they make more waterproof dams than temperate glaciers (Chapter 19).

3 Depositional processes

The processes by which till is deposited have been studied both on present-day glaciers and in the deposits of previous ones. R. P. Goldthwait (1971) has recently reviewed the occurrence and formation of till. It is derived from debris in basal ice layers, and may include lumps of frozen outwash. The proportion of rock debris to ice in Greenland glaciers is usually only 0·03 to 3 per cent by weight, but locally reaches 50 per cent in Spitsbergen, and 64 per cent has been recorded in the dirtiest part of the Casement glacier. Dirty ice layers only extend 1 to 3 m above the marginal rock. Rock debris can, however, extend 6 to 30 m above the valley floor. There is often an amber ice layer above the sharp limit of the debris zone. The dirt in this layer is fine, may only amount to 1cm³ per cubic metre of ice and is deployed by upward diffusion. Debris brought up along shear planes in the marginal zone comes from a local source, but far-travelled material indicates that other processes also carry material up into the ice. Regelation not only helps to break down rocks but also freezes material into the base of the ice. Part of this material may represent unweathered rock flour from abrasion, but some weathered material may also be included. Plucking (see p. 197) is an even more effective process providing basal debris, especially on lee slopes. Material is also incorporated from the valley side by mass movement, and is important in lateral moraine formation. Such material is more angular than that in the basal debris and can include very large boulders. The roundness of rock particles in the basal debris increases rapidly in the first kilometre of transport and attrition occurs through breakage, clasts breaking according to their initial shape. Crushing of englacial and subglacial debris increases the content of fines. Lateral dispersal of debris by the ice from small sources shows spreads of 2° to 10°, but greater ranges of dispersal of 20° to 60° have been recorded elsewhere.

   Deposition of transported material can take place either at the base of the ice or as a result of surface ablation. Basal melting is active beneath substantial areas of large ice sheets today (Chapter 3), releasing basal debris, and a 'plastering' or 'smearing' process spreads material on the bedrock. Continued ice flow will mould or groove these deposits.

   G. S. Boulton (1971) has examined the processes of till deposition in Svalbard

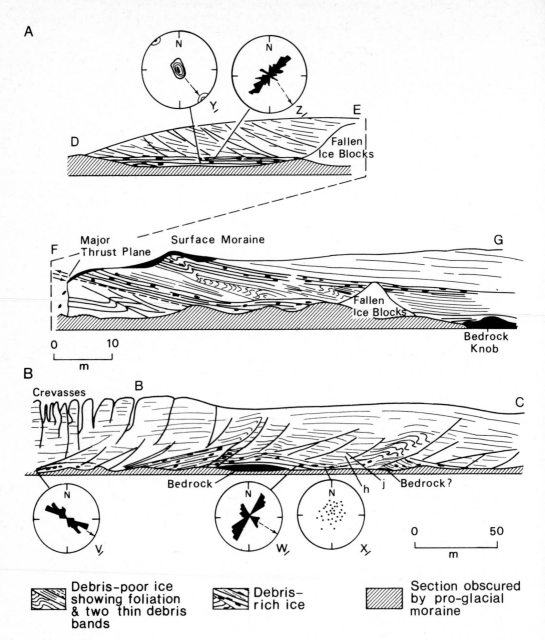

Fig. 13.1
A Section along the southern ice-front cliff of Makarovbreen, Spitsbergen
B Section along the northern ice cliff.

V, *W* and *Z* are rose diagrams showing the orientations of the long axes of prolate- and blade-shaped stones, projected on to the plane of ice foliation. *Y* is an equal-area stereographic plot of poles to *a-b* planes of blade- and plate-shaped particles larger than 10 cm. The plane of projection is the plane of ice foliation. *X* is an equal-area projection, in the plane of ice foliation, of the optic axes of 36 ice crystals. Arrows on *V*, *W*, *Y*, *Z* show direction of ice movement (G. S. Boulton, *J. Glaciol.*, 1970, by permission of the International Glaciological Society)

glaciers. Three main types of till are now forming and a new classification is proposed:

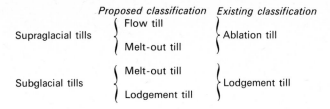

This classification specifies the different types of process whereby till accumulates.

Flow till is released as a fluid mass from englacial debris when exposed by down-wasting of the glacier surface. Melt-out till is deposited from the melting of dead ice covered by a stable overburden. Subglacial lodgement till is released from basal ice either by pressure melting against rock obstructions or from patches of stagnant ice beneath the moving glacier sole. Englacial stone orientations are rarely preserved except in some melt-out or subglacial tills. Flow tills have an upper, more fluid, element with fabrics both transverse and parallel to the flow (see p. 382 for a discussion of till fabrics). The lower part has a parallel fabric with up-slope a-axis imbrication. Melt-out tills have fabrics with a/b axes in the plane of deposition. Subglacial fabrics (Fig. 13.1) vary according to the form of the subglacial bedrock. Where it is flat, parallel fabrics are dominant, but the fabric at one site can rarely be taken as representative of the direction of glacier movement. The englacial stone orientation appears to be affected by the tensile or compressive stress system. The a-axis orientation is controlled by the direction of maximum extension of a triaxial strain ellipsoid. Flow-till fabrics vary more in the vertical than the horizontal direction. Melt-out tills occur where the overburden is thick enough to inhibit flow and where the debris content in the ice below the overburden reaches 50 per cent. Original foliation dipping at 45° will melt to 26° in freshly deposited till, and where the slope is 10°, the dip will become 6° in the a/b plane. Melt-out tills can form at the surface or the base of a glacier. Subglacial deposition is related to the moving glacier sole (Fig. 13.2). The latter shows 'striations' or fluting where it leaves the bedrock floor and the a-axis of stones in the sole lies parallel to these flow striations. Boulton also notes that, where the sole loses contact with the floor, till oozes out like toothpaste and slumps to the floor. The till was more consolidated on the up-glacier flank of bedrock obstacles. Subglacial cavities can be filled in by this process. Pressure melting is important in till deposition around drumlins and roches moutonnées, while basal melting may be responsible for widespread till deposition by lodgement. A plane of décollement separates active from stagnant ice; the latter undergoes folding by shear stress and this can change the fabrics. Fluted lodgement till occurs both on the lee sides and the up-glacier sides of boulders. The flutes are up to 2 m high, but not very long. The long axes of 10 to 30 cm stones tend to parallel the fluting, but pits reveal more complex structures which may be due to flow of till under pressure into subglacial cavities.

Debris-laden stagnant ice underlies mixed material beyond the zone of active ice (Boulton, 1970). Stagnant ice is preserved by the extrusion of flow till and by other material that comes to rest upon it. Melt-out debris from the underlying ice maintains

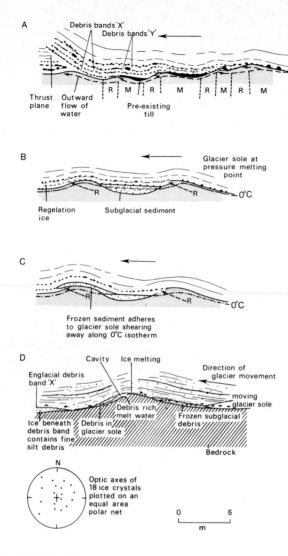

Fig. 13.2

A, B and **C** show possible mechanisms for the englacial incorporation of subglacial debris at bedrock projections.

R and M refer to zones of regelation and melting respectively. Arrows with continuous lines show direction of ice motion; arrows with broken lines show movement of subglacial meltwater, from the glacier interior and from pressure melting against obstructions.

A Debris eroded by the glacier adheres to its moving sole and is then incorporated englacially by the formation of regelation ice beneath it on the down-glacier flanks of bedrock projections. Most of the regelation ice forms from water moving subglacially outwards under a pressure gradient. This accretion of water raises the included debris bands above the regelation zone.

B, C show how masses of subglacial sediments could be incorporated to form debris bands (which may be stratified) with a low ice content.

D A section of the glacier-bedrock contact exposed in the southern ice cliff of Makarovbreen, Spitsbergen (G. S. Boulton, *J. Glaciol.*, 1970, by permission of the International Glaciological Society) its structure if it is undeformed during the melt-out process. Boulton describes how the upper flow till overlies sands and gravels. The basal till, which has a uniform thickness of 0·4 m, appeared to have melted out from the stagnant ice after the sand and gravel had been deposited. The fabric of the basal till shows lower dips and greater dispersion. Both deposits showed a preferred orientation parallel to flow. Where the ice at the surface is active, till can only be deposited subglacially, and this probably takes place from stagnant ice concealed below the active ice. There is more debris in stagnant ice from which ice will already have been melted out. If ice is melted out of the basal till, overriding active ice can produce low-angle shear planes and high-angle joints. Delicate shells can be preserved between the fractures, and stones in the shear planes lie parallel to the movement. Some types of drumlin could form by the overriding of more plastic active ice.

The consolidation of clay under glacial pressure has been considered by R. Aario (1971). He shows that conditions in Finland were favourable to consolidation by glacier loading. Preservation of consolidated clays in this area was favoured by marine inundation following deglaciation, but in some environments, consolidation could be prevented or lost, for example, by periglacial freeze-thaw. Lack of consolidation does not necessarily mean, therefore, that there has been no readvance, although the reverse holds true: consolidated clays do indicate a readvance. In Finland, the ice generally advanced over dry land subjected to permafrost, and this prevented initial consolidation, but as the ice thickness increased, the temperature régime at the base of the ice changed, and when temperatures reached pressure melting point, consolidation resulted.

The thermal régime of the glacier is very important in determining the transport and deposition of sediment by the glacier (Boulton, 1972). Some aspects of this have already been discussed in Chapter 3, and it was noted that four different sets of thermal conditions can exist at the base of a glacier or ice sheet (Fig. 3.8, p. 91). These are: A, a zone of net basal melting; B, a zone in which there is a balance between basal melting and freezing; C, a zone in which freezing of meltwater at the glazier base is sufficient to maintain the ice at pressure melting point; and D, a zone in which the glacier base is below the melting point temperature. In zone A the glacier can slip over its bed, and debris can be entrained by regelation freezing or by the interaction of particles in the ice with those on the bed. The latter process provides only a thin layer of debris, rarely extending more than 1 m above the glacier sole. Regelation ice may be up to 1 m in thickness. Under these conditions debris tends to move down the glacier. Thicker layers of subglacial material in zone A are largely derived from the sole by lodgement. Rates of deposition will be higher where the basal water layer is absent owing to permeability of the bed. A rough bed will also favour deposition: deposits will first accumulate behind obstacles but later may become more uniformly spread.

In zone B where melting balances freezing, depositional processes are basically similar to those in zone A. Lodgement of material will be greater if no meltwater is present, but similar to zone A if meltwater is present, for example if zone B lies down-glacier from zone A. In zone C the glacier will slip over its bed, but plucking may be more effective as loose or jointed material becomes frozen to the glacier sole. Lodgement is not ruled out in this zone, but freezing at the base is more likely to incorporate the material into the glacier sole, therefore causing net erosion. Surface melting will release englacial material as ablation till, which if sufficiently fluid may become flow till, usually when the glacier is retreating. An advancing glacier under these conditions may deposit material proglacially from a steep front, where it will be overridden and incorporated as basal material. In zone D the bed is frozen and no movement occurs between ice and bedrock. Some plucking of large rocks may occur, and any thrusting within the ice will cause material to move upwards as it is transported, thus producing englacial debris bands. Deposition at the glacier base as lodgement till will not be nearly so effective as the deposition of ablation till and flow till supraglacially. Drumlins may possibly form under these conditions by lodgement which, if it does occur, will take place against larger obstructions. Important variables are the bed roughness, the discharge of subglacial water and the permeability of the subglacial strata.

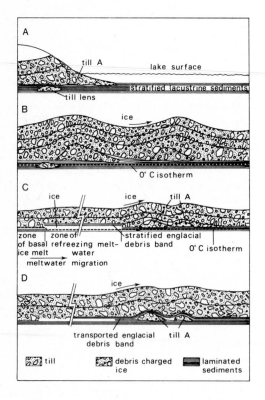

Fig. 13.3
Hypothesis of origin of compound till and stratified sediment sequence formed during one glacial episode in Shropshire, England (J. Shaw, *J. Glaciol.*, 1971, by permission of the International Glaciological Society). The hypothesized sequence of events is as follows:

Deposition of till **A** and the transport of till **B** and stratified sediment across it, during advance of a cold-based glacier. At stage **C**, the increased thickness of ice up-glacier from till **A** caused the basal ice to warm to pressure melting point; stratified sediment is incorporated into the glacier in the regelation zone down-glacier from this. Stage **D** illustrates the transport of the now englacial stratified sediment across till **A**. Note that there is a major change of scale between stages **A/B** and **C/D**.

Thermal conditions in relation to glacial deposition have also been considered by J. Shaw (1971), examining glacial deposits in a gravel pit at Mousecroft Lane near Shrewsbury in England (Fig. 13.3). Here, till overlies sands and gravels, which in turn overlie glacigenic sediments above further sands and gravels. The lower suite is from the Irish Sea ice and the upper one is of Welsh origin. Lenses of till (A) occur in the stratified sediments below the main till (B). Till A differs from till B above it. The former has a regional up-glacier dip. The advancing ice incorporated lacustrine deposits into the basal layers. The gravels were deposited in a braided stream environment, while the laminated deposits were laid down in a proglacial lake. Till lenses and bands were deposited as flow till. Till B was deposited by ice that overrode till A, which was probably frozen at this stage. Basal melting up-glacier would allow ice formation below the stratified sediments of the frozen till (A) deposits. This situation would enable the frozen stratified sediments to be carried over the till (A) lenses. The ice at this stage was polar in type, with basal melting restricted to zones of flow up-glacier. The final deposit of till B caused collapse and both undermelt and surface melting occurred in an ameliorating climate.

The effect of bed irregularities on till deposition have been considered by L. H. Nobles and J. Weertman (1971). They consider that till will be deposited basally:

1 If there is englacial debris in the lower layers, and
2 If the temperature gradient at the base is not sufficient to drain the geothermal heat and sliding heat, because deposition will take place if the basal ice melts.

The larger the bottom temperature gradient, the smaller the rate of melting and thus the smaller the rate of deposition. The temperature gradient will change horizontally as the glacier slides over its bed if the bed is irregular, with wave lengths of the order of, or smaller than, the ice thickness. In an active ice sheet the thermal gradient is larger over a hill than a hollow, and therefore deposition is more rapid over a hollow. Thus bed irregularities tend to be smoothed out. This theory does not account for drumlin formation.

The presence of abundant meltwater is responsible for some special depositional processes. N. C. Hester and P. B. DuMontelle (1971) describe low lobate features consisting of 1·5 m of loess over 3·7 m of mudflow deposit and sand and gravel in front of the Shelbyville Moraine in Illinois. The mudflow deposit is similar to the till of the moraine, the loess being a later deposit. The lobate feature is about 3 km wide and a little shorter, and rises with a surface gradient of 0·5° through about 10 m, where the moraine is about 15 m high.

Unarmed till balls, described by W. A. Pettyjohn and R. W. Lemke (1971), are exposed in a gravel pit near Minot, North Dakota. They lie among crudely stratified layers in a late-Wisconsin channel fill. Chunks of frozen till must have fallen into a melt stream and were carried only a short distance. The till balls have no structure and were formed by erosion and not accretion.

The process of debris sliding over a glacier surface has been observed by J. R. Reid (1969) on the Sioux Glacier in Alaska. The debris covered a third of the glacier surface at the lower end, spreading over a distance of 4·2 km. The material included blocks 15 m in diameter, and the mean debris thickness was 2 m. The debris-covered area of the ice was up to 28 m higher in elevation than the normal glacier surface, although part of this excess elevation could be due to the passage of a kinematic wave down the glacier in 1965–6. Such occurrences indicate how exceptional morainic features could originate.

A good example of the way in which englacial debris is carried by a glacier has been described by Boulton (1970) in some Svalbard glaciers. These glaciers contain a considerable amount of englacial till. Makarovbreen ends in cliffs along the sea and the debris can be seen to be concentrated in the lowest 5 to 7 m of ice. The uppermost 2 to 3 m represents material falling on to the surface, while the lower part is subglacial in origin. Fabrics in the till tend to be transverse in zones of compression and parallel to the ice flow in zones of tension. Active thrust planes were observed within the ice; they occur where compression is high, and material is carried along them to the surface in places. The debris is derived from sub-horizontal debris bands close to the base of the ice, having been incorporated either by thrusting or regelation. The latter method is the most important in these glaciers, and the process of regelation is helped by an irregular glacier bed. Boulton considers that a thick layer of debris is more likely to be incorporated in a cold-based glacier. In the Svalbard glaciers more material is derived subglacially than from the surface above the glacier ice. That basal freezing is important is supported by the erratic content, as those erratics lying higher in the ice were derived from rocks that outcrop farther upstream. In general it seems that cold glaciers carry more englacial material than temperate ones, but glaciers that are entirely frozen to their beds, as opposed to sub-polar glaciers, will carry little debris; the temperature régime is a very important controlling factor in debris load.

A useful account of the characteristics of modern glacial deposition by Boulton (1972) provides a means of interpreting sections through Pleistocene glacial deposits. He also proposes a new classification of glacial sediments in terms of their genesis, and shows that the temperature régime of the depositing ice may be inferred in some circumstances. An important class of glacial deposition is the supraglacial debris, which includes hummocky till moraines and till plains. Examples of these are described from Spitsbergen. Here, the glaciers carry much englacial debris which is exposed by ablation as supraglacial debris near the glacier snout. Till accumulates by ablation to a depth of 1 to 2 m, the depth of summer thaw, but greater thicknesses can be reached as the unstable material slumps down steep slopes to form flow till. Variation in till thickness leads to differential ablation and a very hummocky surface, and hummocky ice-cored moraines can develop with intervening valleys where the till is thinner. The pattern can be either controlled or random according to the till distribution and its relation to ice movement. As the buried ice melts, either a hummocky surface or a till plain can result. Eventually the relief usually becomes inverted, the sites of valleys becoming hummocky ridges when the ice cores that provided the original hillocks have melted. Much of the material collecting behind the ice-cored moraines is water-lain stratified material, usually mixed with flow till and some supraglacial till. The amount of water available will usually determine whether the till surface remains hummocky or becomes flattish on melting, according to the extent of flowage. Flow tills may become stratified and the mode of flow is revealed in the stratification. The association of stratified sediments with till occurs particularly where drainage is impeded and bodies of standing water occur behind an ice-cored morainic ridge on the ice.

As downmelting takes place and the buried ice slowly melts out, structures are formed in the collapsing stratified deposits. Melting can produce collapse, often beneath lakes and streams, and several forms of structure can be recognized, some formed as the surrounding ice melts, others as interstitial ice melts. The structures include both faults and folds, the structures being the result mainly of slumping rather than pushing. Melt-out tills occur as thick englacial debris slowly melts out from the top down, usually accumulating beneath an upper layer of supraglacial till, which holds the melt-out till in place. Stratified sediments can accumulate in englacial or subglacial tunnels.

Basal deposits of subglacial till are not extensive in Spitsbergen, being less so than the supraglacial till. The situation is, however, reversed in Icelandic glaciers. The difference reflects the temperature conditions of the respective ice masses, the Iceland glaciers being temperate in character while those of Spitsbergen are cold or sub-polar. Temperate glaciers tend to produce more lodgement till than supraglacial till because of the low level at which they carry their load, but cold glaciers on the other hand tend to carry basally-derived material at a relatively high level, allowing the formation of extensive flow tills owing to subsequent melting and other types of supraglacial material. Thus position of transport can be related to thermal character, and in turn to type of deposit. Thus former glacier characteristics can to some extent be inferred from the nature of the glacial deposits. Many of the complex sequences of glacial deposits revealed in sections of Pleistocene drifts have been interpreted as the deposits of several glacial advances. Such complex sequences, however, closely resemble those now accumulating at the margins of Spitsbergen glaciers, and probably represent the combination of lodgement, melt-out and flow till elements and outwash accumulating around the wasting margin of one ice mass.

Table 13.2 (Boulton, 1972) summarizes the characteristics of ice-contact stratified deposits. The point is made that eskers and kames, which are usually distinguished on morphological evidence, can be formed by the same processes. The distinctions can give useful data on the type of deglaciation. Supraglacial forms are associated only with a zone of stagnant ice at the glacier margin, and hence give an indication of ice-front position. This cannot be obtained from subglacially derived forms, which only indicate the presence of ice.

4 Methods of analysis of glacial drift deposits

Several techniques have been developed to study the character and origin of glacial deposits: the study of erratics, till fabric analysis and stone shape analysis. Another method of investigation that is useful is the load test in till, for compressibility tests can often differentiate between lodgement till and ablation till. These tests are designed to assess the weight of ice under which the till was deposited; and lodgement tills show the effects of compression by a heavier load of ice than the ablation tills.

4.1   *Erratics*

Erratics were long ago recognized as glacial phenomena. They can provide valuable evidence concerning the movements of now-vanished ice masses, but they can also pose difficult problems, such as that of the origin of the large erratics that occur on the north coast of Devon and the south coast of Cornwall. It has been suggested that these rocks reached their present coastal position on icebergs driven ashore when sea level was at about its present level during one of the earlier glaciations. However, it is the erratics in the tills that provide the most useful evidence of ice flow.

Erratics vary greatly in size. Some rocks, such as Skiddaw slates, do not yield large

Table 13.2 The characteristics of supraglacial, englacial and subglacial ice-contact stratified deposits occurring at the margins of modern glaciers (Boulton, 1972)

	Supraglacial		Englacial	Subglacial
	Ice-floored	Ice-walled		
Topographic expression	Continuous or discontinuous ridges, single or in groups, which may be parallel or apparently randomly distributed. Often beaded. Also mounds, single or in complexes. The cross profiles of the ridges or mounds are smooth with rounded summits. Occurrence is not generally controlled by pre-existing surface relief. Often considerable relief and great volumes of material. Kettle holes may be common.	As ice-floored deposits, except that a flat top is generally retained on ice melting. If, however, the bounding ice walls were steep and the trench in which it accumulated was narrow, collapse or ice-melting could destroy the flat-topped form. The flat-topped forms are particularly common on higher ground and near the margins of valleys, where they will be equivalent to 'kame-terraces' under the existing nomenclature.	Generally single ridges. When underlying ice melts out, relief tends to be low. May have small kettles and surface pits. Relief and volume of material tend to be smaller than in supraglacial forms. Tend to be sinuous in plan with constant wavelength and amplitude.	Essentially sharp-crested ridges which may also be single or associated with others in dendritic or anastomosing patterns. Marked break in slope at foot of ridge flank, ridges tend to be separated by ground of different character. In plan, no obvious regularity.
Component materials	Coarsest to finest grain sizes. Products of high and low discharges may be intimately associated, fluvial gravels with lacustrine silts and clays. Some ridges may be composed dominantly of coarse materials, others of fine materials.	As ice-floored deposits.	Coarsest materials dominate. Sands and gravels deposited in narrow channels.	
Bedding	The full range of fluvial and lacustrine bedding structures.	As ice-floored deposits	Planar bedding common. Internal core flanked by slipped material in which bedded is inclined at the angle of rest.	

Association with till	Till commonly occurs above and below these accumulations. It may occur within them, especially if the outwash sediments are lacustrine.	If the accumulation has built up after a period of erosion down to the glacier bed, an underlying till will be absent. If the underlying ice was destroyed by melting beneath a cover of outwash sediment, an underlying till will remain.	An underlying till to be expected. An overlying till tends not to occur unless morainic debris lies at a high level within the glacier.	Any underlying till would be likely to be destroyed. It may be overlain by a till which is likely to be involved in superficial slumping.
Deformation Structures	Tendency for original bedding to be anticlinally flexed, reflecting the surface, although local synclines may occur. Folding often less acute at higher levels. Normal faulting parallel to lateral flanks. Solifluction structures common.	If the accumulation is built up after erosion to the glacier bed, sediments beneath the summit of the deposit will be undeformed and in the attitude of deposition. If the accumulation was lowered on to the bed after sedimentation had commenced, sediments beneath the summit will be synclinally flexed. In both cases, collapse at the flanks will be similar to that in ice-floored deposits.	Anticlinal warping of inner core, together with normal faulting with downthrows away from ridge crest. Some faults trend across the ridge.	No deformation of inner core.
Orientation	Linear ridge most commonly parallel to the glacier margin, although they occasionally develop normal to it.	As ice-floored deposits.	Most commonly occur trending normal to the glacier margin.	

recognizable erratics, as the rocks break up too easily. The ability of erratics to survive is proportional to their size. Some very large erratics have been recorded: for example, many erratics in Scotland are over 100 t in weight, and one in Caithness, of Cretaceous rock, is 200 m by 137 m by 8 m. In Northumberland, an erratic raft of Great Limestone is 800 m long; at Great Ponton, there is one of Lincolnshire limestone over 300 m long, and near Melton Mowbray, a Lincolnshire Oolite block 275 m by 90 m. There are also Carboniferous Limestone erratics 90 m long in Anglesey and a grit one 180 m long near Abergavenny. These blocks are, however, small compared with the vast 'Schollen' of Germany, the largest of which are 4 km by 2 km by 120 m in thickness. They are formed mainly of Tertiary and Cretaceous clays and sands. Large erratics also occur in North America; for example in south-west Alberta, one quartzite erratic measures 24·5 m by 12 m by 9 m and weighs 18,150 t.

These vast erratics illustrate clearly the immense transporting power of ice, but there remains the problem of how they got into or on to the ice sheet. Some may have slipped on to the ice from steep rock walls, and some may have been quarried off the lee faces of scarps, for example, those of Marlstone and Lincolnshire Limestone in eastern England. In soft rocks, freezing of the ground is a necessary preliminary and the thin blocks appear to have been sheared off along bedding planes or joints in many instances. A good example of the prising-off of erratics is seen in north-west Yorkshire where Silurian boulders lie on Carboniferous Limestone at Norber Brow. If the erratic train is traced back to the Silurian outcrop, the boulders become more angular and the nearest have clearly only been moved a short distance from their outcrop. The beginning of the process is seen in enlarged rectangular cracks developed in the Silurian bedrock as the prising process started to operate. The Norber boulders show that ice can also move erratics uphill, as they lie at a higher elevation than their outcrop. This is most likely to occur where the ice-shed does not coincide with the earlier watershed. In western Scotland, for example, Torridonian sandstone has been carried up 450 m west of the Fannich Mountains. Uplifts are sometimes steep, such as 300 m in 4 km east of Loch Maree. Uplifts of up to 1000 m occurred in the Baltic area and 1525 m has been recorded on the stoss side of Mount Washington and 1615 m in the Winnipeg area.

The ability of ice to move uphill is also well demonstrated in the distribution of Shap granite erratics. Their distribution pattern in the neighbourhood of the outcrop of the granite was studied by S. E. Hollingworth (1931) who deduced interesting patterns of ice movement based on this distribution and drumlin alignment. He shows that the Shap granite erratics do not extend north of a line from Penrith to Melmerby. The form of the drumlins in this area, which is discussed in Chapter 14, shows, however, that basal ice was moving north throughout this part of the Eden valley. To explain this anomaly, Hollingworth suggests that the basal ice, forming the drumlins and carrying the Shap granite erratics northward, must have moved up along shear planes in the ice until it reached the upper layers of the ice sheet. The movement of the upper ice is said to have been determined by the slope of the ice surface. The ice appears to have reached its maximum elevation along a line extending from near Helvellyn to Cross Fell across the Eden valley and to have moved north and south from this position. Thus the Shap granite erratics appear to have been carried first by the basal ice and then to have moved up into the upper layers by the time they reached the area around Penrith. From here they moved southward with the upper ice towards the Stainmore depression. This col was a large funnel into which ice from 15 km around was channelled. The evidence for the passage of ice by this route from the Lake District is clearly indicated by the large trail of Shap granite boulders that stream away from the outcrop north-eastward to the Stainmore gap. The Shap granite erratics do not penetrate north of the position at which the powerful glacier from the Ullswater valley met the northward-moving ice stream. It is possible that this glacier was one of the agents that forced the northward-moving ice to override it and thus enter the upper ice circulation. A basal ice-shed exists in the neighbourhood of Appleby from where the southward- and eastward-moving ice flowed uphill towards the Stainmore col (Fig. 13.4). Through the gap, the train of Shap granite boulders can be followed down to the limit of the Newer Drift ice advance that built the York moraine. Shap granite erratics were also moved southward down the Lune gorge near Tebay, although there is evidence here too of northward-flowing basal ice in the form and erratic content of the drumlins between Shap and Tebay. Ice moving over Shap from the south must, therefore, have moved up from the basal layers into the upper part of the ice sheet to follow the slope of the ice surface southward through the Lune gorge in which the ice would once more reach the ground level. Diagrammatic sections, shown in Fig. 13.4, illustrate the type of ice movement reconstructed from the evidence of Shap granite erratics and the form of drumlins.

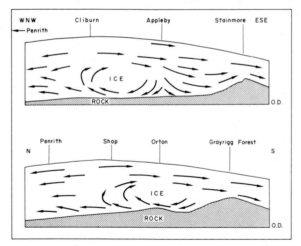

Fig. 13.4
Hypothetical ice flow in north-west England to account for the distribution of Shap granite erratics and drumlins (S. E. Hollingworth, *Q. J. geol. Soc. Lond.*, 1931)

Other erratics indicate extensive ice movement. The riebeckite eurite of Ailsa Craig off the west coast of Scotland has been found in abundance along the east and south-east coast of Ireland and on the north and west coasts of Wales and around Anglesey, showing clearly the movement of the Irish Sea ice sheet. The rhomb porphyry stones found on the east coast of England in Yorkshire and Durham could only have come from Norway and give good evidence of the spread of the Scandinavian ice sheet on to the British coast at at least one stage in the glaciation.

On the west coast of Baffin Island in Arctic Canada, limestone erratics, derived from Foxe Basin, were carried eastward to the mainland of Baffin Island. They give evidence of the existence of a large ice centre over Foxe Basin from which ice spread out in all directions. The maximum extent of this ice sheet occurred about 8000 to 9000 BP. Over the central western part of Baffin Island, the direction of flow at this time was the reverse of that which set in later. By about 7000 BP, an enlarged Barnes ice cap was flowing westward to Foxe Basin in the opposite direction to the earlier ice. This later ice mass only reached the position of the present coast in many parts of the western side of the island.

4.2   Till fabric

The fact that stones in till are not randomly distributed was noted as early as 1859 and was discussed by H. Miller in 1884. The analysis of the direction and dip of elongated stones in till is being increasingly used to study both the direction of former ice flow and the processes of glacial deposition. It is generally agreed that the elongated stones in till become orientated parallel to the direction of ice flow, although this is not always true and the problem of till fabric analysis is a complex one. It is necessary to treat the data statistically to obtain a valid idea of the orientation pattern. It is also desirable to obtain several sets of samples from any one site as variations over short distances can be considerable.

Some valuable work on till fabric analysis was carried out by C. D. Holmes (1941) who attempted to analyse the effect of shape on the orientation of stones in till. He divided stones into six shape classes, discoid, ovoid, tabular, wedge-form, rhombohedroid and varihedroid, and four roundness classes, sharply angular to well rounded. The measurements were plotted on the lower hemisphere to indicate both direction and dip of the stones in a 360° pattern. A composite plot of all the data collected showed that the greatest number of stones lay parallel to the direction of flow of the ice and that they mostly had a low angle of dip, one-third of the stones dipping at less than 10°. In many till fabric diagrams, there is a minor concentration of stones transverse to the flow direction. The stones parallel to the flow move by sliding, but the transverse stones move by rolling about their long axis which is transverse to the ice flow. Many stones with a steeply dipping intermediate axis become orientated parallel to the flow. The ovoids that are well rounded usually maintain the best orientation parallel to the ice flow. A useful means of assessing the method of movement of the stones in the ice is to examine the striae on the measured stones. On wedge-shaped stones, they run parallel to the edges and not to the long axis. This suggests that these stones move mainly by sliding. Large, not very elongated, stones often have a strong transverse pattern. More elongated, well-rounded stones of rhombohedroid shape are particularly adapted to rotational movement, and hence also tend to lie transverse to the flow. The very well-rounded and shorter stones of this form tend to lie parallel to the flow. Because of the wide variety of stone shape and the many factors influencing their emplacement in the till, the fabrics usually show a considerable scatter. From this extensive study, Holmes reached the following general conclusions. He considered that ground moraine accumulated gradually and that stones carried above the glacier bed were subject to rotation, generally about their long axes, but sometimes about the other axes. The stones in contact with the floor moved generally by sliding with both their long and intermediate axes parallel to the floor plane and the long axis parallel to the movement except for wedge-shaped stones. The degree of stone roundness may thus have a profound effect on the preferred orientation of the stone, while shape also plays an important part.

More recently, P. W. Harrison (1957) has studied a clay-till fabric near Chicago, Illinois. He measured the maximum projection plane of shale particles 3 to 40 mm in length. Disk- and blade-shaped particles were mainly selected. Harrison came to the conclusion that much of the fabric in this clay-rich till is inherited from the fabric that developed when the ice was moving. The mean dip of the stones seemed to be related to vanished shear-planes in the heavily laden basal ice, the planes dipping upstream. Along these shear planes, ice tends to override the basal ice that has become static. Stones in such planes would become orientated parallel to them. This orientation also occurs in the ice around the plane, as observed in Greenland today. In the till fabric as measured when the ice has melted, the strength of the orientation is probably partly lost by melting but in basal ice which is laden very heavily, the proportion of ice to debris may only be about 25 per cent. The clay also contains a considerable amount of water and consolidation tests on the till showed that the overlying ice was not more than 90 m thick. Thus it is possible for the ice to retain its fabric from the time when it was moving, if it becomes stagnant and decays slowly *in situ*. End moraines show similar till fabrics which also appear to have been inherited from the time when the ice was moving up thrust-planes as it deformed internally to move over already stagnant ice. Some of the moraine fabrics studied by Harrison showed no up-glacier dip and the stones were dipping with the surface slope of the moraine. These fabrics, he thought, had probably been induced by mudflows down the slope under periglacial conditions, but they may also have originated in the manner of flow tills as described by Boulton (1971).

The bulk of the ground moraine does not appear to have been disturbed since its deposition and Harrison considers that the till fabric indicates that the till accumulated slowly as the ice stagnated by melting from below. The till does not appear to have accumulated by the plastering-on of material beneath moving ice. Thus if this pattern is widespread, it suggests that most till deposition takes place during decay of the ice during deglaciation. The till does, however, retain the pattern imposed upon it when the ice was able to move, giving the common up-glacier dip of the stones, related to the upward shear of ice along planes near its margin where the load is high.

Till fabric analysis was used by R. G. West and J. J. Donner (1956) to differentiate the Lowestoft and Gipping till sheets claimed by D. F. W. Baden-Powell (1948) in East Anglia. West and Donner measured both the dip and direction of 100 stones at each site in undisturbed till. Most of the samples showed a strong preferred orientation when plotted on a symmetrical diagram showing only orientation. The dips on the whole were balanced on either side of the diagrams. Those samples which did not show this balance appeared to have been disturbed by gelifluction. Many of the undisturbed

stones had a very low dip and many were horizontal. As a result of this till fabric analysis, West and Donner attempted to confirm that the ice depositing the Lowestoft till
advanced generally from the north-west over the Midlands and from the west over
East Anglia, while the succeeding Gipping ice came from the north or the north-east
over the Midlands, but east of Cambridge it came from the north or the north-west.
More recent work has disputed these findings (e.g. C. R. Bristow and F. C. Cox, 1973)
and weaknesses in the sampling methods and statistical analysis pointed out.

J. W. Glen, Donner and West (1957) have discussed the processes that lead to the
preferred orientation of stones in till. On theoretical grounds, elongated pebbles placed
at random in a flowing liquid should rapidly become orientated with their long axes
parallel to the flow, but, after a long period, the particles turn so that their long axes
are transverse to the flow. This is the position of theoretical minimum energy requirement. They suggest that parallel orientation is produced by free flow in the ice,
by collision of oblate stones and by dragging on a stationary layer or over a shear-
plane. Transverse orientation should be developed by protracted flow and by collision
of prolate stones. Orientation unrelated to ice flow can be induced by the disturbance
of the deposit when the ice melts or gelifluction occurs.

Observations on currently developing tills in Nordauslandet (Spitsbergen) suggested
that thick till which showed a parallel peak could most likely have been formed by
the plastering-on process. Narrow bands of till between ice layers showed a transverse
peak that could have been produced by long-continued flow or by collisions between
suitably shaped stones. This method of analysis, which takes into account only the long
axis direction and not the dip of the stones, although justified by these authors, omits
some valuable evidence concerning the nature of the fabric. The value of the dip
measurement was apparent in the analysis by Harrison and it provides a means of
eliminating fabrics disturbed by gelifluction as indicated by West and Donner.

J. T. Andrews and B. B. Smithson (1966) have shown the value of dip measurements
in addition to orientation values in their study of the fabric of the cross-valley moraines
of north central Baffin Island (see Chapter 15). These moraines consist of till which
is unstratified, but which has been emplaced subaqueously at the margin of the retreating ice cap by a pressure or squeezing process. The orientation pattern is such that
the preferred direction lies normal to the length of the moraine in most types studied.
The dip pattern is much weaker on the steeper distal slope of the moraine. It is argued
that the strong up-glacier dip, characteristic of the proximal side, indicates that liquid
till may have been forced up into ridges at the edge of the ice cap. This forcing-up
of the ridges would have occurred when the summer melt produced a large volume
of water which liquefied the subglacial material, allowing the ice to settle into it and
to push up the marginal material into a moraine ridge.

Valuable information may also be obtained much more readily by measuring the
orientation of the long axes of surface boulders in a till sheet or moraine. Andrews
(1965) has shown that the orientation pattern of the surface stones shows a preferred
value that agrees closely with the direction of ice-flow. The orientation strength on
the till sheets was consistently strong, and stronger than those found on the surface
of moraine deposits. The weakest orientation patterns were found on outwash plains
where the boulders tended to show a weak orientation perpendicular to the water-

flow direction. Samples of moraine stone orientation measured in west Baffin Island showed a statistically significant orientation at right-angles to the moraine crest in many instances. The stones, therefore, lay parallel to the ice-flow direction.

The study of till fabrics and orientation of surface stones provides a valuable method of reconstructing former ice movement as well as elucidating the processes of till deposition, but it is necessary to analyse the data statistically if valid results are to be obtained. Three-dimensional till fabric analysis is discussed by Andrews and K. Shimizu (1966). The problem of operator variance in till fabric studies must also be considered. To investigate it, A. R. Hill (1968) used a laboratory-simulated till and four operators on different days. He found that, although the operators varied significantly, the variance was less than 5 per cent of the total fabric variability, which suggests that for most purposes it can be ignored.

As more till fabric studies have been undertaken, the more problems in interpretation of the data have become apparent. More sophisticated techniques of statistical analysis have been developed, but the complexity of variations within fabrics and between different sites, and the variety of depositional histories involved, mean that general models of till fabric development have not yet been established. The following case-studies illustrate some of the different approaches used and the varied conclusions reached.

A detailed study of till fabric variation is reported by J. Krüger (1970) from Fakse Banke, Denmark, 50 km south-west of Copenhagen. The orientation of fourteen sets of stones was generally parallel to the ice flow and the stones dipped up-glacier. The influence of shape, length of long axis and roundness on the orientation of the stones was analysed. The larger angular pebbles of low sphericity and symmetrical about their long axes gave the flow direction most precisely. In this area all but three of the fourteen samples were significantly orientated at the 99·9 per cent level, two at 99 per cent and one at 90 per cent. The dip values are presented three-dimensionally by Krüger on an equatorial projection. The composite results show a dip of 7° to the southeast, with individual samples varying from 1° to 11° with reference to the plane of striated limestone on which the till rests and which slopes down 5° to the north-west. The direction of the fourteen observations ranged from north 118° east to north 154° east, while the striae on the limestone ranged from north 124° east to north 150° east,

Table 13.3   Effect of stone shape on till fabric (Krüger, 1970)

Shape class	E	R	W	RH	T	V
Number of stones in each group (total 700):	74	140	167	126	72	121
Mean deviation from orientation parallel to flow:	18	24	25	26	29	32
Per cent of observations where deviation from flow direction exceeds 34°:	12	28	25	29	38	40

For explanation of shape class, see text.

both giving evidence of the final flow of ice in the area. The pattern suggests deposition at the base of an active ice sheet.

Shape studies divided the pebbles into six shape classes: E, elliptical; R, rectangular; W, wedge-shaped; RH, rhombic; T, triangular; V, varied. Six roundness classes were used and the sphericity was calculated by D_n/D_s, the ratio of the true diameter to the minimum diameter of the circumscribing sphere. Roundness and sphericity are related, both to each other and to the different rock types, the angular pebbles being mainly flint. The effect of shape is shown in Table 13.3. Krüger concludes from this that:

1 Streamlined pebbles show strong preferred parallel fabrics, while increasing asymmetry reduces this preference
2 When shapes are pointed down-glacier, the length of the long axis influences the orientation more than the sphericity. Increasing length of long axis raises the chance of parallel deposition
3 Blunt shapes pointing down-glacier are more influenced by sphericity than length of long axis. Decreasing sphericity raises the chance of parallel deposition
4 Increasing roundness decreases the tendency to parallel orientation.

The best pebbles for observations are:

1 Elliptical
2 Rectangular with sphericity of less than 0·70
3 Wedge-shaped, very angular to sub-rounded with sphericity of less than 0·70, and rounded to well-rounded, with sphericity less than 0·60
4 Rhombic, very angular to sub-rounded, with long axis more than 1·5 cm, rounded and well-rounded with long axis more than 2·5 cm
5 Triangular, more than 2·5 cm long
6 Length more than 1 cm and ratio of long axis to intermediate axis at least 1·5.

S. A. Harris (1969), in a study of till fabrics near Waterloo in southern Ontario, introduces the term Minimum Significance Orientation Count (MSOC), based on the strength of the fabric at the 95 per cent confidence level, of the number of pebbles in the modal class of the chi-square test. The number is often as low as 30. Deformation of fabrics may occur through post-depositional stress, for example readvance of the ice, which may result in a mode normal to the former one. Frost action can cause the upturning of pebbles, but while this does not upset the direction of the primary mode, it could create a third vertical mode. Frost action may cause polymodal fabrics, when pebbles are moved up into another till. Mass wasting may completely destroy fabrics. The MSOC value does not correlate with pebble size, shape, roundness or lithology, but data from Arapaho glacier in Colorado suggest that it may be possible to use it to assess rates of flow. Measurements of till fabrics and stone rounding in the moraines of this glacier (Harris, 1968) showed that there is a relationship between the surface velocity and the number of pebbles in the primary mode of 20° classes, suggesting a stronger orientation where the velocity is highest (Table 13.4). The roundness also increases with the speed of flow. The glacier is, however, a small and relatively inactive one, and further studies on other glaciers are needed to support the relationships suggested.

Table 13.4   Relationships between speed of ice flow, till fabric and stone rounding for Arapaho Glacier, Colorado (S. A. Harris, 1968)

Speed of ice flow, m/year	No. of pebbles in the primary mode of 20° classes	Index of pebble rounding*
0·4	19	5·5
1·4	33	1·9
1·7	36	0·1

* Pebble rounding according to G. Luttig (*Eiszeitalter Gegen.* 7 (1956), 13–20)

The variability of till fabrics over quite short distances is stressed by many workers. Andrews (1971) shows how a single till fabric site is unlikely to give a reliable indication of direction of ice flow. On cliff sections in Yorkshire (Andrews and D. I. Smith, 1969[1970]), an array of sites each measuring 3×3 m showed significant variation. The results of analysis showed considerable dispersion and a low order of precision within individual sites. Variations between sites ranged from low to very high. Stone shape was an important factor in explaining the variance, for tabular stones tended to lie parallel to the probable direction of ice motion while rods, ovoids and wedge-shaped stones lay transverse. J. A. T. Young (1969) also comments on short-distance variation, especially great in the vertical direction. At a site south-east of Edinburgh, he found that variation between till units was often less than that within the units, significant changes within units sometimes occurring in as little as 0·6 m vertically.

Till fabrics can be altered by later ice advance. J. Ramsden and J. A. Westgate (1971) give evidence of reorientation of a till fabric in the Edmonton area of Alberta. A lower till shows mainly north-west to south-east pebble orientation, while an upper till has a north-east to south-west pattern. In places, however, the lower till has been reorientated by the ice depositing the upper till and shows the same pattern as the upper till. The north-east to south-west direction is parallel to grooves in the upper surface of the lower till. Reorientation appears to have been achieved by shearing along closely spaced shear planes.

Till fabric analysis has been used to estimate rates of basal till deposition in the stagnating ice mass of Burroughs Glacier in Alaska (D. M. Mickelson, 1973). The ice changed direction of flow around 1892 when hills started to emerge from the ice, deflecting its flow through about 90°, and this change is recorded in the till fabrics. The thickness of till accumulating in the recent phase of ice flow can then be divided by the time involved to give a mean rate of deposition, which varied between 0·5 and 2·5 cm/year.

Most studies of till fabrics have naturally been carried out on present-day or Pleistocene tills, but attention has also recently been directed to pre-Pleistocene tillites. J. F. Lindsay (1970) has analysed fifty till fabrics in tillite of Permian age, outcropping in the central Transantarctic Mountains. Some of the fabrics evolved englacially, and others during settling-out. The former, which developed in moving ice, have only partially been re-organized during settling-out, depending on the level of shear stress at the ice-sediment junction. Of the fifty samples, 26 showed a fabric parallel to the ice flow and 15 a transverse fabric. Three types of parallel fabric were recognized: *P1* had a sharp mode and 9°–20° up-glacier dip, *P2* was less strong, also with an up-glacier dip, and *P3* had a well-defined horizontal mode but weak secondary modes. Of the transverse fabrics, *T1* had a strong horizontal mode, while *T2* had a less strong mode and sometimes a weak parallel mode, which resulted in a broad, well-defined sub-horizontal girdle on the equal-area projection on which the fabrics were plotted. Lindsay compares the fabrics with englacial fabrics measured elsewhere in areas of present-day temperate and polar ice (Casement Glacier, Alaska, and Greenland), finding marked similarities between the Permian and the contemporary fabrics. Where shearing is an important process in the deposition of modern tills, up-glacier dips are common. He suggests that fabric type *P1* developed englacially with one shearing movement, while type *T1* developed englacially with two shearing movements. The larger stones are readily re-orientated during deposition, but the smaller ones may not be. The resulting fabric depends in part on the relationship between controlling obstacle size (p. 141) and the size of the stones in the till. The intermediate types, *P2* and *T2*, represent partial re-orientation. Where basal melting was rapid amd the time of adjustment thus reduced, intermediate-type fabrics are likely. There was no evidence that transverse orientation occurred at the plane of deposition. Results of the analysis not only throw useful light on Permian glacial events and environments, but also on the origins of contemporary till fabrics.

A common problem in studying superficial deposits is that of distinguishing glacial till from other apparently similar deposits, especially those resulting from periglacial activity (gelifluction or other mass movements). Lindsay (1968) show how fabric studies may be applied to differentiate mudflow deposits, for example, from till. He suggests that mud-flow fabrics go through a cycle in their formation. When the point in the cycle is reached where the parallel pattern is strongest, they may be indistinguishable from till fabrics, but as the cycle continues, the strength of the fabric in the mudflow diminishes. The maximum strength occurs when the mode is horizontal. At this point a girdle forms that distinguishes the mudflow fabric from that of a till.

4.3   *Shape and size analysis*

A study of the roundness of stones can give useful information concerning the nature of the feature in which the stones are incorporated. Some information is also provided about processes of deposition. There are several methods of measuring the shape of a stone, but an analysis of three methods indicates that A. Cailleux's formula (1945) for roundness is the most useful. The first of the methods is based on Cailleux's roundness formula, which is given by $2R/a \times 1000$, where R is the minimum radius of curvature in the principal plane and a is the long axis of the stone. Secondly, W. C. Krumbein (1961) has suggested a sphericity measure

$$\sqrt[3]{\frac{bc}{a^2}},$$

where a, b and c are the long, intermediate and short axes respectively. This measure is closely and inversely related to the third measure, Cailleux's flatness formula, given by $(a+b)/c \times 100$. Cailleux's roundness formula was applied to samples of 50 stones taken from a variety of different glacial and fluvioglacial features in west Baffin Island. The deposits sampled included material from fluvioglacial outwash deltas, eskers, kames, moraines and gelifluction material. Analysis of variance tests showed that the roundness of stones in the different deposits was significantly different. Pairs of samples were also tested by using Student's t test to ascertain whether the roundness values were significantly different. The difference was not related to rock type as all the samples tested consisted of granite-gneiss from the Basement Complex which outcrops in this part of Baffin Island. Some of the results are shown in Table 13.5.

Table 13.5   Analyses of stone roundness for various glacial deposits

Pairs	Means mm		Differences mm	t	Degrees of freedom	Significance level
Deltas, eskers	334	332	2	0·14	22	Not significant
Deltas, kames	334	238	96	5·65	16	0·001
Eskers, kames	332	238	94	6·86	13	0·001
Moraines, kames	138	238	100	5·44	10	0·001
Moraines, eskers	138	332	194	12·60	13	0·001
Moraines, deltas	138	334	196	10·90	16	0·001
Moraines, gelifluction	138	185	47	1·68	7	Not significant

It is interesting to note that deltas and eskers cannot be differentiated by roundness. This indicates that their stones have undergone a similar amount of rounding by melt-water streams. The eskers must have developed in close proximity to the ice and their stones cannot have been carried very far by meltwater. It is unlikely that meltwater can exist far beneath cold-based ice caps such as the Barnes ice cap in Baffin Island. Rounding in these fast-flowing, heavily-laden melt-streams must, therefore, be very effective to produce such a marked degree of rounding in such a short distance. The size of the largest boulders also points to this.

Table 13.5 also suggests an important difference in the methods of formation of eskers and kames. Some rounding of stones has taken place in the kames, as they are significantly rounder than the moraine stones, but they can only have been carried a short distance by meltwater before being deposited at the ice margin. The meltwater was probably not concentrated into powerful streams where the kames were formed.

The lowest mean roundness was found in the moraines. This low value suggests that the morainic material was carried either on top of the ice or was frozen beneath the base of the ice. Most of it probably only started to be abraded a short distance and time before it was deposited as moraine. In temperate glaciers, rounding is probably more effective. This is indicated by a significant increase in stone roundness over a distance of about 1 km in recent moraines laid down in front of Blea Water under High Street in the Lake District. More active movement of ice over its bed would be expected in these circumstances and this would produce more effective stone rounding. The material in this case was from the Borrowdale Volcanic Series.

A study of the shape characteristics of pebbles in basal till by L. D. Drake (1972) attempts to relate roundness to distance from the pebble source. Pebbles first undergo rounding to a point of about 0·5 on Krumbein's visual scale of roundness, but are then liable to crushing, causing fracturing. Several cycles of abrasion and crushing affect the pebbles for each kilometre that they are carried, as indicated by the reduction in size away from the source. There are difficulties in an analysis of this type with respect to the lithology, weathering and shape of the pebbles.

In addition to shape analysis, particle-size analysis has also proved useful in differentiating drift deposits of various origins. For instance, P. M. B. Landin and L. A. Frakes (1968) show that the mean (m_z), standard deviation (σ_1), skewness and kurtosis in granule to clay fractions can distinguish till, alluvial fan outwash and mudflow deposits. Plots of σ_1 and m_z show that tills are poorly sorted and finer than mudflow and alluvial fan deposits, there being little overlap. Outwash is better sorted than till and generally finer. Skewness is not particularly diagnostic, although fans tend to be positively skewed and tills vary. Discriminant analysis allows the pairs of sediments to be differentiated. A. Dreimanis and U. J. Vagners (1971) have analysed basal till material and have found that all rock types provide a till with a bimodal size distribution; one mode occurs in the clast size and the other reflects the mineral fragments in the matrix of till. The clast mode is larger nearer the source of the debris. With increasing distance from the source, the matrix mode increases and the clast mode may disappear. Each mineral has a typical matrix mode.

4.4 *Sedimentary structures*

Sedimentary structures provide valuable evidence of depositional environments. The laminations discussed by A. V. Jopling and R. G. Walker (1968) in a kame delta of clays, silts and sands, deposited 13,000 years ago as the Wisconsin ice wasted away, provide a good example. The cross lamination is of ripple drift type, characterized by rhythmic superposition of ripples, or the climbing of one ripple up the stoss slope of the ripple immediately downstream. In type A, the rippled form of cross lamination, the stoss side of the laminae are not preserved, and hence only the climbing sets of lee-side laminae are found. Another type (B) is the sinusoidal ripple lamination with symmetrical sine waves and continuous laminae, in which both climbing lee-side laminae and sandy stoss-side laminae are preserved. The morphology is related to the ratio of sediment fall-out from suspension to bed-load movement. In type A, bed-load is dominant, while sinusoidal ripples (type B) represent dominant fall-out suspension. The different types are due to small fluctuations of current velocity and sediment type. Sinusoidal ripples form when the bed is stabilized by high fall-out of fines and is characteristic of a cohesive load. Type A occurs with non-cohesive bed material. Underflow currents of 10 to 20 cm are capable of initiating rippling in fine sand and silt. The deposits indicate fairly continuous, though pulsating, flow in the depositing water.

J. A. Westgate (1968) has drawn attention to current-produced sole markings in Pleistocene continental ice-sheet tills in Alberta. Grooves, striations and 'prod moulds' occur on the undersurface of the till. Where sand underlies dense loamy till all three types of marking occur. As the glacier slid over a sandy bed, it incorporated stones which formed grooves and other marks. Large rocks formed the grooves and smaller stones were responsible for striations. Prod marks occur where rocks were only in contact with the sand for a short time. The till was probably deposited soon after the marks formed and the sand may well have been frozen at the time.

5 Drift stratigraphy and its analysis

A great deal of work has been accomplished in the analysis of drift stratigraphy. The aim of these studies is to ascertain the nature and sequence of Pleistocene events and to correlate these from place to place. The analysis of glacial deposits is not only concerned with the stratigraphy but the form of the deposits. Their attitude must also be taken into consideration. For example, it is recognized that the deposits of the Newer Drift (Würm, Weichsel, Wisconsin) period have much fresher forms than the older deposits. Newer Drift deposits show original constructional forms and occur within the valley floors. The deposits of the older glaciations tend to show mainly erosional forms and they are often missing from the valley bottoms because of later river downcutting. They therefore tend to have interfluve distributions. However, the most valuable evidence used to build up Pleistocene chronology is derived from a study of the drifts themselves. Data may be obtained from exposures, boreholes and augering. In studies of this type, the sequence of different types of deposits is important and valuable evidence is obtained by studying the weathering of the drifts (Chapter 1). It is not intended to discuss drift chronology, but merely to consider the possible interpretation of special drift sequences, to comment briefly on weathering of drifts and to give one example of the analysis of a sequence of drifts which reveals a complex pattern of events of lasting importance.

5.1   The undermelt drift sequence

R. G. Carruthers (1947–8 and 1953) has described a series of drifts which he considers reveals a complex process of deposition, occurring when an ice sheet becomes stagnant and melts *in situ*. The sequence of drifts contains both stratified and unstratified deposits. The stratified undermelt deposits can be differentiated from lacustrine deposits by certain special characteristics. A complex layer of drift often contains a basal till overlain by stratified deposits and capped by another layer of till. The base of the stratified series frequently contains clay deposits. The term 'shear clay' has been used to describe these clays, which consist of extremely sharply divided but very thin layers of clay, giving a laminated effect, which occurs without rhythm and can be as fine as 0·01 mm. Small particles in the clay show equal flow around them, the underlying and overlying laminae passing under and over the particle respectively. Some of the layers show crumpling or minute faults and some patches are very contorted. There is no trace of organic material in the clays, but they may contain very thin sand partings. The thicker sand layers may show current bedding. The fine definition of the clay layers and the lack of graded bedding suggest deposition by moving water. Where the sand layers show no structure, they are probably secondary, being laid down as the ice slowly melted. The sand usually increases in amount upwards.

The suspension structures are an interesting feature of the deposits. These developed where the depth of flowing water was rather greater. The clay was sometimes held

firm until a late stage in the melting process and, when it did melt out, the lower edge
remained unbroken, but the upper one developed a flame-like appearance. In this form
it became incorporated in the rapidly accumulating sand. At times, roof-falls produced
hour-glass-like forms which provide a valuable clue in identifying the nature of the
deposit, because they do not appear to form under other conditions.

At the top of the sequence there is often a gravel and upper till layer. The contact
between the tills and the sand zone beneath is important in their interpretation. The
contact sometimes forms a clean, straight line, while at other times there are wisps
of till incorporated into the sand. Sometimes there are thin lenses of sand within the
lowest layers of the till. Such forms do not suggest that the upper till was laid down
by a readvance of the ice, but that the two deposits are intimately associated at their
junction. The upper till is not always ablation moraine, as it may have the character
of ground moraine. Carruthers suggests that it could have been introduced by an overriding glacier or be material that has moved up from the base of the ice sheet. Ice
in which a large amount of material was incorporated would resist melting while the
sand accumulated beneath it. Where the roof remained firm, a straight contact would
result. However, if the till was also melting, parts of it would be incorporated into
the sand below. Finally, the till would melt out at the top of the section.

This type of sequence of deposits could be interpreted, therefore, as evidence for
the gradual melting of a stagnant ice sheet. Shear clays only occur on the lower ground
as the fine clay requires special conditions that are not often present in areas of greater
relief. Good examples may be seen in the area between Darlington and Stockton in
north Yorkshire. Carruthers suggests that the deposits are formed by shearing action
in the lower layers of the ice where it is heavily loaded. The silts and sands are not
always present and only occur where meltwater streams were concentrated. Thus in
many areas, only till is found. The intermediate sand deposits, which are very extensively developed in the drift area of the Cheshire plain, may have formed in a similar
way.

5.2   *Examples of superimposed till sequence*

G. Gilberg (1969) describes seven large till profiles in Sweden which show two and,
in one place, three layers of local till. All the tills are thought to have been deposited
by lodgement from basal ice, but accumulation, erosion and redeposition have
occurred. The various tills differ in colour and have undulating contacts, some showing
erosion. Till II sometimes contains coarse material from till I. Till III varies in grain
size, but till II is homogeneous. Each till has unique characteristics but there are also
some common characteristics. All three lie near the area of source rocks. Accumulation
of till I was followed by erosion of both it and the bedrock, so that the new till contains
some different material. There may have been an ice-free period between the deposition
of the tills, as ablation till occurs on the top of till I. The transition between tills II
and III is difficult to define; some features show continuous deposition, but others
suggest an interval. There is more far-travelled material in till III. Changes in ice movement, as it moved up from the Vänern basin, could account for the different till types.

It is possible that all the tills belong to the last ice advance, but Gillberg considers it more likely that there were intervals between their deposition.

L. Drake (1971) has shown, that, in central New Hampshire, there are two super-imposed tills, the lower being lodgement till and the upper ablation till. Seventy-eight properties of 42 samples of the tills were analysed. The results showed that the basal till has strongly aligned pebbles, parallel to the bedrock striations, and it lacks any sign of washing. The upper till possesses weak, inconsistent fabrics, shows evidence of washing, and the pebbles are angular and decomposed by frost action. Not all super-imposed tills are of this type, and F. Pessl (1971), discussing the tills of west Connecticut, shows that two texturally and structurally distinct tills are found. The lower till is sub-glacial and deposited by ice coming from north-north-west, while the ice depositing the upper till came from the north-east. The upper till probably includes both sub-glacial and supraglacial material.

R. P. Kirby (1969) describes glacial deposits in Midlothian where basal till is overlain by sands and till in succession, the lower layers grading into each other. Fabric analysis of four different layers of till shows the same direction and strength of orientation and the same dip. The complex sequence is interpreted as a simple depositional unit, with fluvioglacial and glacial processes alternating in a subglacial environment. The whole section is 14 m thick above the bedrock. From the top downwards, it consists of sand 1·8 m, till 6·1 m, sand 10 cm, till 10 cm, sand 15 cm, till 45 cm, sand 60 cm, gravel 1 m, sand 1 m, grading into 3·5 m till at the base. At one point a gully has cut through the thin till layer and has been refilled with sand. The gully was cut into newly develop-ing till and overlain by fresh till, indicating a minor phase of fluvial action.

A trench section through the Escrick moraine at Wheldrake, Yorkshire, has been described by G. D. Gaunt (1970). The summit of the moraine ridge is formed of boulder clay and minor ridges of till were overlain by glacial sand and gravel on their north slope. The summit ridge probably consists of supraglacial and englacial debris derived from a residual ice-cored moraine. The minor till ridges and glacial sand and gravel on the north slope are denuded remnants of supraglacial flow tills and fluvioglacial deposits. Later phases of downslope sludging also occurred and have filled hollows. On either side of the ridge, 7·5 m of laminated clays occur, passing laterally into sand and gravel to the north, which is thought to be a beach deposit. The till is stiff, grey and stony, and rests on sand, gravel and clays. The ridge is too high to be a squeeze-till feature and is more likely to be formed by upward transport of englacial debris to form thick supraglacial flow tills with an ice core. The 7·5 m drift post-dates the glacial deposits and was deposited in standing water.

Till and sands also occur together in the Blaby brick pit in Leicestershire described by R. J. Rice (1969). The till overlies the sand mainly, although in places they interdigi-tate. The sand varies in thickness from zero to 6 m. Coaly layers indicate the structures in the sand which are very complex and contorted in places. The sand and till are considered to be contemporaneous. The structures were probably induced by slumping, which may at times have been triggered by ice movement, in an ice-marginal environ-ment with abundant meltwater.

5.3   Weathering of tills

The study of the weathering of the drift is an important aspect of drift stratigraphy in establishing glacial chronology. Different types of drift will weather in different ways. Two main processes are important in the weathering of tills. Oxidation occurs first and extends deepest. It consists of the alteration of ferrous compounds to oxides, giving the soil a yellow, red or brownish colour. Ancient soils, sometimes called 'palaeosols', are those that are not related to current processes. They provide a valuable means of establishing glacial chronology (p. 17). The Sangamon palaeosols in North America provide a good example: they are extensive and well preserved, and vary with different conditions as do modern soils covering the same area. Thus they vary from podsols south of the Great Lakes to chernozems in the Great Plains area, and in suitable conditions gumbotil also occurs.

If the tills are calcareous, they will become progressively more decalcified by leaching with time and the depth of decalcification will be at least a partial measure of their age. In more clayey tills, such as those of parts of the United States, weathering produces special characteristics in the till. These deeply weathered clay tills were defined and called gumbotils by G. F. Kay (1916). Their recognition in a number of sections has enabled the till sheets of Iowa to be differentiated and associated with the main advances of the Nebraskan, Kansan, Illinoian and Wisconsin glaciations. Their characteristics have been discussed more recently by J. C. Frye and others (1960) and by A. C. Trowbridge (1961). The latter lists the characteristics of gumbotils by which they can be recognized and their genesis identified. Gumbotil is a compact sticky clay when wet, and hard when dry. It is completely leached and contains no easily weathered materials except at the base. When it is wet, it breaks up into small, shiny-sided polyhedral pellets. It is massive, not stratified, and rarely exceeds 5 m in thickness. The pebbles that it contains are mainly of materials that do not weather readily, such as quartz and chert, and they become more numerous towards the base of the deposit. The base of the gumbotil is not sharply defined, but it grades down through oxidized and leached till, to oxidized and unleached till, and finally to unoxidized and unleached till. Sometimes there are ghost pebbles of crystalline rocks in the lower part of the gumbotil. At its upper limit the layer of gumbotil is overlain by unweathered till or loess. These characteristics indicate that the gumbotil is the product of a long period of weathering of clayey till and its presence indicates an interglacial period. It also provides a method of differentiating the older and newer drift because the Wisconsin tills never have a gumbotil layer at their upper limit. The recognition of weathering in drifts thus provides a valuable tool in chronological analysis.

5.4   An example of drift stratigraphy analysis

The example of drift stratigraphy selected for brief review is from the work of F. W. Shotton (1953). The area lies between Coventry, Rugby and Leamington and the deposits of this area provide evidence for a complex series of events that led to a complete change in the relief and drainage. The main sequence of deposits is shown in Table 13.6. The Bubbenhall clay is very restricted in outcrop and is largely a stoneless, red

Table 13.6   The sequence of Pleistocene deposits around Coventry, Rugby and Leamington

	Modern alluvium	
Newer Drift	River terraces	Avon No. 1 / Avon No. 2 / Avon No. 3 / Avon No. 4
	Long time interval	
	Dunsmore gravel	
	Wolston series	Upper Wolston clay / Wolston sand / Lower Wolston clay
Older Drift	Baginton sand Baginton-Lillington gravel	
	Long time interval	
	Bubbenhall clay	

silty clay, laminated in at least one auger hole and containing pebbles in some places. It contains Bunter pebbles that denote a northerly source. Its deposition was followed by a long period of erosion during which much of it was removed. The overlying Baginton-Lillington gravels were derived almost entirely from the Bunter. The deposit thins towards the south-west. These gravels were deposited on an uneven surface and are overlain, without break in some places, by the Baginton sand. The two deposits show current and deltaic bedding in the lower parts, but in the upper part the bedding is level, which suggests that ponded conditions were beginning. The sand passes undisturbed into the Wolston clay, supporting the view of ponded conditions. The deposits contain a cold steppe fauna. They become thicker to the north-east and were laid down in a valley that drained in this direction. The stones in the gravels indicate that several feeder rivers probably carried them into the area. The reduction in calibre of the deposit upwards shows increasing influence of ponded conditions as ice advanced from the north.

This ice advance is clearly seen in the Wolston series. The deposits are up to 16 m thick. At their base is a stoneless clay laid down in still water. Above the lowest metre, there are occasional pebbles, but the bulk of the deposit is a stoneless red or pink clay, with evidence of varves in places. The clay overlaps and covers a larger area than the Baginton-Lillington deposits and is interpreted as a lake deposit. Nevertheless in the upper part, a small amount of true till also occurs, owing to the spasmodic advance of the ice over its own lake deposits. The rock types contained in the deposit suggest that a northern ice mass was displaced by one coming from the east, as flints increase in number in the upper part, becoming of equal importance with the Bunter pebbles. This tendency continues into the overlying Dunsmore gravels. The Wolston sand between the two clay layers indicates a temporary retreat of the ice with more summer melting carrying coarser deposits. The Dunsmore gravel is the highest of the Older Drift deposits and caps a flat plateau between the Avon and the Leam. It slopes gently to the west-north-west to 98 m OD. The deposit is waterlain and contains 8 cm pebbles, mainly of flint. It nowhere occurs in the same place as the eastern till and these two deposits are mutually exclusive. The Dunsmore gravel appears to be the outwash from the eastern till sheet.

The terrace deposits are everywhere at a lower elevation than the Older Drift Dunsmore gravel. The terraces are related to a river flowing south-westward, while the Older Drift river deposits were laid down in a valley draining north-eastward. The terrace deposits will not be considered in detail.

In interpreting the events that led to the deposition of these sediments, Shotton recognized the presence of a large lake, called Lake Harrison, in the present Avon valley (p. 334). The Wolston clays give evidence for this lake. Clays outside the area studied in detail show that the lake was extensive, reaching from just north of Leicester to Moreton-in-the-Marsh in the south, its maximum length being 90 km. The water level reached a maximum height of 125 m when it overflowed the Jurassic scarp to the east, as its southern outlet was dammed by western ice. The present-day watershed between the Soar and the Avon lies at the site of a deep pre-glacial hollow where 66 m of drift have been proved by boring. The pre-Chalky till surface has been mapped by using the evidence of boreholes and augering. The map shows a broad valley rising to the south-south-west to about 88 m near Bredon Hill, with a gradient of 12 m in 40 km. The present Soar is, therefore, very much shorter than it was preglacially and the Warwickshire Avon is a very recent river. The Avon was initiated by the outwash streams of the retreating eastern ice that eventually overran the lake deposits and advanced as far as Moreton. The detailed study of the drift stratigraphy in this relatively small area, using borehole data and augering as well as examining the available sections, has revealed a complex and important sequence of events.

6 Drift tectonics

Some drift deposits show tectonic disturbance producing either folding and contortion of the strata or faulting. Some of the structures, such as over-thrusting or compressional faults, may be the result of pressure by advancing ice against earlier deposits. Examples of this type of structure are described in Chapter 15, in discussing push moraines. Some small-scale faulting may also be the result of the melting-out of lumps of ice incorporated in deposits, as for example in the kamiform deposits near Carstairs in south Scotland. Other types of fault, particularly normal faults, may require different explanations, such as those put forward by Shotton (1965).

Some of the faulting described by Shotton is on a fairly small scale. At Strett-on-Fosse, a section and auger holes have revealed a succession of normal faults giving rise to small horsts and graben. The displacements are up to about 3 m. At the base of the graben, the underlying Lias clay bulges upward. In this type of fault the vertical force is the dominant one. The pattern of the faulting in relation to the deposits shows that dislocation occurred at several periods. Ice loading does not appear to be a satisfactory explanation for the faulting because it would have to vary over such short distances. The force that Shotton considers could have produced the observed features is the tension induced by the cooling of frozen ground in winter. The fault character suggests that the ground was frozen when the movement occurred. The tension resulting from shrinkage in the winter would be reversed by expansion and compression during the summer. This latter movement appears to have been taken up by movements between the grains of the sands and gravels. The orientation pattern of the faults, which appear to meet at 60° angles, also suggests that tension resulting from contraction has been important in their formation. They appear to be fairly common in the Older Drift.

Other examples of normal faulting and shear plane development have been described by B. J. Taylor (1958) in the Middle Sands of Lancashire and Cheshire. These are water-lain sands lying between two till sheets, belonging to the Irish Sea Glaciation in the Newer Drift period. The disturbance of the sands is of two types. First, near the top, the sand beds have been folded and tilted and small-scale thrusting has occurred. The movements are thought to have been caused by lateral shear-stress imposed by advancing ice. Below these features, the second type of disturbance consists of reticulate shear planes penetrating several metres into the sands. Some of these shear planes penetrate 10 m into the sands and are inclined at 60° either side of the vertical. Most of the faults are normal. The pressure of the ice, which must have been a few hundred metres thick, compressing the sands beneath, which were probably water-logged and frozen, caused shear planes to develop. These planes shortened the vertical dimension. Most of the planes lie between 30 and 60° to the vertical. These features are not present in the Upper Sands which overlie the upper till. This interpretation differs from that already suggested to explain the character of the Middle Sands (p. 392).

Some of the fault structures described by Shotton in a section exposed in motorway construction near Kilsby, Northampton, are considerably larger. The glacial sequence in this area is at least 30 m thick and is of Saale (Gipping) age. It consists of coarse gravel separating two tills. Normal faults have let down the upper till at least 12 m into the gravels. Similar examples of large-scale normal faulting have been revealed in cuttings made along the M1 motorway near Narborough in Leicestershire. The fault throws in this section are from 7·5 to 21 m, and the faults continue down into the Keuper Marl beneath. It seems likely that these faults are reactivating earlier fault patterns, as those near Narborough are fairly closely connected to the fault pattern in the nearby Charnwood Forest area. The weight of the ice sheet that deposited the tills is thought to be a possible cause of the reactivation of old faulting. The ice must have been of considerable thickness in this area and it seems likely that it caused isostatic depression of the ground. The faulting took place after the uppermost of the drift deposits was laid down. The updoming as a result of the reduction of pressure when the ice sheet melted seems to be the most likely cause for these faults. The faulting probably took place quite quickly after the retreat of the ice because there is no indication of the faults at all in the present landscape, neither do they appear to affect the terraces that were formed soon after the ice retreated and in the subsequent interglacial. Tectonic movement in drifts can, therefore, have a variety of origins.

Structures in the stratified drift of the Lac Saint-Jean area of Quebec have been described by J.-C. Dionne (1970). The features include bowl-shaped deformations with vertically placed pebbles and small faults. There are also wedges with infill. The structures are thought to result from the melting of buried ice and are not periglacial.

They could have been formed with soil temperatures near 0°C and are not evidence of a cold climate after the area was deglaciated.

S. R. Moran (1971) describes three types of glaciotectonic structure in north American drift:

1 simple *in situ* deformation
2 large-scale block inclusion
3 transportational stacking of simple till sheets.

The first is caused by ice push and bed shear. The second incorporates bedrock or lumps of old drift into younger till. Shearing of this type occurs most readily where flow is compressive near the ice margin, and as ice flowed up scarps. Ridges produced in this way could be mistaken for end moraines. The third process may produce disturbed sequences by sporadic differential movement along shear planes in zones of debris-charged basal ice. Faulting and stacking may then occur, producing complex patterns.

7 Conclusions

Glacial deposits provide valuable evidence concerning processes operating in and beneath ice masses, and also play an important part in modifying the landscape of areas formerly covered by ice sheets and glaciers. Some of the ways in which these deposits can be analysed have been mentioned and the following chapters enlarge upon some of the specific forms and types of deposit. An example of the analysis of drift stratigraphy has been given, showing how profound can be the changes wrought by the ice sheets and their deposits.

These deposits cover about one-third of Europe, nearly one-quarter of North America, and 8 per cent of the earth's land surface. They have produced much new land in Britain (K. M. Clayton, 1963). The eastern parts of Yorkshire, Lincolnshire and Norfolk, for example, would be under the sea if it were not for their thick and extensive glacial deposits. Drift thicknesses vary enormously over the glaciated area. Flint (1971) gives the following figures for different parts of North America: Great Lakes area—12 m, Illinois—35 m, Iowa—45 to 60 m, Central Ohio—average 29 m, maximum 231 m, New Hampshire—average 10 m, maximum 121 m, Spokane valley, Idaho-Washington—180–400 m. Charlesworth (1957) records a wide variety of estimates of drift thickness. For example, the mean drift thickness in north Germany has been given as 58 m from 470 boreholes, Norrland (Sweden) 4 to 7 m, Denmark 50 m. In some places, exceptionally thick deposits have been recorded, such as 470 m near Lubbendorf in Mecklenburg, 288 m near Rostock, about 400 m near Grenoble, 397 m near Heidelberg, and about 800 m near Imola in the Po valley. In East Anglia, 143 m has been recorded and 175 m in the northern part of the Isle of Man. The greatest thickness recorded in North America is 670 m in the Fraser Delta area, though this consists mostly of outwash.

In all the areas that were covered by ice, the soils have been modified or completely changed. In some areas, former soils have been swept away, but in many areas new drift soils have been laid down. The character of the drift is usually very important in determining the soil type, which in turn has an influence on vegetation and many other aspects of the geography of the area concerned. A. G. McClellan (1971) has discussed the relation between sand and gravel extraction in central Scotland and fluvioglacial geomorphology. The area produced 8 million t in 1964. A survey of 1966–7 showed that 95 per cent of the sand and gravel of Lanarkshire is fluvioglacial, and only 5 per cent is fluvial. Glacial deposits are of interest for their own sake, but also for their widespread influence on many other facets of the landscape. Whole new landscapes have been created by the burial of the solid rocks beneath the thick mantle of glacial deposits. Sometimes this occurs with intriguing surface forms, such as drumlins or eskers. Even the smooth till sheets hide much valuable information within them.

8 References

AARIO, R. (1971), 'Consolidation of Finnish sediment by loading ice sheets', *Bull. geol. Soc. Finl.* **43**, 55–65

AGASSIZ, L. (1837), 'Des glaciers, des moraines, et des blocs erratiques', *Verh. schweiz. naturf. Ges.* **22**, v–xxxii

ANDREWS, J. T. (1965), 'Surface boulder orientation studies around the north-western margin of the Barnes ice cap, Baffin Island, Canada', *J. sedim. Petrol.* **35**, 753–8

(1971), 'Methods in the analysis of till fabrics' in GOLDTHWAIT, R. P. (ed.), *Till —a symposium* (Ohio St. Univ. Press), 321–7

ANDREWS, J. T. and SHIMIZU, K. (1966), 'Three-dimensional vector technique for analysing till fabrics: discussion and Fortran program', *Geogr. Bull.* **8**, 151–65

ANDREWS, J. T. and SMITH, D. I. (1969 [1970]), 'Statistical analysis of till fabric: methodology, local and regional variability (with particular reference to the north Yorkshire till cliffs)', *Q. J. geol. Soc. Lond.* **125**, 503–42

ANDREWS, J. T. and SMITHSON, B. B. (1966), 'Till fabrics of the cross-valley moraines of north-central Baffin Island, North West Territories, Canada', *Bull. geol. Soc. Am.* **77**, 271–90

BADEN-POWELL, D. F. W. (1948), 'The chalky boulder clays of Norfolk and Suffolk', *Geol. Mag.* **85**, 279–96

BOULTON, G. S. (1968), 'Flow tills and related deposits on some Vestspitzbergen glaciers', *J. Glaciol.* **7**, 391–412

(1970), 'On the origin and transport of englacial debris in Svalbard glaciers', *J. Glaciol.* **9**, 213–29; 'On the deposition of subglacial and melt-out tills at the margins of certain Svalbard glaciers', ibid., 231–45

(1971), 'Till genesis and fabric in Svalbard, Spitzbergen' in GOLDTHWAIT, R. P. (ed.), *Till—a symposium* (Ohio St. Univ. Press), 41–72

(1972), 'Modern Arctic glaciers as depositional models for former ice sheets', *J. geol. Soc. Lond.* **128**, 361–93

BRISTOW, C. R. and COX, F. C. (1973), 'The Gipping till: a reappraisal of East Anglian glacial stratigraphy', *J. geol. Soc. Lond.* **129**, 1–37

CAILLEUX, A. (1945), 'Distinction des galets marins et fluviatiles', *Bull. Soc. géol. Fr.* **15**, 375–404

CARRUTHERS, R. G. (1947–8), 'The secret of the glacial drifts', *Proc. Yorks. geol. Soc.* **27**, 43–57 and 129–72

    (1953), *Glacial drifts and the undermelt theory* (Newcastle)

CHAMBERLIN, T. C. (1894), 'Proposed genetic classification of Pleistocene glacial formations', *J. Geol.* **2**, 517–38

CHARLESWORTH, J. K. (1929), 'The South Wales end-moraine', *Q. J. geol. Soc. Lond.* **85**, 335–55

    (1957), *The Quaternary Era* (London, 2 vols)

CLAYTON, K. M. (1963), 'A map of the drift geology of Great Britain and Northern Ireland', *Geogrl J.* **129**, 75–81

DIONNE, J.-C. (1970), 'Structures sedimentaires dans du fluvio-glaciaire', *Rev. Géogr. Montréal* **24**, 255–63

DRAKE, L. D. (1971), 'Evidence for ablation and basal till in eastern central New Hampshire' in GOLDTHWAIT, R. P. (ed.), *Till—a symposium* (Ohio St. Univ. Press), 73–91

    (1972), 'Mechanisms of clast attrition in basal till', *Bull. geol. Soc. Am.* **83**, 2159–65

FLINT, R. F. (1929), 'The stagnation and dissipation of the last ice-sheet', *Geogrl Rev.* **19**, 256–89

    (1971), *Glacial and Quaternary geology*

FRYE, J. C., SHAFFER, P. R., WILLMAN, H. B. and EKLAW, G. E. (1960), 'Accretion gley and the Gumbotil dilemma', *Am. J. Sci.* **258**, 185–90

GAUNT, G. D. (1970), 'A temporary section across the Escrick Moraine at Wheldrake, east Yorkshire', *J. Earth Sci.* (Leeds) **8**, 163–70

GEIKIE, A. (1863), 'On the phenomena of the glacial drift of Scotland', *Trans. geol. Soc. Glasg.* **1**, 1–190

GILLBERG, G. (1969), 'A great till section on Kinnekule, west Sweden', *Geol. För. Stockh. Förh.* **91**, 313–42

GLEN, J. W., DONNER, J. J. and WEST, R. G. (1957), 'On the mechanism by which stones in till become orientated', *Am. J. Sci.* **255**, 194–205

GOLDTHWAIT, R. P. (1971), 'Introduction to till today' in GOLDTHWAIT, R. P. (ed.), *Till—a symposium*, Ohio St. Univ. Press, 3–26

GRAVENOR, C. P. (1955), 'The origin and significance of Prairie mounds', *Am. J. Sci.* **253**, 475–81

HARRIS, S. A. (1968), 'Till fabrics and speed of movement of the Arapaho Glacier, Colorado', *Prof. Geogr.* **20**, 195–8

    (1969), 'The meaning of till fabrics', *Can. Geogr.* **13**, 317–37

HARRISON, P. W. (1957), 'A clay till fabric: its characteristics and origin', *J. Geol.* **65**, 275–308

HESTER, N. C. and DUMONTELLE, P. B. (1971), 'Pleistocene mudflows along the Shelbyville Moraine front, Macon County, Illinois', in GOLDTHWAIT, R. P. (ed.), *Till—a symposium* (Ohio St. Univ. Press), 367–82

HILL, A. R. (1968), 'An experimental test of the field technique of till macrofabric analysis', *Trans. Inst. Br. Geogr.* **45**, 93–106

HOLLINGWORTH, S. E. (1931), 'The glaciation of western Edenside and adjoining areas, and the drumlins of Edenside and the Solway basin', *Q. J. geol. Soc. Lond.* **87**, 281–359

HOLMES, C. D. (1941), 'Till fabric', *Bull. geol. Soc. Am.* **52**, 1299–354

JOPLING, A. V. and WALKER, R. G. (1968), 'Morphology and origin of ripple drift cross lamination, with examples from the Pleistocene of Massachusetts', *J. sedim. Petrol.* **38**, 971–84

KAY, G. F. (1916), 'Gumbotil: a new term in Pleistocene geology', *Science* **44**, 637–8

KAY, G. F. and PEARCE, J. N. (1920), 'The origin of gumbotil', *J. Geol.* **28**, 89–125

KIRBY, R. P. (1969), 'Till fabric analysis from the Lothians, central Scotland', *Geogr. Annlr* **51**A, 48–60

KRÜGER, J. (1970), 'Till fabric in relation to direction of ice movement. A study from Fakse Banke, Denmark', *Geogr. Tidsskr.* **69**, 133–70

KRUMBEIN, W. C. (1961), 'The analysis of observational data from natural beaches', *Tech. Memo. Beach Eros. Bd U.S. (Washington)* **130**, 58 pp.

LANDIN, P. M. B. and FRAKES, L. A. (1968), 'Distinction between tills and other diamictons based on textural characters', *J. sedim. Petrol.* **38**, 1213–23

LINDSAY, J. F. (1968), 'The development of clast fabrics in mudflows', *J. sedim. Petrol.* **38**, 1242–53

—— (1970), 'Clast fabric of till and its development', *J. sedim. Petrol.* **40**, 629–41

LYELL, C. (1830–3), *Principles of Geology* (London, 3 vols, 1st Ed.; 1840, 6th Ed.)

McLELLAN, A. G. (1971), 'Some economic applications of research methods used in glacial geomorphology' in YATSU, E. *et al.* (eds), *Research methods in geomorphology* (1st Guelph Symposium on Geomorphology, Canada 1969), 57–72

MICKELSON, D. M. (1973), 'Nature and rate of basal till deposition in a stagnating ice mass, Burroughs Glacier, Alaska', *Arct. alp. Res.* **5**, 17–27

MILLER, H. (1884), 'On boulder-glaciation', *Proc. R. phys. Soc. Edinb.* **8**, 156–89

MORAN, S. R. (1971), 'Glaciotectonic structures in drift' in GOLDTHWAIT, R. P. (ed.), *Till—a symposium* (Ohio St. Univ. Press), 127–48

NOBLES, L. H. and WEERTMAN, J. (1971), 'Influence of irregularities of the bed of an ice sheet on deposition rate of till' in GOLDTHWAIT, R. P. (ed.), *Till—a symposium* (Ohio St. Univ. Press), 117–26

PESSL, F. (1971), 'Till fabrics and till stratigraphy in west Connecticut' in GOLDTHWAIT, R. P. (ed.), *Till—a symposium* (Ohio St. Univ. Press), 92–105

PETTYJOHN, W. A. and LEMKE, R. W. (1971), 'Unarmed till balls in unusual abundance near Minot, North Dakota' in GOLDTHWAIT, R. P. (ed.), *Till—a symposium* (Ohio St. Univ. Press), 383–94

RAMSDEN, J. and WESTGATE, J. A. (1971), 'Evidence for re-orientation of a till fabric in the Edmonton area, Alberta' in GOLDTHWAIT, R. P. (ed.), *Till—a symposium* (Ohio St. Univ. Press), 335–44

REID, J. R. (1969), 'Effects of a debris slide on "Sioux Glacier", south-central Alaska', *J. Glaciol.* **8**, 353–67

RICE, R. J. (1969), 'Contorted drift at Blaby Brickpit, Leicestershire', *Proc. Geol. Ass.* **80**, 283–91

SAUSSURE, H. B. DE (1779–96), *Voyages dans les Alpes* (Neuchâtel), 4 vols

SHAW, J. (1971), 'Mechanism of till deposition related to thermal conditions in a Pleistocene glacier', *J. Glaciol.* **10**, 363–73

SHOTTON, F. W. (1953), 'The Pleistocene deposits of the area between Coventry, Rugby and Leamington, and their bearing on the topographic development of the Midlands', *Phil. Trans. R. Soc.* **237**B, 209–60

(1965), 'Normal faulting in British Pleistocene deposits', *Q. J. geol. Soc. Lond.* **121**, 419–34

TAYLOR, B. J. (1958), 'Cemented shear planes in the Pleistocene Middle Sands of Lancashire–Cheshire', *Proc. Yorks. geol. Soc.* **3**, 359–65

TROWBRIDGE, A. C. (1961), 'Discussion: accretion gley and the gumbotil dilemma', *Am. J. Sci.* **259**, 154–7

WEST, R. G. and DONNER, J. J. (1956), 'The glaciations of East Anglia and the East Midlands: a differentiation based on stone-orientation measurements of the tills', *Q. J. geol. Soc. Lond.* **112**, 69–91

WESTGATE, J. A. (1968), 'Linear sole markings in Pleistocene till', *Geol. Mag.* **105**, 501–5

YOUNG, J. A. T. (1969), 'Variations in till microfabric over very short distances', *Bull. geol. Soc. Am.* **80**, 2343–52

14

Drumlins

The whole County is remarkable for a Number of small Hills, which are compared with wooden Bowls inverted, or Eggs set in Salt; and from thence it is said to have taken the Name of Down, which signifies a hilly Situation. (C. SMITH, 1744)

The term 'drumlin' is of Gaelic origin, being derived from *druim*, a word for a mound or rounded hill, and has been used in glaciological literature since 1866. It refers to the small round, oval or elongated hills, which diversify some areas of glacial deposition and which are largely, if not entirely, composed of till. The term has been used to describe quite a wide variety of features, from isolated rounded hills up to 60 m in height to low, gentle swells. Drumlins can occur as isolated features, but more often they develop in swarms. Often, thousands or more together cover considerable stretches of country, as for example in Ireland and New York State.

Normally drumlins have a length to breadth ratio of about 2·5:1, extending up to 3 or 4:1. But in some instances very long narrow features have also been referred to as drumlins. The term has, therefore, been used with some lack of precision. Until the method of formation of drumlins is known with more certainty, it seems best to use the term for a specific form rather than in a genetic sense. A drumlin could be defined as a low hill, having an oval outline and not exceeding 60 m in height. It is formed mainly of till, but sometimes contains stratified material or a rock core. It is not desirable to call features of similar form, but composed entirely of solid rock, drumlins, as the main essential of a true drumlin is that it should consist of glacial deposits, shaped to create a streamlined form. There are certain intermediate forms that may contain a considerable amount of solid rock. Crag and tail features are well-known examples. These also show the streamlined form associated with a moving medium.

G. Glückert (1973), from a survey of two large drumlin fields in central Finland covering 25,000 km² and containing 14,500 drumlins, classifies drumlins into five categories:

1 rock drumlins
2 rock end drumlins
3 typical drumlins
4 drumloids
5 drumlin shields.

The most typical drumlins occur in the centre of the field and the rocky ones near the margins, where the relief is uneven and rock outcrops. The drumlins were moulded by south-easterly-moving ice, which reshaped till to form the drumlins. Esker systems run from south or south-east to north or north-west and overlie the drumlins. The pattern shows that the eskers postdate the drumlins and, in some parts, that the ice front retreated in a different direction from its direction of advance. Dimensions vary from 0·1 to 7 km in length and 50 m to 2 km in width, with a height range of 1 to 120 m. The largest drumlins are in the form of shields in which several drumlins are linked. Most of the drumlins are formed of hard sandy till, which is unstratified and contains stones, although some of the narrow ridges contain stratified material. The till thickness ranges from 5 to 30 m, thinning in the proximal areas to 0 to 5 m. It is interesting to note that most of the drumlins have a rock core. The form of the drumlins is related by Glückert to the speed of ice flow, the amount of till and the relief. They appear to have formed where the ground is fairly flat, the till cover is fairly thick, and where the ice was advancing strongly, and they become shorter where the flow rate was reduced by a reversed gradient of the land slope.

1 The shape of drumlins

Various studies have been made of the shape of drumlins in relation to specific geometrical figures. Drumlins are particularly easily treated in this way as they show more symmetry than many geomorphological forms and occur in large numbers in close proximity, so that a large sample can easily be obtained. They are revealed clearly on aerial photographs and often on good, closely-contoured topographical maps, such as the Ordnance Survey 1 : 25,000 series and the US Geological Survey topographic maps, such as the 1 : 24,000 series, some of which have a contour interval of 10 feet (3 m).

B. Reed *et al.* (1962) have used these maps to study the drumlins of three areas, near Boston (Mass.), central-western New York State, and the Voltaire area of North Dakota. The drumlin form was compared with an ellipsoid, the shape of which can be defined by use of the formula

$$\frac{x^2}{a^2}+\frac{y^2}{b^2}+\frac{z^2}{c^2}=1,$$

where a, b, c are the long, intermediate and short semi-axes respectively. This gives the x, y and z co-ordinates. Each contour should be an ellipse if this formula is used and the ratio of major to minor axes must be constant for each contour.

The results of plotting the values for major and minor axes of a series of drumlins shows that, at their lower contours, the drumlins approximate closely to the elliptical form. However, at their summits, they depart slightly from the ideal shape. The elongation of the drumlins is also apparent in the steepness of the curve plotting length against breadth. Those in New York are more elongated than those near Boston. A study of the orientation and spacing of the drumlins was also carried out. The spacing both parallel to and normal to the mean direction of orientation was measured, using adjacent drumlins for the measurements. The results show a wide variation in the spac-

ing with a multi-modal curve resulting. There is, however, some evidence for a preferred spacing and some indication of periodicity.

Another shape model has been applied to drumlins by R. J. Chorley (1959). This
shape emphasizes the streamlined form of drumlins and their formation by a moving
medium. They are compared with the shape of other streamlined forms, such as aero-
plane wings and snow drifts. These shapes are such that the air, which is the moving
medium causing their formation or to which they are adapted, can flow smoothly
over them. The form is not symmetrical as an ellipse about two axes. It is only symmetrical about the longitudinal axis, having a rounded end facing into the flow and an elongated tapering end on the lee side. The profile is also not symmetrical, but has a
steeper slope on the exposed side and a gentler lee slope.

It is interesting to see how closely a drumlin corresponds to the ideal streamlined
form. If a close correspondence can be shown, then it is necessary to investigate the
processes that can cause such a streamlining effect. In comparing the shape of drumlins
with that of aircraft wings, Chorley makes the point that the maximum wing width is
situated about three-tenths of the distance from the leading edge of the wing. A more
important point is that the greater the air speed that the wing has to withstand, the
greater the length relative to the breadth. This means that the form is more elongated.
Greater elongation of drumlins may well be associated with more powerful ice flow.

The type of curve that can most suitably be used is the lemniscate loop. It is adaptable
to the varying forms of actual drumlins, and also provides the ideal streamlined form.
The formula for the lemniscate loop is given by

$$\rho = L \cos k\theta$$

where L is the length of the long axis (this being the value of ρ when $\theta = 0$); k is a
dimensionless number expressing the elongation of the lemniscate loop, such that k
is equal to unity when the form is circular, and as the loop increases in elongation,
so k also increases. The value of k is given by

$$k = \frac{L^2 \pi}{4A}$$

where A is the area of the loop. The only measurements that are needed to fit the
loop to the actual outline of the drumlin are the length, L, and the area, A. These
values can be used to obtain k in the second equation, and then k can be used in the
first equation to plot the form of the loop. The tapering of the drumlin is often not
so extreme as that given by the lemniscate loop when the ice flow is relatively slow,
and the lee end becomes slightly rounded relative to the plotted curve. Chorley points
out that the smaller drumlins tend to be more symmetrical, having a form approaching
more closely to that of an ellipse than a lemniscate loop. As the size of the drumlins
increases, so does the asymmetry about the long axis.

An interesting comparison is made between this tendency to become more rounded
and the shape of birds' eggs. A field of drumlins is often referred to as 'basket of eggs'
relief. This analogy is a useful one as it appears that there may be some genetic connection between the shape of eggs and drumlins in this respect. A hen's egg is always laid
with the blunt end first, so that the pressure at this end is greatest and the shape of

the egg is adjusted to this. The blunt end of the hen's egg thus corresponds to the stoss end of the drumlin, which also experiences the greatest pressure of the moving ice. This relationship can be carried further, in that there is a relationship between the shape of a bird's egg and the size of the bird laying it. Birds that lay large eggs relative to their size tend to lay the most asymmetrical and pointed or tapering eggs. Birds laying relatively small eggs lay much rounder, symmetrical ones. The less the pressure

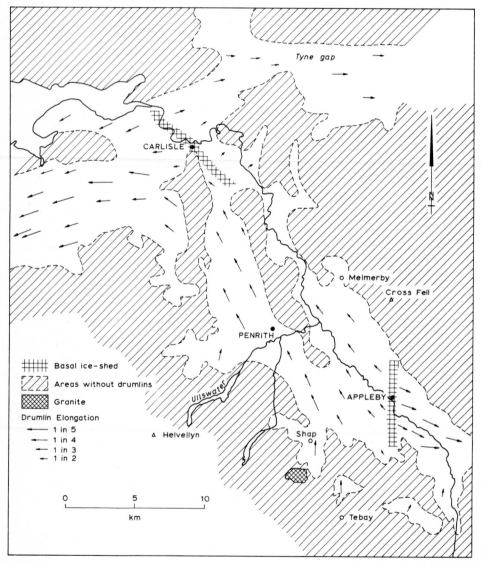

Fig. 14.1
The distribution and elongation of the drumlins in the Eden valley and Solway lowlands, northern England

on the eggs as they are laid, the rounder they should be. By analogy, the more slowly the ice is flowing, and hence the lower the ice pressure, the more rounded the drumlins should be. This analogy should not be carried too far, as the supply of till and other factors clearly play a part. Nevertheless it does suggest that the shape of drumlins can be related in part to the pressure exerted by the ice in their formation. The factor k gives a useful inverse indication of the relative resistance of the equilibrium form of the drumlin to the moving ice. This resistance depends both on the consistency of the material and the nature of the force of the moving ice.

It is important to note the great variety of elongation of the features that have been described as drumlins. At one extreme there are the narrow, linear drumlins described by R. W. Lemke (1958) in north-central North Dakota near Velva. These features are normally 1·6 to 4·8 km in length, 46 to 60 m wide and 1·5 to 4·6 m high. They have steep sides and sharp crests. Their ratio of length to breadth is about 60 to 1. They are a little higher at the stoss end and have a symmetrical transverse profile. The longest of this set is 21·5 km long. It is 15 m high at the north-west stoss end, falls ·to 10 m in height throughout most of its length, and tapers to only 1·5 m in height at the south-east or lee end. This particular feature has an elongation of 240 to 1. Similar-shaped features have also been described by J. L. Dyson (1952) in front of the snouts of the Grinnel and Sperry glaciers. They are 150 to 180 m long, 0·6 to 3·7 m wide and 0·3 to 0·9 m high. Such long, low features appear to have a fairly wide distribution in North America, but not all drumlins in this area are of this shape. Many have the much lower elongation normally associated with drumlins.

Although it is generally true that, in any one drumlin field, the variation in elongation is limited, there are nevertheless interesting differences within some drumlin fields. These variations give useful information concerning the formative processes. The Eden Valley and Solway Lowland drumlin belt may be taken as an example. This area has been studied by S. E. Hollingworth (1931). A further analysis of the drumlin elongation in this area has revealed interesting variations within it as shown on Fig. 14.1. The elongation of 100 drumlins varied from 1·1:1 to 9·7:1, with a mean elongation of 3·0:1. A series of 20 drumlins was measured on the 1:25,000 map, using the lowest closed contour to give a measure of the drumlin length and breadth, in five separate localities within the area. The first group is situated near Appleby, the second in the Eden Valley around Penrith, the third in the neighbourhood of Carlisle, the fourth in the Tyne Gap, and the fifth set in the area to the west of Carlisle. The mean elongation for each of the five areas is:

$$
\begin{array}{rl}
\text{Area 1} & 2\cdot55:1 \\
2 & 3\cdot03:1 \\
3 & 1\cdot92:1 \\
4 & 2\cdot85:1 \\
5 & 4\cdot66:1
\end{array}
$$

The data were then used in an analysis of variance test. The resulting value of F was 22·2, which is highly significant. The F value shows that the variation between the groups of drumlins is considerably greater than that between the drumlins within the individual groups. The change in elongation within the whole drumlin field is, there-

fore, statistically significant. These differences in elongation must be accounted for
in terms of the formative processes in relation to the environment in each of the five
groups of drumlins selected. A discussion of the factors that may be responsible for this
variation in drumlin elongation will be considered when their genesis is discussed
(see p. 421).

F. D. Hole (1970) discusses some measurements of drumlins made by R. J. Allen.
Measurements of 320 drumlins give an average length of 930 m (standard deviation
537 m), mean width 290 m (S.D. 136 m), crest height 14·6 m, average north–south spacing 1860 m, the trend being mainly north–south. Of the material in the drumlins, 27
per cent had particle size greater than 2 mm, while, of the finer material, 72 per cent
was sand, 17 per cent silt and 11 per cent clay. The patchy distribution of drumlins
in parts of the field probably resulted from modification of the original pattern by meltwater erosion, or by burial under outwash or other deposits.

Most measurements of drumlin dimensions have been made from maps or air photographs, but H. F. Barnett and P. G. Finke (1971) report field measurements of drumlins
in southern Germany, in north-central Massachusetts and central-west New York
State. The field measurements include slope gradients and lengths. The selection was
not random; slopes were measured on 55 oval-shaped drumlins, making 74 km of traverse. Four traverses were made for each drumlin and angles were measured to 1°.
The results are summarized in Table 14.1. The slopes of the North American drumlins
were rather less steep than those of the German ones, the median slope being nearly
5° for the former and between 8° and $9\frac{1}{2}°$ for the latter.

Table 14.1   Field measurements on fifty-five oval-shaped drumlins (Bennett and Finke, 1971)

	Hudson, Mass.	Weedsport, NY	Cato, Cayuga	Eberfing, Germany	Bodanruck, Germany	Rosenheim, Germany
No. of drumlins	17	9	3	13	10	3
Mean length, m	586	544	1129	549	571	745
Mean width, m	391	252	241	268	236	279
Length/width ratio	1·54	2·19	4·66	2·18	1·98	2·80
Mean height, m	30	24	33	29	29	16
Azimuth of long axis	347°	347°	344°	335°	315°	034°
Longitudinal asymmetry (ratio of lee slopes/ stoss slopes)	1·00	1·29	1·63	1·65	1·22	1·09
Transverse asymmetry (ratio of RH/LH slopes)	1·18	1·18	1·16	1·37	1·34	1·09

The relationships between length (L), width (W) and height (H) of drumlins for
an area in southern Ontario have been analysed statistically by A. S. Trenhaile (1971).
The following six equations define the best-fit relationships in this area:

$$L=2·68W+256·9(r=0·65) \qquad W=0·167L+116·16(r=0·65) \qquad H=0·03L+1·65(r=0·65)$$
$$L=13·1H+516·5(r=0·65) \qquad W=4·71H+141·1(r=0·73) \qquad H=0·11W-0·43(r=0·73)$$

The scatter for all three relationships was large. The lee slope gradient varied mainly from 1 : 10 to 1 : 40, with a modal stoss value of 1 : 12. The lee slopes were most variable and may be significant in adjusting to ice flow. Most of the drumlins were ellipsoid in shape. Their height did not affect the shape; for about a quarter, the L/W ratio increased as the summit was approached. The lemniscate shape gave a mean k value of 4·2, with a standard deviation of 2·1. Polynomial curves were fitted to seventeen surveyed profiles; the quartic polynomial fitted best, defined by

$$L = 0·02915\,H + 0·00234\,H^2 - 0·00142\,H^3 + 0·00008\,H^4.$$

The results generally substantiate those found elsewhere.

2 The material of drumlins

It has been suggested by some authorities (e.g. R. F. Flint, 1971) that features having the elongated, streamlined form typical of drumlins can consist of a wide variety of material, ranging from almost entirely solid rock to entirely drift. The glacial material of which most drumlins are formed can vary widely. Some drumlins include stratified, water-laid deposits, although many consist mainly of unstratified till, deposited directly by moving ice. The till can vary according to the rocks exposed in the path of the approaching ice. Thus some drumlins consist of mainly sandy material, while others are made up mainly of clay. Some are formed of true boulder clay, consisting of a mixture of subangular and striated boulders set in a clayey matrix, such as many of the drumlins in the north of England. The narrow, linear drumlins of North Dakota that have already been mentioned, consist mainly of stratified sand. This sand includes balls and lenses of till; there are also layers of gravel, and the whole is shrouded in a thick layer of till. The till also underlies at least part of the sand. Finally a layer of wind-blown sand is banked up against one side of the drumlins. This complex structure is found in the larger ridges, while the smaller ones are often composed only of till. The included lenses of till within the sand lie with their long axes parallel to the axis of the ridge.

The internal composition of a drumlin in Upper Bavaria is described by I. von Schaefr (1969). The drumlin, which is 800 m long, 300 m broad and 32 m high, is dissected by a gravel pit. This shows, from the top downwards, 3 to 5 m ground moraine, 0·5 to 1 m moraine gravel, 3 m fine sand above 2–20 cm dark brown lake clay and black peat. Beneath the peat is a deposit of clay and fine sand, 1 m thick, on grey, grey-blue or grey-yellow bedded gravels.

J. Lundqvist (1970) has described the drumlin tracts in central Sweden. Some of the best developed occur in the low part of the country in a sub-aquatic environment, but most in this area are crag-and-tail features or rock drumlins. True drumlins occur at higher levels, above the high late-glacial shoreline. Well-developed groups consist of two or three drumlins, 500 to 700 m long, 100 m wide and 10 m high. Some are narrower, rising from low basements. They consist of loose till with little gravel or stones and about 15 per cent clay. They represent the only patches of glacial drift on otherwise bare, grooved rock, and they occur where the slope is convex rising to the high plateau. Rather broader drumlins occur at the foot of the high plateau. Another large cluster occurs at the south-west rim of a broad plateau at 700 to 800 m. These are about 500 m long, 200 m broad and 10 m high and are formed of ground moraine, which is dominantly sand and silt, with less than 5 per cent clay. The terrain where the drumlins occur is flat transverse to the direction of ice flow and the drumlins tend to occur on the edge of the plateaux. Some of the drumlins become incomplete towards the depressions in the high plateau. They become crescentic in form and linked transversely, eventually merging into Rogen moraine, as shown in Fig. 14.2. Some smaller drumlins occur, only about 100 m long, 20 to 30 m wide and a few metres high. The drumlins occur both in areas of thick drift, very thin drift or almost no drift, and consist of ordi-

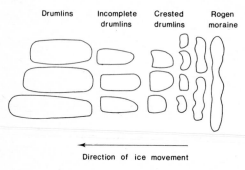

| Drumlins | Incomplete drumlins | Crested drumlins | Rogen moraine |

Direction of ice movement

Fig. 14.2
Sketch to show the transition from Rogen moraine to drumlins in part of central Sweden (J. Lundqvist, *Acta Geographica Lodziensia*, 1970)

nary sandy ground moraine. The narrower drumlins appear to occur where the ice flow was faster, and some types are associated with erosion. The elongation of the drumlins follows the direction of the flow of the last ice movement. Over most of the drumlin area, the curvature of the ground slope was convex in the direction of the ice flow. The change from drumlins to the Rogen-type moraine occurs where the slope changes from convex to concave. Lundqvist considers that the drumlins formed in the initial phase of subglacial accumulation. A heavy load of till is not essential, as the same effect can be achieved by stagnation of the basal ice. Convexity of ground slope causes the basal ice to move less far than the upper ice and therefore accumulation takes place. Both erosion and deposition occur during drumlin formation. Some drumlins are formed during recession but they need not be either low-level or marginal features. Drumlins may form where the ice became thinner, so that horizontal pressure was greater than the vertical, a process that only took place at high levels in Jämtland, owing to the stagnation of ice in depressions where drumlins may be hidden by dead-ice features.

In some areas, studies have been made of the orientation of stones included in the till of the drumlins. G. Hoppe (1959) has shown that the drumlins of Norrbotten in northern Sweden, which are composed mainly of sandy till, contain stones that show a preferred orientation in the direction of the ridge, except when the stones are too numerous to enable them to move freely. These particular drumlins are irregular, elongated ridges, usually less than 10 m high and up to 500 m long. The stones in them show no preferred plunge and the drumlins show no longitudinal asymmetry as many drumlins do. Hoppe suggests that these particular features may have been formed by the squeezing of water-soaked till into elongated cavities within or under the ice or by direct subglacial deposition. The fact that the stones near the surface of the features also show a preferred orientation suggests that the amount of supraglacial till must have been small where the features were formed. Nearly all these ridges have no rock core, although a few do contain a rock core. Their orientation is parallel to the latest striations and thus to the final direction of ice movement in the area. This suggests that they were formed near the front of the ice, when it was receding in the area. Such data conflict with characteristics of drumlins in other areas to be considered, and it may be that these ridges are not true drumlins.

Detailed study of the orientation of stones in till has given some valuable evidence concerning the character of drumlins. Work by H. E. Wright (1957) was based on observations made in the Wadena drumlin field of Minnesota. Measurements of stone orientation in drumlins were made in the area to the west of Lake Superior, where the drumlin field is 100 km long and covers an area of 5000 km². To the west, it becomes buried by outwash, to the south by drift of younger ice, and on the east lies a younger moraine. The younger ice did not destroy the drumlins in the area where it overlapped them in the north-east. In the north of the drumlin field, the drumlins trend south 45° west, and they fan out to south 30° east and north 80° west to the south and west respectively. The drumlins range from an elongated form, such as 12 km by 0·5 km, to oval ones about 0·7 by 0·5 km, and they vary from 5 to 10 m in height. They are formed of yellowish-brown sandy calcareous till.

The stones that were measured had a long axis of 1 to 8 cm and an elongation ratio of at least 3 to 2, fifty stones being measured at each site. The strike and plunge of the long axis were both measured. Nearly all the sites show a major orientation trend parallel to the elongation of the drumlin and to the assumed direction of ice flow. Some have a minor transverse concentration as well. The evidence suggests that the drumlins were formed by accretion of basal material from the lowest layers of the ice, and that the stones were carried by basal sliding. Stones carried in this way tend to be orientated parallel to the flow, whereas stones carried by rolling are usually arranged with their long axis at right angles to the direction of movement (Chapter 13).

Measurements of stone orientation in drumlins are of value as the actual form of the drumlin gives an independent indication of the direction of ice flow. It is, therefore, possible to relate stone orientation to direction of ice movement with more confidence in these circumstances. In a smooth till sheet, orientation studies are usually made in order to deduce the direction of ice movement. This is a topic about which there is still much to be learnt and it is significant that in drumlins it is not always true that the stone orientation is parallel to the elongation and flow direction. For example, in a drumlin in Wensleydale (Yorkshire), the preferred orientation was found to be at a considerable angle to the elongation of the drumlin. This drumlin lies in the valley floor and is elongated down-valley and hence presumably parallel to the ice flow. The mean orientation of ten samples of fifty stones each, taken from the exposed interior of the drumlin, makes an angle of 33° to the elongation of the drumlins. The maximum deviation is 54° and the minimum 12°. All but two of the ten samples show a statistically significant preferred up-glacier dip of the long axes in a direction perpendicular to the mean orientation of the pebbles. The observations were made on the south side of the drumlin, where till is exposed by river erosion (Fig. 14.3). The orientation, there-fore, points in towards the centre line of the drumlin. The two samples that have no

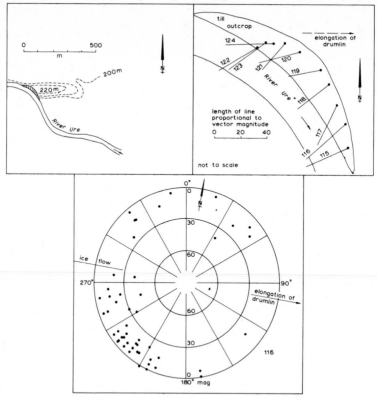

Fig. 14.3
Till fabrics from a drumlin in Wensleydale, Yorkshire

significant dip are near the crest of the drumlin. There is a gradual increase in divergence between elongation of the drumlin and direction of the mean vector from the base upwards. The drumlin is elongated in the direction 280–100° and the mean vector directions from the base upwards are 226°, 232°, 237°, 242° and 245°. The values for the summit samples have means of 258° and 268°, but are less consistent and show a much wider scatter. These orientation values suggest that ice was pressing material against the side of the drumlin, with an element of lateral stress which was missing near the summit of the drumlin. The suggestion that drumlins could occur when the ice flow has subsidiary lateral stresses may help to explain the occurrence of these features.

Wright has also measured the plunge of the stones in the Wadena drumlin field. Of the 565 stones analysed, 34 per cent of the long axes plunge at 10° or less, the average plunge being 23°. This value is higher than earlier studies made on other types of till deposits. 64 per cent of the stones plunge up-glacier, 27 per cent plunge down-glacier and 9 per cent plunge transversely. The up-glacier plunge can be compared with the up-glacier plunge of shear zones, for example the plunge at 10 to 30° on the margin of the Barnes ice cap in Baffin Island. It has been shown that stones associated with such shear zones normally dip with them, parallel to the shear plane. As a result of these studies, Wright came to the conclusion that the Wadena drumlins must have been deposited by active ice. The probable thickness of the ice at the time of the formation of the drumlins can be estimated by comparison with the Greenland ice sheet. The breadth of the marginal zone at the time of drumlin formation was about 140 km. The thickness of the Greenland ice sheet at this distance from the front is 1000 m and it is about 300 m thick over the outer 40 km zone. If similar conditions existed at Wadena, it is apparent that the drumlins must have formed under a fairly thick ice cover. It would certainly have been sufficiently thick to make it unlikely that crevasses would extend to the base. Although these drumlins appear to have formed fairly close to the limit of the ice sheet, squeezing of material into cavities in the ice would not seem to be a likely method of formation in this instance.

Another interesting study of the internal structure of a series of drumlins has been carried out near Dollard (Saskatchewan) by W. O. Kupsch (1955). Some of the drumlins studied appear to be composed of till alone, but others have till and stratified sand and boulders. Both forms, however, show clear evidence of having been streamlined by moving ice. The drumlin field lies in a topographical depression with a south-west trend. It is 29 km long and has a maximum width of 10 km and contains 100 drumlins. Their density is therefore much less than in many drumlin fields. They lie about 160 km north of the drift limit. Their average length is 230 m and their width 10·7 m. They vary from 75 to 750 m in length and 23 to 30 m in width in the more elongated ones. Most of them have a length to breadth ratio of about 2·5 : 1, and a height of 46 m. An interesting point concerning their distribution is that they only occur on the sandy Upper Cretaceous strata and not on shale. The maximum thickness of the drift is 20 m but in places it is as little as 6 m around the drumlins.

The till of which they are partially composed is a silty clay with few boulders, but containing irregular pockets of sand and gravel. A sand pit has been dug in one of the drumlins, and this reveals an interesting internal structure (Fig. 14.4). This drumlin, 490 m long and 335 m wide, consists of a layer of unstratified till on the gentle distal slope. A line of boulders stands vertically beneath the highest portion. The steep proximal side consists of stratified sands and gravels. The whole feature is overlain by about a metre of unstratified sandy drift. The wall of boulders is perpendicular to the long axis, and the boulders are rounded and range up to between 0·3 and 1 m in size. Many of the boulders are fractured and have three sets of joints. These fractures are clearly related to the formation of the drumlin, as they can be traced at times through one boulder into a subjacent one. Most of the fracture planes are nearly vertical. The preferred orientation of the joints is 355°, with a secondary preferred orientation at 92° and a minor set at 34°. The orientation of the drumlin is 33°. The pattern of the joints can be explained if they were formed by the pressure of ice against the drumlin exceeding the strength of the rocks within it. The joints at 355° and 92° are shear joints. They would occur if the top and bottom of the drumlin were confined and pressure were relieved by lateral movement. The set of joints at 34° represents extension joints. The pattern of the joints suggests that the wall of boulders was already in place and was overridden by active ice. The joints agree with an experimental test pattern.

The formation of the drumlin is thought to have been carried out in the following

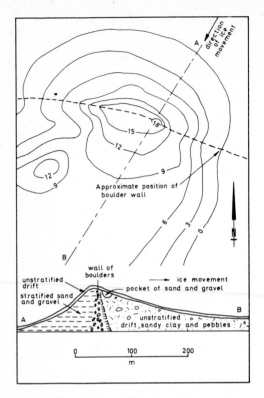

Fig. 14.4
Internal structure of a drumlin near Dollard, Saskatchewan (W. O. Kupsch, *Bull. geol. Soc. Am.*, 1955)

stages. First, the ice thinned and became stagnant. Then the boulders collected in linear crevasses, where they were concentrated by meltwater washing away the finer material. The finer material remained in the rest of the deposits and now forms the proximal side of the drumlin. The ice then readvanced after the kame, formed of the boulders, had frozen into a solid block. The advancing ice streamlined the deposits into drumlin form. The boulders were fractured by the ice pressure and a tail of unstratified till was left on the distal side of the original kame, that is, in the lee of the obstruction. This particular drumlin, in the details of its sedimentary character, shows evidence of both erosion and deposition, the final overall covering being in the form of ablation moraine. The thickness of the ice when the drumlin was shaped could have been as much as 400 m at this point. It was probably much less at the stage when the fluvioglacial material was being deposited to form the original kame, which was later modified by active ice to form the present drumlin.

The drumlin field of south-east Wisconsin has been described by N. P. Lasca (1971). It was in his area that T. C. Chamberlin applied the term 'drumlin' as long ago as 1893, by which time over 2500 drumlins had been reported. Each set of drumlins seems to be related to a set of marginal or terminal moraines; in some places, the drumlins merge into the moraines with kame-and-kettle relief, in others marginal deposits rest on drumlins. The drumlin long axes often overlap, while laterally they are usually separated by distances greater than the drumlin width. The drumlin length tends to increase towards the axis of movement of the glacial lobe. They consist of clayey and sandy till with patches of stratified drift in some cases. Occasionally the stratified beds are folded, but the overlying material is undisturbed. Few drumlins here have rock cores, and 85–90 per cent of the material is of local origin, suggesting deposition by basal ice. The drumlins formed in a zone several kilometres wide behind the ice front and Lasca considers that they can be satisfactorily explained by Smalley and Unwin's theory (1968) (p. 426). E. B. Evenson (1970) found that till fabrics in this area were affected by short transport, high frequency of stone collision to produce parallel orientation of prolate stones, and possible dragging over a stationary or slow-moving layer. The differential pressure theory of till shear, in his opinion, accounts for the fabrics in the drumlins of Jefferson County. There is little correlation between slope angle and fabric plunge, so that ice-shear or till-shear theories must be considered in this area, and the latter better fits the data. Overriding ice can reorientate fabrics to a depth of 10 m. The till fabrics suggest that drumlins in this area were formed beneath the ice as frozen or partially frozen but viscous till sheared in response to overriding ice. The drumlins were formed by ice readvancing over previously deposited till and outwash. The overfolded stratified beds were injected by subglacial lateral pressures at the time of formation. The up-glacier plunge of the fabrics is related to the movement from high to low pressure zones, and to up-curving shear zones in the overriding ice.

A. R. Hill (1970, 1971) has examined the material and pattern of the drumlins of County Down and parts of Antrim in Northern Ireland. The drumlins are related to the direction of ice movement in County Down. There are two till sheets, the lower having shells dated at 24050 ± 650 BP. The lower till was deposited by ice moving from Scotland while the upper till was laid down by ice coming from the Lough Neagh basin in Ireland. There are 3900 drumlins in an area of 1600 km² in Co. Down, and these form part of an even larger field. In the area near Bangor their mean orientation is $18 \cdot 9° \pm 22 \cdot 69°$; in the rest of northern County Down the orientation ranges from $290 \cdot 7° \pm 17 \cdot 82$ to $336 \cdot 2° \pm 15 \cdot 04°$. The drumlins of the Bangor district are composed of lower till, but the others are of upper till. The orientation strengths indicate that the upper till was deposited by ice flowing less fast in the Bangor district than elsewhere, where the orientation strengths are higher.

The drumlins of the north Down and south Antrim area consist of a variety of material, some being composed of one till, others of two tills, while yet others have rock cores or consist of sand and gravel. Most till fabrics are nearly parallel to the drumlin trend, but some were very different. The drumlins made of till are mainly depositional in type. Their fabric and the direction of plunge of the stones appear to differ from those of the ground moraine. In some areas the two separate tills, from the north and west respectively, are separated by sands and gravels, although the tills may be contemporaneous. Most of the drumlins formed by the Lough Neagh ice are orientated north-west to south-east. They vary from 200 to 850 m in length and are 100 to 350 m wide.

The drumlins of the northern Ards Peninsula are more subdued and lie north–south or north-east to south-west. They were formed earlier by North Channel ice and have survived Irish ice movement across the area. The seventy-six sections examined indicated that most of these drumlins were built of till, but a few had rock cores. Four drumlins had three concentric till zones. Half the fabrics had orientations within 15° of the long axis direction, but in some there was a change in direction vertically upwards through the drumlin from west of north to east of north. The fabrics of the lower till core of some drumlins were unrelated to the present drumlin trend, but the upper till fabrics mostly lay within 20° of the drumlin trend. There was no consistent plunge direction, although the drumlins had stronger fabrics than the surrounding ground moraine.

The variations in the material of which different drumlins are composed requires a variety of possible methods of formation. Individual drumlins and drumlin fields need not all have been formed by the same mechanism. The presence of both stratified and unstratified material makes it necessary to consider both erosional and depositional processes in drumlin formation (p. 423).

3 The distribution of drumlins

Drumlins are widely distributed in those areas that were glaciated during the last major ice advance (Würm, Weichsel, Wisconsin). Although they are widely scattered, their occurrence is often very localized in any one area. A few are found outside the areas of the last major ice advance: for example, there are drumlins of Riss age in the Danubian part of the Rhine glacier area and of Illinoian age in North America. One of the main problems concerning drumlins is to account for their localized occurrence in any one area. It is also worth noting that, within any one area, the drumlins tend to be of fairly uniform type and dimensions.

Over much of glaciated Europe, drumlins are rare, while in North America they are found infrequently in Pennsylvania, New Jersey, Ohio, Indiana and south Michigan. However, where they do occur they are often found in very large numbers. In east-central Wisconsin, there are about 5000 drumlins, in south-central New England there are 3000, central-western New York State has 10,000 and Nova Scotia 2300. In Canada, the drumlins tend to be long and narrow in the north.

The North American drumlins are found in five main areas:

1 Manitoba and Athabasca
2 New England, Ontario, New Brunswick and Nova Scotia
3 Michigan and Wisconsin
4 British Columbia
5 central-western New York State.

This last area probably contains the most extensive drumlin belt in the world. The drumlins of this area lie in a belt 56 km broad and 225 km long, between Lake Ontario and the Finger Lakes. Generally, the drumlins of North America tend to occur in belts behind morainic barriers on flat till plains.

In Europe they also occur behind and between morainic ridges, for example, in Switzerland. Drumlins occur in western Pomerania, where they have an elongation of 3·75 : 1 on average and number over 3000. Many of these contain a fluvioglacial core. Another small but typical drumlin area lies in northern East Prussia between Memel and Nimmerstatt. There are about 100 drumlins in this area and they are formed entirely of till. Drumlins are numerous in parts of the Alps, for example, the drumlin field of the Bodensee and Eberfiner areas (P. Woldstedt, 1954).

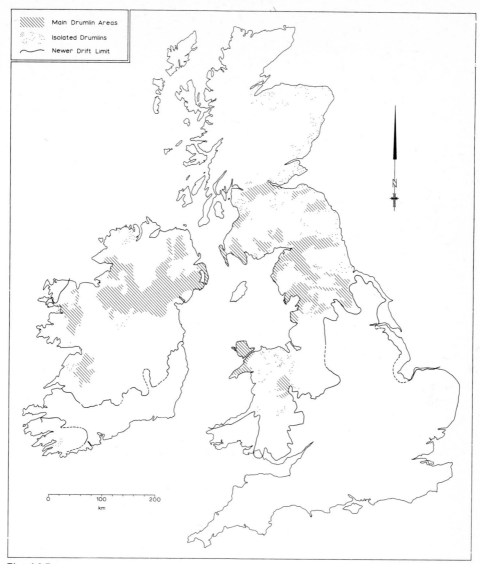

Fig. 14.5
The distribution of drumlins in Great Britain

In Britain (Fig. 14.5), drumlins are also common in some parts of the Newer Drift areas. The great belt of drumlins, extending across nearly all of the northern half of Ireland, is one of the largest in the world. It extends from County Down in the east across Armagh, Cavan, the Erne area, and into Donegal and Sligo Bay in the west. The drumlins extend out into the Atlantic in Clew Bay, County Mayo, although this set of drumlins is not directly continuous with the more northerly belt. Drumlins also occur farther south in a great belt in County Clare to the north of the Shannon. These drumlins lie inside the southern Irish end-moraine, which marks the limit of the Newer

Drift ice advancing from the north. In the extreme south-west of Ireland, drumlins occur within the limits of the lesser Kerry-Cork glaciation, which also belongs to the Newer Drift period.

In Scotland, drumlins occur in the Midland Valley, in Galloway and the Merse area of the lower Tweed. Some not very well developed drumlins have also been described in Skye by J. J. Donner and R. G. West (1955). They had previously been described as dead ice features, but examination of stone orientation in them suggests that they are in fact drumlins. The preferred stone orientation is parallel to their elongation. They occur north of the Cuillin Hills and are remarkable in that they are small in size, being only 2 to 5 m high and 10 to 20 m long, and their lee side is the steepest.

One of the most extensive drumlin fields in England is that in the Eden Valley and Solway. This belt continues into south-west Scotland and across the Tyne gap into north-eastern England. It almost links with another set of drumlins to the south of the Lake District around Kendal, Barrow-in-Furness and Ulverston. To the west of the Lake District, the belt swings round past Wigton and Aspatria towards Maryport. Drumlins also occur in the Ribble valley between Skipton and south of Settle. A very well developed set occurs in the Ribblehead region. This group is almost linked with another set that stretches into Wensleydale which, in turn, nearly link with the Eden valley set at Kirkby Stephen. The high ground of the Lake District and North Pennines is thus almost entirely surrounded by drumlins. They also occur in Anglesey and in the Welsh Border country near Wrexham.

Other areas in the world with drumlins include central France, the Dinaric Alps, Tienshan, Siberia, Novaya Zemlya and China.

3.1   Spacing of drumlins

Regularity in spacing of drumlins has frequently been demonstrated, each drumlin field appearing to have its own preferred wavelength in a direction transverse to the longer axes. For the Wadena drumlin field in Minnesota, S. Baranowski (1969) found that 30 per cent of the drumlins are spaced 400–600 m apart, and for a drumlin field near Zbojno in Poland, the equivalent spacing is 90–120 m.

A. S. Trenhaile (1971) has analysed the spacing of drumlins in southern Ontario. The hypothesis of a negative binomial distribution was tested using three quadrat sizes of 4, 9 and 16 km². The hypothesis of clustering at the 0·05 per cent level of significance was rejected in all cases. The Poisson and Dacey 'more regular than random' distribution was also used, and the results suggested that the drumlin distribution lies between a random and a more regular distribution. Nearest-neighbour analysis of distances measured to the four nearest drumlins showed in all cases a significant difference at 0·05 per cent between the computed random distribution and the observed distribution. The orientation pattern was normally distributed with a small standard deviation, except where bedrock control is effective.

S. Aranow (1959) considered that six possible factors may influence the distribution of drumlins within a drumlin field. The area he examined was the Warwick-Tokio area of North Dakota, where drumlins on the whole are not plentiful. The drumlins are confined to the Heimdal moraine and its southern continuation. There are some drumlins on the proximal side of the moraine and on patches of the moraine that extend up through the outwash. Most of these drumlins, however, are not very regular, only twenty being true drumlin-shaped features out of 160. The drumlins are elongated in a north-east direction and are formed of clayey till. Six aspects of relief and geology were selected as having a possible effect on drumlin distribution:

1 regional slope
2 topographic grain
3 character of the drift
4 configuration of the bedrock
5 character of the bedrock
6 relief prior to the last glaciation.

None of these six factors appears to control the distribution of drumlins in the area and several of them are fairly uniform throughout the area. The conclusion is reached that there are no easily discernible differences in the factors enumerated that could account for the variation in drumlin distribution. It is more likely that the cause is to be sought in the character of the now-vanished ice that deposited the drumlins. Where conditions in the ice are suitable for drumlin formation, features showing stream-lining will form regardless of the material over which the ice is flowing. True drumlins will occur under circumstances in which till is available for drumlin formation.

3.2   *Variation of drumlin form within drumlin fields*

The spatial variation of drumlin form within a drumlin field has also attracted considerable attention. Many possible factors may control this variation, and investigations of the spatial trends often help to throw light on the problems of drumlin formation. J. W. Miller (1972) identified three belts, each characterized by distinctive drumlin forms, in central New York State between Lake Ontario and the Allegheny mountain front. The northern belt lies north of the Niagara scarp and consists of large, poorly streamlined forms. The central drumlins are smaller and better streamlined. Drumlins in the southern belt, lying against the Allegheny front, are still smaller and steeper, and show more intricate streamlining patterns. Variables measured included drumlin crest elevation, inter-drumlin elevation, drumlin length, width, orientation and steepest gradient. Three traverses across the three belts were surveyed. Statistical analysis to compare the three belts showed a significant difference in length and width, but not in length-to-width ratio. The slopes steepened southwards. The proportion of the area covered by drumlins declined from 45 per cent in the north to 33 per cent in the centre and 27 per cent in the south, while density values were respectively, from north to south, $0.51/km^2$, $1.08/km^2$ and $1.26/km^2$. Orientation did not differentiate the three groups, nor did composition. One possible hypothesis to explain the variation is in terms of till volumes and rates of ice flow. The volume of till may have a significant effect on the latter, so that a decrease in volume of till at the glacier sole would cause faster flow and drumlins would become more elongated. Slow ice movement in the northern region would result in the formation of large drumlins around large obstructions. In the central zone where the volume of drift was less, the ice was able to flow faster, a process that would be accentuated as the ice moved up the Allegheny front in the third zone. At the Allegheny front, loss of load caused the drumlin-forming process to cease. Numerous features round which drumlins could be streamlined occur in the southern zone, accounting for the greater density in this area. Thus variation in the speed of ice flow could account for the variations, and the changes in flow were themselves probably related to changes in bed load. Streamlining and elongation were greatest where the flow was fastest, as in the Green Bay lobe of south-east Wisconsin.

A. R. Hill (1973), in an analysis of part of the drumlin belt of northern Ireland, finds a non-random distribution of drumlins on three scales by means of trend-surface analysis and block-size distribution studies. At a regional scale, the density of drumlins rises to a maximum in the centre of the field, declining towards the terminal zone. There are also alternating zones of high and low density, about 3 km in width, perpendicular to the direction of ice movement. Finally there is a tendency for drumlins to cluster at a linear scale of 1 km. The analysis covers an area of 1600 km² in a much larger drumlin field, involving in the study area a total of 3900 drumlins. The drumlins range from 180 to 850 m in length, 100–350 m in width, and are most elongated in the central and southern parts. The mean drumlin density is 3·29/km². The alternation of zones of drumlin density may result from mechanisms of movement in the ice sheet, possibly related to waves within the ice causing variations in pressure. Like Miller (1972), Hill also believes that the availability of till may play a part, the drumlins occurring at greater density where more till is present.

3.3   *The drumlins of the Eden Valley and Solway area*

The variation of the elongation of the drumlins in this area has already been noted (see p. 407). The distribution of the drumlins is shown on Fig. 14.1 which shows how their orientation changes in different parts of the area. This gives an indication of their probable date of origin. Some of the drumlins occur in the area covered by the Scottish Readvance as defined by Hollingworth, but they occur mainly in the area glaciated by the Main Lake District ice sheet during the main Newer Drift advance. However, there is still controversy over the extent of the Scottish Readvance glaciation. L. F. Penny (1964) adheres to the limit suggested by Hollingworth (1931), but J. B. Sissons (1964) suggests that most of these drumlins were formed during the Lammermuir advance, which occurred after the maximum of the Newer Drift but before the Perth advance of Zone I. The ice at this stage, according to Sissons, extended well south of the Lake District and the ice front linked with that forming the Kells moraine in Ireland. This moraine separates the major drumlin area to the north in Ireland, from the esker area to the south. Whichever ice sheet formed the drumlins, it can be shown that most of the drumlins must have formed when the ice sheet was fairly extensive. The reason for this is that the Lake District ice, when the drumlins were formed, must have moved continuously from the Lake District and Howgill Fells north down the Eden Valley and then it must have come into contact with Scottish ice in the inner part of the Solway Firth. The combined ice masses must have been diverted south-west and south down the Irish Sea along the western flanks of the Lake District.

The pattern of the drumlins conforms exactly to the direction of basal ice movement. This can be inferred from other evidence, such as the distribution of erratics, which include distinctive marker rocks such as the Shap granite (Chapter 13). The ice must have been thick and moving actively during the deposition of the drumlins as their elongation shows that the ice was escaping over every available col during this stage. The drumlins around Appleby are almost circular because a basal ice-shed existed at this point. Ice was diverted northward down the Eden valley in one direction and eastward across the Stainmore pass in another. Around Carlisle, the elongation falls to very low values, and in this area also a basal ice-shed existed (Fig. 14.1). It separated ice of both Scottish and Lake District origin that was moving westward round the edge of the Lake District into the Irish Sea from that moving eastward across the Tyne gap to the low ground east of the Pennines.

The elongation of the drumlins is found to be greatest where the ice was moving fastest and was thickest. This occurred in the area to the north of the Lake District around Wigton and Aspatria. In this area, the drumlins trend east-north-east to west-south-west and their stoss end is to the east. The elongation also increases in the Tyne gap where the ice was being squeezed through the gap and was flowing fast. The greater velocity increased the pressure and the drumlin form became more elongated in response. The evidence of this area points to the fact that these drumlins must have been formed by actively moving ice and not in the sub-marginal zone in which some drumlins seem to occur.

When the Lake District ice parted from the Scottish ice during deglaciation, the ice front to the north of the Lake District lay in an east–west direction almost parallel to the drumlin elongation. Had the drumlins been formed close to the margin of the ice, which would presumably have been moving northward at this stage, then they should have had an elongation at right angles to their actual one. The drumlins that lie within the Readvance limit as defined by Hollingworth do not show an alignment radial to the ice front, as would be expected had they been formed by this later ice. This advance has left no terminal moraine so that it was probably too short-lived and the ice moving too slowly to form drumlins or seriously to modify the earlier ones. However, there are some drumlins 8 km east of Carlisle that must have been formed by later ice moving down the Tyne gap, as they rest on the Middle Sands and include some of this material in their till. The Middle Sands in this area also contain clay and it is possibly the presence of the clay that allowed drumlins to form in this area. According to Hollingworth, in some parts of the area covered by the Scottish Readvance in northern England and southern Scotland there are two distinct sets of drumlins with different orientations. Most of the drumlins, however, belong to the more vigorous ice sheets.

Most of the drumlins of the north of England are composed of till, a true boulder clay with striated boulders set in clay. Some of them contain sand and gravel in a central core, suggesting deposition of material before the shaping by moving ice. Other drumlins have a rock core, while a few consist mainly of rock. It is not possible to distinguish the material from the surface form of the drumlin.

Another point of interest concerning the distribution of drumlins in the Eden valley and neighbouring areas is their restriction to the lower levels. They are found in the valley and lowlands and rarely occur above 300 m. They are common, however, in cols, for example the Tyne gap, Stainmore, the Lune gorge near Tebay, the cols at the head of the Ure valley, Ribblehead, and the broad col linking the Lune, Ribble and Aire valleys near Settle. In areas where the ice comes down into a valley or crosses a col, its thickness or speed will be increased. Both these factors would increase the ice pressure on the bed, and, where suitable material is available, this could well be shaped into drumlin form. It is in these areas of relatively gentle slope that, in the early stages of glaciation, deposition would be most likely to occur.

3.4 Drumlinoid drift tails

A rather modified form of drumlin development is well seen in Wensleydale (Fig. 14.6). The ice moving down the valley from the west was of purely local origin and its speed of flow was partially checked by the great width of the valley and the presence

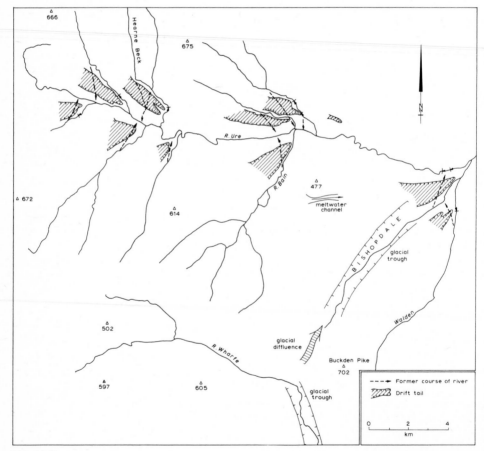

Fig. 14.6
Drumlinoid drift tails and other glacial features in Wensleydale, Yorkshire

of a large ice mass in the Vale of York at the outlet of the dale. Thus a considerable amount of till was deposited along the valley sides. In the valley floor, some well-developed true drumlins occur. They are elongated down-valley and are composed entirely of a tough boulder clay. The drumlins along the valley sides have become linked together and are joined on to the solid spurs of the tributary valleys. The spurs are thus elongated down-valley by these series of linked drumlins, which are often arranged *en échelon*, so that swampy hollows develop between the smooth drumlinoid mounds.

The development of these drumlinoid drift tails can be seen on nearly every spur in the valley. One effect has been to divert all the tributary streams eastward down the valley. The tributaries now flow eastward behind the drift tails in abnormally gentle thalwegs until they can find a low enough route across the drift tail. The stream then rejoins the main river often along an over-steepened course in which waterfalls are common. The best known of these waterfalls is Hardrow Scar, 30 m high. This has been caused by the diversion of Hearne Beck and Fossdale Beck by a long drift tail. Other examples include Whitfield and Mill Gill Forces farther east, both lying on the same stream as the stream finds its way round a double drift tail. The Bain at Bainbridge has been pushed out of its pre-glacial valley by a drift tail, which can be seen to fill the old valley where this is exposed in the flanks of its post-glacial gorge. Part of this gorge is cut through the solid rock (W. B. R. King, 1935).

4 Theories of drumlin formation: conclusions

In describing the material and distribution of various drumlins, some suggestions concerning the development of drumlins have already been made. It has been shown that drumlins vary considerably in most of their major characteristics, such as elongation, spacing, size and material, but they have one essential feature in common. This feature is their streamlined form, and it is on the basis of this feature that their genesis must be sought. It is not necessary to postulate that all drumlins have been formed in the same way. In fact their great variety points to variations in the formative processes. These depend not only on the nature of the ice forming the drumlins but also on the type and quantity of drift available, the relief and other factors. One point that nearly all those interested in the formation of drumlins agree upon is that they have been formed by moving ice, but here the uniformity of opinion ceases. One of the problems of explaining drumlins is that it is difficult to examine the base of an active and moving ice mass although it is in this situation that the drumlins are formed.

C. P. Gravenor (1953) has assembled and analysed some facts concerning drumlins. He states that drumlins on the whole tend to be sand-rich rather than clay-rich, and clay does not appear to be necessary for their formation, although they sometimes consist of a clay-till. They can also consist of sandy or loamy till, rock or pre-existing drift. They often contain lenses of stratified drift. Rock drumlins may occur with other drumlins but many glacial areas do not have any drumlins at all. They exist in fields that are often wider than moraine belts. They are streamlined, with the stoss end facing up-glacier. Some drumlins have cores, but most do not. Their long axes are parallel to the ice flow.

Theories of drumlin formation may be classified into two main groups: those that support an erosional origin for drumlins, and those that support a depositional origin. The erosional theory considers a two-stage formation and can account for stratified material in drumlins. The unstratified material often does not resemble kame deposits and often the drift of which drumlins are composed is the same age as that of the surrounding drift. The basis of the erosional theory is that pre-existing till or stratified glacial deposits are shaped into drumlins by the ice of a readvance. Some evidence for this has been put forward by S. Jewtuchowicz (1956). The drumlins he describes in Poland have a core of ground moraine or till, which has been covered by fluvioglacial deposits. These stratified deposits have suffered cryoturbation and glacial tectonic disturbance. The ice must first have advanced to deposit the till, then it retreated, allowing the fluvioglacial deposits to accumulate on the till. The retreat must have lasted long enough for frost action to produce the cryoturbation features. The ice must then have readvanced, producing the glacio-tectonic features in the deposits already in place and shaping them into the characteristic streamlined drumlin form by partial erosion of the deposits.

R. F. Flint (1971) points out that there is a complete range of forms from features of pure rock to those composed entirely of drift, all of which have the same drumlin shape. The solid rock features are clearly erosional, so the similarity of form suggests that the others may be as well. It is worth noting, however, that frequently the solid rock features, in the form of roches moutonnées, show a reversal of the normal stoss and lee slope of the typical drift drumlin. The drumlin normally has the steeper slope on the proximal or stoss side, while the roche moutonnée often has a relatively gentle proximal or stoss side and a steeper, more angular and quarried distal or lee slope. This difference suggests that the processes forming drumlins are not identical to those shaping the solid rock. There is also a difference in location. Drumlins usually occur on the lower ground, while smoothed rock features are also characteristic of the steeper slopes and higher areas. Both are, however, essentially streamlined at least in part.

Some authorities consider that drumlins are formed submarginally. There are examples of drumlins associated with moraines. The suggestion that drumlins could form under stagnant conditions has been put forward by F. M. Synge (1952). He pointed out that the drumlins on the north of Ireland lie mainly north of the Kells moraine and they have been destroyed by the Antrim Readvance. He therefore suggests that drumlins may not be able to survive active glaciation. This view, however, does not agree with the streamlined form and orientation of the drumlins.

Gravenor considers that drumlins could form during phases of uneven advance. He suggests that material deposited at the front of a glacier during a halt could be shaped into drumlins when the ice advanced again. The morainic belts would be expected to be wider during ice advance than during retreat when eskers and such forms develop. The bedded material found in drumlins resembles more closely that found in ground moraine than that found in kames. Gravenor contends that this modified erosion theory accounts for many of the characteristics of drumlins.

The drumlins described by Kupsch (1955) in Saskatchewan, which have already been mentioned, show characteristics of both erosion and deposition. These particular drumlins, consisting of stratified sands and gravels on the steep, proximal slope, with a line of boulders in the centre and a tail of unstratified till, must also have formed

in two stages. The stratified sands and gravels and the boulders must have been deposited near the ice margin, under almost stagnant ice conditions, but the tail of till
indicates a renewed advance of ice. This advance must have been responsible for the
streamlining of the pre-existing drift as well as adding its own contribution on the lee
side of the drumlin. Both erosion and deposition must have taken place in the formation
of this particular drumlin. The active movement of the ice at the time of the shaping
of the drumlin is indicated by the fracturing of the boulders in the centre of the drumlin.
This fracturing must have required considerable pressure from thick, fast-moving
ice.

The long low tail is characteristic of the streamline form. It is designed to create
the minimum of disturbance in the flow of the moving medium, whether it be air,
water or ice. This form is more likely to result from deposition than from erosion
because it is this part of the drumlin where the pressure would be least, in the lee of
the main body of the feature. Material could therefore accumulate in this position.
On a similar argument, the lack of pressure in this position would minimize erosion.
It is significant that in roches moutonnées, the lee side is steep and rugged, and is eroded
by plucking rather than by grinding and smoothing. In so far as drumlins resemble
crag-and-tail features, it seems likely that their shape is partly depositional in character
at least, especially the long low tail.

The purely depositional theory has been advocated for drumlins formed entirely
of till. It has been suggested that they form around a nucleus of frozen till or rock.
A few drumlins show concentric layers, suggesting that they have been built up by
accretion of layer upon layer. The depositional view of drumlin formation is supported
by J. K. Charlesworth (1957) and Hollingworth (1931). The drumlins that are composed entirely of till show no evidence of two stages of development. The distribution
of drumlins on the lower ground suggests that they occur in areas that are predominantly depositional. In detail it has been suggested that the drumlins of northern England form in areas where the pressure of the ice is locally increased by thickening or
by increase of velocity. The extra pressure could provide the conditions necessary for
drumlin formation in ice in which there is sufficient basal till to form them. The orientation of the stones in the till of drumlins is a strong argument in favour of their deposition
by moving ice, which is also responsible for shaping them into drumlin form. Both
the drumlin elongation and the stone orientation have clearly been determined by the
same ice flow and, where this can be shown to be true, those particular drumlins can
confidently be stated to be depositional in character.

Some observers have suggested a different method of emplacement of the till in drumlins. Hoppe and V. Schytt (1953) have suggested that elongated hollows occur in the
ice as it moves past boulders on the bed and that in relatively thin ice and stagnant
ice these cavities would remain in front of the boulder. Water-soaked till would be
squeezed up into the hollow to form long linear drumlin features. This explanation
cannot, however, be of very wide application because it demands special conditions
and does not agree with many of the characteristics of drumlins.

I. J. Smalley and D. J. Unwin (1968) have put forward an interesting new theory
of drumlin formation, related to the dilatancy of the material forming them. The basic
conditions necessary for till-drumlin formation are said to be: (1) the material at the

base must have been being continuously deformed by shearing, and (2) the material must consist of boulders in a clay-water system, the large particles forming a dilatant system. Dilatancy is a property of granular masses. Smalley suggested in 1966 that dilatancy may be a possible cause of drumlin formation. They form in a layer of boulder clay separating a glacier from certain types of terrain. When flow in this layer is obstructed, part of the layer packs into an obstruction and the rest flows around and moulds it. When a granular mass is disturbed, expansion results, the dilatant materials being then more resistant to shear stress. It is this high resistance that causes drumlins. Fig. 14.7 shows that the load on the till first leads to deformation by expansion to a

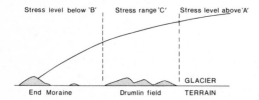

Stress level below 'B'   Stress range 'C'   Stress level above 'A'

GLACIER

End Moraine   Drumlin field   TERRAIN

Fig. 14.7
Cross-section of the edge of an ice sheet showing critical stress regions (I. J. Smalley and D. J. Unwin, *J. Glaciol.*, 1968, by permission of the International Glaciological Society)

maximum value at A, and further deformation leads to collapse, C–B. The load leading to deformation is reduced after this stage, i.e. at B. Experiments with pure sand show that this responds in a similar way to till, but the latter shows less regularity. The stress caused by glacier movement in the range C is critical to drumlin formation. If the stress is greater than A, no drumlin can form; values below B do not allow continuous deformation, and so drumlins cannot form under these conditions either. In general the stress is likely to be greater than A, but towards the edge of the ice, stresses drop and drumlins can form until the lower limit is reached, at which point end moraines will form instead.

The large drumlin fields in Ireland are attributed to fairly small local glaciers covering a large area, while elsewhere the belt in which the stress lies within the appropriate range is narrow so that a narrow drumlin belt results. The large particles in the till make the material dilatant. The clay may be thixotropic, which probably helps in the shaping process. The growing till accumulation causes a local rise of pressure at the base of the ice and the fine material is carried round more easily to form streamlined forms. Thixotropic materials are less resistant and therefore react in the opposite way to dilatant materials. The argument can be applied to erosional drumlins as well as depositional ones. Indeed, both processes can act simultaneously.

Drumlin distribution on this hypothesis ought to be random. A random placement model was made and the nearest-neighbour test for random spacing was used, $R = D/0.5(A/N)^{-1/2}$, where D is the linear distance, A is the area and N is the number of points (drumlins) in the area. R ranges between 0 and 2·1491 for hexagonal spacing. Random spacing gives a value of unity. The results for measured values in different areas are shown in Table 14.2. The R values for the random models were 1·1500, 1·0441, 1·2766 and 1·1367, which are not significantly different from the actual spacing.

In Wisconsin, the drumlin fields are in zones 44 to 80 km behind the ice margin, where the estimated ice thickness would range from 130 to 440 m. Drumlins tend to

Table 14.2   Analysis of drumlin spacing and density

Location	Area, km²	Density, N/km²	N	D, m	R
Co. Clare	24·08	1·33	32	563·9	1·3096
Co. Clare	24·08	1·53	37	554·2	1·3842
Co. Clare	56·16	1·62	91	497·9	1·2784
Eden Valley	40·04	0·92	37	583·7	1·1295

form in basins where the ice is moving slowly or dying. They are probably formed only beneath temperate glaciers, and they are unlikely to form if the ice is unable to slip on its bed. The actual till is the dilatant material and there must be shear deformation in the till at the glacier-terrain boundary. This may occur where ice is advancing over already deposited till.

Baranowski (1969) has attempted to relate drumlin formation to a change from temperate to cold ice at the base of a glacier. Under temperate ice conditions, water-soaked material could accumulate at the base, but as cold penetrated, the material would freeze and expansion could cause heaving. The spacing of the resulting 'hummocks' would depend on the hydrostatic pressure, i.e. on the ice thickness and the amount of water. Drumlins could form as the zone of cold ice extended down-glacier and their formation would continue until the movement of the thermal front became slower than the movement of the ice at its base.

Some of the observations on drumlins suggest that there is some periodicity in the ice flow, which induces zones of deposition. These cores can then be shaped by the moving ice into drumlins. This would apply particularly to the drumlin fields situated on open areas of low relief. The suggestion that variations in the ice flow itself are a factor in the pattern and occurrence of drumlins is worth close investigation. The approximately equal height of the drumlins in any one drumlin field probably reflects the thickness of the debris concentrated in the basal ice.

It would seem reasonable to conclude that drumlins can be formed either by the modification of previously existing drift sheets, the readvancing ice causing erosion and deposition in varying proportions, or by the direct deposition of till by active ice in areas where till is available and the pressure of the ice is sufficient to produce the stream-lining characteristic of drumlins. Dilatancy of the drumlin-forming material is an important pre-requisite. The absence of drumlins from most of the areas of Older Drift throughout the world may reflect the great time that has elapsed since this drift was deposited. During this long period, the till surface has been modified and any drumlins that were present have been lost by infilling or erosion. Drumlins remain an intriguing and enigmatic feature of many areas of glacial deposition.

5 References

ARANOW, S. (1959), 'Drumlins and related streamline features in the Warwick-Tokio area, North Dakota', *Am. J. Sci.* **257**, 191–203

BARANOWSKI, S. (1969), 'Some remarks on the origin of drumlins', *Geogr. Polonica* **17**, 197–208

BARNETT, H. F. and FINKE, P. G. 'Morphometry of landforms: drumlins', *U.S. Army Earth Sci. Lab. Tech. Rep.* ES–**63** (Natick Labs, Mass.), 41 pp.

CHARLESWORTH, J. K. (1957), *The Quaternary Era* (London, 2 vols)

CHORLEY, R. J. (1959), 'The shape of drumlins', *J. Glaciol.* **3**, 339–44

DONNER, J. J. and WEST, R. G. (1955), 'Ett drumlinsfält pä ön Skye, Skottland', *Eripainos Terresta* **2**, 45–8

DYSON, J. L. (1952), 'Ice-ridged moraines and their relation to glaciers', *Am. J. Sci.* **250**, 204–11

EVENSON, E. B. (1970), 'The relationship of macro- and micro-fabrics in till and the genesis of glacial landforms in Jefferson County, Wisconsin', Unpubl. M.S. thesis, Univ. of Wisconsin, Milwaukee, 91 pp.

FLINT, R. F. (1971), *Glacial and Quaternary geology* (New York)

GLÜCKERT, G. (1973), 'Two large drumlin fields in central Finland', *Fennia* **120**, 37 pp.

GRAVENOR, C. P. (1953), 'The origin of drumlins', *Am. J. Sci.* **251**, 674–81

GRAVENOR, C. P. and MENELEY, W. A. (1958), 'Glacial flutings in central and northern Alberta', *Am. J. Sci.* **256**, 715–28

HILL, A. R. (1970), 'The relationship of drumlins to the directions of ice movement in northern County Down', in *Irish geographical studies* (ed. STEPHENS, N. and GLASSCOCK, R. E.), 53–9

  (1971), 'The internal composition and structure of drumlins in northern Down and southern Antrim, Northern Ireland', *Geogr. Annlr* **53**A, 14–31

  (1973), 'The distribution of drumlins in County Down, Ireland', *Ann. Ass. Am. Geogr.* **63**, 226–40

HOLE, F. D. (1970), 'Drumlin morphology and soil relationship in Jefferson County, Wisconsin', in *Pleistocene Geology of South Wisconsin (Geol. Soc. Am. Ann. Meeting* (Milwaukee, 1970); Field trip guide: *Wisc. Univ. Geol. Nat. Hist. Surv., Inf. Circ.* **15**, F1–F9)

HOLLINGWORTH, S. E. (1931), 'The glaciation of western Edenside and adjoining areas, and the drumlins of Edenside and the Solway basin', *Q.J. geol. Soc. Lond.* **87**, 281–359

HOPPE, G. (1959), 'Glacial morphology and inland ice recession in northern Sweden', *Geogr. Annlr* **41**, 193–212

HOPPE, G. and SCHYTT, V. (1953), 'Some observations on fluted moraine surfaces', *Geogr. Annlr* **35**, 105–15

JEWTUCHOWICZ, S. (1956), 'Structure des drumlins aux environs de Zbojno', *Acta geogr. Univ. lodz.* **7**, 1–74

KING, W. B. R. (1935), 'The Upper Wensleydale river system', *Proc. Yorks. geol. Soc.* **23**, 10–24

KUPSCH, W. O. (1955), 'Drumlins with jointed boulders near Dollard, Saskatchewan', *Bull. geol. Soc. Am.* **66**, 327–38

  (1962), 'Ice-thrust ridges in western Canada', *J. Geol.* **70**, 582–94

LASCA, N. P. (1970), 'The drumlin field of south-east Wisconsin', Pt. E. E1–E13 in HOLE, F. D. (op. cit.)

LEBEAU, R. (1954), 'Forme mineure du relief sous-glaciaire', *Revue Géogr. Lyon* **29**, 227–56

LEMKE, R. W. (1958), 'Narrow linear drumlins near Velva, North Dakota', *Am. J. Sci.* **256**, 270–84

LUNDQVIST, J. (1970), 'Studies of drumlin tracts in central Sweden', *Acta geogr. Lodziensia* **24**, 317–26

MILLER, J. W. (1972), 'Variations in New York drumlins', *Ann. Ass. Am. Geogr.* **62**, 418–23

PENNY, L. F. (1964), 'A review of the Last Glaciation in Great Britain', *Proc. Yorks. geol. Soc.* **34**, 387–411

REED, B., GALVIN, C. J. and MILLER, J. P. (1962), 'Some aspects of drumlin geometry', *Am. J. Sci.* **260**, 200–10

SCHAEFR, I. VON (1969), 'Der Drumlin von Hörmating in Oberbäyern', *Eiszeitalter Gegenw.* **20**, 175–95

SISSONS, J. B. (1964), 'The glacial period' in *The British Isles, A Systematic Geography* (ed. WATSON, J. W. and SISSONS, J. B.)

SMALLEY, I. J. (1966), 'Drumlin formation: a rheological model', *Science* **151** (3716), 1379–80

SMALLEY, I. J. and UNWIN, D. J. (1968), 'The formation and shape of drumlins and their distribution and orientation in drumlin fields', *J. Glaciol.* **7**, 377–90

SYNGE, F. M. (1952), 'Retreat stages of the last ice-sheet in the British Isles', *Ir. Geogr.* **2**, 168–71

SYNGE, F. M. and STEPHENS, N. (1960), 'The Quaternary period in Ireland—an assessment, 1960', *Ir. Geogr.* **4**, 121–30

TRENHAILE, A. S. (1971), 'Drumlins: their distribution, orientation and morphology', *Can. Geogr.* **15**, 113–26

VERNON, P. (1966), 'Drumlins and Pleistocene ice flow over the Ards Peninsula–Strangford Lough area, County Down, Ireland', *J. Glaciol.* **6**, 401–9

WOLDSTEDT, P. (1954), *Das Eiszeitalter* (Stuttgart)

WRIGHT, H. E. (1957), 'Stone orientation in the Wadena drumlin field, Minnesota', *Geogr. Annlr* **39**, 19–31

15

Moraines

These fragments they [the glaciers] gradually transport to their utmost boundaries, where a for-midable wall ascertains the magnitude, and attests the force, of the great engine by which it was erected. (J. PLAYFAIR, 1802)

The term 'moraine' covers a wide range of depositional features associated with glaciers and ice sheets. Moraines normally consist of unstratified material, although at times stratified deposits laid down in front of ice sheets have been referred to as moraines. The well-known Salpausselka moraines in Finland, for example, consist mainly of water-lain deposits. This type of feature is best referred to as a 'delta-moraine' and in many respects resembles one type of kame. It will therefore be dealt with when fluvioglacial features are described. This chapter is concerned with features that have been deposited directly by the ice.

Moraines can be classified in several ways, according to their position, their state of activity, and their method of formation. In the first classification, moraines are divided into terminal, lateral and medial moraines according to their position relative to the glacier. Terminal and lateral moraines are of great interest to the geomorpho-logist, as they indicate halt or readvance positions, at which the ice margin lay for a considerable time during deglaciation. In the second classification, moraines may be termed 'active' if they are in contact with active ice. Ice-cored moraines constitute a special group in this classification, usually occurring in a marginal position along the side or front of a glacier or ice sheet. Another group comprises inactive moraines, which have lost all contact with the ice forming them as a result of glacier retreat or decay. It is these moraines that provide evidence for stages of glacier retreat, glacial chronology and correlation.

The third classification subdivides moraines according to their method of formation. Two groups may be differentiated: ablation moraine and ground moraine. Ablation moraine is material that has accumulated on the surface of the glacier by gradual down-wasting of the surface ice and can be distinguished by its relatively coarse nature, because the fine material has usually been washed out by meltwater. Ground moraine is deposited subglacially and can accumulate in several different ways. The second method of classification will be adopted for a primary description of morainic types, although the other classifications will be used for secondary subdivision.

The problems of moraine nomenclature have been discussed by V. K. Prest (1968) in connection with the glacial map of Canada. On this, moraines are grouped into three: first, those lying transverse to the direction of ice flow, secondly, those parallel to the ice flow, and thirdly, non-orientated moraines. The first group includes seven different forms:

1 *Corrugated ground moraine* that is often made conspicuous on air photographs by ponds. The corrugations are 0·3 to 3 m high and 0·1 to 1·5 km long; their transverse spacing is 70 to 270 m. They consist mainly of till and sometimes form the longitudinal elements in a reticulate pattern.

2 *End moraines* that delimit former frontal positions and vary from 3 to 180 m in height.

3 *Ice-thrust moraine.* This type can consist of ice-thrust bedrock structures, often mantled by hummocky drift. Stratified sediments are often incorporated.

4 *Ribbed moraine ridges* of bouldery till. These are about 10 to 30 m high with crests 100 to 300 m apart. The ridges are steep and can be 3 to 5 km long. They are common in Newfoundland. This type may have a fluted or drumlinized surface and may form by different processes.

5 *de Geer moraines*, sometimes called 'winter moraines'. The ridges are narrow and occur in former lake or sea-covered areas. Cross-valley moraines are included in this category; they are more irregular and more closely spaced, owing to the presence of valley walls.

6 *Inter-lobate kame moraines.* This is a type of end moraine in which there is a predominance of fluvioglacial deposits rather than till.

7 *Linear, ice-block ridges.* This type of moraine is probably formed under stagnant ice, where material slumps into a crevasse system. The ridges are less than 300 m long and 7 m high. Their exact method of formation is in doubt.

The second main group of moraines, those lying parallel to the ice flow, may occur both in overridden older till or newly deposited till. They may consist of drumlinoid ridges 3 to 30 m high, with narrow fluting and shallow furrowing. The features are closely spaced with a local relief of 1·8 to 10 m, and are most commonly formed of till. Crag-and-tail features occur with thin drift and knobbly bedrock. A second type consists of marginal and medial moraine in glaciated mountain areas, a third consists of interlobate and kame moraines, and a fourth type comprises linear ice-block ridge forms, possibly originating by ice pressing material into cracks parallel to the flow.

The group of non-orientated moraines is divided into three. Overall, this group consists of hummocky moraine with pitted surfaces in places. Some forms have high relief and others low. Prest distinguishes

1 ground moraine
2 disintegration, dead-ice or stagnation moraine
3 interlobate and kame moraines.

The first type is commonly deposited from basal ice that became inactive during recession, through ablation and disintegration. This type may include prairie mounds and plains plateaux with raised rims, consisting of till and about 100 m in diameter, 3 to 4·5 m high. Some of these features probably developed as frozen ground phenomena after the glaciation but various other formative processes have been considered. The second type consists of hummocky moraine with a local relief up to 60 m. It may contain relatively few pits and may not always be hummocky, consisting instead of flat-topped moraine plateaux.

1 Active moraines

1.1 *Medial moraines*

The medial moraine is the most conspicuous form of active moraine in valley glaciers, but it is least likely to form a permanent, stable feature. Medial moraines can be seen prolonging a spur between two confluent valley glaciers in a down-glacier direction (Fig. 3.9). Such moraines are sometimes figured as continuing down through the ice to the glacier bed. In many instances this does not, in fact, occur. The medial moraine is usually an entirely superficial feature, consisting often of only a metre or so of coarse stony debris lying on the ice. It consists of angular and unsorted material and usually stands out as a conspicuous elevated ridge on the glacier surface. In reality the ridge is mainly formed of ice. In this respect the medial moraines of many temperate mountain glaciers resemble dirt cones in having a core of ice. They stand up as ridges because the layer of debris, once it exceeds a critical thickness, protects the ice beneath from melting. The thickness required to prevent or slow down melting is only of the order of 0·5 m.

If the spur from which the debris is derived lies above the firn-line, then the moraine may not appear on the ice surface until lower down the glacier. In other instances, the moraine is hidden under avalanche debris until this melts off as the glacier moves down-valley. Some medial moraines show a beaded structure, for example on Austerdalsbreen in Norway (Plate V). The beading appears to have an annual spacing and can be accounted for by the more active production of waste during the summer season. More of the rock face is exposed at this time to frost and other destructive processes such as avalanches.

Differential ablation on a medial moraine has been studied by S. R. Loomis (1970). The observations were made on the Kaskawulsh Glacier, Yukon, where a medial moraine has formed at the confluence of two tributaries. The medial moraine is nearly 1 km wide near the junction, narrowing to 60 m at 0·75 km down-glacier. The debris is of supraglacial origin, the height of the moraine increasing from 2 m at the junction to 20 m, 2 km down-glacier; and it then decreases to 7 m at 5 km, partly because of extending flow farther down-glacier. The till thickness increases in inverse proportion to the ablation rate. Debris-free ice ablated at 5·4 cm/day, but the rate was retarded with more than 1 cm of debris. With 22 cm the ablation rate was only 0·6 cm/day. Maximum slopes of the debris-covered ice were 39° to 43°. The moraine narrows and becomes higher down-glacier, forming a prominent ice-cored ridge, the size becoming constant where the tendency to increase in height is balanced by extending flow, the moraine then becoming stable with mossy growth on its rocks. Slope, orientation and relief affect the ablation rate, the relief being lowest under the thinnest cover of debris.

Some medial moraines are very straight and orderly, such as those of the Barnard glacier in Alaska. Others, such as those of the Alaska Range glacier, show considerable contortions in the morainic pattern of the glacier surface. The contorted pattern is evidence of the action of surges (Chapter 2) which do not affect all the tributary glaciers that join to form the main glacier at the same time. Individual ice streams thrust forward into the main valley at different times, causing the irregular bulges indicated by the medial moraine pattern (Fig. 2.7). These moraines, therefore, give useful information concerning the character of the glacier flow.

The effect of glacier surges on medial moraines is demonstrated by A. Post (1972) on the large Bering Piedmont Glacier in Alaska, using photographs taken in 1957, 1963 and 1967. During two surges the ice was displaced up to 13 km. The displacement of medial moraines during surges is lateral when the main glacier surges past non-surging tributary glaciers. These irregularities are converted into 'accordion'-type folds by compressive flow and lateral, transverse expansion of previous minor irregularities. A large debris band composed of repeatedly folded medial moraine extends across the centre of the Bering glacier lobe. The zone of intensive shear at the glacier margin plays an important part in creating this pattern (see also Fig. 2.7).

N. W. Rutter (1969) has compared the moraines formed by surging glaciers with those formed by normal glaciers. The surge moraine is thin, discontinuous and irregular, mostly till, but with some stratified ice-contact material. Subdued lateral ridges occur. The non-surge moraine is more continuous and formed of thicker till, with prominent ridges marking the upper ice limits. Fabrics are weak in surge moraines, but oriented in the flow direction in normal moraines. The surge moraine has a coarser fraction in the size range less than 2 mm, compared with the normal moraine.

1.2 *Lateral moraines*

Lateral moraines of present glaciers can vary greatly in character. Some consist of only a thin covering of surface debris resting on the ice and derived from the rock walls of the valley. The lateral moraine can take another form when the glacier is retreating or thinning. Because of the general climatic amelioration of the last few decades, many glaciers have been thinning, resulting in the formation of the trim-lines that are characteristic of many glaciers at present. The trim-line separates an area of bare moraine below from vegetated moraine above. It is often noticeable that the vertical distance between the trim-line on the lateral moraine and the ice surface gradually increases towards the glacier snout. This fact can be explained in some instances by the passage of a kinematic wave down the glacier. The wave frequently reaches its greatest dimensions at the snout of the glacier, owing to the deceleration of flow causing compression in the down-glacier direction. If this factor is greater than the diffusion of the wave, then the change in ice elevation at the snout will be greater than farther up-glacier.

In some areas of temperate valley glaciers, such as south Norway, parts of the Alps and the Southern Alps of New Zealand, lateral moraines form well defined ridges. They are separated from the valley wall by an ablation valley. Good examples of lateral

Plate XXIII
Lateral moraine, Glacier de la Grande Casse, France. *(C.A.M.K.)*

Plate XXIV
Moraines below the terminus of the Cameron Glacier, New Zealand. *(C.A.M.K.)*

moraines of this type occur on the flanks of the Tasman glacier and Cameron glacier (Plate XXIV) on the eastern side of the New Zealand Alps. These glaciers, draining an area of easily frost-shattered greywacke rocks, have a thick cover of surface and lateral moraine. They are now down-wasting slowly beneath this cover. In order to form the lateral moraine ridge, it must be assumed that at one time the glacier had a higher and more convex cross profile than it has at present. Debris must have accumulated between the valley wall and the higher ice surface. Then as the ice slowly melted down in the valley to form the generally flat or concave surface that it has at present, the valley side of the moraine has maintained its slope away from the ice. The steeper side, facing the glacier, has developed as the ice surface has gradually been lowered. The valley-side slope often lies at the angle of rest of the material, but the glacier side is often much steeper approaching the vertical in places. The over-steepened slope often develops a corrugated form, and it rarely contains an ice core. W. B. Whalley (1974) suggests that quartz cementation could account for the cohesive strength shown by the unconsolidated moraine slopes. The process can operate rapidly under suitable conditions.

At times, small ridges of lateral moraine can form along the flanks of retreating glaciers. A small example was observed to develop in one year along the lower edge of Austerdalsbreen in Norway. It is thought that, during the winter, the ice thickened and debris moved off the ice to lie against the ice in a pile. Then as the surface was lowered by melting during the summer, the ice-side of the pile collapsed to form a small ridge nearly 1 m high. This type of small ridge had no ice core and was symmetrical in form, each flank lying at the angle of rest.

1.3   *Terminal moraines*

Few valley glaciers have well-defined terminal moraines in contact with the ice at present. This is because most glaciers are retreating and terminal moraines can only form effectively when the glacier is advancing or stationary in position and when the ice is active. An advancing active glacier often has a fairly steep, high front, against which a bank of material can accumulate. When the ice thins or retreats a ridge of terminal moraine will form in the same way as the lateral moraine already mentioned. Terminal moraines sometimes dam up water to form marginal lakes particularly when the ice retreat is rapid. Examples of lakes of this type are seen in front of Skeiðarárjökull in Iceland and the Steingletscher and Feegletscher in the Alps. When the glacier is retreating rapidly, after a period of still-stand and moraine formation, the material released from the melting ice is not sufficient to fill up the area between the glacier snout and the terminal moraine. Lakes accumulate temporarily in this position until their outlets are cut down through the moraine and the lake is drained.

The morphological effects of surges on terminal moraines are described by P. G. Johnson (1972). The Donjek Glacier is the largest in the St Elias Mountains of Yukon and is 55 km long. The glacier snout was reactivated by a surge, creating a steep unstable front, although its position did not change markedly, the forward movement being dissipated by the lobate form of the snout. The old stagnant front was pushed forward by the active ice, forming push structures with ice cores. Previous surges have caused the Donjek River to shift repeatedly, leaving dry gaps in the proglacial area.

Plate XXV
The 1890 moraine, Breiðamerkurjökull proglacial area, Iceland (see Fig. 15.6). The ice margin now lies several kilometres to the right. *(C.E.)*

1.4   *Ablation moraine*

Ablation moraine is often a conspicuous form of active moraine on the surface of a valley glacier. This surface debris usually increases in amount towards the snout of the glacier, where it is concentrated both by the deceleration of the ice and by the increased ablation near the snout. Where glacier caves are available, however, it can usually be seen that this layer of ablation moraine is only a surface layer and the ice beneath is normally very clean.

A layer of ablation moraine is most likely to be present on a glacier that is moving slowly and which is slowly down-wasting while its snout remains more or less static. In this respect, the contrast between the glaciers draining to the west of the Southern Alps of New Zealand and those draining east is of interest. The Tasman, which flows eastward and receives a considerably smaller accumulation as it lies in the lee of the mountains, moves more slowly and its snout is at 700 m. In an area of easily broken rocks, it receives a large supraglacial load. This has become concentrated by compression of the slowly moving snout into a thick layer of ablation moraine, covering the whole of the lower 8 to 10 km of the glacier. Slow down-wasting of the ice by about 60 m has also helped to increase the thickness of the ablation moraine. This active ablation moraine merges almost imperceptibly into ice-cored moraine and this again merges into the proglacial moraine and outwash. The precise location of the glacier snout is, therefore, difficult to determine.

On the wetter western side of the Southern Alps, the Franz Josef glacier is very different. This glacier flowed actively to within 210 m of sea-level in 1956 and ended in a clean, steep ice front, despite a rapid retreat which brought its snout to 300 m above sea level in 1961. The cleanness of the ice is partly the result of the large precipitation which falls as rain in summer when the abundant rain and meltwater can evacuate much of the glacier load. The ice flows fast all the way to the snout and this decreases the accumulation of ablation moraine. The snout of the glacier responds much more readily to changes in accumulation and ablation. This glacier has no deposits comparable with the ablation moraine of the eastern glaciers, such as the Tasman.

G. S. Boulton (1967) has described a complex supraglacial moraine sequence at the margin of Sorbreen in west Spitsbergen. The glacier has a bi-lobate front of largely stagnant ice, with a series of supraglacial ice-cored ridges. The sediment consists in large part of sand and gravel, much being bedded, and the ridges superficially resemble kames and eskers. Most of these features, however, appear to have formed subaerially, the material coming from the englacial debris as the ice melted. Bands of englacial debris up to 1 m thick dip up-glacier at more than 70° on the north lobe, but have lower dips on the south lobe. The debris was probably brought to the surface along shear planes, or by being frozen to the base of the ice and moving up where the flow lines ascend near the margin. Deposition of flow till (Boulton, 1970) is also taking place on the flatter areas. Other features include small ridges with ice cores on the flanks of which slumping produces complex structures, arch-bedded ridges forming arcs parallel to the glacier margin, and terminal ridges at the actual margin where the terminal face is steep. The features resemble controlled disintegration features (see Chapter 17). The abundant debris is partly the result of easily weathered material, and in this characteristic the glacier resembles some of the Pleistocene ice sheets. The complex interleaving of till with fluvioglacial material is instructive. Pleistocene sequences of this sort have often been interpreted in terms of successive advances and retreats of an ice margin, on no more evidence than such lithological alternations (see also p. 492). Here, Boulton shows how such sequences can be explained in terms of a single episode of glacier retreat.

1.5 *Ice-cored moraines*

In areas marginal to existing ice masses, some interesting ice-cored moraines have been studied recently, particularly by G. Østrem (1963, 1964). Most of his observations have been carried out in the Jotunheim district of Norway, Kebnekajse (north Sweden), and Baffin Island (arctic Canada). Experiments were carried out to determine the effect of different thicknesses of dirt on the ablation of the ice. Ablation of clean ice at the time of the experiments was taking place at a rate of 4·5 cm/day. The ablation rate decreased when the dirt cover exceeded 0·5 cm, falling to 3 cm/day with a dirt cover of 6 cm and to less than 1 cm/day when the dirt cover exceeded 20 cm. The studies of the effects of varying thicknesses of artificial dirt cover were extended to an examination of the distribution of ice-cored moraines. The study included the nature and age of the ice core and interesting results concerning the length of time the ice remains buried have emerged.

Geophysical techniques were used to investigate the presence and thickness of ice cores. When some ice-cored moraines have been identified in the field, it is possible to continue the study on aerial photographs. Air photographs have been used extensively to locate ice-cored moraines in various parts of Scandinavia. The ice-cored moraines can be identified on the aerial photographs by comparing the size of the moraine and the size of the glacier forming it. When the moraine is large compared with the glacier, it can safely be assumed to contain an ice core and the reverse is also true. This leaves a doubtful middle category that may or may not have an ice core. In all moraines in which ice exists, a snow bank is found on the distal side of the moraine. Many of the moraines consist of a series of small curved ridges of rounded outline. Ice-cored moraines do not normally have a sharp crested ridge.

In the past, it has normally been assumed that the buried ice in the ice-cored moraine was derived from the glacier. From the crystallographic structure of the ice in the core, Østrem showed that it was possible to determine whether the ice had originated as glacier ice or as snow. He found that although some of the buried ice was of glacial origin, in many instances the ice had originated as a snow bank, which had been piled up against the glacier margin.

The results of the study of the distribution of ice-cored moraines show that they are unevenly spread through Scandinavia. They are concentrated in the Jotunheim district, none occurring in the Jostedalsbreen area except for a few on the east. A few occur in Dovre, but the majority are found farther north, beyond the Arctic circle, around Sarek and Kebnekajse. In the extreme north, they are found fairly close to sea level. The type of bedrock does not appear to affect the distribution of ice-cored moraines.

It is striking that many of the smaller glaciers have ice-cored moraines, while the large ones often do not. This may be accounted for by the relatively small changes in the position of the glacier snout in a small cirque glacier compared with a longer valley glacier. Once a large moraine has formed in front of a small glacier, it will help to slow down its forward movement, thus helping to maintain the relatively stationary position of the ice front. Glaciers with a short, broad tongue will also tend to remain fairly static, and would be expected to have an ice-cored moraine.

Studies of the ice of the core that have been carried out have shown that both glacier ice and snow-bank ice are incorporated in some of the ice cores. On the whole in Scandinavia the snow-bank ice cores are the more common and the presence of snow banks around the outer end of the moraines is also common. Conditions that favour the formation of ice-cored moraines are those where accumulation and ablation are small, that is in areas where the climate tends to be more continental.

A close connection is found between the altitude of the glaciers and the occurrence of ice-cored moraines. The glacial limit in Norway and Sweden gradually rises eastward from 1200 m on the coast near the Lofoten Islands to 2200 m in the Jotunheim area. Two main factors account for the relationship between the glacial limit and the ice-cored moraines. First, the high-lying glaciers in the eastern part of the area are associated with a more continental climate. The effect of this on ice-core formation has already been mentioned. The glaciers are at a high altitude and, therefore, ablation will be slow and the ice front will be static. This will allow time for the formation of ice-cored moraines. Secondly, the winter temperatures will be very low in the areas of the highest glaciers. This factor and the thin snow cover will allow great cooling of the exposed moraine ridge. Only a thin layer will thaw in summer as a result and once the ice core has formed, it will be preserved in the moraine.

Snow-cored moraines are formed of material carried to the end of the glacier and dumped on the frontal snow patch either directly or by slumping. It then becomes mixed with snow, allowing the supersaturation that is sometimes noted in the material of these moraines. When the dirt cover is thick enough, it will protect the snow bank beneath. During this accumulation phase the ice front must remain stationary.

The analysis of the ice from ice-cored moraines can give valuable evidence of the age of the moraine. In order to obtain sufficient organic matter from the snow-bank, it is necessary to extract a large sample. Østrem (1961) has shown that 200 kg of ice may be required to produce sufficient organic matter for radiocarbon dating. The organic particles had blown on to the snow before it was covered by debris and, there-fore, the snow must be older than the organic material. It is thus possible to obtain a maximum age for the moraine, as the morainic material must have covered the snow which included the debris. When the organic material had been separated from the ice, it contained 3 g of carbon, half of which was used for dating purposes. One sample of dated ice was obtained from the outermost moraine ridge of Grausbreen in Jotun-heim, Norway. The result of the radiocarbon analysis gave a date of 2600 ± 100 years BP. The result shows that the moraine is very much older than was formerly thought. K. Faegri (1948) had suggested that these moraines belonged to the great glacier advance in the middle of the eighteenth century in Norway. However, the date given by the carbon-14 analysis agrees with the climatic deterioration that set in at the time of the cool, wet Sub-Atlantic phase. Further observations will show if this was a period of general glacier advance in this part of Scandinavia. This evidence does not neces-sarily negate Faegri's dates for non-ice-cored moraines, for which there are historical data. This will be mentioned later.

The ice-cored moraines of Scandinavia have been compared with those of Alberta and southern British Columbia by Østrem and K. Arnold (1971). The proportion of ice-cored to non-ice-cored moraines in both areas increases inland, and is greatest where the glacier limit is highest, that is, in the most continental type of climate. Ice-cored moraines are large relative to the size of the glacier and have rounded outlines, sugges-tive of slumping. There are no fracture zones and no fresh scree is visible, and they appear to be stable. Fresh moraines with angular outlines are believed to be ice free, although some may contain glacier ice. The ice-cored moraines often have marginal snow banks that survive the summer.

A problem that has been the centre of some recent controversy concerns the distinc-tion between ice-cored moraines and rock glaciers. A fuller account of rock glaciers is given in Embleton and King (1975, Chapter 4); here, it may be briefly noted that the ice in active rock glaciers may include glacier ice, snowbank ice and interstitial ice formed by the refreezing of melting snow. When the ice content in the rock debris is sufficiently high, the lower layers become plastic and the rock glacier flows slowly, rounded ridges and furrows characteristically forming on its surface due to movement. Rock glaciers are akin to small glaciers in that both end near the snow line where permafrost and interstitial ice can form. The arcuate ridges on rock glaciers superficially resemble moraines, but their origins are quite different. Ideally, it should be possible to base a distinction on movement: many rock glaciers have been shown to be in motion, whereas ice-cored moraines are normally fixed in position. However, some rock glaciers have lost their interstitial ice through climatic amelioration and have become inactive, while on the other hand, some ice-cored moraines exhibit slow mass movement. The distinction between the two types of feature is therefore often blurred in practice, and there are obvious difficulties if identification of features is based on air photographs rather than detailed field study. D. Barsch (1971) has claimed that some of the ice-cored moraines of Østrem and Arnold (1971) are in reality rock glaciers, but in defending his interpretation of the features as ice-cored moraines, Østrem (1971) argues that movement could not be established, and that no melting occurred below the summer melt level. Rock glaciers are said to be rare at present in Scandinavia, usually existing on steeper gradients where movement is more likely.

P. G. Johnson (1971) has described the formation of ice-cored moraines in front of the Donjek glacier in the Yukon. As noted already, this glacier experiences repeated surges. Its neo-glacial moraines, formed 300 years ago, are ice-cored. Those forming at present have cores of glacier ice and not of snow-bank ice. They form by debris from ablation and meltwater accumulating over a stagnant wedge of ice, which the glacier pushes forward during a surge. The wedge is structurally distinct from the active ice. A smaller type is formed by the shearing of stagnant ice from beneath a cover of outwash or moraine across which the surging glacier was advancing. These ridges may not be permanent features. Some fluvioglacial and glaciolacustrine deposits have ice cores buried under 6 m of deposits. Several processes lead to the degradation of ice-cored moraines. The most common is the melting-out of ice below a simple ridge to produce kettle topography. Where the core remains covered, kettles, cracking, slips and mud-flows occur, the last requiring saturation of till with water or a favourable aspect. Degradation is fastest once the ice core is exposed; in one instance, it then melted at 60 cm/week with accompanying slumping and mud flow.

The ice-cored moraines of south-west Ellesmere Island have been studied by R. A. Souchez (1971). The debris in the ice-cored moraines can be traced to a subglacial source because of the distinctive rock outcrops. The ice cap ends in a ramp, sloping at 20 to 35°, although locally there are cliffs 10 m high. The ramps lead up to ice-cored moraine ridges, the debris in which has come from within 3 km of the ice margin. Stone counts of different rock types indicate that the inner ice-cored moraines receive material from nearer the ice margin than those in the outer zone, and the stone counts are not inconsistent with the basal freezing hypothesis. The existing surface debris is calculated to have accumulated by ablation over a period between 65 and 200 years.

2 Inactive or stable moraines

Moraines no longer in contact with the ice that deposited them adopt a variety of forms, suggesting that they have developed by different mechanisms. They can be broadly divided, first, into those formed marginally to the ice as lateral and terminal moraines, and second, into those laid down over a wider area as ground moraine or ablation moraine. The first category itself includes a range of moraine types. It is important to differentiate between moraines deposited by temperate glaciers and those deposited by cold, polar glaciers. Most of the moraines that have been considered so far belong to temperate valley glaciers. There is also a significant difference between moraines laid down in water (but not *by* water) ponded in front of the ice, and those deposited subaerially. There is, too, a difference between moraines deposited by ice sheets and those formed by valley and cirque glaciers. Examples of some of these different types will be mentioned. These types include the terminal moraines of valley glaciers in temperate areas, the type of cold ice-sheet moraine that has been designated the Thule-Baffin type, and the cross-valley moraines that have been formed in water.

2.1   *Lateral and terminal moraines*

Lateral moraines lacking ice cores are usually much less conspicuous than those already discussed adjacent to active glaciers. In mountain areas this type of moraine sometimes only forms an amorphous deposit of debris on the valley side. The lateral moraine usually becomes more distinctive when it leaves the valley side and merges with the terminal moraine to swing across the valley as an arcuate ridge. Examples in the Conway valley of North Wales are given by C. Embleton (1961).

R. W. Galloway (1956) has studied the fabric of lateral and terminal moraines in an area of north Norway and has shown that the preferred orientation of the elongated stones lies at right angles to the lateral moraine ridge crest. This suggests strong radial movements when the moraine was formed. In the terminal moraine the stones lie mainly parallel to the moraine crest. These stones were probably emplaced by rolling and thrusting.

Terminal moraines are also considerably less imposing when they no longer contain an ice core. However, moraines without an ice core can attain considerable heights under suitable conditions. The terminal moraine of the Franz Josef glacier in New Zealand reaches 430 m in height on the narrow coastal plain beyond the mountains. This is an active glacier. Other high terminal moraines include the arcuate hills that mark the limit of the extension of the Alpine glaciers on the north Italian plain. The ice in this area was also active and built up large moraines from the abundant debris carried out of the mountains. Similar high terminal moraines can be built by ice sheets under favourable circumstances. In the east of Denmark, for example, the ice sheet of the Weichsel maximum remained stationary for long enough to build up morainic hills reaching an altitude of 173 m. Exceptionally large lateral moraines 700–900 m high are said to occur in the Bassin de Laragne, southern French Alps (J. Gabert-Delay, 1967).

Moraines are not always high or continuous. This may result from several possible causes. On the east of the New Zealand Alps, for example, some of the terminal moraines have been considerably eroded by subsequent fluvioglacial meltwater activity. The lower part of the moraine has also been buried in some areas under the rising mass of fluvioglacial outwash. The absence of terminal moraines does not necessarily mean that they were never laid down. This absence of moraines in some valleys may make the correlation of moraines in neighbouring valleys rather difficult if the number of moraines is used.

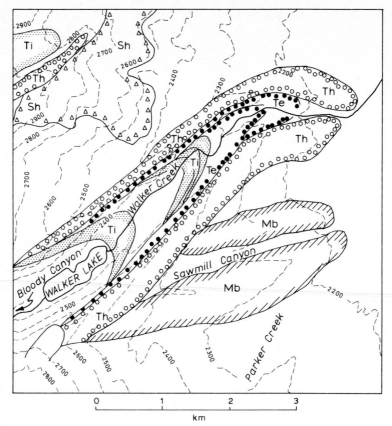

Fig. 15.1
Moraines of Walker Creek (Bloody Canyon) in the Sierra Nevada, California (R. P. Sharp and J. H. Birman, *Bull. geol. Soc. Am.*, 1963)
Ti—Tioga;   Te—Tenaya;   Th—Tahoe;   Mb—Mono Basin

A series of well-defined lateral and terminal moraine loops in the Walker Creek area of California (in the Sierra Nevada) illustrates stages of glacial advance and moraine formation. These moraines are quite substantial features and their distribution is shown on Fig. 15.1. The moraines have been differentiated by studying their position and character. The earliest of the moraine loops was formed during the Mono Glaciation, probably of Illinoian age. It now consists of subdued ridges of lateral moraine. The stream changed course before the deposition of the next set of moraines, the Tahoe series, which probably belong to an early Wisconsin glaciation. These are lateral moraines up to nearly 100 m high and in places they bury the older moraines. The older moraines were preserved by the change in stream course. This change suggests a considerable time interval between the two moraines. There are two younger moraines (late Wisconsin), belonging to the Tenaya and Tioga glaciations, and lying within the arc of the large Tahoe moraines (Plate XXVI).

These moraines have been differentiated by interesting semi-quantitative methods (R. P. Sharp and J. H. Birman, 1963). The criteria used were:

1 The total number of surface boulders.
2 The ratio of granodiorite to hard metamorphic boulders (the former are much more susceptible to weathering than the latter).
3 The ratio of weathered to unweathered rocks on the moraine, unweathered being defined as a boulder with a recognizable portion of abraded surface.
4 The proportions of material of different sizes.

In the last case, samples are taken from holes dug in the ground. One gallon of material is taken from the upper 15 cm and divided into three size categories, over 5·6 mm, 5·6–0·8 mm, and less than 0·8 mm. A second sample is taken from the next lower 15 cm of material and is similarly sieved in the field. The results shown in Table 15·1 indicate the striking contrasts between the different moraines.

Table 15.1 Semi-quantitative investigations of moraines of different ages in California

	Mono (oldest glaciation)	Tahoe	Tenaya	Tioga (youngest glaciation)
No of boulders over 0·3 m diam in 6×30 m strip	60	115	180	300
Granite/Resistate ratio	17/83	29/71	50/50	no count
Weathered to unweathered ratio	95/5	80/20	50/50	10/90
Hole-in-ground data:				
1st Interval—				
Coarse	24	40	67	79
Medium	26	24	7	4
Fine	50	36	26	17
2nd interval—				
Coarse	32	40	53	65
Medium	24	23	13	12
Fine	44	37	34	23

Sharp (1969) has further developed his semi-quantitative methods of differentiating moraines in an area near Convict Lake, Sierra Nevada. The weathering of granite boulders is one of the more sensitive tests, and the results suggested that a lateral moraine 300 m high was mainly a Tahoe feature, while a series of lobate moraines belongs to the Tioga glaciation. Both are late Pleistocene and no older deposits have been recognized, contrary to previous opinion.

A quantitative analysis of Pinedale (late Wisconsin) morainic landforms in the Beartooth Mountains of Montana and Wyoming has been carried out by W. L. Graf (1971). A moraine index was developed, $M = S.N/C$, where S is a shape index, the length from the major semi-axis divided by the width from the minor semi-axis, N is an orientation index, the angle of intersection of the major semi-axis and the general valley trend, and C is the cross-valley distance from the centre of the outer margin in metres. End moraines had large M values (up to 5), and lateral moraine low values. Boulder frequency counts and a weathering index were also used, and slope angles and soil development were recorded. Two Bull Lake (early Wisconsin), four Pinedale (dated at 23,000, 14,000, 11,500 and 8,500 years), and two neoglacial (2800 and 400

Plate XXVI
Tioga Moraine (last glaciation), Walker Creek Valley, Sierra Nevada, California. The recent nature of this moraine is reflected in the coarseness and lack of weathering of the surface material. *(C.E.)*

years BP) moraine groups were recognized. Regression analyses were carried out between related variables with the following results:

Log Y = 1·7190 − 0·721 X (Y = mean boulder frequency, X = age in thousands of years, $r = 0·9017$)

Log Y = − 3·3626 + 0·1849 X (Y = weathering index × 100, X = age in thousands of years, $r = 0·3488$)

Log Y = 2·1185 − 1·0984 X (Y = slope expressed as a percentage, X = age in thousands of years, $r = 0·9360$)

Terminal moraines formed directly of glacial deposits are usually found in front of stationary or advancing glaciers. The material forming these moraines is derived from the valley up-glacier and consists partly of material pushed forward by the glacier in sliding over its bed. Material is also added by the sliding of superficial debris down the steep end-slope of the glacier, which is the characteristic form during advance, and by the movement of debris up shear planes near the snout of the glacier. None of these processes operates so effectively when a glacier is retreating. Thus terminal moraines determine stages of advance during general deglaciation.

The character of the moraine depends on the material available to build it. In general, its slopes will be fairly symmetrical, both sides tending to lie at the angle of rest of the material. The coarser the material, the larger the moraine will tend to be, as meltwater cannot transport the larger blocks. A moraine of very large blocks, about the size of a small house, was formed in the first decade of this century during a short advance of Austerdalsbreen in Norway (Fig. 15.7).

The climate also affects the size of the terminal moraines as it determines to a considerable extent the amount of frost activity around the glacier. If the rocks are suitable, this may be a major source of debris. More important, however, the climate determines the activity of the glacier in large part. In a very cold dry climate, the glaciers are less active than in a wetter, warmer one. On the other hand, the abundant meltwater in the latter type may remove much of the glacial debris. Climatic factors can therefore have opposite effects, in this instance. The active glacier promotes moraine formation but the abundant meltwater either washes away the moraine or buries it beneath fluvio-glacial deposits. A cold climate, with active glaciers, should produce the largest moraines.

V. Okko (1955) has discussed the moraines of Iceland and suggested that moraines be divided into two types, which may be called 'push' and 'dump' moraines. He suggests that most of the moraines of the Vatnajökull glaciers belong to the first type. Evidence for this is seen in the peat layers that have been folded and now underlie the moraines of Breiðamerkurjökull and Kviárjökull. Sometimes these push-moraines are overlain by a layer of ground moraine brought up along thrust planes. The Breiðamerkurjökull moraines are 30 km long and 1 to 2 km broad. Orientation studies of the stones in the moraine give a weak maximum parallel to the last ice flow and perpendicular to the ridge elongation. In places the moraine consists of several parallel ridges. The fabrics in these double ridges are clearly parallel to the flow in till, which overlies stratified sand, mixed with the peat beds. This succession suggests a double ice advance with sedimentation taking place in the interval between the advances. Only the proximal slope of such moraines is the result of glacial deposition, the distal one being an erosional feature. The Kviárjökull moraines are more than 100 m high and consist mainly of sand with very large boulders. The Morsárjökull moraines show a stone orientation parallel to the ice flow and this suggests addition of basal till to the moraine. These moraines appear to be of the first type, but they have been subsequently overridden by thicker and more active ice. The end moraines of Hoffellsjökull are mainly dump moraines. The fabric analysis at a depth of 1 m shows no certain orientation and the material is stratified in places. Evidence of the presence of marine shells indicates that the moraines were deposited in water.

K. Straszewska (1969) has described some pseudo-morainic features formed mainly of till which she terms 'boulder-clay festoons'. They are tens of kilometres long and a few kilometres wide, comprising asymmetrical ridges with a short, steep proximal side and a long, gentle distal side. They occur within the extent of the Saale glaciation in Poland and are connected with its recession. One feature has a steep slope 13 m high facing north-east. To the south lies moraine upland, and to the north, ice-margin features and meltwater kames are situated within an outwash plain. The largest festoon is the Opinogora escarpment, which is 30 km long and 40 m high, running north–south, unlike all the others. The features are thought to be caused by the flow of ablation till from the ice front. Similar features are now forming around Spitsbergen glaciers. The steep slope is formed partly by meltwater erosion flowing along it as the ice melts from its contact zone.

2.2 *Push-moraines*

The term 'push-moraine' has been used to describe features different from those described by Okko in Iceland (p. 445). Some of these push-moraines occur in front of an advancing ice sheet, rather than a glacier, and there must be suitable deposits for the ice sheet to make into the push-moraine. Recently formed moraines of this type can be differentiated from other terminal moraines by their form. Those in Schleswig, for example, are more convex in cross-profile. Their slopes are steeper, especially on the distal side, where the frontal concavity is not developed as in normal terminal moraines. There may be a break of slope between the moraine and the proglacial deposits in front of it.

Push-moraines of this type can be identified best by their material and internal structure. The material shows signs of tectonic disturbance by faulting and thrusting. This causes the development of imbricate structure. A good example of this type of moraine is seen in the Itterback region of north-west Germany. Another example of this type of tectonic structure is seen in the drift exposed in the cliffs near Dinas Dinlle in North Wales. Glacial tectonic structures are well displayed in this coastal section. The material lying in the path of the advancing ice sheet must have been in a rigid, frozen state in order to fracture abruptly when subjected to high lateral pressure by the advancing ice. The tectonic features revealed in the structure of push-moraines of this type are formed by this strong pressure. Sometimes waves and folds are formed as well as low-angle thrusts and faults.

Where the material shows fracturing throughout a considerable thickness of up to

100 to 150 m, in the Itterback area for example, the depth of permafrost must have been considerable. It can be inferred, therefore, that the Scandinavian ice sheet advanced into an area that had previously been deeply frozen. This deep freezing would require a prolonged cold period before the ice advanced to form the push-moraine. Several tens of thousands of years would be required to produce a permafrost layer of 100 to 150 m depth. It can be concluded that the ice only advanced after a period of general cooling of considerable duration. The push-moraines of the last major ice advance were not so extensive as those of the earlier Saale period, to which the Itterback moraine belongs. This conclusion agrees with the known colder conditions in the earlier glacial period.

Similar types of push-moraine have been recorded in North America, for example, in the New York area and in southern Iowa, Alberta, and on the Yukon coast. There are ice-pushed moraines in Holland, which have been described by M. G. Rutten (1960). They consist of early Pleistocene and Tertiary sediments, sand, gravel and clay. The tectonic structures have a dip of 50 to 70° and strike parallel to the length of the push-moraine ridges, dipping away from the outer faces. The ridges are only covered by a veneer of drift. They are 50 to 200 m high and up to 100 km long, arranged in regular sequence following the valleys. It appears that they form best in those areas where the advancing ice ponded the drainage, since this allowed more effective permafrost development. The push-moraines are the result of advancing ice pressure but they are not formed of glacial deposits except for the thin surface veneer.

2.3 Thule-Baffin moraines

The moraines of cold ice sheets differ in several respects from those of temperate valley glaciers. The areas in which they have been studied most are around the Barnes ice cap in Baffin Island and in the Thule area of Greenland. J. Weertman (1961) has proposed, therefore, that they be called the Thule-Baffin moraine type. This name is preferable to 'shear' moraine, as this latter term suggests a genetic origin that has not been generally accepted, and for which alternatives exist.

These moraines were described in detail by R. P. Goldthwait (1951). It is on the basis of his explanation of the features that they have been called 'shear' moraines, although he used the term 'end moraine'. The moraines that he described lie around the edge of the Barnes ice cap, which is a cold ice sheet, having a mean temperature of about $-10°C$. The material forming the supraglacial debris is thought to have come to the surface along shear planes from beneath the glacier, where the ice is thinning near its edge. The ice up to 30 to 60 m above the steep toe of the ice cap is layered by many thin dirt bands, striking nearly parallel to the ice edge. Although much of the material is fine dirt, there are some pebbles and boulders as well. Goldthwait suggested that the debris moved up along shear planes dipping down towards the bed of the ice cap. There is good evidence for the existence of shear planes of this type near Flyway Lake. The material accumulates on the surface of the ice near the ice margin to form a dirty zone about 135 m wide. The concentration of dirt on the surface only takes place where the ice thickness is less than 85 m.

The ice-cap surface is clean beyond the 135 m wide strip of dirty ice. Near the edge, there is a black ice surface, which is covered by a thin film of dirt. This layer is usually not more than 2·5 mm thick and the surface slopes at 15°. The dirt-covered ice melts more rapidly than the white ice above it. The ablation rate measured was 20 cm/week on the clean ice and 32·5 cm/week on the dirty ice. It was estimated that 150-200 cm of the dirty ice was lost in a year's ablation, but only 100-150 cm of the white ice. This difference results in a steepening of the dirty ice slope. Material moving down this steepening slope accumulates at the bottom to form a thicker layer of debris. Thus an accumulation of till 0·6 to 1·2 m thick forms on the last 30 m of the slope, and a depression develops at the junction of the dirty ice slope and the thicker till

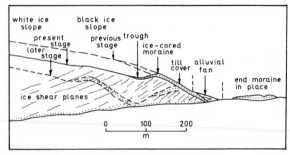

Fig. 15.2
Thule-Baffin cold-ice moraine formation at the retreating margin of the Barnes Ice Cap (R. P. Goldthwait, *J. Geol.*, 1951, University of Chicago Press). Debris is transported up inclined shear planes, gathering on the surface as ablation moraine, and finally deposited as end moraine (right)

layer on the outer margin. This is caused by the more rapid melting of the ice under the thin dirt cover than under the thicker till cover. The trough separates the developing ice-cored moraine from the dirty ice beyond. In time, the ice core may melt and a low terminal moraine will be left in place (Fig. 15.2).

Goldthwait estimated that it would take about 25 years to accumulate 0·9 m of debris on the outer ice-cap margin. The ice-cored moraines have a relatively steep slope of 30° and are about 15 m high, but when the ice core has finally melted, the end moraines are no more than 1·5 m to 4·6 m in height and 15 m to 150 m broad. Such features can only form where the ice is moving actively to within 30 m to 60 m of its margin, and they are, therefore, characteristic of active cold ice caps.

Although the later stage of development of this type of moraine is clear, there is more doubt concerning the method by which the debris reaches the surface. It is this part of Goldthwait's theory that has been modified by the later work of Weertman (1961), who calls these moraines the Thule-Baffin type. The main problem that must be solved is how the debris becomes concentrated in the lower layers of the ice. Weertman points out that the character of the debris does not support the shear hypothesis, as the fine debris can be distributed thinly throughout a considerable thickness of ice. There are also examples of thick layers of debris concentrated in bands up to 0·5 m thick. These layers consist of stones, sand and other debris, while the slightly dirty layers may be up to 1 to 2 m thick. A new suggestion is put forward to account for the dirt in the ice.

It is suggested that the debris becomes frozen into the base of the ice when basal meltwater freezes. There is a considerable amount of evidence to show that the cold glaciers do not slide over their beds (Chapter 4). In the Thule area, tunnels have shown that there is no movement between ice and bedrock, although there is considerable shearing in the lower layers of ice (p. 131). Where the ice is temperate, on the other hand, water can exist beneath the ice, and in these conditions debris can be moved as the ice can slide over its bed. Weertman suggests that only the outer part of the ice cap in Baffin Island and at Thule in Greenland is frozen to its bed. Farther in from the edge, the base of the glacier may be at pressure melting-point. The heat of sliding and geothermal heat may exceed the rate at which heat can escape upwards by conduction from the inner part of the ice cap. The basal ice would melt to form subglacial water. The pressure gradient would force the water out towards the edge of the ice cap, where it would enter a zone of the ice cap where the thinner ice can allow the extra heat to escape to the atmosphere. The basal water will then refreeze at the bed of the glacier, while the base of the ice will still be at melting-point owing to the latent heat of freezing (Fig. 3.8). If it can be shown that the base of the ice cap in its central parts is at the pressure melting-point and yet no water is emerging from the basal margin of the ice cap, then it may be assumed that basal water is refreezing near the edge of the ice cap.

An ice cap in a steady state would not introduce much debris, as distinct layers, into the lower part of the ice. Ice caps, however, change with time. Variations in thickness, and the accompanying change in pressure, will cause variations in sliding velocity. Thus the amount of heat generated at the base of the ice will vary with time, and the point at which the ice refreezes on to its bed will also vary. If the freezing zone extends outward towards the edge of the ice cap, the water will freeze at the bottom in an area where before it was liquid. This may occur below a zone into which material had previously been frozen. The overlying debris is thus incorporated into a layer of ice with clean ice above and below. If this cycle is repeated, with freezing following a renewed melting, then the layers of debris can increase in number with each cycle.

Once the debris has been included in the ice, it flows with it and can be carried into the ablation zone where the flow lines must come to the surface. Where the edge of the ice sheet is in the ablation zone, as in the Barnes ice cap and the Greenland ice sheet near Thule, the debris will arrive at the surface and can form terminal moraines in the way already described. At the moment, there are no direct measurements to support the view that the centres of these ice sheets are at the pressure melting-point. However, from theoretical considerations this appears likely; moreover, at Byrd Station in the Antarctic, deep boring proved that the base of the thick inland ice here was at pressure melting-point (see p. 49). It seems, then, that Weertman's method of moraine formation, if not proved, is at least possible.

R. Le B. Hooke (1970) recognizes three types of ice-sheet margin near Thule:

1 A gently sloping ice margin leading to an ice-cored moraine ridge
2 A gently sloping ice margin with little moraine on the ice, and
3 Ice cliffs rising above moraine ridges.

The first type includes shear moraines, termed ice-cored moraines by the author. The moraine ridges (Fig. 15.3) may be either sharp-crested or broad and rounded, the latter being older and more stable. There are dirt bands in the ice, and Weertman's theory of the method of debris incorporation in the ice is accepted. The ice is flowing compressively in the area at a maximum rate of 4·5 m/year. Along the margin of the ice sheet

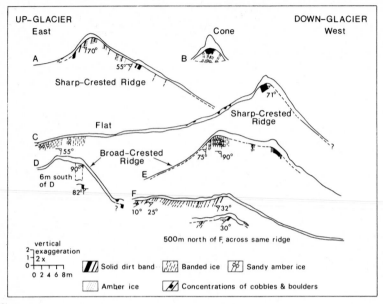

Fig. 15.3
Morphological and sedimentary features of the ice-sheet margin near Thule, Greenland, as revealed by trenches cut through moraines (R. le B. Hooke, *J. Glaciol.*, 1970, by permission of the International Glaciological Society)

and in front of the ramps, there are some stagnant zones of wind-drift ice (Fig. 15.4). This appears to have been buckled as indicated by foliation within it, owing to pressure from the ice behind the moraine. Ice cliffs occur where the orientation of the ice front is such that wind-drift ice cannot accumulate. The time needed for moraine formation in this area is estimated at 42 to 86 years. It is a sporadic process, only occurring when thick dirt bands are exposed, under which conditions a moraine can form in a few years. Wind-blown snow is an essential element in differentiating the types of margin. All types could form under equilibrium conditions at the ice margin, or with either slow advance or retreat.

   Hooke (1973) has also looked at the processes operating to form moraines at the edge of the Barnes ice cap on the basis of bore hole and tunnel data and fabric studies. His evidence suggests that a wedge of deformed superimposed ice thinning up-glacier is present beneath the glacier ice. Debris occurs within the glacier ice in places, and, where it is found, an ice-cored moraine ridge develops 100 to 150 m from the ice margin. A general advance of the glacier is necessary to incorporate the superimposed ice which consists of refrozen meltwater and some snow forming at the edge of the ice; it is overridden as the ice cap advances. The moraine can continue to grow in height and move

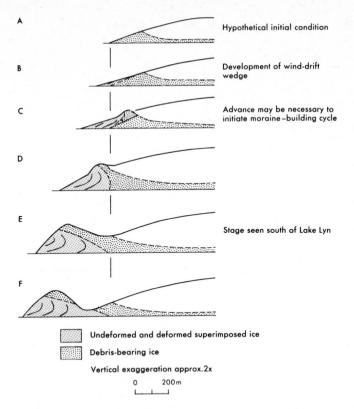

A   Hypothetical initial condition

B   Development of wind-drift
     wedge

C   Advance may be necessary to
     initiate moraine—building cycle

D

E   Stage seen south of Lake Lyn

F

☐ Undeformed and deformed superimposed ice

☐ Debris-bearing ice

Vertical exaggeration approx.2x

0   200m

Fig. 15.4
Possible stages in the formation of moraines at the margin of a cold ice sheet. Progressive overturning of original sedimentary bands in superimposed ice is indicated. Foliation may later obscure this banding
(R. le B. Hooke, *Bull. geol. Soc. Am.* 1973)

outwards without further ice advance. A depression forms, separating the moraine from the glacier, as melting continues. Debris may be recycled by slumping down the front face of the growing ridge and becoming incorporated in newly formed superimposed ice. The material is then deformed to add to the moraine again, if and when the ice readvances. Complex structures result from this process. No discrete shear zones were found in the marginal zone in the tunnel or bore holes. The moraines would probably take a few hundred years to form by this process.

2.4   *Cross-valley and washboard moraines*

The Baffin-Thule type of moraine is formed in front of an ice cap that terminates on dry land. Another type of moraine has been studied that demands that the ice cap should end in ponded proglacial lakes or the sea. Moraines of this type have been variously named 'De Geer', ribbed and 'cross-valley' moraines. The De Geer moraines, described by G. Hoppe (1959) in Sweden, have also been called 'annual' moraines. However, the suggestion of annual formation cannot be proved, according to

Hoppe. Until they have definitely been shown to be annual, this is not a desirable name.

The De Geer moraines, described by Hoppe in Norrboten, are about 6 to 7 m high; they have a breadth of 8 to 40 m and may be up to 1 km long. They are asymmetrical, having steeper distal or down-glacier slopes, on which many loose boulders occur. They are usually formed of till, which occasionally shows evidence of pressure structures and sometimes they contain stratified lenses. These features are only found below the highest marine shoreline. They occur below sea level in places. It is generally agreed that the features were formed at or very near to the ice front, when this was in the form of an ice cliff. They were explained by De Geer as a type of push moraine, formed as the glacier advanced in winter. If this were the correct method of formation they would be annual features. However, it has been shown, by comparison with varves, that in fact they are not always annual and more than one can form in one year. G. Lundqvist (1948) showed that the pebbles on the moraines were oriented perpendicular to the ridge direction. Although this direction is often parallel to the last ice movement, it appears to be little related to it, for the pebbles remain oriented at right-angles to the ridge direction even when this is not parallel to the ice movement. Hoppe suggests that water-soaked moraine under the ice has been squeezed into crevasses near or against the snout, lying perpendicular to the ice front. Irregularities in the pattern can be explained by the calving pattern of the glacier front. The pattern also reflects the rate of retreat of the ice front, which tends to be more rapid where the water is deeper in broken country. The resulting pattern is usually uneven.

The cross-valley moraines, described by J. T. Andrews (1963a, b) from Baffin Island are similar in many ways to the moraines described by Hoppe. The cross-valley moraines were formed in lakes held up in front of the Barnes ice cap on its north-west margin. Similar lakes now occur on the north-east side of the ice cap, such as Conn and Bieler Lakes. Andrews undertook a detailed study of the till fabric and the spacing of the moraines. The moraines occur in the valley of the Isortoq River and their spacing and number were studied on aerial photographs. Unlike normal terminal moraines, they tend to be concave rather than convex away from the ice front. In form, the moraines are asymmetrical, having a steep distal face with a slope of 34 degrees and a proximal slope that varies between 15 and 24 degrees. The moraines extend from near the valley bottom up towards the lake shoreline, but are best developed in their middle section, degenerating towards their lower limit to a rather irregular mass of ridges and hollows. At their largest, the moraines are up to 15 m high and are composed of 20 per cent gravel, 60 per cent sand and 20 per cent silt and clay.

Analysis of the fabric of the till in the moraines has yielded valuable information concerning their possible method of formation. Both the orientation and the dip of the longer stones in the till were analysed, using pairs of samples from the distal and proximal moraine slopes. The fabrics were found to be very different on the two sides of the moraines. Those on the proximal side of the moraines show a stronger dip and orientation pattern. It is clear, therefore, that the force forming the moraine must have come from the up-glacier or proximal side. The orientation of the pebbles shows that they lie at right-angles to the moraine orientation and not necessarily normal to the direction of the latest ice movement. This is sometimes not at right-angles to the ridge crest. The dip of the pebbles shows a very strong up-glacier plunge on the proximal side. There is little preferred dip on the distal side of the moraines (Fig. 15.5).

The fabric is not associated with the basal movement of the ice, but is imposed by another process of moraine emplacement. In studying the spacing of the moraines, it was shown that there is a correlation between the spacing of the moraines and the distance from the watershed from which the ice is retreating down-valley. Gradually deepening lakes are thus impounded until the exposure of a lower col allows sudden shallowing of the lake by reducing its level. The moraines are closer together farther from the watershed. A graph of moraine spacing against depth shows a striking linear

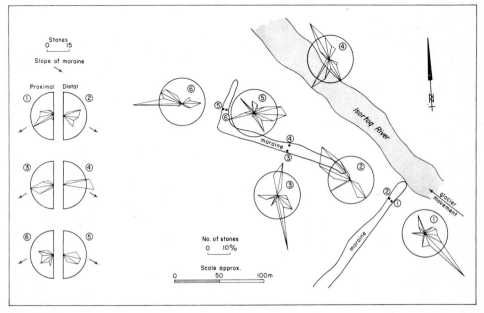

Fig. 15.5
Till fabrics in cross-valley moraines in Baffin Island (J. T. Andrews and B. B. Smithson, *Bull. geol. Soc. Am.*, 1966)

correlation, which is highly significant statistically. The depth of the lake in which the moraines formed varied from 120 to 200 m, the ice thickness being about the same as the water depth. The mean distance between the moraines is 36 m and 33 m in two areas. There is a tendency in part of the valley for two moraines to be closer together, followed by a wider distance before the next pair.

A possible method by which these moraines could form is associated with the hydro-static pressure present at the base of a high ice cliff in a lake. The hydrostatic pressure of the water at the base of the cliff would tend to cause seepage under the ice foot. The seepage would tend to thaw the basal ice and the moraine-covered ground on which it was resting. This moraine would then become soaked with water, reducing its strength until it became squeezed up in front of the ice face. The pressure of the ice would induce the till to flow and the stones in it would take up a preferred orientation and dip pattern based on the stresses in the till. The flow on the distal side would not be confined to the same extent as on the proximal side and hence the less strong till fabric on this distal side can be accounted for. The double moraines may form as a result of the ablation and forward movement of the ice acting together to give two periods in the year when the ice front would tend to be stationary for a short period. There is some evidence that 650 to 800 moraines had formed in a 700-year period. This suggests that they may be annual in character, although this is by no means proved.

The washboard moraines of North Dakota, described by D. N. Nielson (1970), are 1·2 to 4·6 m high, and 80 to 170 m apart. They occur mainly in the southern marginal part of the Mankato drift; nearly half the glaciated part of North Dakota has washboard moraines. Similar features are described from Manitoba by J. A. Elson (1957), where they are formed of sandy till. The ridges are subdued, discontinuous, irregular and parallel, but their preferred orientation is generally unrelated to the regional direction of ice flow. Individual ridges extend up to 2·4 km in length, though most are 0·5–1 km long. The relationships between the washboard moraines and eskers are complex. Some moraines change alignment near large eskers but are not affected by small ones. Small eskers may be offset by the moraines, but others cross them without change. The slopes of the moraines range from 2° to 6°, and there is no systematic variation between the proximal and distal sides. They appear to be remnants of shear moraines deposited in a supraglacial position by shearing of active over stagnant ice at the margin. Debris-laden shear planes are formed, which develop into ice-cored shear moraines by ablation. Most fabrics in the washboard moraines are bimodal or polymodal, and a few trend parallel to the moraine elongation. The disturbance of the fabrics is probably due to slumping, and this process probably accounts for the fabrics, while till flowage may also exert an effect. If the shear-moraine theory of their origin is correct, the ridges are evidence of active ice movement. An alternative view is that they are linear disintegration features associated with decay of stagnant ice, as discussed in Chapter 17.

R. J. Price (1969) has studied the proglacial area revealed by the retreat of Breiðamerkurjökull since 1890. Rapid retreat has resulted from ice thinning, and has exposed ground moraine, moraine ridges, sandar, kames and eskers (Fig. 15.6). The outermost moraines are the largest, being 5 to 10 m high, and there are four systems of younger moraines 1 to 3 m high. The moraines are formed of material that is less than 30 per cent finer than 0·06 mm, and the coarser material is well rounded. Orientations of elongated stones are strongly normal to the ridge crest and dip towards the proximal side. Price considers that the moraine ridges are formed by the squeezing of water-soaked till from under the ice to form ridges. This would explain the lobate pattern and fabric. The moraines of Fjallsjökull (Price, 1970) are similar in some respects. The glacier has also retreated during the last 70 years from moraines 5 to 10 m high, leaving smaller moraines 1 to 4 m high. Their pattern, interval and surface form suggest squeezing of semi-liquid till from beneath the glacier. The slope of the ice surface plays a part in determining the position and orientation of the moraines which may have formed annually, thus being similar to washboard moraines. Most of them have formed in the last 30 years. The ice margin is steep, crenulated and highly crevassed, and is fronted by proglacial lakes. The distal slopes of the moraine ridges are steeper than the proximal

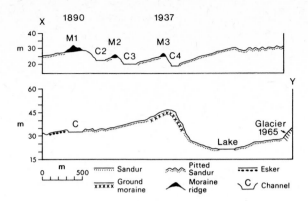

Fig. 15.6
Profile, X-Y, across the proglacial area of Breiðamerkurjökull, Iceland, from the 1890 moraine (M1) to the 1965 glacier margin. The ice margin at this point is now 4·4 km inside the 1937 moraine (R. J. Price, *Institute of British Geographers, Transactions*, 1969)

slopes, with angles of 25 to 35° and 20 to 25° respectively. The moraines are sub-parallel to each other and to the ice front. They are spaced more closely on the western than the eastern-facing slopes. Most of the fabrics showed a preferred orientation normal to the ridge crest, but one is parallel. Their pattern suggests linked arcuate segments. In one place at least fifteen ridges have formed in 16 years.

Features described as 'ribbed moraine' are reported from an area south-east of Schefferville, Labrador, by W. R. Cowan (1968). Lengths of ridges vary from 300 m to more than 1·5 km, widths from 100 to 300 m and heights from 3 to 15 m; slopes range from 2° to 15° and spacing from 100 to 300 m. The till of the area is mostly 6 to 15 m thick. The ribbed moraines trend normal to the last direction of ice movement, and there are ridges curved concave to the ice movement. Fabrics on the proximal side were found to be imbricated up-glacier into the slope, while on the distal side the fabric was conformable to the slope. Both the ribbed moraines and curved ridges had directional components additional to those of the regional glaciation. Recurrent ridges of ribbed moraine transverse to the regional ice flow were thought to have resulted from a reactivation of a retreating ice front, leading to bulldozing and piling-up of proglacial debris that was then overridden. The curved and straight types of ribbed moraine appear similar and the hardness and fissility of the till suggests a subglacial origin, and a fairly late formation under the glacier seems likely. The spacing of the ribbed moraine appears to be related to ice thickness, and no temporal periodicity is involved. The formation of ribbed moraine is thus quite different from the formation of washboard or De Geer moraines, though there are morphological resemblances.

2.5   *Fluted moraine*

Fluting of till surfaces on various scales has been widely reported. In North Dakota and Saskatchewan, for instance, shallow parallel grooves and intervening ridges cover vast areas. The fluting is best developed on till (or sometimes stratified drift) but softer solid rock formations are also moulded in places. The ridges range from less than 2 m in height up to 25 m, and each extends for many kilometres. In origin, they are akin to very elongated drumlins, and their alignment provides good evidence of the former ice movement.

On a much smaller scale are the parallel flutings on newly-exposed clay- or silt-rich till first described by J. L. Dyson (1952) and V. Schytt (1959). S. Baranowski (1970) has examined those now forming in front of the Werenskjold Glacier in south-west Spitsbergen. The furrows and ridges lying parallel to the ice flow are 80 to 150 m long with a width of a few decimetres to 2 m, and a relative relief of up to 30 cm. There were more than 150 ridges in front of the Werenskjold glacier in 1962. The features were wider and better developed where there was more silty material. The biggest distance between ridges was 11 m, and the average spacing was 4·5 m, with 52 per cent between 3 and 5 m, and 78 per cent between 2 and 6 m. The fluted moraine was seen to exist in a frozen state underneath the glacier, where the ridges were much steeper; some had vertical or overhanging sides. Such micro-flutings have been interpreted as small-scale versions of the larger fluting and elongated drumlins referred to above.

Hoppe and Schytt (1953) suggested an erosional origin, where boulders embedded in the ice are dragged over a till surface. Schytt (1959) has put forward another hypothesis based on ice pressing. Ice-pressed forms will be considered more fully in Chapter 17; here, the essential elements of the idea are that water-soaked till beneath a glacier may be squeezed, by hydrostatic pressure, into subglacial cavities or grooves, the latter forming down-glacier from fixed boulders or bedrock obstacles and lengthening with the ice motion. The fact that elongated stones in the level parts of the ground moraine are aligned parallel to the direction of ice flow, but those in the ridges are not, is in accordance with such a theory of ice pressing, as is also the fact that when the ridges are traced under the glacier, they are found to contain much refrozen meltwater.

Baranowski's (1970) hypothesis also starts with subglacial water-soaked till, and considers the effects of the freezing of this material when it moves from a zone of temperate basal ice to a zone of cold ice. Freezing results in subglacial frost-heaving, causing evenly-spaced bulges to develop where the cryogenic pressure is lowest, the distance between bulges depending on the water content. The upheaved patches groove the moving ice as do boulders, forming cavities on the lee side of the bulges. The semi-liquid material squeezed into these cavities by pressure of the ice and by cryogenic pressure prevents the cavities closing up. If the freezing belt advances down-glacier, then long ridges can form. This theory attempts to explain the regularity of the fluting, which other theories do not.

Fluted moraine has also been described by J. L. Andersen and J. L. Sollid (1971) around Midtdalsbreen near Finse, Norway. Block orientation in and on the fluted moraines shows a preferred orientation parallel to the long axes of the ridges and to striations, which showed a similar pattern.

3 Moraine landscapes

The more distinct examples of morainic forms have been discussed, but there are some other features of ground moraine and ablation moraine that demand brief consideration.

Hoppe (1957, 1959) has described various types of hummocky morainic landscape in Sweden. One of these occurs around Lake Rogen in Härjedalen, and has been referred to as the Rogen type. It covers large areas, stretching from Dalarna in the south to Norrbotn in the north. The other type is known as Veiki moraine, from an area near Gällivare. The Veiki-type moraine consists of three elements: moraine plateaux, hollows, and ridges marking the edges of the plateaux. Stones in these rim ridges show a strong preferred orientation. Hoppe suggests that moraine was deposited unevenly beneath actively moving ice giving the plateaux and hollows. Ice movement then ceased, but water-soaked moraine was squeezed up from the hollows against the plateaux, giving the rim ridges. The presence of drumlins associated with the moraine plateaux and hollows supports the contention that these two elements, unlike the rim ridges, formed beneath active ice.

Ablation moraine, sometimes called kettle moraine, can give rise to distinctive landscapes under some circumstances. A good example of an ablation moraine landscape can be seen in the south-east of Ireland in County Wexford. The moraine consists of a thick layer of sandy till, forming a hummocky landscape. Amongst the rounded hummocks are small round lakes. The whole landscape is the result of the melting away of an ice sheet that just impinged on the south-east coast of Ireland from the Irish Sea. Large lumps of ice must have been incorporated in the sandy material and these melted out slowly to form the kettle-holes, now filled with water. The whole landscape appears to be of recent origin, as the features are fresh, and it probably represents a late ice advance on to the coast of Ireland. Other features associated with a decaying ice sheet will be considered when dead-ice features are dealt with in Chapter 17.

The effect of glaciers or ice sheets overriding older moraines is described by S. M. Totten (1969) in north-central Ohio. The overridden moraines show partial removal by scouring and some covering with drift. Inter-moraine areas have been partially filled and some new moraines deposited giving a composite landscape. The moraines belong to a single glaciation, probably pre-Woodfordian (early Wisconsin).

4 Chronological significance of moraines

It has been shown that different types of moraine require different conditions for their formation. Some only form in water, others only if the ice sheet is cold and frozen to its bed at least at the margin, others again form at the snout of a glacier when it is stationary or advancing.

It is the latter type of terminal moraine which has been used a great deal to assist chronological analysis of glacial action. The well-known terminology of Alpine glaciations in Europe was partly based by A. Penck and E. Brückner (1909) on the main terminal moraines in the area surrounding the Alps, to which the enlarged glaciers extended. In the same way, the maximum advance of the Newer Drift in Britain is marked over part of its length by terminal moraine formation. The moraines at York and Eskrick in the Vale of York are well-known limits of the Newer Drift maximum as are those of South Wales (for example, at Glais in the Tawe valley). Another subdued morainic ridge has been identified by A. Straw (1958) on the eastern margin of the ice sheet where it was pressing into the Fenland gap near Stickney. This moraine is low partly because of its subsequent partial burial by Fenland deposits.

According to H. D. Foster (1970) the erratics of the mostly submerged ridge of Sarn Badrig, in Cardigan Bay, support the view that it is a medial moraine between Welsh ice moving out into Cardigan Bay and Irish Sea ice moving south. Its age is of the order of 17,000 to 32,000 years BP.

From the outer moraines of the Newer Drift ice advance, a series of moraines can be traced back towards the ice source in the hills. These moraines were formed by intermittent ice advances during a period of general deglaciation. It is noticeable that the moraines become progressively fresher and more hummocky as they are traced towards the hills. The latest moraines in the sequence are found in the cirques of North Wales, the Lake District and Scotland, where they date from the post-Allerød cold phase (Zone III of the Late-glacial period). These moraines show little post-glacial modification, as they were formed only about 11,000 years ago. The ones marking the maximum extent of the Newer Drift ice probably date from between 20,000 and 30,000 years ago.

The moraines in the valleys around the Jostedalsbreen in Norway provide a sequence that continues into the more recent past. The outermost of a series of recent moraines formed by Austerdalsbreen (Fig. 15.7) and Nigardsbreen can be associated with the advance in the mid-eighteenth century. Historical evidence, studied by Faegri (1948) for Nigardsbreen, is valuable in this respect. There is evidence that the glacier advance of this period invaded farm land. A similar moraine in Austerdalsbreen has also been dated to the same period by tree-ring analysis. Between this outer moraine, which formed around 1750, and the glacier snout is a series of low symmetrical moraines. The moraines nearest to the glacier snouts are dated at 1928 and 1930 in Austerdalsbreen and Nigardsbreen respectively. The next nearest moraines in each glacier were deposited around 1909 and 1906 respectively. For this period climatic data are available and also annual measurements of the movement of the glacier snouts. The measurements are complete for Nigardsbreen and nearly complete for Austerdalsbreen. The measurements show quite distinctly that these particular moraines were formed during a period of ice advance and climatic deterioration. Nigardsbreen advanced from 1905 to 1910, when it remained stationary for a few years. It advanced again between 1922 and 1925 and then remained stationary until 1932 when a rapid retreat started. This retreat has continued uninterruptedly until 1960. This example shows how changes in the position of a glacier margin can be related to the formation of terminal moraines of this particular type, which are low symmetrical moraines.

5 Conclusions

Moraines can be divided into active moraines, ice-cored moraines and stable moraines. The first type rests on or alongside active glaciers or ice sheets. This type can be further subdivided into medial, lateral, terminal and ablation moraines according to their position relative to the ice mass. The second category of ice-cored moraines can yield useful

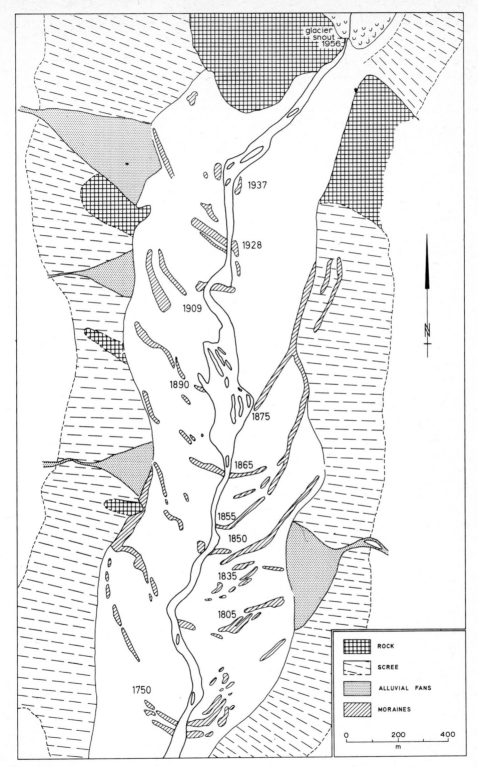

Fig. 15.7
Terminal moraines in Austerdalen, Norway

information when the ice that they contain (partly glacier ice and partly snow-bank ice) has been dated. Where the ice has retreated, moraines become stable and in this form they provide valuable evidence of past glacier advances. This type can also be subdivided into (1) lateral and terminal, and (2) ground moraine. Another form of terminal moraine is the push-moraine, formed of material lying in front of an advancing ice sheet. It gives evidence of the conditions in front of the advancing ice sheet, but it is not primarily a glacial deposit, although it may consist of glacial deposits of an earlier glacial period.

Moraines formed by cold-based ice sheets show different characteristics and have been termed Thule-Baffin moraines from the areas in which they have been studied. Another special type of moraine that forms only in proglacial water bodies, but which nevertheless consists of till, is the cross-valley or De Geer moraine. This type of moraine should be differentiated from the delta-moraine type, which is basically a fluvioglacial form and will therefore be discussed in the next chapter, although it also forms in proglacial water bodies.

The deposition of moraine has given rise in places to distinctive types of landscape, while moraines also give valuable evidence of phases of intermittent ice advance during general deglaciation. This is especially true of the more recent phases of the last major glaciation.

6 References

ANDERSEN, J. L. and SOLLID, J. L. (1971), 'Glacial chronology and glacial geomorphology in the marginal zones of the glaciers Midtdalsbreen and Nigardsbreen, south Norway', *Norsk geogr. Tidsskr.* **25**, 1–38

ANDREWS, J. T. (1963a), 'Cross-valley moraines of the Rimrock and Isortoq River valleys, Baffin Island, North West Territories', *Geogr. Bull.* **19**, 49–77
(1963b), 'Cross-valley moraines of north-central Baffin Island: a quantitative analysis', *Geogr. Bull.* **20**, 82–129

ANDREWS, J. T. and SMITHSON, B. B. (1966), 'Till fabrics of the cross-valley moraines of north-central Baffin Island, North West Territories, Canada', *Bull. geol. Soc. Am.* **77**, 271–90

BARANOWSKI, S. (1970), 'The origin of fluted moraine at the fronts of contemporary glaciers', *Geogr. Annlr* **52** A, 68–75

BARSCH, D. (1971), 'Rock glaciers and ice-cored moraines', *Geogr. Annlr* **53** A, 203–6

BOULTON, G. S. (1967), 'The development of a complex supraglacial moraine at the margin of Sørbreen, Ny Friesland, Vestspitsbergen', *J. Glaciol.* **6**, 717–36
(1970), 'On the deposition of subglacial and melt-out tills at the margins of certain Svalbard glaciers', *J. Glaciol.* **9**, 231–45

COWAN, W. R. (1968), 'Ribbed moraine till-fabric analysis and origin', *Can. J. Earth Sci.* **5**, 1145–59

DYSON, J. L. (1952), 'Ice-ridged moraines and their relation to glaciers', *Am. J. Sci.* **250**, 204–11

ELSON, J. A. (1957), 'Origin of washboard moraines', *Bull. geol. Soc. Am.* **68**, 1721

EMBLETON, C. (1961), 'The geomorphology of the Vale of Conway, North Wales, with particular reference to its deglaciation', *Trans. Inst. Br. Geogr.* **29**, 47–70

EMBLETON, C. and KING, C. A. M. (1975), *Periglacial geomorphology*

FAEGRI, K. (1948), 'On the variation of western Norwegian glaciers during the last 200 years', *Gen. Assembly int. Un. Geod. Geophys., Oslo (1948)*, **2**, 293–303

FOSTER, H. D. (1970), 'Sarn Badrig, a submarine moraine in Cardigan Bay, North Wales', *Z. Geomorph.* **14**, 475–86

GABERT-DELAY, J. (1967), 'Moraines, terrasses et glacis de versant du bassin de Laragne', *Revue Géogr. alp.* **55**, 521–48

GALLOWAY, R. W. (1956), 'The structure of moraines in Lyngsdalen, north Norway', *J. Glaciol.* **2**, 730–3

GOLDTHWAIT, R. P. (1951), 'Development of end moraines in east-central Baffin Island', *J. Geol.* **59**, 567–77

GRAF, W. L. (1971), 'Quantitative analysis of Pinedale landforms, Beartooth Mountains, Montana and Wyoming', *Arct. alp. Res.* **3**, 253–61

GRESSWELL, R. K. (1962), 'The glaciology of the Coniston basin', *Lpool Manchr geol. J.* **3**, 83–96

HARRINGTON, H. J. (1952), 'Glacier wasting and retreat in the Southern Alps of New Zealand', *J. Glaciol.* **2**, 140–4

HEWITT, K. (1967), 'Ice-front deposition and the seasonal effect: a Himalayan example', *Trans. Inst. Br. Geogr.* **42**, 93–106

HOOKE, R. LE B. (1970), 'Morphology of the ice-sheet margin near Thule, Greenland', *J. Glaciol.* **9**, 303–24

(1973), 'Flow near the margin of the Barnes ice cap and the development of ice-cored moraines', *Bull. geol. Soc. Am.* **84**, 3929–48

HOPPE, G. (1957), 'Problems of glacial geomorphology and the Ice Age', *Geogr. Annlr* **39**, 1–16

(1959), 'Glacial morphology and inland ice recession in north Sweden', *Geogr. Annlr* **41**, 193–212

HOPPE, G. and SCHYTT, V. (1953), 'Some observations on fluted moraine surfaces', *Geogr. Annlr* **35**, 105–15

JOHNSON, P. G. (1971), 'Ice-cored moraine formation and degradation, Donjek glacier, St Elias Mountains, Yukon Territory, Canada', *Geogr. Annlr* **53** A, 198–202

(1972), 'The morphological effects of surges of the Donjek glacier, St Elias Mountains, Yukon Territory, Canada,' *J. Glaciol.* **11**, 227–34

KING, C. A. M. (1959), 'Geomorphology in Austerdalen, Norway', *Geogrl J.* **125**, 357–69

LOOMIS, S. R. (1970), 'Morphology and ablation processes on glacier ice', *Proc. Ass. Am. Geogr.* **2**, 88–92

LUNDQVIST, G. (1949), 'The orientation of block material in certain species of flow earth', *Geogr. Annlr* **31**, 335–47

MANLEY, G. (1959), 'The late-glacial climate of north-west England', *Lpool Manchr geol. J.* **2**, 188–215

NIELSON, D. N. (1970), 'Washboard moraines in north-east North Dakota', *Compass* **47**, 154–62

OKKO, V. (1955), 'Glacial drift in Iceland, its origin and morphology', *Bull. Commn géol. Finl.* **170**, 1–133

ØSTREM, G. (1959), 'Ice melting under a thin layer of moraine and the existence of ice cores in moraine ridges', *Geogr. Annlr* **41**, 228–30

— (1961), 'A new approach to end moraine chronology', *Geogr. Annlr* **43**, 418–19

— (1963), 'Comparative crystallographic studies on ice from ice-cored moraine, snow banks and glaciers', *Geogr. Annlr* **45**, 210–40

— (1964), 'Ice-cored moraines', *Geogr. Annlr* **46**, 282–337

— (1971), 'Rock glaciers and ice-cored moraines. A reply to D. Barsch', *Geogr. Annlr* **53** A, 207–13

ØSTREM, G. and ARNOLD, K. (1970), 'Ice-cored moraines in southern British Columbia and Alberta, Canada', *Geogr. Annlr* **52** A, 120–8

PENCK, A. and BRUCKNER, E. (1909), *Die Alpen im Eiszeitalter* (Leipzig)

PENNY, L. F. (1964), 'A review of the Last Glaciation in Great Britain', *Proc. Yorks. geol. Soc.* **34**, 387–411

POST, A. (1972), 'Periodic surge origin of folded medial moraines on Bering Piedmont Glacier, Alaska', *J. Glaciol.* **11**, 219–26

PREST, V. K. (1967), 'Nomenclature of moraines and ice-flow features as applied to the Glacial Map of Canada', *Geol. Surv. Can. Pap.* **67–57**, 32 pp.

PRICE, R. J. (1969), 'Moraines, sandar, kames and eskers near Breiðamerkurjökull, Iceland', *Trans. Inst. Br. Geogr.* **46**, 17–43

— (1970), 'Moraines of Fjallsjökull, Iceland', *Arct. alp. Res.* **2**, 27–42

RAISTRICK, A. (1927), 'Periodicity in the glacial retreat in west Yorkshire', *Proc. Yorks. geol. Soc.* **21**, 24–8

RUTTEN, M. G. (1960), 'Ice-pushed ridges, permafrost and drainage', *Am. J. Sci.* **285**, 293

RUTTER, N. W. (1969), 'Comparison of moraines formed by surging and normal glaciers', *Can. J. Earth Sci.* **6**, 991–9

SCHYTT, V. (1959), 'The glaciers of the Kebnekajse massif', *Geogr. Annlr* **41**, 213–27

SHARP, R. P. (1969), 'Semi-quantitative differentiation of glacial moraines near Convict Lake, Sierra Nevada, California', *J. Geol.* **77**, 68-91

SHARP, R. P. and BIRMAN, J. H. (1963), 'Additions to the classical sequence of Pleistocene glaciations, Sierra Nevada, California', *Bull. geol. Soc. Am.* **77**, 1079-86

SLATER, G. (1931), 'The structure of the Bride Moraine, Isle of Man', *Proc. Lpool geol. Soc.* **15**, 284–96

SOUCHEZ, R. A. (1971), 'Ice-cored moraines in south-west Ellesmere Island, N.W.T., Canada', *J. Glaciol.* **10**, 245–54

STRASZEWSKA, K. (1969), 'Boulder clay festoons accumulated in front of the ice sheet', *Geogr. Polonica* **17**, 161–72

STRAW, A. (1958), 'The glacial sequence in Lincolnshire', *E. Midld Geogr.* **2**, 29–40

TOTTEN, S. M. (1969), 'Overridden recessional moraines of north-central Ohio', *Bull. geol. Soc. Am.* **80**, 1931–46

WEERTMAN, J. (1961), 'Mechanism for the formation of inner moraines found near the edge of cold ice caps and ice sheets', *J. Glaciol.* **3**, 965–78

WHALLEY, W. N. (1974), 'A possible mechanism for the formation of interparticle quartz cementation in recently deposited sediments', *Trans. N.Y. Acad. Sci.,* Series II, **36** (1), 108–23

YATES, E. M. and MOSELEY, F. A. (1967), 'A contribution to the glacial geomorphology of the Cheshire plain', *Trans. Inst. Br. Geogr.* **42**, 107–25

16

Fluvioglacial ice-contact features: eskers and kames

Eskers consist of washed and sorted, usually water-rolled and stratified materials.... Esker ridges are usually irregular, often very sinuous.... (M. H. CLOSE, 1866)

In this chapter some of the features deposited by glacial meltwater in contact with the ice will be considered. One of the problems in discussing these features is that of nomenclature, for terms such as 'esker', 'ose' and 'kame' have not been used systematic-ally. The first two terms have been used to describe the same features by different workers. The word 'esker', of Celtic origin, has been used more frequently in the English literature than the Scandinavian 'ose'. The latter term has been used more in Europe, and is adopted by J. Tricart and A. Cailleux (1962). The term 'esker' will be used in this chapter as it is best known to English-speaking people. The term 'kame' is useful, but again there are other terms that to a certain extent overlap with it. A method by which eskers and kames can sometimes be differentiated is to use the term 'esker' only for those features that in general run at right-angles to the ice edge (G. Hoppe, 1961). The term 'kame' can be used for features lying parallel to the ice front. Often, however, the features are amorphous and cannot be easily grouped in this way. The term 'delta-moraine' has been suggested for some of the larger-scale features that are probably formed in a similar way to many kames, which are usually fairly small. Kame terraces should be differentiated from kames. Both may lie parallel to the ice edge, but kame terraces usually form along the side of a valley and not at the terminal ice edge. Kame terraces may thus lie at right-angles to those kames that form along the snout of an ice-sheet, and parallel to eskers. They can be differentiated from eskers by their position along the valley side and their different form.

The most important criterion by which fluvioglacial deposits can be differentiated from those laid down directly by the ice is the stratified nature of the meltwater deposits. However, not all stratified drift deposits need occur in one of the forms associated with the fluvioglacial features that will be described. For example, it has already been shown that some drumlins contain stratified deposits. These were laid down by fluvioglacial agencies but were later shaped by moving ice into the typical drumlin form. There are also other fluvioglacial deposits that may not have the form normally associated with ice-contact meltwater deposits. These may have been laid down beneath the ice when it was stagnant and melting. Some of these deposits have been referred to as undermelt features and often they show interesting structures. Other features associated with stagnant ice will be considered in Chapter 17.

1 Sedimentology and sedimentary structures

The sediments constituting eskers, kames and associated forms were laid down in environments dominated, first, by variable but often powerful meltwater flow, and secondly, by collapse of the adjacent ice through melting. Thus characteristic structures of normal fluvial sediments, such as cross-bedding, ripples and dunes, are combined in these fluvioglacial sediments with contorted bedding, faulting and collapse structures, making interpretation often very difficult and complex.

The characteristics of the ice-contact sediments in the Shrewsbury area of England have been discussed by J. Shaw (1972a, b). The sediments under investigation are formed into ridges parallel to the current direction. The fluvioglacial deposits revealed in boreholes showed a wide variety: gravel was dominant in one borehole, clay in another, and sand predominated between. The gravel consists of steeply dipping foreset beds, and horizontal or gently dipping imbricate gravels, cross-bedded units, about 1·5 m thick and lense-shaped, with some gravels, and cross-laminated fine sand and silt. There is some faulting between the gravels and sands, the latter being in a zone of subsidence. The sands show large-scale trough cross-stratification (Fig. 16.1C). There are also thin beds of horizontal stratification and small-scale cross-stratification. The sands merge into parallel-bedded silts. The upper part of the section is contorted with injection structures of fine material, which have moved up from below, possibly by squeezing, and collapse structures in coarser sediment. Kettles occur among the

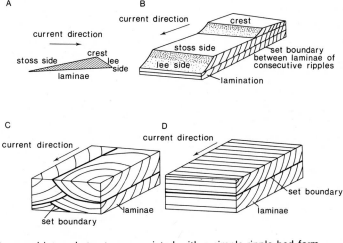

Fig. 16.1
A & B The form and internal structure associated with a simple ripple bed form
C Large-scale trough cross-stratification
D Large-scale tabular cross-stratification
(J. Shaw, *Field Studies*, 1972)

ridges trending north-west to south-east, indicating accumulation in ice-walled channels. The vertical distribution of deposits in such an ice-contact environment is very different from the normal fluvial distribution, since lateral movement is prevented by the ice walls and the normal fluvial cyclothem fining upwards is not found. Gravels were being actively carried in the channels as indicated by the cross-bedded gravel units, reflecting a high flow régime. The presence of large-scale trough cross-stratification adjacent to the gravels indicates a fall in stream power. During periods of low flow, the dune bedform, which generates this type of stratification, is replaced by small-scale ripples, which generate small-scale trough cross-stratification. The flow is likely to be reduced with time as the channel becomes wider by melting and as water leaks to lower levels, as well as by the formation of marginal lakes where clays can collect. Layers of saturated sand overlying saturated silts and clays are unstable, leading to collapse into them of overlying tills, and injection phenomena can arise owing to the lower bulk density of silts. Variable flow is indicated by the intervening ripple and parallel-laminated fine beds, and the depth of flowing water was probably small.

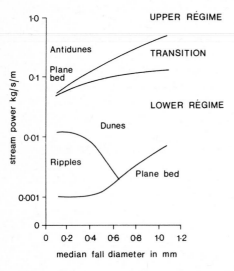

Fig. 16.2
The relationship between stream power, bed grain size and bed form (J. Shaw, *Field Studies*, 1972)

The sedimentary structures can be related to the flow régime (Fig. 16.2). Ripples (Fig. 16.1A) are triangular in cross-section parallel to the flow. Dunes range in wavelength from a few centimetres to several metres, but have much the same shape. Large-scale asymmetrical ripples or dunes can have either tabular or trough structure (Fig. 16.1C, D). Plane beds are horizontally stratified, which may occur at low-power flow or upper-flow régime; in this case, the latter applies. Antidunes are long-crested, symmetrical and sinusoidal, and occur at very high régime flow. They show upstream-dipping and poorly developed cross-strata, with dips less than 10°.

Both macro- and micro-studies can give evidence of the nature of former fluvioglacial water flow (A. V. Jopling, 1971). The micro-scale evidence includes sedimentary structures. The sharpness of bedding tends to diminish as the flow régime increases in intensity. The degree of definition is related to the effectiveness of sorting and is optimal in the dune range. The plane bed develops at $4 \cdot 5 \, V_{\text{crit}}$, while dunes are stable from 1·5 to 3·5 V_{crit}, where V is the velocity of flow. It is possible to calculate the stream depth, if some other variables are known, from the equation

$$\frac{V.D}{\gamma}F=\frac{V.D}{\gamma}\cdot\frac{V}{gd^{1/2}}=300$$

where F is the Froude number, γ is the kinematic viscosity for a smooth bed ($c.$ 3×10^{-6} m/s) and $d=$ stream depth. If $V=1\cdot37$ m/s and $D=0\cdot4$ mm, $d=0\cdot86$ m.

Another complex sequence of fluvioglacial deposits at Hendre near Pentraeth, Anglesey, has been examined by D. G. Helm (1971). The nature of the bedding provides significant evidence of the method of formation. The deposits in general become coarser upwards, corresponding to lacustrine, deltaic and fluvial sediments, and the features are not now thought to be eskers. Six lithostratigraphic units have been identified. The first consists of gravels, and the second of well-bedded sands with cross-laminated units, ripples and ripple-drift bedding. The third shows most variation of grain size and largest cross-bedding, the maximum being 6 m with a mean of 3–4 m. The fourth unit fills isolated erosion channels, cut into the foresets of the third, while the fifth contains a wide range of grain size and changes facies rapidly, being poorly stratified. The sixth unit is fine and has features of wind-blown loess, being silty and structureless. The sediments are mostly undisturbed although some fracturing has taken place. The material accumulated in a depression and water currents were directed to the south or south-east. Seasonal control of sedimentation is indicated by the range of material from coarse to fine. The unconformity below the fifth unit marks the change from glacio-lacustrine to fluvial processes. Morphologically, the feature is a deceptively simple ovoid with a long axis trending north-east to south-west, transverse or oblique to the flow of water. Surface features of a great many fluvioglacial deposits give little clue to their true origin, and great caution is needed, in the absence of proper exposures and sections, if the origins of fluvioglacial forms are to be correctly identified.

2 Eskers and related features

An esker is essentially an elongated ridge, formed of stratified fluvioglacial deposits, usually including much sand and gravel. The ridge may be sinuous or straight. Occasionally an esker consists of a number of separated elongated hills of bedded sand and gravel, forming together an elongated but discontinuous feature. The term 'beaded esker' has been used to describe this rather unusual type. G. de Geer in 1897 used beaded eskers as evidence of still-stands in the ice margin retreat. Eskers also show occasional widening and other irregularities. Esker slopes are variable; at times they are steep, lying at the angle of rest of their material, but sometimes they have been modified to gentler slopes of 5–10°. Some of the largest eskers occur in central Sweden, where they attain hundreds of kilometres in length, with only short gaps. Typical eskers have a fairly constant relationship between their different dimensions. Large eskers, even though they are very long, are never wider than 400–700 m and 40–50 m high. Smaller eskers may be 200–300 m long, 40–50 m wide and 10–20 m high. The material within them is always well worn and sorted. Some contain pebbles 5–20 cm long and although many have smaller material, some in western Baffin Island consist of large stones up to 1 m in size. Many eskers consist of bedded sand, dipping at 10–20° or even more. The dip is normally outward from the centre of the esker and sometimes coincides with the surface slope. An esker fabric measured in an esker in Baffin Island showed a strong preferred stone orientation in the direction of esker elongation. There was also a strong preferred dip in a downstream direction.

2.1   *Observations of eskers forming currently*

Several eskers have been described in the process of formation. W. V. Lewis in 1949 described an esker forming near the snout of Böverbreen in Norway. The glacier ends on a gentle slope on which much mixed outwash has been deposited. The esker, as it emerged from beneath rapidly retreating ice, was nearly 37 m long and 3·6 m high. It had a flat, almost horizontal top, contrasting markedly with the sloping gelifluction area around it. The surface carried several small and still-active kettle-holes. There were more or less horizontal topset beds on the upper surface. Pits revealed foreset beds beneath 30–45 cm of topset beds, and this bedding suggests that the whole feature must have formed in standing water. It was apparent, however, that no lake could have existed in front of this glacier while the esker was forming. The steep side-slopes

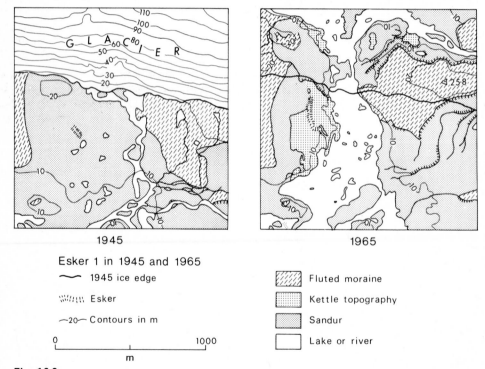

Fig. 16.3
Maps comparing the same area of the ice margin and proglacial zone of east Breiðamerkurjökull, Iceland, in 1945 and 1965. The 1945 ice edge has been superimposed on the 1965 map to aid comparison (P. J. Howarth, *Arctic and Alpine Research*, 1971)

of the esker were caused by two meltwater streams that were issuing from the glacier on either side of the esker. The esker still contained an ice core, at least in part, whose melting would disturb the stratification and the form of the esker. Well-preserved eskers are therefore only likely to form where ice is unlikely to be incorporated in the deposits. The material forming the esker must have been deposited in subglacially ponded water near the point at which a stream emerged from the glacier snout. The ponded water subsequently escaped as the ice front melted back.

J. C. Stokes (1958) has described an esker-like ridge extending on to dry land in front of the retreating Svartisen ice cap in north Norway. The ridge emerged from a tunnel, and consisted of sand, boulders and rock flour, with an ice core near the snout

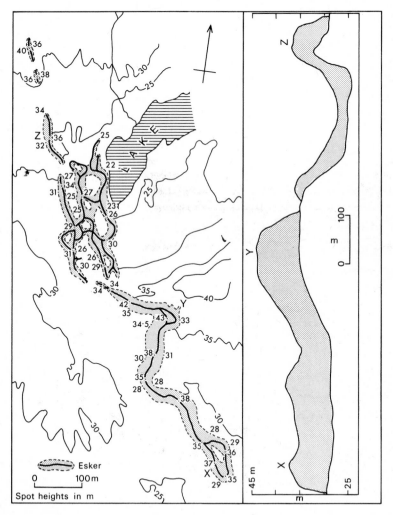

Fig. 16.4
Map and long profile of an esker near Breiðamerkurjökull, Iceland. The ice margin is to the north
(R. J. Price, *Institute of British Geographers, Transactions*, 1969)

where it was 16 m high. It decreased in height to 1–2 m away from the glacier as its ice core melts. The ice in the vicinity was almost stagnant, although there were signs of thrusting in the dirty ice as it pressed against a marginal buffer of dead moraine-covered ice. In tunnels beneath the glacier, debris was found piled up against the walls, ready to form ridges as the ice melted. This process would form small un-stratified esker-like ridges, similar to genuine eskers in that they form in a subglacial tunnel, but unlike them in their lack of stratification.

Eskers forming at the margin of Breiðamerkurjökull in Iceland have been examined by P. J. Howarth (1971) (Fig. 16.3). An area shown as sandur (outwash) on an air photograph of 1945 has since then been modified by the melting of buried ice to reveal eskers flanked by kettles. One esker was 350 m long, 5 to 15 m wide and 1 to 5 m high, possessing minor ridges which looped around and joined the main ridge at both ends. Its proximal end was discontinuous and, in the centre, it split into two ridges which rejoined. The form and height of the eskers was affected by the melting-out of buried ice in them. Gravels in at least one esker were underlain by glacial ice, indicating a possible englacial origin. Another esker, which started to appear in 1960, melted out of the glacier and increased in length from 420 m in 1965 to 470 m in 1966. It was then 15 to 20 m wide and 5 m high close to the ice, but only 5 m wide and 1 m high at the distal end. A shear-plane carrying material into the upper part of the esker was revealed in a section cut by a stream. Part of the esker runs parallel to the ice margin and part at right-angles to it, probably reflecting control by structures in the ice. Eskers were clearly able to survive melting-out without losing their internal structure; only a small proportion of the gravel and sand slumps down the steep outer slopes. As the ice core melts, the main body of the esker is slowly let down on to the ground. The rate of melting of the ice core is normally greater than the rate of slumping of sands and gravels from the crest. Similar processes can, of course, operate if the esker is of subglacial or supraglacial origin.

Further examples of eskers currently and recently visible near Breiðamerkurjökull (Figs. 16.4 and 16.5) are analysed by R. J. Price (1969). As well as the particular examples described by Howarth, there are some 5 km of sharp-crested, steep-sided sand and gravel ridges (Plate XXVII) in parts of which stratification is visible despite the unstable nature of the loose material. These ridges are typical eskers, representing the former courses of streams in, under or on the ice. Pebbles in the gravels are well-rounded except in the layer of ablation moraine which sometimes caps the eskers. The base of the eskers falls generally towards the distal end, by some metres, but it is irregular, as is the crest profile. In 1965, the largest esker extended up on to the ice surface, and a stream cutting through it in 1966 revealed an ice core. This and other evidence strongly suggests that the eskers were let down from a position on or in the ice to the subglacial surface: this would most readily account for the undulating base profile (Fig. 16.6). There is no need to invoke the flow of subglacial streams under hydrostatic pressure (p. 473) to explain these eskers: an englacial origin presents fewer problems. The work of Price and Howarth in this area of southern Iceland clearly shows the value of sequential air photography in studying rapidly changing proglacial and ice-marginal areas.

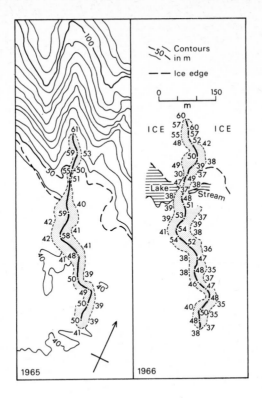

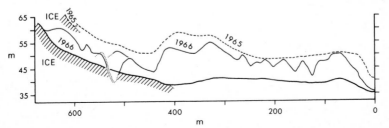

Fig. 16.5
Maps and profiles of an esker near Breiðamerkurjökull, Iceland, showing changes 1965–66. The 1965 map and profile are based on photogrammetric data; those for 1966 on tachyometry (R. J. Price, *Institute of British Geographers, Transactions*, 1969)

2.2 *Methods of esker formation*

Two main hypotheses of esker formation have been put forward. First, there is the view that the esker deposits have been laid down by subglacial or englacial meltwaters as they emerge from the glacier into ponded water. It has been suggested that eskers will form in certain conditions by the annual addition of small mounds of sediment as the ice front retreats. This leads to the formation of the beaded esker type. The summer period of rapid water-flow and abundant material adds each bead, while the

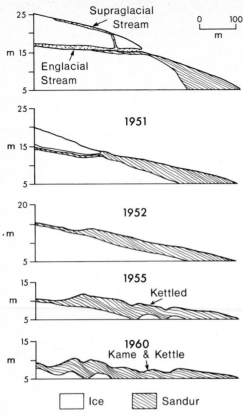

1950

Supraglacial
Stream

0   100
m

Englacial
Stream

1951

1952

1955
Kettled

1960
Kame & Kettle

☐ Ice   ▨ Sandur

Fig. 16.6
The forms resulting from deposits laid down
by supraglacial and englacial streams on top
of stagnant ice. The sequence is inferred
from air photographic evidence for the Breiðamerkurjökull area, Iceland, though the
diagram is schematic (R. J. Price, *Institute
of British Geographers, Transactions,*
1969)

Plate XXVII
An esker ridge crossing the proglacial area of Breiðamerkurjökull, Iceland. (*C.E.*)

winter is represented by the intervening gap. Eskers of this type seem rather rare. The second method of formation is the most favoured. This hypothesis suggests that the eskers are formed subglacially or englacially by deposition from meltwater. The fact that meltwater streams can flow uphill under hydrostatic pressure has been used to explain the fact that some eskers run uphill in places. However, a simpler hypothesis, as noted already in the discussion of Price's and Howarth's work on the eskers of Breiða-merkurjökull, is that the eskers formed first in normally sloping englacial tunnels and that the esker gravels were later superimposed on the underlying irregular ground surface as the ice melted. In general, eskers are aligned down the ground slope and roughly parallel to the direction of ice flow, but there are exceptions.

A subglacial or englacial origin of the eskers of the Rondane district of Norway is also supported by the fact that the stratified esker material is overlain by a thin layer of unstratified ablation till, which must have been let down on to the esker as the overlying ice melted away. Such eskers are often rounded in cross-section. The streams depositing some of these eskers in the Rondane first flowed down the open hillside, and then under or through a lobe of decaying ice in the valley, before emerging again from the ice. The reduction of gradient under or in the ice could account for the deposition of material in this instance. The blocking of stream channels by the collapse of decaying ice may at times be instrumental in initiating the deposition of material in the tunnels. C. M. Mannerfelt (1945) has described similar eskers in parts of Sweden.

The necessity for the ice to be almost dead in order that true eskers may be formed has restricted their occurrence mainly to areas covered by large ice caps rather than by valley glaciers. Even rapidly retreating valley glaciers such as Austerdalsbreen move actively right to their snouts and this prevents esker formation. Eskers can also only form in those areas where meltwater can penetrate readily through the ice. Eskers are, therefore, normally found in association with temperate ice caps, where the ice is at the pressure melting-point.

It has been suggested that a rare type of esker could form by deposition of material in supraglacial meltwater channels. The material could then be let down on to the surface beneath the ice as the surrounding ice melts. Such a feature could form on a cold glacier with surface meltwater drainage, but as most glacier surface streams contain very little debris, eskers formed in this way are likely to be uncommon.

A model to demonstrate the formation of eskers has been devised by G. F. Hanson (1943). The experiment was designed to simulate a glacier stream discharging into a lake. A pipe, discharging sand and water into standing water, was slowly moved backwards on a platform, representing the ice margin. This sand emerging through the pipe produced an esker in the lake. This deposit remained uncollapsed if the water in the lake was allowed to drain away slowly. The 'esker' formed both with the pipe outlet just above the lake level, and also when the pipe was lowered below the lake level to simulate a subglacial or englacial stream. A flat-topped ridge formed when the lake level was lowered below the level to which sand could be deposited. The height and width of the ridge appeared to be related. These experimental eskers illustrate some interesting similarities with genuine eskers. The flat top, in particular, is seen on some of the Trim eskers (pp. 477–9).

2.3　Descriptions of eskers in different areas

R. J. Price (1965) has described eskers left in front of the retreating Casement glacier in Alaska, some of which have been revealed since the glacier became land-ending about 1907. A complex set has been exposed from the retreating ice since 1935. The eskers are ridges 3–9 m high, formed of well-rounded gravel, cobbles and boulders. They occur on ground that slopes both away from and towards the glacier. Those formed on the slope down towards the glacier must either have been formed by sub-glacial streams under hydrostatic pressure or by englacial streams. In the latter event they must have been let down subsequently on to the reverse slope. A complex series of ridges is thought to have developed supraglacially in lakes, which have since drained, leaving shoreline features on the ridges. The largest ridges were probably on the site of meltwater streams and they appear to be true eskers which were underlain by ice at one time. Some near the glacier have ice cores, and photographs showed them extending on to the ice in places. They could either have formed in englacial tunnels or on the ice. Subsequent melting must have produced the uneven crest line. The lakes must have formed after the ice melted out beneath and around the eskers. Some of the eskers were 40 m high and 33 m wide at their base when they were still on the ice. These eskers have been destroyed by meltwater streams subsequently.

Hoppe (1961) has described some of the eskers of northern Sweden. These eskers occur both above and below the highest marine limit. They often extend for 5 to 10 km between breaks, which normally occur where the relief is highest. The largest eskers have rather flat, broad crests, supporting the view that the deposits were built up in successive layers in subglacial or englacial meltwater channels. Hoppe agrees with V. Tanner (1932) that the sharp-crested narrow variety of esker is the result of subsequent collapse of the sides when the supporting ice walls melt away. The eskers formed above the water-level consist mainly of stratified sand and gravel. Sometimes silt occurs within the esker or on its surface, and occasionally heaps of boulders are found.

One particular esker, Hammesharju, shows some unusual features. It is situated south-east of Gällivare and is over 4 km long, 50 to 80 m wide and up to 16 m high, although its crest is broken by gaps. The esker was formed probably in a subglacial or englacial tunnel; later, a supraglacial stream was let down across it and eroded the gaps that occur along it. Ridges of till flank the esker on either side. These ridges contain stones preferentially orientated transverse to the ridge elongation. It is thought that they were formed by the squeezing of water-soaked till into the area on either side of the tunnel. There are also examples of a similar type of feature that forms when the till is pressed beneath and into the central core of the esker rather than alongside it.

Another interesting example is the Vessö esker in southern Finland. It is in part a continuous ridge, but in places it is broken into separate mounds. It is situated about 50 km east of Helsinki, is 8 km long and varies in width from 300 m to 2 km, with a maximum height of 38 m (Fig. 16.7). It runs south-east to north-west radially to the large Salpausselka moraines to which it is linked. Its outer part consists of several islands, while it has been modified by the sea as it has risen isostatically since its emergence from the ice. The esker as a whole is asymmetrical and has a steeper eastern slope with a stony surface, continuing in places below sea level. In places, there are depressions, which are steep and circular, along the ridge. Along part of its length, the esker forms a broad ridge with associated boulder-covered mounds. Elsewhere, the esker spreads out and merges into an even, sandy plain. The fluvioglacial material of which the esker is composed lies directly on the rock with no intervening till. The bedrock, an easily weathered granite, was eroded to form a depression along which the esker is situated.

The structure of the esker has been revealed in cuttings at a point where the eastern slope is 13° and the western only 3°. The esker is covered with boulders and its internal structure is shown in Fig. 16.7. The sloping layer of clay (4), with a dip of 8°, runs

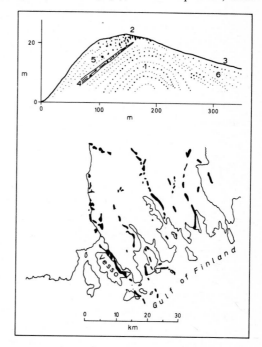

Fig. 16.7
The Vessö esker, Finland (O. Granö, *Fennia*, 1958, by permission of the Johnson Reprint Corporation)

almost parallel to the steeper eastern side. This layer probably formed the original esker surface, and the overlying sand and stones (5) consist of later marine additions. Beneath the clay and silt, sand (1) occurs in alternating layers of finer and coarser particles, showing cross-bedding. In the centre, the material is coarser. Some large angular boulders occur in the silt layer and on the upper surface (2), where the gravel and stone layer is 1–3 m thick.

The form of the esker and its constituent material give indications of its mode of formation. O. Granö (1958) suggests that the fluvioglacial material of the esker was deposited close to the ice margin in deep water. He considers that the character of the esker, particularly its core, agrees with the views of De Geer concerning esker formation in submarine environments. The original core was probably deposited as a small delta at the mouth of a subglacial stream emptying into the sea. A clay and silt layer represents a remnant of the covering of fine material deposited when the esker was still in deep water. It includes large ice-rafted angular blocks. The boulders covering the clay could have been derived from a covering of stony till from which the fine material has since been washed away. The sand (6) above the unconformity on the western side of the esker was deposited by the waves as the esker emerged above the sea. The steeper eastern slope is the result of movement of material to the lee side by the waves whose maximum fetch lay to the west. The feature has been considerably modified by the sea since its formation at the edge of the retreating Scandinavian ice sheet.

The Lammi esker, described by E. Wisniewski (1973) stretches north-west from the Salpausselka I moraines. It consists of a series of proglacial deltas linked by short ridges, and lies in a bedrock trough. The deltas formed successively in the Baltic Ice Lake. In places, the esker ridge is 200 m wide and 35 m high, and is composed of gravel and pebbles up to 30 cm in size. Its sides slope at 25° to 30°. Some boulders up to 2 m in size indicate an ice-contact position; they overlie bedded material whose stratification is deltaic. Deformation of the finer layers in places has resulted from melting of buried ice, and there are also kettle holes. The esker now has the form of hillocks of varying size, formed during deglaciation in a sub-aqueous environment. Each delta marks a pause in the retreat of the ice front, and the esker deposits accumulated in crevasses or channels along which fluvioglacial transport was taking place. It is a classic example of a De Geer-type esker. Most of it was submerged on deglaciation, resulting in some loss of fines and abrasion through wave action.

P. Fogelberg (1970) analyses the features of deglaciation associated with the Salpausselka II moraine in an area of southern Finland. In the moraine foreland, there are eskers whose southern ends rise to ridges or hills 135 to 140 m above sea level, northwards from which a lower tail continues at 100 to 110 m. The steep hillock at the southern end was deposited between ice walls where the ice-channelled stream or tunnel opened into a lake. The tail is part of the feeding esker formed in the ice tunnel. In the moraine hinterland, the fluvioglacial deposits include more continuous eskers. No De Geer-type eskers occur either in the foreland or the hinterland of this area.

Farther north in Finland, the deglacial history of the Tana valley provides further illustration of the complex environments with which esker systems are associated. The ice sheet melted rapidly from this area, its margin melting back at an average rate of 230 m/year until it became ice-free in 9500 BP (H. Mansikkaniemi, 1970). Subglacial sedimentation was extensive during this phase, and bottom eskers and subglacial valley trains formed. The largest eskers occur where the valley is narrow and deep; these include bottom eskers 10 to 40 m high that follow the valley floor and that were formed in ice tunnels. They have been trimmed by later marine incursion. During deglaciation, short-lived lakes formed in the narrow parts of the valley and these were filled with fine sediments, at a level above that of the river deposits. The sorted sediments attain thicknesses of 20 to 70 m on the valley floor. There are also 'slope eskers' descending the valley sides, which are smaller and less than 10 m high, the infillings of subglacial chutes (p. 346). Some of the large bottom eskers are continuous for 7 km. Their width varies from 100 to 300 m—25 to 50 per cent of the width of the valley bottom. Deep kettle holes are associated with them. The finest deposits in the eskers are well-sorted, and the stones rounded, having travelled 25 to 35 km. Coarse sediments, on the other hand, are poorly sorted and only slightly stratified. Some of the bedding in the bottom eskers is nearly horizontal. The eskers probably formed under dead ice 100 to 200 m thick, whose downwasting led to the plateaux and uplands being exposed while ice remained in the valleys. Fine sediments accumulated as melting ice caused blockages in the valley at some places. Coarse sediments were deposited when the flow was rapid in subglacial or englacial tunnels. These were enlarged by melting until the marginal ice walls collapsed, together with the roof, and the fallen blocks of ice were then covered by further sediment. The slope eskers are less numerous because rapid flow of water down the subglacial chutes tended to prevent deposition.

The deglaciation of the Holy Cross Mountain area west of the Vistula in Poland has been discussed by C. Radłowska (1969). The Tarłow esker is associated with the Vistula ice lobe of Riss age. Its roots lie, in part, in a valley, although not at the lowest point. Its crest reaches 192 m where the plateau level is 210 m, and its course is sinuous, consisting of long hillocks and mounds, whose irregular relief is said to be partly the result of subsequent erosion. Coarse material occurs in the esker core, overlain by bands of gravel-pebble mixture and some sand. The structure is complex and in places disturbed, particularly where the material is coarsest in the north. Boulders up to 0·5 m across are tightly packed at the esker base. Material is more uniform towards the south. The lower eastern slope is thinly mantled with till up to 5 m from the crest, suggesting that the esker was formed in a tunnel. Large soft-rock boulders indicate short distances of transport, and there is little sign of sorting. Strong cross-bedding occurs and there are lenses of sand and silt. Three stages of deglaciation are inferred:

1 the esker developed in a subglacial tunnel and was then partially covered by till and ablation moraine as the ice decayed
2 a free space developed between the esker and the dead-ice wall, and water flowing along this channel added a kame terrace to the esker
3 after the dead ice decayed, a valley formed adjacent to the kame-esker, followed by settling and gelifluction on the slopes. Eolian activity finally added some sand dunes.

In central Poland, Z. Michalska (1969) differentiates true eskers from crevasse fillings. Esker materials are well washed but vary in particle size and bedding. Accumulation occurred in a marginal tunnel or, more rarely, in a decay crack near a tunnel outlet. There were probably short breaks in sedimentation, a series of layers representing separate depositional episodes with different sources. Till layers several metres thick are found near the bases of some eskers; at some later stage, channels up to 30 m deep were cut in the till. Following channel erosion, up to 30 m of fluvioglacial sediments were deposited in several episodes, to be overlain in turn by sandy deposits laid down in continuously flowing water. The tunnel then increased in size and deposits 15 m thick accumulated along the channel axis. Finally, the tunnel roof melted, letting down a mantle of till 1–2 m thick.

Mention should be made of some of the eskers of Ireland, as it was from this country that the name 'esker' originated. An interesting series of eskers and associated features has been described by F. M. Synge (1950) in the area around Trim, north-west of Dublin. The area lies in the drainage basin of the River Boyne, which flows north-eastward to the north of the main esker area. The area is drained by streams flowing north-westward to join the Boyne (Fig. 16.8). When the ice sheet associated with the eskers lay over the area, its margin ran roughly from north-east to south-west and was

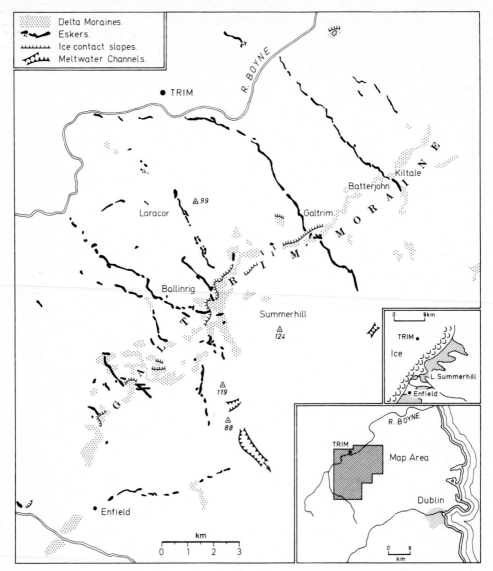

Fig. 16.8
The eskers near Trim, Eire (F. M. Synge, *Proc. R. Ir. Acad.,* 1950)

retreating north-west. As it was retreating downslope, it ponded up the drainage in front and lakes were formed which overflowed across the watershed to the south-east. It was in these lakes, according to Synge, that the eskers and associated features were formed. The eskers lie in a strip of land between the Boyne and the elongated ridge of sand and gravel, named by Synge the Galtrim moraine, which will be discussed in more detail on p. 491. The eskers trend from north-west to south-east within this strip. The bedrock is hidden beneath superficial deposits of drift, including sands, gravels, silts and till, which form a gently undulating area. There are twelve eskers in this area, including the impressive Trim esker. They all trend in about the same direction and join the Galtrim moraine at their eastern or south-eastern ends. The eskers are morphologically similar, consisting of long, narrow, steep-sided ridges which are sinuous in plan.

All of them consist of bedded sand, gravel and some silt. There appears to be relatively little connection between their internal structure and their external form. One esker of uniform shape may change internally from fine sand to coarse gravel. Structurally also they are varied. Some consist of apparently horizontally bedded sands and gravels, often showing slumping and disturbance by faulting on their flanks. Some show arched bedding. Others consist of unsorted, though washed gravels, containing very large striated erratic blocks of limestone. Some may show delicate deltaic bedding, with some bands of lacustrine clay. The bedding may be graded either from coarse to fine upwards or the reverse. The main impression is one of variety.

The Trim esker is one of the best of the series. Its total length is 14·5 km and it increases in size towards the south-east. It starts as a small ridge 4·6 m high on the west bank of the Boyne. As it is traced to the east of Trim, it runs parallel with the Boyne and increases to 9 m in height, and becomes a well-marked ridge 15 m high for the last 8 km of its length. Where it reaches the Galtrim moraine, it extends up on to it and ends beyond the moraine crest at a height of nearly 100 m. It is unusual in that it does not end in a delta formation as many of these eskers do.

Another rather unusual esker is the Laracor esker, consisting of two parallel ridges which show a tendency to beading along part of their length. It is a small esker, only about 7·5–9 m high and 4 km long. Two of the other eskers approach one another, but then swing apart, one ending in the Galtrim moraine and the other ending in a small delta.

Three important generalizations concerning this series of Irish eskers can be made. First, they show winding courses like rivers. Secondly, most of them act as feeders to the Galtrim moraine. Thirdly, many of them show some tendency to beading. The pattern and character of these eskers agree with the hypothesis that they formed under lacustrine conditions against or within walls of stagnant ice. Two of the small eskers continue over the crest of the Galtrim moraine. These probably formed where a cliff of dirty ice overhung the developing Galtrim moraine. Two parallel tunnels formed beneath this ice cliff, which later collapsed on to the surface of the moraine. The tunnels became choked with gravel, and eventually formed the eskers. Where the eskers show beading, it is suggested that the beads represent stages in the retreat of the ice front, the material being deposited marginally. The absence of eskers north of the Boyne valley shows that ponded water is necessary for their formation. As the ice retreated beyond the valley line, the water could escape freely from in front of the ice margin. However, it has been shown that small eskers can form where water is ponded up temporarily within the marginal zones of a glacier or ice sheet. It is probable that features of this type can form fairly rapidly. Fluvioglacial streams carry a large load of material and where conditions are suitable, deposits can build up fast. Incision can be equally rapid where the streams are acting erosively.

The Tullamore eskers have been described by A. Farrington and F. M. Synge (1970).

The Kilcormac esker is 32 km long and runs almost from Shannon to Screggan. It appears to be a ridge formed by a subglacial (? englacial) stream. It is narrow, symmetrical and continuous, ending in a gravel spread covering 2·5 km², with horizontal bedding. The esker rises 15 to 18 m above its surroundings; where it merges into the gravel spread, there is a moraine perpendicular to it and a hollow with a steep slope that is probably an ice-contact slope. Other moraines perpendicular to the esker suggest that it formed as the ice front retreated. The Geashill esker south-east of Tullamore is 13 km long but only 4·6 to 6 m high. It is similar to the other eskers and is also crossed by moraines. The Ballyduff esker is larger, up to 31 m high. It runs east–west for 17·6 km. At one place it is a steep-sided ridge with a level upper surface, pitted with kettle holes, and it is crossed in places by other smaller ridges. Arched symmetrical bedding is revealed, with no sign of slumping. The whole assemblage of features is similar to that described in the area around Trim.

Some interesting observations have also been made of the stone types and mineral grains in eskers. H. A. Lee (1965) has described a long esker in Canada called the Munro esker (latit. 48°N, long. 80°W). The esker can be traced easily on aerial photographs over a distance of 400 km trending north–south. It consists of a winding plateau 1·6 to 6·4 km wide, with depressions along its axis. Bell-shaped ridges occur between the depressions, and these correspond to subsurface depressions, giving a total thickness of more than 88 m of sediment. The sediment is dominantly sand, but there are layers of coarse gravel or boulders. Sheet bedding is characteristic and cross-bedding suggests sedimentation in braided streams. Foreset beds are not prominent, but deep sections are uncommon. The sides of the esker have been modified by wave action in lake waters.

Samples of the esker material were taken at intervals to assess the distance that different sizes and types of minerals had been carried in the esker. The maximum abundance of any mineral does not occur at the point of outcrop of the rock containing the mineral but at some distance downstream. The distance between the outcrop source of the mineral and its maximum abundance has been called the K distance. This distance is the same for particles that behave in a similar way hydrologically but it differs with the size and density of the particle. The results of the sampling in the Munro esker indicate that dunite in the size range 3·35–8 mm had a K distance of $13 \pm 3·2$ km, while particles in the size range 8–16 mm had travelled less far. Trachyte of 8–16 mm had travelled $4·8 \pm 3·2$ km, and gold grains of 10 μm and over had travelled only $3·2 \pm 3·2$ km. These results show that transport of material in eskers is not very great, although some mineral grains are carried far beyond the K distance, as the curve of percentage number of grains tails off slowly from the K distance peak.

A. Hellaakoski (1931) has made similar observations on an esker at Laitila, Finland. The esker is rather sinuous and rarely exceeds 20 m in height. It runs roughly parallel to the striations and is approximately perpendicular to the ice front direction. It is more than 27 km long and is not influenced by the relief, although its shape changes with the altitude. It is ridge-like in the higher parts and forms rows of hills in the lower parts. At one time, it was under 150 m of sea. It runs across an outcrop of Rapakivi granite for 27 km of its length. The stone types were counted both in the drift alongside the esker and in the esker itself from the point where it crossed on to the Rapakivi outcrop. At a distance of 1·5 km from the proximal contact with the Rapakivi granite, this rock gives 50 per cent of the stone count in the drift. However, in the esker the Rapakivi granite only starts to appear 5 to 8 km from the proximal contact. Thereafter the percentage rises steadily and reaches a peak between 15 and 35 km from the proximal contact, where it is twice as high as in the adjacent morainic drift. The percentage of Rapakivi granite stones begins to fall at a similar distance from the distal contact of 5 to 8 km. The morainic drift was the source of the esker material and this was carried about 4 km or more before being deposited. The stones are destroyed in the esker, but preserved in the morainic drift. The stones also become rounded in the esker, but remain angular in the morainic drift. The agency forming the esker, however, could not drag the large boulders, which were lying close to it. The esker material was well worn, mixed and washed. It is suggested that the deposit was laid down by a subglacial river, but the results of the observations indicate that the stones forming the esker have not been carried very far.

Two different types of esker were examined in Baffin Island, where they occur in association with dead-ice phenomena and fluvioglacial outwash deltas. Some of the eskers feed into deltas in the same way as those associated with the Galtrim moraine in Ireland. These short feeding eskers in Baffin Island sometimes contain large boulders (the mean length of fifty was 89 cm) and merge into the ice-contact slope of the deltas. They must have been formed when the ice front remained stationary for long enough for extensive outwash spreads to build out into the sea or lake as deltas.

The other type of esker is more sinuous and occurs in dead-ice areas among the outwash deposits in association with kettle-holes and kames. Some of the eskers are sinuous and others bifurcate. At least some of these sinuous ridges indicate by their internal structure that they have been largely deposited by flowing water. The fabric of one esker shows a bimodal orientation of the long axes of the stones within it. The main direction is parallel to the esker crest with a marked downstream dip. The secondary mode is at right-angles to the crest with a dip down the flank of the esker and was probably induced by later collapse and slumping of the esker sediment. The variable down-slope dip of the beds also suggests collapse as the retaining ice walls melted away. The sand also shows characteristics of flow in its good sorting $(\sigma_1 = 0·70)$.

The form of another esker in the same dead-ice area is shown in Fig. 16.9. The ridge is sinuous and rises slowly in crest elevation by about 40 m along its length before it finally merges into a kame terrace. The material is coarse as shown in the Figure. The ridge is sharp crested and flanked on both sides by kettle-holes. Its sides slope at about 25°. These characteristics suggest that the final form of the ridge is the result of slumping. It seems unlikely that the ridge could have been formed entirely by water flowing in a subglacial tunnel owing to the very coarse material and the upward trend of the ridge in the direction of flow. It is more likely that the finer material was deposited by water flowing in ice-walled channels, possibly open, and that the larger stones rolled down into the deposits as the surrounding ice surface melted down. The similarity in roundness value (see p. 389) between delta and esker stones supports the view that most of the smaller esker stones and the finer sediment have been deposited by running water. The esker stones are significantly rounder than those of kames and moraines.

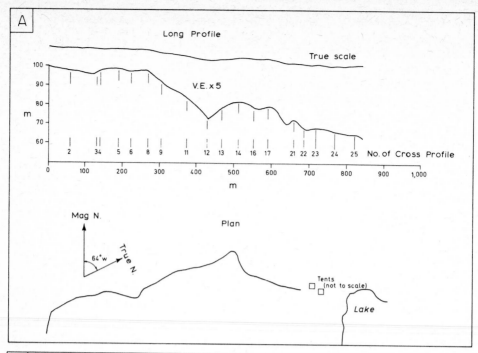

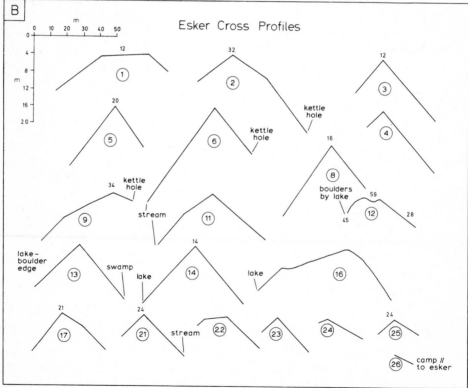

Fig. 16.9
An esker near Reflection Lake, Baffin Island, in profile **A** and section **B**. The figures above the cross profiles indicate the mean stone size in cm on the esker surface.

Plate XXVIII
Kettle-holes near Gillian Lake, Baffin Island. (*C.A.M.K.*)

This rounding can apparently take place very quickly in the powerful meltwater streams that issued from the retreating ice cap in this area.

It is likely that not all the ridges that appear superficially similar to the eskers in the dead-ice areas were deposited mainly by flowing water. A detailed survey of one dead-ice area, with large kettle-holes and ridges, in the large outwash spread near Tikerarsuk Point in west Baffin Island is shown in Fig. 16.10. The esker-like ridges bifurcate round small kettle-holes. The largest ridge has a beaded crest, the humps being 2–6 m above the hollows. The ridge is composed of sand and boulders. The sand has a much lower sorting value ($\sigma_1 = 1 \cdot 275$) than the other one mentioned and it is also mixed with stones of a mean size of 4·6 cm. These ridges probably were formed mainly of material washed into crevasses or hollows between decaying masses of ice, although they may have a foundation of water-deposited material. The ice masses finally melt out to form the kettle-holes and the sediments slump into the sharp-crested ridge form as the supporting ice melts away.

R. F. Flint (1928) has drawn attention to the differences between the characteristics of true eskers and esker-like features that are formed in crevasses in dead ice. True eskers may be up to 240 km long, and 4–40 m high. They trend parallel to the direction of ice movement, are discontinuous and sinuous, trend up and down slope, and rarely have level crests. They often show a tributary pattern. Coarse material predominates;

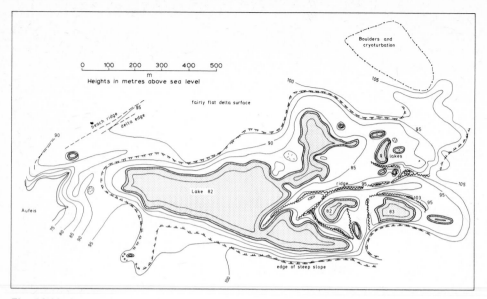

Fig. 16.10
A plane-table survey of a dead ice area near Tikerarsuk Point, west Baffin Island, showing esker-like ridges and kettle-holes

fine sand or clay is rare. Their bedding is variable, and they frequently have a coating of till. A cross-section often shows that the marginal bedding is parallel with the side slope as a result of slumping, with an average dip of 20°. These characteristics may be contrasted with those typical of the esker-like ridges formed as crevasse fillings. These ridges are associated with pitted outwash plains, and are small compared with true eskers, being 250 m to 1·6 km in length and 4–4·5 m high. Their trend is variable; they are normally not sinuous but they are continuous. They have no distributaries and do not pass over divides. They contain both fine and coarse material, which is horizontally bedded. There is no till coating, but marginal slumping occurs as in eskers. This type of feature is best called a crevasse filling and not an esker. The true esker is formed by water flowing in glacial tunnels.

3 Kames

The term 'kame', 'cam', or 'kaim' is of Scottish origin and has two meanings. It means either crooked and winding, or alternatively, a steep-sided mound. The term has been used to describe a great variety of forms. Both kames and eskers have a common relationship to stagnant ice and both are normally composed of sand and gravel. Because kames can form in a variety of ways, the meaning of the term should not be too closely associated with the genesis of the form. A kame can be conveniently defined as follows, after C. D. Holmes (1947): 'A kame is a mound composed chiefly of sand and gravel, whose form has resulted from original deposition modified by any slumping incident to later melting of glacial ice against or upon which the deposit accumulated'. T. F.

Jamieson in 1874 was the first person to interpret kames correctly. In 1894, J. Geikie used the term 'kame' to describe both isolated mounds of stratified sand and gravel and also other more complex features of morainic origin. His typical kame consisted of a single hillock, made up of stratified and cross-bedded sand and gravel near the centre, while the marginal parts showed indications of slumping. Kame deposits are sometimes poorly sorted and they may even contain some unstratified material and finer deposits. A. B. Keeble (1970) notes how flow tills (p. 370) are incorporated near the crest of a kame near Edmonton, Alberta; the kame also has a capping of ablation moraine that must have slumped on to the ridge as the ice melted away.

While kames consist of more or less isolated mounds (Plate XXIX), kame terraces form more continuous features along a valley side (Plate XXX). Both kames and kame terraces consist mainly of stratified material. Kame terraces are deposited by meltwater flowing along the edge of an ice mass between the ice and the valley wall (Chapter 12). The side of the deposit near the ice slumps down as the ice melts away to form the steep ice-contact slope that is characteristic of the outer side of the kame terrace. The surface of the kame terrace sometimes slopes gently in towards the hillside, as in the case of some kame terraces in western Baffin Island. Kame terraces may be discontinuous, only forming where the meltwater spreads out and can deposit its load. They may be absent where the water is concentrated in flowing round spurs and other obstructions along the valley side (A. Jahn, 1969).

S. Jewtuchowicz (1969) has described a kame at Zieleniew in Poland. It is composed

Plate XXIX
Section in a kame, Pentraeth, North Wales. Ablation moraine rests on water-bedded sands and gravels. (*C.E.*)

Plate XXX
Kame terraces (left) of Tioga age at Sonora Pass Junction, Sierra Nevada, California. *(C.E.)*

of sand, gravel and silt. Dips of the bedding vary greatly in direction and amount from almost horizontal to vertical. Several nearly vertical beds of till are also built into it, their strikes not always coinciding with the axis of the kame. They are 5 to 25 m apart and separated by sands and gravels. There is evidence of faulting and slumping—some of the deposits are disturbed and others are not—but no sign of upward squeezing of sediment. The distribution of materials is consistent with the melting of dead ice below the kame and associated collapse.

Observations in 1966 and 1967 of a collapsing kame terrace in south-eastern Alaska are described by G. D. McKenzie (1969). The deposits were laid down over the stagnant Adams Inlet glacier. Three to 6 m of gravel covered the dead ice, and the surface of the collapsing area had depressions up to 5 m deep. A cover of Dryas vegetation suggested that a period of 20 to 25 years had elapsed since colonization by vegetation began. The stagnant ice was at least 35 m thick. Water emerged from the junction of the gravel and ice, resulting in gullying. Slumping and sliding of debris were continually in progress. Heat, transferred downward through the gravel by rain water, raised the gravel temperature. The average rate of melting in an unvegetated gravel-covered area in July and August was 0·8 cm/day compared with an annual ablation rate of 24 cm/year (=0·07 cm/day). This rate of melting was, however, much lower than that of the stagnant ice where it was not covered with gravel; the exposed ice melted at an average rate of 2 cm/day.

J. B. Sissons (1958a) has described the ice-contact deposits of the Eddleston valley near Edinburgh. The ice in this valley gradually decayed *in situ*, providing suitable conditions for kame formation. Kames occur at two main elevations, at 268–274 m and 258–262 m, the latter being the more numerous. These kames were deposited by meltwater, which was flowing freely at the base of decaying ice. They must have formed in large irregular openings in the ice and the presence of kettles indicates the inclusion of lumps of ice among the deposits. Many of the kames have steep ice-contact slopes, which are up to 15 m high on at least one side. Eskers are associated with these kames, and formed as the continuations of subglacial meltwater channels. The heights of the crests of both kames and eskers are similar, suggesting that they were both related to the level of the englacial water-table, which was in turn controlled by a col to the north.

Various hypotheses have been put forward to explain kames. One view holds that the kames are formed where a meltwater stream, carrying a heavy load, debouches into a proglacial lake. The material is deposited in a deltaic form, and the side of the feature adjacent to the ice later collapses to form a steep ice-contact slope on the proximal side. Where several streams enter a lake close together, the small deltaic deposits may merge into a longer feature aligned parallel to the ice front and at right-angles to any eskers present. Normally, however, kames show a disorganized pattern and often exist as isolated mounds, which Synge calls 'lone kames'. These features can originate in several ways.

One possibility is the perforation hypothesis. This idea, put forward by J. H. Cook (1946), considers that the debris forming the kame collected in pools on the surface of a stagnant ice sheet. Sufficient debris must have been exposed on the surface of the ice sheet by ablation partially to fill the pool. As the pool warmed, it gradually melted its way down to the ground beneath. The material that had accumulated in it then became the kame mound as the surrounding ice slowly melted away. It must be assumed that ablation is largely achieved by evaporation. This process appears to be going on to a certain extent at the present time on stagnant ice masses such as the outer marginal area of the Malaspina glacier. However, the theory raises several problems. If ablation takes place largely by evaporation, there is no means whereby the finer sediment in the till can be dispersed. In some areas where the underlying till is clayey, a layer of fine clay should accumulate at the base of the hole and insulate it from further down-melting. The resulting deposits should contain more fine material than is normally found in kames. This is true of parts of New York State, which was the area discussed by Cook. It also seems unlikely that evaporation was the major means of ablation during the decay of the Pleistocene ice sheets in areas that are now temperate. Glaciers such as those in Spitsbergen do not now lose more than 3 per cent of their ice by evaporation. There is also evidence of abundant meltwater in other features, so that it is reasonable to associate kame formation with meltwater.

Another hypothesis put forward by Holmes (1947) suggests that kames develop where heavily laden melt-streams flow into small ponds on or within a stagnant ice mass. At the point of inflow the stream would deposit some of its load to form a small deltaic deposit. This would form an isolated mound when the surrounding ice melted away. Glacial meltwater streams often develop moulins (Chapter 6), and as these deepen, the material accumulating in them may be let down on to the glacier bed.

The original current bedding will then have a better chance of surviving when eventually the material is exposed as a kame mound. These mounds have some affinity with eskers and sometimes a thin band of deposits linking the kame mounds may be an embryo esker. Kames may at times develop as crevasse fillings in stagnant ice. Such deposits are usually elongated but probably show relatively little internal structure.

One of the best-known series of kames is that situated near Carstairs in southern Scotland. In 1926, J. K. Charlesworth interpreted this system of steep-sided fluvioglacial ridges, in places over 15 m high, as an end-moraine formed along the margin of ice from the Scottish Highlands. The ridges extend over a distance of 7 km and consist of coarse sand and gravel much of which shows current bedding. J. B. Sissons (1961) has shown that the features are not consistent with Charlesworth's views, and that they were probably the result of deposition by a complex series of sublacial meltwater streams, flowing beneath decaying Southern Uplands ice.

Sissons (1958b) has also described kames and kame terraces that formed during the deglaciation of East Lothian in association with stagnant ice. The kame terraces attain a maximum length of 1·6 km and a width of 200 m, but most of them are only a few hundred metres long and a few tens of metres wide. Their surfaces are mostly flat in transverse profile, but some slope slightly up towards the ice-contact slope, where the supply of debris from the ice must have been greatest. The longitudinal slope varies from almost horizontal to a few metres per kilometre at a maximum. They are mainly formed of sand and gravel, but in some, clay and silt also occur. Some of them show by their bedding that they were built up by streams draining along the edge of the ice or out from the ice. Where the beds are horizontal and the material is fine, the kame terraces were deposited in elongated marginal lakes. At their upstream ends, the kame terraces sometimes merge with thick deposits of valley till, deposited in the tributaries to the main valley in which the ice lay. At their downstream end, they frequently merge into meltwater channels.

As the ice shrank a series of kame terraces developed one above the other. For example, 6 km south of Gifford, a series of four kame terraces has been formed at levels between 244 and 320 m. Both the longitudinal and the frontal slopes of the uppermost terrace are steep, and it is composed of coarse gravel. The melting-out of detached blocks of ice has led to this terrace forming a series of separated mounds and ridges.

Most of the kames of this area lie in a belt about 3·2 km wide at 168–228 m on the northern flanks of the Lammermuir Hills. A ring of kames formed around a small lobe of isolated and stagnant ice south-east of Gifford. These kames form flat-topped mounds around a series of marshy hollows about 12 m deep, and formed as material accumulated around the edge of the decaying ice and in crevasses within it. The marshy hollows represent kettle-holes. The internal structure of one of the isolated kames, situated about 5 km south of Gifford, is revealed in a section that runs from south-south-east to north-north-west along the length of the feature. The lowest beds consist of thick layers of sand, which have been partially eroded before the next layer, consisting of gravel, was laid down. The gravel is overlain by alternating bands of reddish clay and brown silty sand or yellow sand, and then by another thick layer of sand, which merges northward into layers of gravel. Similar gravels occur in two lenses within the uppermost layer of alternating sand and silt. These deposits show that, at times, a

Plate XXXI
Crevasse in buried ice. The buried ice lies under 1–2 m of sandur gravels which are collapsing into the chasm, showing how crevasse-fillings originate. A short linear ridge will be left after the ice melts. Active glacial ice of Breiðamerkurjökull, Iceland, can be seen behind. *(C.E.)*

powerful stream fed material into a small lake, held up in front of the ice, while at other times finer material could accumulate. The flat top of this kame suggests that the lake was filled to the surface at least at the northern end from which the material was derived. In places, the gravels consist of foreset beds dipping at about 20°.

Both kames and eskers have a rather similar distribution. They occur in those areas where the ice sheets stagnated, melting away slowly and in such a way that much fairly coarse material was available. Much meltwater must also have been present to redistribute the debris and deposit it in and around the margins of the static and decaying ice masses. Kames and eskers are found in large numbers in central Ireland, south of the Kells moraine, and in those parts of Scotland where ice masses became stagnant. They are also well developed and numerous in the lower parts of Sweden and Finland, and in similar areas relative to the ice sheets in parts of Canada. In both Canada and Scandinavia, large ice sheets wasted away rapidly and released much meltwater into lakes ponded near the ice margins or into the sea, the level of which was then higher owing to isostatic depression. Eskers and kames also occur in western Baffin Island in association with dead-ice phenomena (Plate XXVIII).

The term 'kame' is restricted by W. Niewiarowski (1963) to forms that occur in association with dead ice; thus he eliminates forms laid down by meltwater around active ice margins. The majority of Polish kames occur in association with drift of the last glaciation between the outermost moraines and the Pomeranian moraines. This same distribution occurs also in Germany. The kames are usually associated with kettle holes and never with drumlins. They occur in four forms: (a) kame hillocks, (b) flat hills and kame plateaux, (c) kame ridges, and (d) kame terraces. The first (a) rarely exceed 10 m in height and occur in groups, while those of the second type (b) occur singly, and their flat uniform summit surface is their most notable character. Both types have slopes of 5°–15°. One kame plateau covers 12 km². The kame ridges (c) are aligned in uniform directions and range from 8 to 20 m in height. They are transitional between kames and eskers, but differ from the latter in structure and origin, being shorter and straighter. Kame terraces (d) are not common in Poland. The kames are made of clay, silt, sand, gravel and pebbles, but sand predominates. The deposits may reach 20 to 30 m in thickness. Some kames are formed by meltwater streams; others form in still lake-water and are poorly sorted. The lake kames usually consist of finer sediments and may show varves. Sometimes the bedding is disturbed owing to melting of dead ice, which causes subsidence. The disturbance may be in the form of small faults or arcuate bedding, giving an outward dip to the beds along the slopes. The ridges formed in crevasses, and the hillocks in hollows on the dead ice or between blocks of dead ice. There is some doubt whether the kames formed supraglacially, or within or beneath the ice, but the manner of disturbance of the bedding supports a supraglacial origin in some examples at least.

The hummocky ridge of water-lain sands, veneered by till, which is called the Waterloo kame-moraine in Ontario, has been described by S. A. Harris (1970). The ridge trends north-north-west to south-south-east at right-angles to other moraines. Its sands are older than the earliest Wisconsin tills of the Erie lobe, and it was probably deposited immediately prior to the Erie ice-lobe advance, which produced the Canning till. The kame has kettle holes and well-developed bedding with better sorting to the west. It may have been produced by stagnation of the Simcoe ice lobe. It is overlain by six thin tills, four deposited by the Erie lobe and two by the Simcoe lobe. The feature consists of outwash sands laid down at the beginning of the Wisconsin period. It was not overrun by ice during the last 40,000 years. Ice movements have varied in the area and the feature is not interlobate: ice lobes have moved from different directions at different times.

4 Delta-moraines

Many morainic features show evidence of the action of meltwater in their formation, in that they contain stratified drift. Push-moraines, however, which were described in the last chapter, should not be included under the term 'delta-moraine', even though they contain stratified material. They can usually be differentiated from true delta-moraines by their structure which often shows the effects of intense pressure, producing imbrication and other similar structural features. Delta-moraines, on the other hand, form where stationary ice fronts remain for some time, ending in a lake or the sea. They have much in common with kames of the type that form at the margin of a stagnant ice sheet, where this ends in ponded water. The delta-moraines are, however, usually larger features.

A good example of a delta-moraine has already been mentioned in connection with the eskers in the vicinity of Trim in Ireland (Fig. 16.8). Many of the eskers merged into the so-called Galtrim moraine which has all the characteristics of a delta-moraine. It formed adjacent to the stagnant ice sheet which dammed up the lake into which the deltaic deposits were carried by the meltwater streams. One of the most significant aspects of this delta-moraine is the well-developed ice-contact slope on its northern margin, which gives clear evidence of the former presence of ice here. The moraine trends from north-east to south-west, in its northern section, from the Hill of Tara for a distance of 10 km, and forms a prominent arcuate ridge up to 15 m high. Southward, near Kiltale, the delta-moraine is crossed by an esker. Its typical deltaic structure is revealed in a series of pits near Batterjohn, where foreset beds dip steeply to the south. Overlying the deltaic strata is a layer of coarse unstratified material which appears to be ablation moraine. The feature has a broad, flat top 1·6 km south-west of this point. At Galtrim, the steep northern slope overlooks flat, marshy ground in which a lone kame 6 m high stands. The slope to the south-east is much gentler. The level top shows that the delta was built up to the lake level during the period when the ice front occupied the position now revealed by the ice-contact slope. As the ice melted, so the marshy hollow developed.

To the west of Galtrim, the delta-moraine splits into two parallel ridges. The southern one is a rather sharp-crested ridge, indicating that the ice did not stay long in this position before it retreated to the ice-contact slope of the north-western ridge. This ridge is wider and was fed by the three eskers that merge into its ice-contact slope on the north-west. One of the eskers climbs up the northern face of the delta-moraine to its top where the esker stops. The esker, therefore, acted as a feeder channel to the delta-moraine. There are kettle-holes, forming enclosed hollows, within the delta-moraine deposits as well as lone kames to the north of it. Some eskers merge with the foot of the ice-contact slope. The largest esker is as high as the 10 m ice-contact slope, and the two merge accordantly together.

At its southern end, the delta-moraine splits up into three separate gravel ridges. All these features have the characteristics of deltas forming where the ice front was stationary for a fairly short period. Synge (1950), who has described these features, believes that they were deposited by heavily laden meltwater streams draining through the ice, the marginal zone of which was stagnant.

The overlying unsorted material on the crest of the ridge is thought to have been derived from an overhanging ice-cliff. This cliff must have contained much englacial material that would collapse into the newly deposited deltaic sediments beneath the water surface. Dirty ice terminating in water frequently forms an overhanging face because the dirty ice has a greater ablation rate in water than in air. Thus the overlying material can be described as ablation moraine. The presence of small eskers on the ridge crest also supports this view.

The two large Salpausselka moraines of Finland are probably the largest examples of delta-moraines. The outer one is 70–80 m high and 2·5 km wide. The inner one lies 20–25 km north of the outer. The Salpausselka moraines form a long arc, stretching right across the southern part of Finland, and making a conspicuous belt of higher ground which marks important halt stages in the deglaciation of the area. The features were laid down as a line of marginal deltas in the sea, which was high at this time since isostatic recovery had barely begun. The material consists of well-washed sand and gravel showing deltaic bedding on the distal side. The proximal side shows some evidence of later glacial advance in the contortions and overfolds in the struc- ture. Thus the proximal part has some of the characteristics of a push-moraine. Till only occurs along part of the northern margin of the ridge, mainly as a sandy ablation deposit.

Associated with the two major Salpausselka ridges is a shorter ridge known as Jaa- mankangas, forming an arc about 50 km long and attached to the innermost of the Salpausselka ridges at its northern end. Jaamankangas ridge is associated with many eskers, one of which is 300 km long, the longest in Finland. The ridge consists of three main elements: a plateau ridge, an esker ridge and an area of knob-and-kettle relief. The plateau area is 2–4 km broad and 110–120 m high, sloping gently south and including deep kettle-holes and hillocks. It consists of deltaic deposits. In parts of the Jaamankangas ridge, the structure is, however, horizontal rather than deltaic. This may be the result of later resorting by the sea. Another possibility is that part of the feature was formed as an enlarged esker between two lobes of ice. Whatever the precise mechanism by which the feature was formed, it was deposited by heavily laden meltwater streams next to stagnant ice.

Features that have already been described were deposited either within or at the margin of the ice sheet. It is often assumed that when fluvioglacial deposits underlie a thick layer of unstratified till that this sequence indicates a readvance of the ice. That this need not be so if the till layer is thin and consists only of ablation moraine, has already been pointed out in discussing the Galtrim moraine. It is also probably true that, even under fairly thick till, stratified deposits need not necessarily indicate a readvance. The decay of a stagnant ice sheet can give rise to a complex sequence of deposits, whose chronological significance needs to be carefully evaluated. The matter will be discussed further in the next chapter.

5 Conclusions

Eskers and kames are formed of stratified deposits laid down in contact with slow-moving or stagnant ice by glacial meltwater. Eskers form elongated ridges, some of which originate on the ice and have ice cores. Others are laid down subglacially or englacially and may have been built up within the ice to the level of the englacial water-table, or to a uniform lake-level, to give an even crest elevation. Eskers are sometimes beaded, some are straight and some are sinuous in form. They are usually formed of sand and gravel, but may contain large boulders. Most eskers were probably deposited in channels within or beneath the stagnant ice.

Kames are more irregular in form than eskers and can be divided into two types. Those formed at or near the snout of an ice mass can be differentiated from kame terraces formed along the valley-side margin of a glacier. These terraces slope gently down-valley and often have a slight transverse slope away from the hillside. They are formed by deposition from meltwater flowing between the hillside and the ice. The other type of kame may be deposited at the margin of a glacier in a proglacial lake, sometimes forming a short irregular gravelly ridge, orientated parallel to the ice front. Isolated mounds of stratified material, forming lone kames, possibly accumulated in hollows on or in the decaying ice.

Delta-moraines may be distinguished by their deltaic bedding and their steep ice-contact slope facing the decayed and vanished ice margin. They are formed by the deposition of meltwater load in standing water in front of the ice sheet and are frequently fed by esker channels. The eskers now merge into the ice-contact slope of the delta-moraine. They mark periods of prolonged halt and decay of stagnant ice sheets.

6 References

CARRUTHERS, R. G. (1947–8), 'The secret of the glacial drifts', *Proc. Yorks. geol. Soc.* **27**, 43–57 and 129–72

(1953), *Glacial drifts and the undermelt theory* (Newcastle)

CHARLESWORTH, J. K. (1926), 'The readvance, marginal kame moraine of the south of Scotland, and some later stages of retreat', *Trans. R. Soc. Edinb*, **55**, 25–50

COOK, J. H. (1946), 'Kame complexes and perforation deposits', *Am. J. Sci.* **244**, 573–83

EMBLETON, C. (1964), 'The deglaciation of Arfon and southern Anglesey, and the origin of the Menai Straits', *Proc. Geol. Ass.* **75**, 407–30

FARRINGTON, A. with SYNGE, F. M. (1970), 'The eskers of the Tullamore district' in *Irish geographical studies* (ed. STEPHENS, N. and GLASSCOCK, R. E.), Belfast, 49–52

FLINT, R.F. (1928), 'Eskers and crevasse fillings', *Am. J. Sci.* **235**, 410–16

(1930), 'The origin of the Irish "eskers"', *Geogrl Rev.* **20**, 615–30

FOGELBERG, P. (1970), 'Geomorphology and deglaciation at the second Salpausselka between Vääksy and Vierumäki, southern Finland', *Soc. Sci. Fennica Phys. Math.* **39**, 7–90

GEER, G. DE (1897), 'Om rullstensåsarnas bildnigssatt', *Geol. För. Stockh. Förh.* **19**, 366–88

GEIKIE, J. (1894), *The Great Ice Age* (London, 3rd Ed.; 1st Ed., 1874)

GRANÖ, O. (1958), 'The Vessö esker of south Finland and its economic importance', *Fennia* **82**, 3–33

HANSON, G. F. (1943), 'A contribution to experimental geology: the origin of eskers', *Am. J. Sci.* **241**, 447–52

HARRIS, S. A. (1970), 'The Waterloo kame-moraine, Ontario, and its relationship to the Wisconsin advances of the Erie and Simcoe lobes', *Z. Geomorph.* **14**, 487–509

HELLAAKOSKI, A. (1931), 'On the transportation of materials in the esker of Laitila', *Fennia* **52**, 1–41

HELM, D. G. (1971), 'Succession and sedimentation of glacigenic deposits at Hendre, Anglesey', *Geol. J.* **7**, 271–98

HOLMES, C. D. (1947), 'Kames', *Am. J. Sci.* **245**, 240–9

HOPPE, G. (1961), 'The continuation of the Uppsala esker in the Bothnian Sea and ice recession in the Gävle area', *Geogr. Annlr* **43**, 329–35

HOWARTH, P. J. (1971), 'Investigations of two eskers at east Breiðamerkurjökull, Iceland', *Arct. alp. Res.* **3**, 305–18

JAHN, A. (1969), 'Kame terraces in the Sudetes', *Folia quatern.* **30**, 17–22

JAMIESON, T. F. (1874), 'On the last stage of the glacial period in North Britain', *Q.J. geol. Soc. Lond.* **30**, 317–38

JEWTUCHOWICZ, S. (1969), 'Struktura kemu w Zieleniewie', *Folia quatern.* **30**, 59–69

JOPLING, A. V. (1971), 'Some techniques used in the hydraulic interpretation of fluvial and fluvioglacial deposits' in *Research methods in geomorphology* (ed. YATSU, E.), 1st Guelph Symposium on Geomorphology, 1969, 93–116

KEEBLE, A. B. (1970), 'The mode and origin and depositional history of kames, south-west of Edmonton', *Albertan Geogr.* **6**, 33–44

LEE, H. A. (1965), 'Investigation of eskers for mineral exploration', *Geol. Surv. Pap. Can.* **65–14**, 1–17

LEWIS, W. V. (1949), 'An esker in process of formation, Böverbreen, Jotunheim, 1947', *J. Glaciol.* **1**, 314–19

MCKENZIE, G. D. (1969), 'Observations on a collapsing kame terrace in Glacier Bay National Monument, south-east Alaska', *J. Glaciol.* **8**, 413–26

MANNERFELT, C. M. (1945), 'Några glacialmorfologiska formelement', *Geogr. Annlr* **27**, 1–239

MANSIKKANIEMI, H. (1970), 'Deposits of sorted material in the Inarijoki-Tana river valley in Lapland', *Ann. Univ. Turku.* (ser. AII–43), **6**, 1–63

MICHALSKA, Z. (1969), 'Problems of the origin of eskers based on the examples from central Poland', *Geogr. Polonica* **16**, 105–19

NIEWIAROWSKI, W. (1963), 'Types of kames occurring within the area of the last glaciation in Poland compared with kames known from other regions', *Rep. 6th Conf. int. Ass. quatern. Res. (Warsaw, 1961)*, Łodz (1963), **3**, 475–85

PRICE, R. J. (1965), 'The changing pro-glacial environment of the Casement Glacier, Glacier Bay, Alaska', *Trans. Inst. Br. Geogr.* **36**, 107–16

—— (1969), 'Moraines, sandar, kames and eskers near Breiðamerkurjökull, Iceland', *Trans. Inst. Br. Geogr.* **46**, 17–43

RADŁOWSKA, C. (1969), 'On the problematics of eskers', *Geogr. Polonica* **16**, 87–103

SHARP, R. P. (1953), 'Glacial features of Cook County, Minnesota', *Am. J. Sci.* **251**, 855–83

SHAW, J. (1972a), 'Sedimentation in the ice-contact environment, with examples from Shropshire (England)', *Sedimentology* **18**, 23–62

—— (1972b), 'The Irish Sea Glaciation of north Shropshire—some environmental reconstructions', *Field Stud.* **3**, 603–31

SISSONS, J. B. (1958a), 'Supposed ice-dammed lakes in Britain, with particular reference to the Eddleston valley, southern Scotland', *Geogr. Annlr* **40**, 159–87

—— (1958b), 'The deglaciation of part of East Lothian', *Trans. Inst. Br. Geogr.* **25**, 59–78

—— (1961), 'The central and eastern parts of the Lammermuir-Stranraer moraine', *Geol. Mag.* **98**, 380–92

STOKES, J. C. (1958), 'An esker-like ridge in process of formation, Flåtisen, Norway', *J. Glaciol.* **3**, 286–90

SYNGE, F. M. (1950), 'The glacial deposits around Trim, County Meath', *Proc. R. Ir. Acad.* B, **53**, 99–110

TANNER, V. (1932), 'The problem of eskers', *Fennia* **55**, 1–13

TRICART, J. and CAILLEUX, A. (1962), *Le modelé glaciaire et nival* (Paris)

TROTTER, F. M. (1929), 'The glaciation of eastern Edenside, the Alston Block and the Carlisle Plain', *Q.J. geol. Soc. Lond.* **85**, 549–612

WISNIEWSKI, E. (1973), 'Genesis of Lammi esker (southern Finland)', *Fennia* **122**, 30 pp.

17

Ice stagnation features

Each glacier loaded with stones from the rocks above it may be regarded as a ship freighted with specimens of its native mountains, which it deposits by thawing in the place where it ultimately rests. (R. BAKEWELL, 1838)

The concept of ice disappearance by stagnation was developed at the turn of the century. F. P. Gulliver hinted at the possibility of ice stagnation in 1899 and in 1904, M. L. Fuller collected evidence in south-eastern Massachusetts that led to the first proposal of ice stagnation during the deglaciation of this area. J. H. Cook (1924) and R. F. Flint (1929) recognized the wide extent of the area of north-eastern North America affected by stagnation of ice during deglaciation. Flint showed that stagnation of the last ice sheet occurred throughout Connecticut. Other areas affected by stagnant ice include Rhode Island, parts of Massachusetts, New York State east of the Hudson River, northern New Jersey and all northern New England. Similar features also occur in parts of western North America, such as Yakima Valley, Washington.

When an ice mass ceases to receive an adequate supply to keep it moving, it will become stagnant and waste away by slow down-melting. Because, under conditions of stagnation, the ice takes longer to melt away where it was originally thicker, it will melt off interfluves while still remaining in valleys. The progress of deglaciation will depend very largely on the geometrical relationships between the slopes of the ice surface and the relief of the bedrock (or drift) surface, as C. M. Mannerfelt set out in 1945, and, if there is sufficient evidence, for instance from marginal meltwater channels (p. 343), to determine the slopes of the ice surface, the theoretical patterns of ice margins at different stages during downwasting can be easily plotted. The ice surface will melt most rapidly where nunataks first appear and lakes form marginal to the rock exposures. These lakes become elongated along the valley side as the ice shrinks. Evidence for their previous existence may be seen in paired terraces of deltaic-bedded outwash deposits. The decaying ice will usually contain crevasses that cannot close up so that a relatively uniform water-table is established in the dead ice mass. For this reason, the terraces tend to be paired at equal heights on either side of the valley, and are often connected with spillways, by which the lakes drained. The terrace margins

Table 17.1A   Descriptive classification of deglacial depositional features (R. J. Fulton, 1967)

Unit Designation	Topographic Expression Relief	Topographic Expression Surface	Areal Shape	Sorting	Material Stratification	Material Grain Size	Compaction
Ground moraine	Low	Gently rolling	Sheet	Poor	Unstratified	Boulder to clay	Compact
Drumlinoid moraine	Moderate	Streamlined ridges and grooves	Sheet	Poor	Unstratified	Boulder to clay	Compact
Hummocky moraine	Moderate	Hummocks Parallel ridges Closed depressions	Irregular sheet and ridge	Poor	Unstratified	Boulder to clay	Compact to loose
Morainic gravel	Moderate to high	Hummocks, sinuous, anastomosing ridges Closed depressions	Irregular sheet and ridge	Moderate to good	Stratified to partly stratified	Pebble to silt	Loose
Collapsed lacustrine	Low	Gently to sharply rolling	Sheet	Good	Stratified (disrupted strat.)	Fine sand to clay	Loose
Kettled terrace	Moderate to high	Flat Closed depressions	Prism Elongate prism	Good	Stratified	Pebble to fine sand	Loose
Rill complex	Low to moderate	Small hummocks Channels	Sheet Ribbon Patchy	Poor to good	Unstratified to partly stratified	Boulder to clay	Loose
Terrace deposit	Low	Flat	Prism Elongate prism	Good	Stratified	Pebble to silt	Loose
Lacustrine deposit	Low	Flat to gently rolling	Sheet Patchy	Good	Stratified	Fine sand to clay	Loose

Table 17.1B Facies classification of glacial deposits (R. J. Fulton, 1967)

Ice Facies	Ice>Water Facies	Water>Ice Facies	Water Facies
Ground moraine	Morainic gravel:	Kettled terrace:	Terrace deposit:
Drumlinoid moraine	Hummocky gravel	Kettled stream	Stream terrace
Hummocky moraine	Esker	terrace	Deltaic stream
	Kame	Kettled deltaic	terrace
	Crevasse filling	terrace	Lacustrine deposit
	Collapsed lacustrine	Rill complex	
	deposit		

represent ice-contact slopes that reveal the pattern of the decaying ice mass. The lake deposits also contain varves at times which point to slow rates of downwasting. The downwasting becomes extremely slow when the ice surface is thickly covered with debris.

In Connecticut, the debris appears to have been washed on to the ice surface from upstream, and for this reason, the southernmost ice masses were often the last to disappear because they had the thickest debris cover washed on to them by the south-draining streams (Flint, 1929). The horizontality of the terraces formed along the flanks of the Connecticut valley is good evidence that they are of lacustrine origin. The terraces show little erosion since their formation and this lack of erosion probably reflects the presence of stagnant ice in the valley bottom throughout the whole period of deglaciation. If the ice from the lower reaches of the valley had disappeared before the ice in the upper part had melted, the meltwater from the upper ice would have eroded the terraces in the lower part of the valley. The great extent of the terraces also points to the widespread presence of dead ice throughout the period of deglaciation.

Relief plays an important part in the distribution of stagnant ice. Features of stagnant ice are often more conspicuous in the broader parts of valleys. These parts may be cut off from the ice source by the narrower reaches upstream in which the smaller bulk of ice melted away more rapidly. This local stagnation can only occur, however, when the whole ice supply has become reduced because of climatic factors. Many stagnant ice features occur in areas of considerable relief or in the lee of large relief features. As the ice thins, it becomes unable to surmount the relief barrier and the ice beyond becomes cut off and stagnates (Mannerfelt, 1945, p. 110).

R. J. Fulton (1967) has examined the deglaciation features in an area of moderate relief in the interior of British Columbia. The valley floors are from 750 to 1500 m above sea level and the hills rise to 1500–1800 m. The area was completely ice-covered at the maximum of glaciation, with ice varying in thickness from about 550 m above the hills to 2100 m above the valley floors. Fulton proposes two types of classification of the glacial deposits, one of which is descriptive and the other environmental. Table 17.1A gives the first, while Table 17.1B gives the second which is based on facies in relation to the amount of water available during deposition. For the group in which ice facies dominate water facies, the relief forms are controlled by ice, which was largely inactive. Where water is dominant, the facies indicate the presence of ice but are otherwise normal aqueous deposits which form mainly beyond the ice.

Table 17.2 Summary of deglaciation in the Kamloops region, British Columbia (R. J. Fulton, 1967)

Model	Process Elements						Results	
	Valley Characteristics				Agent			
	Size	Shape	Area rel. to upland	Orientation (relation to ice mvmt)	Ice	Water	Deposit or Feature (order of abundance)	Distinguishing Characteristics
Broad Longitudinal (Minnie L. area)	Broad Shallow	Open	Large	Parallel	Stagnant	Marginal, proglacial, englacial, and subglacial drainage; Valley drained	Rill deposit; Lat. overflow channel; Kettled terrace; Morainic gravel; Lacustrine deposit	Open slope, broad valley: free ice-marginal drainage; Erosion and deposition of rill deposit early stage; Marginal terraces, morainic gravel, lacustrine later
Narrow Longitudinal (Guichon Cr. area)	Narrow Deep	U-shaped	Small	Parallel	Stagnant	Marginal, englacial, and subglacial drainage; Valley drained	Morainic gravel; Kettled terrace; Lat. overflow channel	Ice-tongue in deep, narrow valley after uplands free; Ice-marginal terraces early stage; Morainic gravel later stage
Broad transverse (Separation Lake area)	Broad Shallow	Open	Large	Transverse	Stagnant	Marginal, proglacial, and englacial drainage; Valley drained	Kettled terrace; Lat. overflow channel; Rill deposit; Morainic gravel	Irregular slope, complex retreat; irregular meltwater drainage; Gravel and coarse material deposited as terraces and rill dep.; fines removed
Broad transverse (Kamloops (Re-entrant area))	Broad Deep	U-shaped	Equal	Transverse	Stagnant	Marginal, englacial, and subglacial drainage; Valley ponded	Lacustrine deposit; Kettle terrace; Morainic gravel; Lat. overflow channel	Major valley blocked meltwater deposits coarse material on uplands, moving fines into major valley; Major lacustrine deposit in valley

As long as the ice retains some plastic flow and has a surface gradient, marginal drainage can occur and kettled stream terraces can form. When the ice is dead and has no significant movement or gradient, englacial and subglacial drainage leads to deposits associated with a preponderance of ice. Ice form-lines can be reconstructed to mark stages of downwasting, indicated by marginal channels where moraines do not occur. The features of deglaciation in four type areas are summarized in Table 17.2. The pattern is influenced by the relief, and in the area studied, water was active in the formation of all features. Where drainage is unimpeded, numerous meltwater channels form; where the valley sides are irregular, local ponding occurs, and kettled deltaic and kame terraces occur. Four major phases of deglaciation are recognized:

1 Active supraglacial drainage dominant—no ice-free area
2 Transitional: supraglacial drainage dominant—highest land ice-free
3 Stagnant: supraglacial and marginal drainage dominant—all uplands ice-free
4 Dead: englacial and subglacial drainage dominant—all areas except valley bottoms ice-free.

Various types of ice-contact deposit associated with stagnant ice, such as eskers and kames, have already been discussed. There are, however, other features of interest associated with the decay and disintegration of stagnant ice. Ice stagnation occurred very extensively in some areas, for example in the Canadian Prairies and parts of Scandinavia. The characteristics of an ice mass that is stagnant and decaying will first be described, and then other features deriving their origin from stagnant ice will be mentioned. These features include prairie mounds, ice disintegration features, crevasse fillings, ice-pressed forms and Kalixpinnmo hills.

1 Glacier karst features

The processes by which some of the stagnant ice features have formed have been likened to those that operate in karst landscapes. In karst country, the limestone is dissolved in a manner analogous to the melting of a stagnant ice sheet. This idea has been pursued by L. Clayton (1964) in describing the features of a present-day stagnant ice mass. Although the removal of ice is by melting, rather than chemical solution as in true karst, the resulting features have much in common. There are not many glaciers that today have the necessary conditions for the development of good examples of glacier karst features. The stagnant and heavily drift-covered parts of the Malaspina glacier provide such conditions. Another nearby glacier, the Martin River glacier, 200 km west of the Malaspina, also illustrates these features clearly (J. R. Reid, 1969). The cover of till is necessary to produce the karst-like features, because it prevents too rapid ablation which would destroy the developing forms. These forms require more localized ablation which is provided by the thick and uneven cover of till. In the case of the Martin River glacier, Reid recognizes four zones—active, intermediate, terminal and glaciated (proglacial). In the active zone, ice flow reaches 200 m/year. Where ice is exposed, supraglacial drainage is common, but where it becomes buried by debris, subglacial or englacial drainage is the rule. Ice-cored medial moraines up to 30 m high occur in this zone. In the intermediate zone, ice is only exposed in the walls of sink holes. Ice flow is much slower in this zone, the debris cover is thicker and vegetation has taken hold. Lakes are to be found, some up to 2 km long, which drain at intervals or change level suddenly. Trees in the terminal zone, where ice flow is from 0 to 30 m/year, show that it was active up to 1650 AD. The debris cover on the ice here is 1–3 m thick on average and, as in the two higher zones, much of it appears to have reached the surface of the glacier from subglacial sources along thrust planes, where the more active ice has been over-thrusting the stagnant ice. Sink holes and lakes are abundant. The lake water here is often clear, in contrast to the lakes of the higher zones containing much rock flour. The outermost 'glaciated' zone is also forest-covered and possesses much deeply buried glacial ice. The moraines here probably date from the late Wisconsin.

The funnel-shaped sink-holes are the most striking morphological features. They are almost circular and are mostly between 60 and 370 m across. They form with a density of about 30 to the square kilometre. Many hold small lakes and are 15 to 90 m deep. At times, the sink-holes coalesce to form compound features. They also enlarge and collapse to reveal the meltwater stream below, forming a type of glacial uvala. Even polje-sized features can form. For example, there is a lake 2·5 km long and 0·75 km wide near the margin of the glacier. The floors of these features may be the solid ground below the glacier. They are ice-walled outwash plains, and resemble features that have been described in North Dakota. The normal sink-holes are probably formed mainly by collapse. However, some are formed by 'solution' or melting by the operation of glacier moulins. The moulins are vertical-walled, cylindrical holes only a few metres in diameter, but probably up to 30 or more metres deep. These moulins, however, only appear in abundance in the active part of the glacier (Chapter 6). Ice caves and tunnels are also common as is indicated by the numerous collapse features. The tunnels were about 3 m high and up to 2 km long. New ones were continually developing as old ones were abandoned. Some of these tunnels were floored with gravel, while the minor ones were often completely clogged with sediment. These would form eskers when the ice melted.

The ablation till on the surface, which is the equivalent of the residual soil on true karst, supports vegetation when it exceeds half a metre in thickness. Where the ablation till is 1·5–3 m thick it is stable, and spruce and hemlock trees over 100 years old are growing on it. In the zone where the ice is still flowing actively, there are moulins, little surface moraine, except medial moraine, and no vegetation. In the intermediate zone, there are funnel-shaped sink-holes, little or no movement, and unstable sediment up to 1 m thick and some vegetation. In the terminal zone, the ice is stagnant, and the till cover is up to and over 3 m in thickness. The karst features include uvalas, poljes and hums. The drainage comes to the surface again and a dense forest can flourish on the deeply buried ice, which only melts away slowly.

The formation of kettle-holes has been discussed by R. P. K. Clark (1969). Some may be formed by a process of non-sedimentation, rather than the melting-out of buried ice—they may represent the positions of former ice pillars or roof pendants in decaying ice under which sedimentation is taking place to form kames and eskers. R. J. Price (1969) illustrates the complexity of kettled landscapes in the proglacial area of Breiða-merkurjökull, Iceland (Fig. 18.8, p. 526); some kettle-holes here represent the spaces or hollows separating anastomosing kame and esker ridges. Kettle-holes may form some distance from the ice margin where a large ice mass becomes stagnant over a wide area. They may have formed in this way up to 30 km from the ice margin in northern England, and possibly as much as 100 km in Norway. Some of the most distant may represent ice masses floated on water away from the ice front and later grounding; the resulting kettles are usually relatively shallow. The stratification of the enclosing sediments may give some clue to the type of kettle: the bedding of the sediment around the deep kettles caused by the decay of ice projecting through the full thickness of the deposits may show evidence of sliding and creep as the ice support melted.

2 Features of ice stagnation

The range of forms attributed to ice stagnation is so varied, and the circumstances of deglaciation differ so widely from one region to another, that valid generalizations are not easy to formulate. To begin with, some case studies will be presented, mainly from Canada and Scandinavia.

Features of stagnant ice in eastern central Alberta have been described by C. P. Gravenor (1955). These are the 'prairie mounds', consisting of ground moraine and covering an area of about 64 km² about 22 km east of Hemaruksa. The mounds are nearly circular and often have a central depression up to 1 m deep. They are about 100 m in diameter and 4·5 m high. Most of the mounds are composed of clayey till, but some contain stratified silt and clay. The mounds are underlain by at least 12 m of clayey till.

To account for these mounds, Gravenor suggests that, first of all, pits develop at the surface of a stagnant ice sheet. These pits might develop either by increased melting at places where the debris cover on the ice was comparatively thin, or by ice collapse into subglacial or englacial cavities, or into enlarged crevasses. Once the pits had formed in the stagnant ice, material would be washed into them from the surrounding ice. Then, as the surrounding ice became cleaner and the cover in the pit thicker, the surrounding area would melt away faster than the ice under the pit, so that eventually there would be inversion of relief and the infillings of the pits would form the prairie mounds. This hypothesis accounts better for the characteristics of these features than that which suggests that they have a periglacial origin (Embleton and King, 1975, Chap. 2). There is no evidence of ice-wedge fill. In this area, the wastage of the ice was probably largely by evaporation so that meltwater would not be so likely to disperse the sediment. The lack of outwash in this area supports the view that evaporation was important and it also accounts for the availability of englacial moraine to form the mounds.

M. Okko and M. Perttunen (1971) have described some morphologically similar mounds in southern Finland. They are 20 to 100 m in diameter and 3 to 5 m high, and occur at the junction of two late-Weichselian ice lobes. One of the mounds, made of stratified silt and fine sand, seems to have been deposited by meltwater in a subglacial cavity in stagnant marginal ice, and is therefore a type of kame, but some of the other mounds consist of washed till and probably represent debris-filled sink holes forming as Gravenor described on the surface of stagnant ice. The second Salpausselka ice-marginal belt thus contains dead-ice deposits that originated as thermokarst.

The ice disintegration features of western Canada have also been described by Gravenor and W. O. Kupsch (1959). Most of the material left by the Wisconsin ice in this area was till and there is relatively little stratified material, in contrast to eastern North America where stratified deposits predominate. The features are also well preserved and easily visible, owing to the dry climate and the sparse vegetation. The term 'disintegration' is used to describe the way in which the ice sheet decayed and disappeared in this region. It appears that the ice must have broken up into numerous separate small blocks. The breaking-up of an ice sheet may take place under uncontrolled conditions, in which case there is no evidence of alignment of the resulting features; alternatively, the ice may break up along aligned or lobate weaknesses within the ice, and these weaknesses may give a pattern to the resulting deposits. This gives rise to the controlled type of disintegration deposits. The patterns that result from this process are inherited from a more active phase of glacier motion, when the ice developed a pattern of crevasses and thrust planes associated with its movement. The linear zones along which the decaying ice fractured are normally aligned parallel or perpendicular to the ice movement. Sometimes the crevasse pattern in the lower part of an ice sheet or glacier is aligned at an angle of 45° to the ice movement (Chapter 4). It also sometimes happens that dead ice features are superimposed on features formed by the ice when it was still active. The patterns can be particularly clearly seen on aerial photographs.

A variety of features occurs in ice disintegration areas. Where the disintegration was uncontrolled, there is no pattern apparent. Ridges, hollows and mounds are distributed indiscriminately. The major elements of the relief are knobs and kettles, moraine plateaux and beaded linear ridges. The moraine plateaux are relatively high areas consisting of clay till, sometimes covered by thin lake clays. Some of the moraine plateaux have kettles and poorly developed rims. The features described by Gravenor and Kupsch in Canada seem to be similar to those described by G. Hoppe (1952, 1957, 1959) in northern Sweden. Hoppe considers that the moraine plateaux, which he calls the Veiki moraine type, have originated subglacially as described in Chapter 15. The evidence for this is the preferred orientation of the pebbles in the raised ridges around the plateaux, the fact that such features are often associated with drumlins, and that some are cut by meltwater channels. Gravenor and Kupsch do not give any information concerning pebble orientation, so that it is possible that the features they described, although similar in relief to those described by Hoppe, are in fact of different origin. Other possible methods of formation will be mentioned later in the chapter.

Another interesting feature of the uncontrolled disintegration conditions is the closed disintegration ridge. The circular features of this type are similar to one of the types of prairie mound, but others are not so symmetrical. They range in height up to 6 m and from 6 to 600 m in diameter. The base of the central depression in some of them lies below the level of the surroundings. In this they differ from the prairie mounds whose central depression is always shallow. Occasionally a small mound is found in the centre rather than a depression. Most of the ridges are formed of till, but some contain stratified material.

Controlled disintegration features are widespread in western Canada and the United

States. They are composed chiefly of till with a thin covering of gravel, but some are of stratified material. Their dimensions vary from 1 to 11 m in height, 7·5 to 90 m in width and from a few metres to 13 km in length. They are usually fairly straight or arcuate. Two sets can intersect at right-angles, or at an acute angle to form a diamond pattern. Their alignment is controlled by the last effective ice movement, which caused lines of weakness to develop along which the ice later broke up. Many different-shaped ridges occur, from hairpin shapes to those like a shepherd's crook. When two ridges meet or cross, one may lie above the other or they may merge together.

Linear intersecting ridges and depressions, or 'minor moraines', either parallel to or at an angle of 45–90° to associated end-moraine series of the Des Moines lobe of Iowa have been described by J. D. Foster and R. C. Palmquist (1969). The ridges are 1·5 to 6 m high and formed of till; there are T-shaped offsets, steps and box patterns. The main ridges are spaced between 30 and 200 m apart. The irregularity and high dip of small cross-bedded sand bodies in the till, small faults and joints suggest an origin by ice disintegration with controlled patterns. The alignment of the till fabric with glacier flow suggests a subglacial lodgement origin; the features are certainly not end-moraines or ice-marginal.

Ice-contact rings and ridges associated with end-moraines, moraine plateaux and outwash plains have been studied by R. R. Parizek (1969) in south-central Saskatchewan. They are also known in many other states from Alberta to Manitoba in Canada and from North Dakota to Illinois and Indiana in the USA. The diameters of the rings vary from 100 m to several kilometres, and they may be tens of metres high. They may be oval-shaped, crenulate or irregular. The ice-contact ridges are straight and of similar height, but a few are sinuous or anastomosing. Some occur in parallel sequences, and most are sharp-crested, rising from bases 30 to 75 m wide. These rings and ridges are thought to be the products of dead-ice wasting involving debris of a great variety of origins; some may be subglacial ice-pressed forms (see next section). Ice-contact ridges may form from crevasse fillings, or from the fillings of marginal or supraglacial meltwater channels during ice disintegration. Ice-contact rings may be analogous to prairie mounds, for the latter often possess central depressions as already noted. The larger the central depressions are, the more the features begin to resemble rings rather than mounds.

A possible mode of origin is as follows. Pits in stagnant ice fill with debris. As the ice wastes, the stage will be reached when the infillings of the pits are let down on to the underlying ground but, between the pits, hillocks of ice remain. Debris slips off these ice hummocks and accumulates around them; finally the ice hummocks melt to leave a hollow surrounded by a bank of debris. Multiple rings are fairly common, and some are breached by escaping lake water. Fabrics in the ridges often show little preferred orientation, indicating an ablational origin or slumping of supraglacial debris. Essentially, the features involve inversion of relief during the melting and disintegration of an ice layer heavily laden with debris.

The term 'wash-board moraine' has been used to describe the pattern that the ridges make when they are parallel and close together. Descriptions of these features were given in Chapter 15 (p. 454), where the shear-moraine hypothesis of origin was noted.

Gravenor and Kupsch (1959) agree with this suggestion, and go on to point out that the ridges owe their preservation to ice disintegration. They also point out that the ridges show evidence of former ice movement and are thus controlled patterns. Their pattern tends to be lobate and in this they differ from the linear disintegration ridges. The linear features, which are sometimes even-crested, probably formed in more open crevasses. They often show a composite pattern, with several different directions super-imposed. Erosional features, formed originally in ice-walled channels, also occur. The term 'ice-walled channel' can be applied both to open channels with ice walls and to closed tunnels within the ice (Chapter 12).

In considering the origin of these features, two major processes appear to have operated. First, some of the material was probably let down by the ablation of the ice. Secondly, some may have been squeezed into the cracks and hollows within the ice as it gradually broke up into stagnating blocks. The disintegration features formed late in the glacial sequence; sometimes they overlie drumlins and other features of active ice movement, and must have formed later than these.

Another possibility is that the features that show alignment and controlled patterns may have received this pattern subsequent to their original formation. If the decayed ice that originally formed them had been rejuvenated, by thickening, to become active again, it might have pushed the material into the pattern in which it is now found. This process, however, appears less likely than the view that the pattern was inherited from a period of more active flow prior to the formation of the disintegration features. Thus inherited flow control, rather than regenerated flow control, appears more likely.

The process of ablation must clearly play an important part in the latest stages of the disappearance of a dead ice mass. The superimposition of ridges one above another can best be accounted for by downwasting, where one crevasse was filled with material, and later, another was opened on top of it, the material in the second being let down on top of the first. Moraine plateaux could form by the accumulation of material in a low part of the glacier, which then becomes a plateau by ablation of the surrounding cleaner ice. This would account for the lake deposits that sometimes overlie the till on the tops of the plateaux. The moraine-filled depressions would have been small lakes on the ice as they were developing. Support for the ablation view is also found in the collapse structures that are common in this type of disintegration deposit.

One of the difficulties of the ablation theory is that ablation moraine is normally loosely packed. It does not contain much fine material, which has been washed away in the meltwater. Many of the features described, however, consist of compact silty till. Another point is that many existing glaciers only carry a relatively small area of supraglacial till close to their snouts. It is difficult to envisage a process whereby huge areas of decaying ice could become thickly covered by supraglacial material. It is known from a study of modern ice sheets that till can be brought up to the surface near the edge of the ice sheet, but this only occurs in live ice or at the junction of live and dead ice. Till ridges parallel to the ice front can form in this way.

S. Jewtuchowicz (1968) has observed accumulation processes on Spitsbergen glaciers which have a wide stagnant zone of debris-covered ice. Differential ablation leads to a hummocky surface with dirt cones, crevasses and kettles. The surface is gradually levelled as material slumps down the steeper slopes. Surface streams find subglacial tunnels which are widened and then collapse, causing deformation of the overlying material. On Keilhaubreen, ridges of material accumulating along glide planes have formed perpendicular to the main glacier axis. There are also parallel ridges, only 2 m high, of material squeezed into bottom crevasses during stagnation. In places, previous glacier structures are revealed in the ridge pattern.

2.1   Ice-pressed features

A. M. Stalker (1960) has described similar features in Alberta and has produced much evidence to show that they probably originated as ice-pressed features. He therefore supports the second major hypothesis. Stalker shows that the same process of ice pressure or squeezing can account for the wide variety of forms that exist in this area. The forms can be classified into the following types: (a) rim ridges of moraine plateaux and plains plateaux, (b) ice-pressed moraine ridges, (c) ice-pressed plains ridges, (d) till-cored eskers and eskers made wholly of till, (e) ice-pressed minor moraine ridges, (f) ice-pressed long drumlinoid ridges. Some of these features have already been mentioned in connection with moraines, eskers, and drumlins. There is no need to describe all these features in detail, but the processes that may have operated in their formation are worthy of comment. Stalker considers that during deglaciation there were holes and crevasses at the base of the disintegrating ice. The till at the base of the ice was either not frozen or only partially frozen as the ice was disappearing. The melting of the overlying ice produced much meltwater, so that the till beneath the ice was full of water, commonly being completely saturated. The weight of the overlying ice would press the saturated till towards the holes and crevasses within the ice. Where the holes were very large, more material would accumulate around their margins than in the central part of the hole. When the ice finally melted, this material round the margins would form the marginal ridges, while the hollows around the ridges occur in those parts from which the material was pressed. A strong argument in support of this theory is that the till forming the ridges around the moraine plateaux is of basal type, as it contains much fine material.

One of the essentials for the operation of the ice-pressing process is an abundance of water. The water was provided in this part of Alberta because the ice was more active to the east and north and would thus prevent the flow of water in this direction, which is the natural drainage direction because of the general slope of the land. Thus there was plenty of water available to soak the subglacial till and fill the holes and crevasses in the basal ice. The long ridges could form in basal tunnels cut by streams, but it is less easy to account for the much larger holes that formed the moraine plateaux. Many of these features are more than 1 km in diameter. Smaller holes probably occurred at the junction of two or more crevasses.

A process that could account for the enlargement of a hole once it had formed is calving. Where the holes were water-filled, the ice would tend to be undermined at the water level, and this would cause collapse of the ice walls by calving and the enlargement of the hole. This process of calving had probably ceased before the formation of the rims by ice pressure. This latter process requires a certain stability of the ice

Plate XXXII
Margin of Breiðamerkurjökull, Iceland. Stratified dirty ice overlies semi-liquid till that has been squeezed up into a fissure in the ice. *(C.E.)*

walls. The cessation of calving would take place as increased melting opened more channels and allowed the water, filling the holes, to drain away. Water in the holes is not essential to the squeezing process. Another advantage of the hypothesis that calving occurred before the formation of the ridges is that it would provide steep ice walls to assist the pressing process. Calving cannot, however, explain the origin of the larger holes, although it probably played an important part in the smaller holes.

A sufficient ice thickness is necessary to provide sufficient pressure to produce the ice-pressed forms. However, the ice thickness must not be too great or the ice will flow actively and destroy the disintegration forms. Under thick ice, the tunnels and holes would also close up and the process could not operate. There is thus an optimum ice thickness required to produce the features by ice pressure. Stalker suggests that, under favourable conditions, with ice 60 m thick, ice density of 0·8 and till density of 2·0, a ridge about 24 m high could be built up. The observed maximum height of the ridges agrees fairly well with this estimate, as the highest ridges are about 21 m high. A greater height could be achieved if a great thickness of supraglacial till were also available to add to the height of the ridge. However, many of the ridges are much lower than these values, which suggests that the ice thickness generally must have been less than 60 m. Another common limiting factor in the growth of the ridge was the supply of semi-liquid till.

H. A. Winters (1961) has described ice disintegration features on the Missouri Coteau in North Dakota. The landforms of this area resemble the hummocky moraine of western Canada described by Gravenor, Kupsch and Stalker. Perched lacustrine plains are a conspicuous element of the landforms of North Dakota. They may have formed in hollows on the ice, to be left perched as the surrounding ice melted. Some of the perched features are surrounded by rims of till or fluvioglacial gravel. The rims are not conspicuous where the former lakes were almost completely filled with lacustrine sediment, but where the lake deposits were thin, the ridges stand up like the edges of upturned saucers around the plateau edge (Fig. 17.1).

2.2 *Stagnant and active ice disintegration*

Many features show evidence of the influence of both active and dead ice. The Kalixpinnmo hills, described by Hoppe (1959), come into this category. These hills are formed of fine sand and coarse silt. They are up to 25 m high and are elongated, aligned parallel to the ice front. They are thought to have originated sub-marginally in areas where the meltwater streams slowed down. The slowing-down could be the result either of loss of hydrostatic head caused by thinning, or of the widening of the channel in which the streams were flowing. The material deposited in this way was mixed with till falling from the tunnel roof. The deposits seem to have been thrust together and formed into ridges by the ice, which appears to have been still in motion. The features, therefore, demonstrate characteristics of both dead and live ice sedimentation.

It is difficult to establish the precise nature of the formative processes merely from the external shape of the deposits. Some hummocky drift is the result of live ice deposition. A study of the internal structure and the fabric of the material may be necessary to reach a conclusion concerning the origin of the features. However, more must be

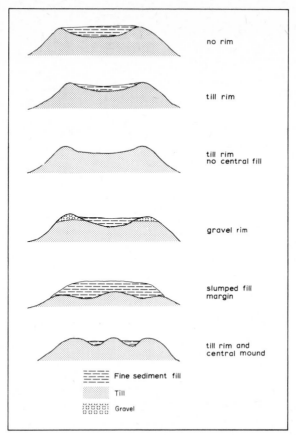

Fig. 17.1
Landforms associated with ice disintegration: diagrammatic cross sections (H. A. Winters, *Prof. Geogr.*, 1961, Association of American Geographers)

learnt about the material fabrics before this technique can provide an infallible method of differentiation. The superimposition of live ice and dead ice features also leads to problems of interpretation, although the different types can frequently be recognized by studying the form of the features and the deposits of which they are composed. Dead ice features must also not be confused with permafrost and periglacial features, which may resemble them in some respects. Rimmed depressions could in fact be either disintegration features or collapsed pingos (Embleton and King, 1975, Chap. 2).

Features that have been interpreted as the result of stagnating ice by M. M. Leighton (1959) occur in the Illinoian drift of south-west Illinois, where the drift consists of many ridges. The ridges trend mainly parallel to the south-westerly ice movement of the Illinoian lobe. The ridges are thought to have originated as crevasse fillings and are 25–30 m high in places. They consist mainly of fluvioglacial gravel. Features referred to as moulin kames are also described. These are mainly 6–12 m high, but at times attain 60 m. They appear to have formed from material choking moulins at a late stage of disintegration.

J. H. Hartshorn (1958) has shown how till may come to overlie stratified outwash deposits as a result of processes associated with stagnating ice. Till derived from supra-glacial deposits is found to overlie sands and gravels near Taunton in south-eastern Massachussetts. Hartshorn envisaged lumps of stagnant ice, covered with a layer of till, separated by zones of fluvioglacial deposition. The supraglacial till must rise to a higher level than the surface of the outwash. As the ice lump melts, the till on its steepening slopes may move off laterally by flow on to the surrounding lower areas of outwash (Fig. 17.2). The deposits of till that have flowed off the wasting ice mass can form deposits, as much as 6 m thick, of what has been called flow till on top of

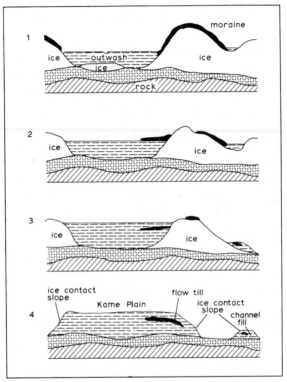

Fig. 17.2
Stages in the development of flow till (J. H. Hartshorn, *Bull. geol. Soc. Am.*, 1958)

the stratified deposits. Sometimes the flow till becomes interstratified with the outwash deposits; this occurs when later fluvioglacial activity covers a layer of till that has flowed down over a lower layer of stratified deposits. The outwash that was originally laid down in contact with the stagnant ice lumps would develop ice-contact slopes as the ice slowly melted away. These ice-contact slopes are always associated with the presence of flow tills. The association of flow till with outwash and the formation of ice-contact slopes along the margin of the outwash is usually accompanied by an inversion of the original relief before the ice lumps start to melt. The process by which flow till forms can be seen in operation on the Malaspina glacier at present.

Flow tills, as noted in Chapter 13, are being actively formed around the snouts of some west Spitsbergen glaciers (G. S. Boulton, 1968). Some of the glaciers of Ny Friesland contain much englacial debris that becomes supraglacial near the snout. This highly mobile till can flow down the glacier surface and cover other deposits. Dunerbreen has an ice-cored terminal moraine and near its margin, flow till overlies material derived from the glacier bed. In front of some other glaciers, flow till overlies stratified deposits. Some of the flow tills were submarine at one stage. Stubendorfbreen has an extensive hummocky ice-cored moraine from which flow till runs down on to fluvial deposits at the ice edge. The till is more liable to flow if it is rich in clay. In the fabrics of flow tills, the preferred orientation is parallel to the surface slope. The fluid movement of the till causes this pattern to develop, apart from a transverse orientation at the front of the flow. Controlled forms occur where debris bands are separated by cleaner ice, while uncontrolled forms occur where the debris is randomly distributed in hummocky ice-cored moraines.

W. Niewiarowski (1963) has shown that large areas of Poland were covered by stagnant ice during deglaciation. These zones of dead ice were up to 250 km wide. Kames, described in Chapter 16, are one of the features characteristic of these zones of decaying stagnant ice. Dead-ice moraines are another type of feature and they occur, for example, on the moraine plateau of Chelmno. These moraines are underlain by ground moraine and are characterized by irregular mounds and ridges and an abundance of kettle-holes. Much of the material is stratified and very mixed in size from silt to pebbles. The dip of the pebbles usually conforms to the direction of slope, but the orientation is often varied. Small faults cut the stratified sediment and occur as a result of subsidence as the ice melts. There is a cover of unstratified drift 1–2 m thick. The stones in this drift do not show any preferred orientation and the material is a type of ablation moraine although it differs from the ablation moraine of active glaciers. Dead-ice ablation moraine contains much fine material and many small fault structures resulting from displacement as the dead ice melts. Another important point concerning the dead-ice moraine is that it often lies a few metres lower than the outwash produced by the ice when it was still active. Marginal valleys also sometimes occur in dead-ice areas and it has been suggested that the very flat areas of ground moraine may be associated with decaying masses of uncrevassed dead ice.

Kettle-holes need not necessarily indicate dead ice. Some kettle-holes form as a result of the freezing of small lakes, melt-streams or ground ice during the accumulation of the deposits. Then when these ice masses melt, kettle-holes form. Kettles associated with dead ice normally occur in combination with kames and eskers.

The length of time that buried glacial ice can survive is not known with any certainty, and obviously varies according to local circumstances. M.-B. Florin and H. E. Wright (1969) discuss evidence for the persistence of stagnant glacial ice in Minnesota. Coarse material was found below a lacustrine organic-rich mud which included wood fragments and probably represented the floor of a supraglacial forest. The mud accumulated in shallow pools formed when the buried Wisconsin ice slowly melted beneath a cover of till or outwash. The biological evidence suggests rapid deepening of the pools on the dead ice once they formed. Radiocarbon ages inferred for the ice range from 16,000 to 12,000 BP. The dead ice lasted for a long time. Wastage proceeds both from the surface and from the base. Melting is delayed by a thick cover of fine-grained deposits which inhibit the percolation of meltwater, and under favourable circumstances the buried ice may have survived for over 2000 years. In North Dakota, ice-contact deposits range in age from 11,650 to 9000 BP, where the last ice was active 12,000 years ago. The thickness of the dead ice is indicated by the height of ice-cored moraines which were probably 200 to 300 m high, while the St Croix moraine is now less than 100 m high.

3 Conclusion

Ice disintegration appears to have been an important method of deglaciation over wide areas in western North America and also parts of Scandinavia. G. Holmssen (1963) has shown that dead-ice features and deposits occur widely in south-eastern Norway. Most of the morainic drift of this area has been left by stagnant ice, for only terminal moraines and sparsely covered rock occur in areas of active ice retreat. Altitude plays an important part in determining the areas characterized by dead-ice deposits. Both Holmssen and Stalker point out that the higher areas tend to produce the earlier dead-ice deposits, because it is in these areas that the ice first thins where they are situated far from the ice source. If ice-pressed forms can be associated with areas of ice disintegration, then they provide a useful means of identifying those areas in which the ice disappeared in this way. It is particularly important that dead-ice hummocky moraine landscapes should be correctly differentiated from landscapes incorporating terminal moraines formed by active ice.

The stagnation and disintegration of the ice seems to have occurred largely in broad, marginal belts of the ice sheet, particularly where the ice was thin over higher areas. Because the zone of disintegration was wide and the hummocky areas extensive, it is not possible to recognize the position of the ice front by studying these deposits. The ice disintegration features do, however, provide valuable evidence of an important and widespread method of ice dispersal.

4 References

BOULTON, G. S. (1968), 'Flow tills and related deposits on some Vestspitsbergen glaciers', *J. Glaciol.* **7**, 391–412

CLARK, R. P. C. (1969), 'Kettle holes', *J. Glaciol.* **8**, 485–6

CLAYTON, L. (1964), 'Karst topography on stagnant glaciers', *J. Glaciol.* **5**, 107–12

COOK, J. H. (1924), 'The disappearance of the last glacial ice-sheet from eastern New York', *Bull. N.Y. St. Mus.* **251**, 158–76

DYSON, J. L. (1952), 'Ice-ridged moraines and their relation to glaciers', *Am. J. Sci.* **250**, 204–11

ELSON, J. A. (1957), 'Origin of washboard moraines', *Bull. geol. Soc. Am.* **68**, 1721

EMBLETON, C. and KING, C. A. M. (1975), *Periglacial geomorphology*

FLINT, R. F. (1929), 'The stagnation and dissipation of the last ice sheet', *Geogrl Rev.* **19**, 256–89

FLORIN, M.-B. and WRIGHT, H. E. (1969), 'Diatom evidence for the persistence of stagnant glacial ice in Minnesota', *Bull. geol. Soc. Am.* **80**, 694–704

FOSTER, J. D. and PALMQUIST, R. C. (1970), 'Possible subglacial origin for "minor moraine" topography', *Proc. Iowa Acad. Sci.* **76** (1969), 296–310

FULLER, M. L. (1904), 'Ice retreat in Glacial Lake Neponsat and south-east Massachusetts', *J. Geol.* **12**, 181–97

FULTON, R. J. (1967), 'Deglaciation studies in Kamloops region, an area of moderate relief, British Columbia', *Bull. geol. Surv. Can.* **154**, 36 pp.

GRAVENOR, C. P. (1955), 'The origin and significance of Prairie Mounds', *Am. J. Sci.* **253**, 475–81

GRAVENOR, C. P. and KUPSCH, W. O. (1959), 'Ice disintegration features in western Canada', *J. Geol.* **67**, 48–64

GULLIVER, F. P. (1899), 'Thames River Terraces in Connecticut', *Bull. geol. Soc. Am.* **10**, 492–5

HARRISON, P. W. (1957), 'A clay till fabric: its character and origin', *J. Geol.* **65**, 275–308

HARTSHORN, J. H. (1958), 'Flow till in south-eastern Massachusetts', *Bull. geol. Soc. Am.* **69**, 477–82

HOLMSSEN, G. (1963), 'Glacial deposits of south-eastern Norway', *Am. J. Sci.* **261**, 880–9

HOPPE, G. (1952), 'Hummocky moraine regions with special reference to the interior of Norrbotten', *Geogr. Annlr* **34,** 1–71

(1957), 'Problems of glacial morphology and the Ice Age', *Geogr. Annlr* **39**, 1–18

(1959), 'Glacial morphology and Inland Ice recession in north Sweden', *Geogr. Annlr* **41**, 193–212

JEWTUCHOWICZ, S. (1968), 'Accumulation in stagnant ice, with the Spitsbergen glaciers as examples', *Geogr. Polonica* **13**, 49–56

LEIGHTON, M. M. (1959), 'Stagnancy of the Illinoian glacial lobe east of the Illinois and Mississippi rivers', *J. Geol.* **67**, 337–44

MANNERFELT, C. M. (1945), 'Några glacialmorfologiska formelement', *Geogr. Annlr* **27**, 1–239 (with English summary)

NIEWIAROWSKI, W. (1963), 'Some problems concerning deglaciation by stagnation and wastage of large portions of the ice-sheet within the area of the last glaciation in Poland', *Rep. 6th Conf. int. Ass. quatern. Res. (Warsaw, 1961)*, Łodz (1963), **3**, 245–56

OKKO, M. and PERTTUNEN, M. (1971), 'A mound field in the second Salpausselka ice marginal belt at Kurhila, south Finland', *Bull. geol. Soc. Finl.* **43**, 47–54

PARIZEK, R. R. (1969), 'Glacier ice-contact rings and ridges', *Geol. Soc. Am. Spec. Pap.* **123**, 49–102

PRICE, R. J. (1969), 'Moraines, sandar, kames and eskers near Breiðamerkurjökull, Iceland', *Trans. Inst. Br. Geogr.* **46**, 17–43

REID J. R. (1969), 'Glaciers "living and dead" ', *Proc. N. Dak. Acad. Sci.* **21** (1967), 42–55

STALKER, A. (1960), 'Ice-pressed drift forms and associated deposits in Alberta', *Bull. geol. Surv. Can.* **57**, 1–38

VIVIAN, R. (1965), 'Glaces mortes et morphologie glaciaire', *Revue Géogr. alp.* **53**, 371–401

WINTERS, W. A. (1961), 'Landforms associated with stagnant ice', *Prof. Geogr.* **13**, 19–23

18

Proglacial features

The first impression of Breiðamerkursandur is that of a desert, ... an almost smooth plain of dark, blackish gravel which from the end-moraines of the glaciers, high as the ramparts of a fort, slopes gently down towards the sea. The water from the ice for ever changes its course; ... the swiftest rivers change their course almost daily on account of the huge quantities of gravel which they carry with them. (H. W. AHLMANN, 1938)

Where glaciers are active and can erode their beds effectively the meltwater streams issuing from their snouts are heavily laden. Their load usually consists of coarse waste as well as fine silt, the latter giving the meltwater streams their characteristic milky appearance. The coarse material is normally deposited in front of the glacier margin to form proglacial features that consist of fluvioglacial deposits. There are two major landform types resulting from this deposition, depending on the character of the local relief. If the glacier is confined within steep valley walls, the outwash must be similarly confined and the resulting deposit is generally known as a valley train. But where there is a wide ice margin, terminating on a broad lowland area, the melt-streams can spread widely and build up an extensive surface. This is usually called an outwash plain, also known by the Icelandic term 'sandur' (Plate XXXV). Valley trains are associated essentially with valley glaciers and the outwash plains with ice sheets. Both landforms, but particularly the outwash plains, can cover a large area.

Before considering these two major types of proglacial landform separately, some points concerning the general characteristics of these features will be mentioned. The debris provided by the ice is normally very poorly sorted and consists of a wide range of size of material. Normally the coarser material is deposited near the ice margin, and the finer, sandy sediment is carried farther. It is this material that gives rise to the Icelandic name 'sandur', which means sandy area.

Because of their large load, the meltwater rivers usually aggrade their valleys and follow extremely braided courses that are continually changing in pattern (Plate XXXIII). Another characteristic of outwash streams is their variable flow. The variation near the glacier snout may be diurnal in character in summer, but it is the seasonal variation that is more important. It is associated with the summer ablation season when

Plate XXXIII
Outwash plain of the Hvitá, Iceland, about 5 km from the margin of the Langjökull. Scale given by person standing in centre foreground. (*C.E.*)

the flow increases very markedly for a few months, diminishing to low values during the winter.

Some glacial rivers are also liable to more violent and short-lived variations in the discharge rate. These result from glacier bursts or jökulhlaups, as they are called in Iceland where they are particularly common. These sudden outbursts are caused by the rapid drainage of lakes dammed up by the ice or of englacial or subglacial water bodies, as described in Chapter 19. The largest sandur in Iceland, Skeiðarársandur, is particularly liable to experience jökulhlaups owing to the periodic draining of Grims-vötn, a subglacial lake on Vatnajökull. From the point of view of the sandur development, it is the volume of water and the load it carries that is of significance. The usual discharge pattern of a jökulhlaup consists of a fairly gradual rise to a peak in the water level, followed by a short crest height, and a sudden rapid fall to normal level (Fig. 19.2). For a short time, the velocity of flow may reach figures of 7–8 m/s. In addition to the large volumes of water suddenly discharged on to the sandur, a great deal of material is also carried down by the water. Near the glacier snout, large lumps of ice may be broken off and left stranded on the sandur as the flood subsides, eventually melting out to form kettle-holes. The kettle-holes often occur within the large channels, which are cut by the powerful streams before they start depositing their load.

The hydrograph of the 1954 jökulhlaup of the Skeiðará shows that the water rose gradually at an accelerating rate from 7 July to reach a peak on 18 July and then fell rapidly to reach its normal level by 21 July (Fig. 18.1). During this period, S. Rist (1955) estimated that a total of 28.9×10^6 t of silt were carried by the river. The load

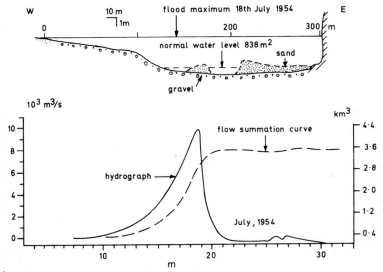

Fig. 18.1
Hydrograph for the Skeiðará jökulhlaup of July, 1954. A cross-section of the main channel, showing water levels, is included (S. Rist, *Jökull*, 1955, Iceland Glaciological Society)

of silt reached a maximum of 9.5 g/l on 18 July, the day of the flood peak. During this day, a total load of 7.8×10^6 t of silt was carried on to the sandur. The total volume for the whole period must clearly represent very rapid removal of material from beneath the glacier. The much larger jökulhlaup of 1934 produced about 150 million t of silt. If this were spread over the area drained by the Skeiðará it would amount to a loss of a layer 2.8 mm thick each year over the whole area (Chapter 11).

When these vigorous flood streams emerge on to the sandur, they at first erode deep, wide channels. One of these measured 200 m in width and varied from 5 to 7.5 m in depth. There were deeper troughs on the margins of a flat central section (Fig. 18.2). Very soon their heavy load causes the rivers to spread out into the typical braided pattern, the accompanying deposition causing rapid aggradation of the sandur.

One result of the observed rapid fall in the discharge at the end of a jökulhlaup is that the river suddenly loses its transporting power and much of its load will be deposited. Such sudden variations in the discharge will help to account for the aggrada-

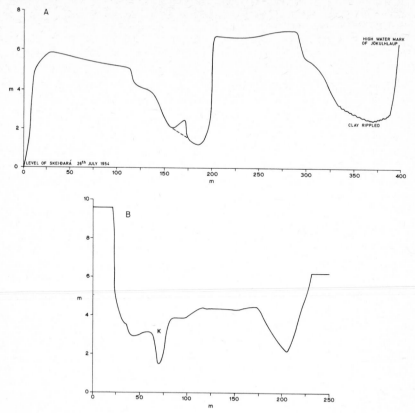

Fig. 18.2
Jökulhlaup channels on Skeiðarársandur, Iceland
A surveyed after the 1954 jökulhlaup
B older channel with kettle hole (K)

tion that goes on under these conditions on the sandur. The same process will also operate to a lesser degree as a result of the annual variations of flow, with the relatively rapid fall-off of discharge in later summer.

1 Valley trains

Valley trains owe their characteristics to the glacial meltwater streams that form them and the fact that these streams are confined by valley walls. The valley train in Morsár-dalur, Iceland, is a good example. This valley is tributary to the Skeiðará, which has already been mentioned in connection with jökulhlaups. Morsárjökull is a small outlet glacier from Vatnajökull and it ends in a steep-sided valley about 7 km from Skeiðará. The valley is 870 m wide and a profile levelled across it, as shown in Fig. 18.3, does not vary more than 0·3 m on either side of the mean height, apart from one or two deeper channels that are about 0·5 m deep. The slope down-valley is, however, fairly steep, at about 9·7 m/km to 6·25 m/km, decreasing down-valley. The floor of the valley consists of coarse gravel with little or no vegetation over most of its width. This lack

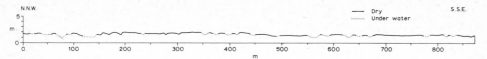

Fig. 18.3
Profile levelled across the valley train in Morsárdalur

of vegetation testifies to the unstable and variable nature of the channels of the braided river.

The rate at which aggradation takes place depends mainly on the discharge and load of the meltwater rivers. Morsá aggrades less rapidly than Skeiðará, into which it drains. The difference in level was apparent during the jökulhlaup of 1954. During the outburst, Morsárdalur became flooded for a long distance upstream of the confluence as it lay at a lower level. The difference in level of the two outwash areas was about 2 m. The distributaries flowing on the west side of Morsárdalur flow along the margin of Skeiðarársandur when they come into contact with it before joining Skeiðará.

Most valleys·that have been glaciated in the past show some evidence of having been modified by valley train development. The great Alpine valleys, such as the Rhine and Rhône valleys in their upper reaches, have extremely flat floors, which have been built up by fluvioglacial deposition. The flatness may be in part the result of deltaic sedimentation in lakes. However, where the stream gradient is considerable, much of the deposition is the result of valley train development. A great deal of material is sometimes deposited: a depth of 120 m has been recorded in the Rhône valley. The low gradient of the transverse profile of the valley train exaggerates the U-shape associated with glaciated valleys. There is often a sharp break of slope at the margin of the valley train.

Valley trains are well developed in the valleys on the east side of the Southern Alps of New Zealand. Those in the Tasman valley show the characteristic flatness and braided stream pattern. In places, the active fluvioglacial streams have almost completely eroded away moraines, of which only small remnants now exist. Other moraines may have been buried completely.

Valley trains in areas where glaciers are no longer present frequently become dissected. As the ice disappears and as hillsides become better protected by vegetation from erosion, the load transported by the stream is gradually reduced. As a result, the stream begins to cut into its former deposits and the valley trains become dissected into terraces. There are many good examples of terraces of this type in the valleys on the eastern side of the Southern Alps of New Zealand. The formation of these terraces depends on the relationship between the load and volume of the river (Chapter 5) and has no connection with changes of base level that may be affecting the lower part of the valleys. The terraces formed by changes in discharge and load are mainly found in the upper reaches of the rivers.

In the Waimakariri valley, studied by M. Gage (1959), the outwash deposits, now forming sharp terraces, can be traced up to moraines. The moraines mark halt stages or readvances of the glaciers, whose outwash built up the terraces, originally in the form of valley trains. One of the most extensive spreads of outwash gravels in the valley

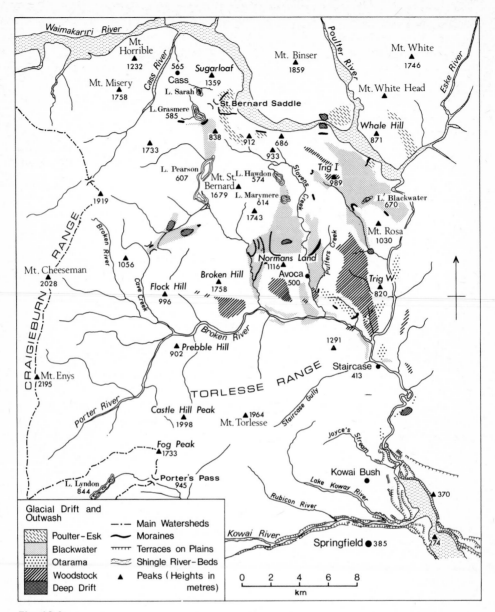

Fig. 18.4
Terrace deposits and moraines in the Waimakariri valley, New Zealand (M. Gage, *N.Z. Jl Geol. Geophys.*, 1958, DSIR, R. E. Owen, Govt Pr., Wellington)

can be traced to the Blackwater moraines (Fig. 18.4). Some of these deposits, which consist of roughly stratified and poorly sorted gravels, are over 150 m thick and now cover an area of about 25 km². Other valley trains were deposited in tributary valleys that were fed by diffluent glaciers. These deposits are in places between 60 and 75 m thick. Some of them show very good sorting and small sized gravel. The periods of aggradation when the moraines and valley trains were being deposited and the glaciers were active were followed by down-cutting and terrace formation when the ice had retreated.

It is worth pointing out that similar occurrences, but not usually so clearly displayed, are well known in Britain. The greater age of the deposits has rendered them less conspicuous in parts of England. The Main Terrace of the Severn, of Würm age, is a good example. The upper Hilton terrace of the Trent, which was aggraded at the beginning of the previous interglacial, shows many of the stratification characteristics of glacial outwash deposits. The stratification is crude and patches of till are included in the deposits. This suggests that dirty lumps of ice were incorporated and that the current, laying down the deposit, was strong and tumultuous. The deposits have beeen mainly removed by later incision to form terraces, during periods when the loads of the rivers were very much reduced after the glaciers had retreated. The case of the Thames flood-plain and terraces is mentioned briefly in Chapter 5.

2 Outwash plains or sandar

The outwash plains or sandar of Iceland have already been mentioned in connection with jökulhlaups. They are a good example of the contemporary development of these features and thus provide a model by which older features, no longer actively developing, can be studied and recognized. Skeiðarársandur is the largest outwash plain in Iceland. It stretches down to the sea from the snout of the large piedmont glacier Skeiðarárjökull, a distance of 20 to 30 km. Over this distance, the Skeiðará, the major outlet river, has a gradient of about 5 m/km. The river follows a braided course and only during the maximum of a jökulhlaup is the whole sandur flooded. The slope of the sandur close to the terminal moraines, which were formed in the mid-nineteenth century, was levelled and found to be about 1 in 77 as shown in Fig. 18.5. This sandur is aggrading rapidly at the present time. The whole sandur has been built up by the slow migration of the major meltwater streams across it from one side to the other. At the present time, the river is at the eastern extremity of the sandur; in the future,

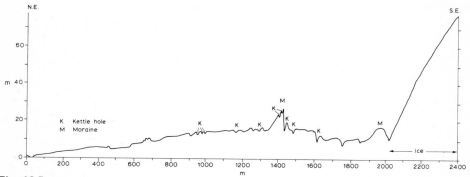

Fig. 18.5
Profile levelled across the upper part of Skeiðarársandur, crossing the moraines to the glacier margin

it may start migrating westward again across it. Skeiðará is not the only stream to emerge on to the sandur, although it is the largest. Towards the seaward limit of the sandur, the material becomes much finer. Instead of gravel and boulders, black basaltic sand becomes the dominant material and a shallow, flat lagoon separates the coastal sand barrier from the dry sandur. At the eastern end of the sandur this lagoon is about 6 km wide and less than 0·3 m deep throughout this width. The gradient of the sandur, therefore, falls off as the material becomes finer and the slope required to carry it is reduced.

Detailed work has been carried out on Hoffellsandur, the outwash of another glacier draining from Vatnajökull (F. Hjulström, 1955). It lies to the east of Skeiðarársandur.

Plate XXXIV
The present shore is being cut into, and some 20 m below the level of, the sandur surface. Horizontally bedded sandur gravels are exposed in a transverse gully. Breiðamerkursandur, Iceland. *(C.E.)*

The work included a detailed study of the hydrology of the fluvioglacial rivers that have built up the sandur and a study of the climate. The character of the sediments on the sandur was studied by boring through the gravel and sand. Studies were also made of the marginal lakes that are associated with the sandur. Repeated surveys have given some indication of the variation of the stream positions with time. During the period 1820 to 1873, the Austerfljot moved gradually north-eastward across the sandur and then by 1937 it had moved most of the way back again. The river appears to have remained in about the same position between 1903 and 1938. During this period, the glacier gradually started to split into two parts. Between 1945 and 1951, a proglacial lake gradually developed in front of the glacier, although the position of its snout did not change very much. The actual width of the area through which the river has fluctu-ated is fairly small, covering less than 400 m.

A study was made of the depth of the proglacial lake which developed as the glacier snout withdrew slowly from the terminal moraine. The sandur slopes down steeply into the lake and the almost vertical ice front in the lake extends far below the level to which the sandur surface has been built up. The greatest depth of the lake, 23 m, was found along the ice margin. This deep point occurs where the meltwater stream emerges from the glacier. The floor of the river slopes up steeply to a depth of 2 m and then more gradually to the sill at a depth of 20 cm. The sandur level is about 4–8 m above the river level, the difference in height decreasing downstream. A horizon-tal line from the bottom of the lake does not cut the sandur surface until 3 km from the ice front. The lake floor character suggests that, at present, sedimentation of fine material is in progress. The lake also acts as a trap for the coarser material brought down by the glacier streams. The river, under these conditions, starts with almost no load and erodes its bed as indicated by the depth of the stream below the sandur surface. Sandar cannot, therefore, be built up near the ice front where proglacial lakes exist.

One problem is to account for the process by which the meltwater streams can build up the level of the sandur above the lower limit of the ice. This aggradation must take place before the formation of the proglacial lake. Such lakes are common features in

Plate XXXV
Skeiðarársandur and (in distance) terminus of Skeiðarárjökull, South Iceland. (*C.A.M.K.*)

front of many of the outlet glaciers of Vatnajökull (Chapter 19). They appear to form most readily when the glacier snout is retreating and thinning after a period of advance, during which frontal moraines are formed.

The profile surveyed from the margin of Skeiðarárjökull (Fig. 18.5) shows how the level of the sandur behind the major moraine has not been built up to the level in front of it. In this vicinity, proglacial lakes have formed during the rapid retreat of the last few decades (see Chapter 19, p. 537). J. Jonsson (1955) has suggested that the proglacial lakes form at points where large meltwater streams issue from a retreating ice front. He suggests that the lake is deepened by the powerful erosive activity of these streams, which flow mainly beneath rather than within the ice. This process can only apply to temperate glaciers. Three proglacial lakes in front of Breiðamerkurjökull were found to be 60 m, 48·5 m and 34·5 m deep. The river issuing from beneath the ice erodes downward. Then, when the river moves its position, the ice settles into the hollow so formed. The old channel thus becomes blocked by ice and erosive activity is transferred elsewhere. Slow sedimentation reduces the depth at the old river site, although the ice blocking the old channel prevents rapid sedimentation at this point.

The level of the sandur frequently seems to be linked with the position of a moraine, to which the outwash is graded. It appears, therefore, that aggradation of outwash material can take place most effectively, if not only, when the ice front is advancing and then remains stationary for long enough for a terminal moraine to form. Under these conditions, the meltwater streams must flow upward from beneath the ice to reach the level of the developing sandur. It is possible to observe that this does in fact take place. The melt-rivers can be seen to be emerging under strong hydrostatic pressure from the glacier. The source of Skaftafellsá illustrates this point well.

Because of the gravelly and sandy nature of the sandur sediments, much meltwater percolates underground near the ice margin where the material is coarser. The water emerges from the sediment farther from the ice front where the material is finer and less permeable. The emerging water is clearer than glacial water, but still carries some fine sediment. Nevertheless much of the fine sediment must have been deposited amongst the coarser material. Some of this ground-water emerges as springs, and some forms shallow lakes in old channels. The ground water-table slopes at almost the same gradient as the sandur surface, which is from 1:66·7 to 1:108·6. The sandur material tends to become finer vertically downward. This increase in fine material also helps to prevent water penetration, so that ground-water movement is probably restricted to the uppermost 10 m of the sediment. An estimate suggests that the ground-water discharge is 0·4 per cent of the total water discharge (Hjulström, 1955).

The proglacial area of Breiðamerkurjökull, Iceland, has been described by R. J. Price (1969) (Figs. 18.6 and 18.7). Two-thirds of the area is covered by fluvioglacial material that has not been carried far, but stones are well-rounded, probably because of a long history of previous transport. The combined thickness of the glacial and fluvioglacial sediments is at least 97 m, much of which may have been deposited when the nearby volcano Oraefi erupted in 1362. The sandar within the 1890 moraine are smaller than those outside, and kettle-holes are developing in them. The gradients of the sandar are from 1:30 to 1:45, and they are covered by anastomosing channels. Kettle-holes are mainly to be found in the proximal part, but some larger ones occur near the 1890

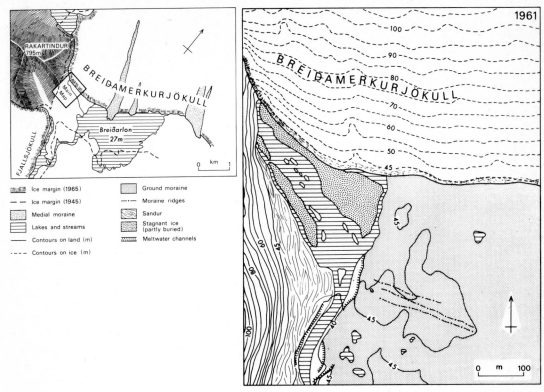

Fig. 18.6
(*Left*) Location map
(*Right*) Photogrammetric map showing part of the sandur between Breiðamerkurjökull and Fjallsjökull, Iceland, in 1961 (R. J. Price, *Arctic and Alpine Research*, 1971)

moraine. Most of them are less than 10 m across. The sandur gravels were spread out over a sheet of glacial ice and, in some places, little of the sandur surface survived the melting of the ice. The very pitted sandar are morphologically almost identical to kame-and-kettle topography, but in this case the 'kames' are remnants of the sandar (Fig. 18.8). A kettled sandur and kame-and-kettle topography really only differ in the number and density of kettles. In 1966, ice could be seen under 4 m of sand and gravel, and also in the walls of the kettle-holes.

Further observations on the sandur formation were subsequently reported by Price in 1971, using air photograph evidence for 1961, 1964 and 1965. Some parts of the sandur were lowered by buried ice wastage and kettle-hole development, while other parts were built up owing to the drainage of an ice-dammed lake. This led to an increase of discharge and sediment load in the stream crossing the sandur after 1961, but by 1965 the stream had abandoned the sandur (Fig. 18.7). The height in a triangular area increased by 1–3·4 m as a result of the lake drainage and sedimentation. After the stream had shifted away, continued melting of the buried ice caused kettle-holes 2–7 m deep to appear.

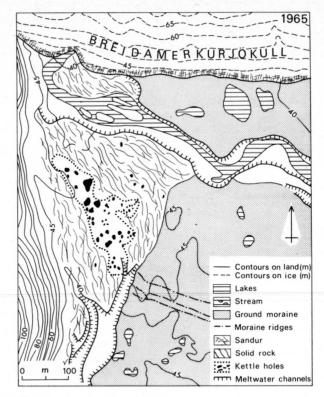

Fig. 18.7
Photogrammetric map, 1965, of the same area as in Fig. 18.6, showing changes in the sandur over a 4-year period (R. J. Price, *Arctic and Alpine Research*, 1971)

Most of the fluvioglacial outwash probably accumulates during advancing stages, when the erosional capacity of the glacier is also enhanced, so that the meltwater streams carry a larger load. The sandur reaches its maximum elevation when the ice front becomes stationary and terminal moraines are forming. The relationship between terminal moraines and outwash spreads is clearly seen in the valleys on the eastern side of the Southern Alps of New Zealand. The features described by J. G. Speight

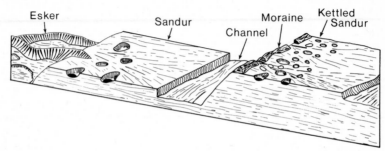

Fig. 18.8
Idealized diagram showing the relationship between sandar, kettled sandar, eskers, moraines and meltwater channels (R. J. Price, *Institute of British Geographers, Transactions*, 1969)

(1963) in connection with the Pukaki stage of the glaciation of the area around Lake Pukaki illustrate these points well. The Pukaki moraine consists of various ridges with steep ice-contact slopes facing up-valley. A series of outwash surfaces form broad channels through an area of kame-and-kettle formation in front of the moraines. These outwash surfaces still show clearly the pattern of braided streams that formed them. The outwash is in the form of a broad fan with a gradient of about 9 m/km. This outwash surface stands about 2 m below that of the former stage, which was formed in relation to moraines a little farther down the valley. The ice-contact slope, facing up-valley and the Lake Pukaki basin, indicates that the glacier must have eroded down well below the level to which the outwash was built up when the ice occupied the position indicated by the moraine. The streams depositing the outwash must have been flowing under hydrostatic pressure if they were flowing under and not on or within the ice. The ice depositing the moraines must have been 300 m thick in the main basin behind the moraine. This gives some indication of the upward component of flow of the streams under the ice.

Some of the areas of older outwash formation can usefully be compared with these recent examples. The Canterbury Plains in New Zealand were at one time thought to be formed by fluvioglacial deposition. This area of smooth but gently sloping land is formed of thick layers of coarse gravel on the east of the Southern Alps. It is now thought to be only partially of fluvioglacial origin. Its surface slope is similar to that of the sandar of Iceland. R. P. Suggate (1958, 1963) has shown that the Plains consist of three major fans of material brought out from the mountains by heavily laden streams. It was thought that the two lower fans were built up as outwash plains in association with two of the major moraines recognized in the foot-hill area, the Otarama and the Blackwater moraines. It was recognized that the upper fan was formed in post-glacial times. It lies on the seaward side of remnants of the earlier fans that form the surface near the mountain area. It is now thought that only the oldest fan is in fact of glacial outwash material and that this fan correlates with the Blackwater stage in the Waimakariri. This stage probably belongs to part of the last glaciation in the Northern Hemisphere.

Fluvioglacial outwash deltas were formed in Baffin Island as the ice retreated from its readvance position near the present central west coast about 7000 BP. These pro-glacial deltas link the ice-front position, which is marked by a steep ice-contact slope, with the sea level or lake level controlling the deltaic deposition. The deltas consist of large boulders near the ice-contact face. The mean value for seven deltas, taking

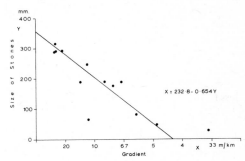

Fig. 18.9
The relationship between the size of the material and the gradient of an outwash delta in western Baffin Island

the mean of 50 stones on each, was 44 cm, with a standard deviation of 25 cm. The size of the material and the gradient falls off from the ice-contact face outwards. There appears to be a linear relationship between the stone size and slope as shown in Fig. 18.9. The two variables correlate significantly at the 99 per cent level, r, the co-efficient of correlation, being -0.858. The deltas built of coarser material usually have large dry channels which show that they were built up by braided streams. At their front faces, the material forming the deltas is much finer and consists of sand and finer sediment. The sand layers show the steeper dip of foreset beds beneath a layer of horizontal topset beds in some instances. Many of the large deltas in this area are 4–5 km across. They indicate halt stages in the recession of the ice, and are also associated with dead-ice features, some of which have already been noted in Chapter 16.

A detailed study of the sandar of eastern Baffin Island has been carried out by M. Church (1972). The nival climatic régime of eastern Baffin Island results in a peak of snowmelt runoff in early July, with later summer storm floods and occasional jökulhlaups. The sandar rivers show predominant bedload transport, although solution and suspended sediment loads are also significant. The sandar studied included a small one on the Lewis River, a meltwater stream from the Barnes ice cap, and Ekalugad sandur, which receives glacial runoff. Normally from 25 to 75 per cent of the total sediment transport occurs on 4 to 5 days of peak flow, although this is not unusual in temperate rivers. The total work is fairly evenly divided between normal and high flows. There is an abundant supply of sediment, more than is being currently created, with much material being stored in the stream beds and channels. The material moves downstream intermittently, the diurnal melt cycle being important as well as the larger, less frequent events. Channels are wide and shallow, and velocities are high. The rapid decline in boundary resistance as flow increases causes rapid velocity adjustment. The sandar profiles are generally concave, indicating recent aggradation, which is concentrated on the distal portion at present. A slight entrenchment now occurs on the proximal part.

Channel bed bars are common, and are spaced at five to eight times the channel width. They form an important part of the resistance to flow. The channel gradient can be related to the material size. Channels in weak cohesionless sediments tend to be wide, in order to distribute the flow stress as widely as possible. The sediment can be passed along the channel most efficiently in a series of steps, as wide stretches are separated by narrower, deeper ones, with lower flow resistance. Fluctuating discharge may be responsible for this characteristic, creating storage elements and conditions suitable for initiating motion as flow increases. The channel bars are the storage areas as well as the resistance elements. The processes result in alternate erosion and deposition, a characteristic of braided streams, which are typical of sandar.

In the aggrading section of Ekalugad sandur, sedimentation occurs at all flows above 6 m³/s, but most occurs at high flood stage. Gravelly deposits show poor sorting and normal imbrication, and the sandur surface shows a veneer of flood sheet deposits. Scouring occurs in channels during the rising stage, but this is more than compensated by deposition in the peak and falling stage of the flood. The sediments are sedimentologically immature, with a rapid change of texture down-sandur in the coarse sediments, the size decreasing and sorting increasing. The fines lack pattern and bedding is crude. Sediment size changes little downstream in the channel bed, showing conditions to be homogeneous; the river is a transporting agent and sediment sorting is achieved by selective erosion rather than deposition. On an active sandur the river profile is concave upwards, while on an inactive sandur the river adjusts to become a rectilinear 'transport surface' owing to aggradation in the lower part and degradation in the upper course. The sandar represent the effects of an abrupt change in geomorphological process as a result of deglaciation, when sediment is in plentiful supply.

Outwash plains are not very conspicuous in Britain, although examples of valley trains have already been mentioned. In continental Europe, on the other hand, they are widely distributed and of great importance in the landscape. In Denmark, for example, there are large areas of outwash associated with the ice front at the time of the Main Stationary Line, the Weichsel maximum. These outwash areas form broad fans cutting through and partially destroying the Old Morainic landscape, dating from the Saale glaciation. The outwash plains, therefore, belong to the Weichsel glaciation and are associated with the high moraines formed at the maximum of this advance. The outwash slopes gently away from the moraines at a gradient of 1 in 700 near the middle of the plains, and 1 in 1000 near the outer margin. These spreads of sandy gravel are analogous to the sandar of Iceland. In places, the moraines are much higher than the outwash, but elsewhere the outwash has built up to a higher level than the moraines with which it is associated. In these areas the proximal margin of the outwash is in the form of an ice-contact slope. In many other parts, the moraine and outwash merge imperceptibly into one another, the only change that indicates the passage from one to the other being a change in soil character. There are kettle-holes in the outwash deposits that show that lumps of ice were included in the deposits. The outwash surface has been dissected by later stream activity, with different load–discharge relationship.

Large areas of submerged glacial outwash have been located off the Lincolnshire coast by A. H. W. Robinson (1968). The Protector Overfalls is formed of coarse sand, gravel and pebbles, up to 10 cm in diameter. It is stable and represents a slightly modified fluvioglacial feature. The Theddlethorpe Overfalls is similar, but some other features built of fluvioglacial material are in fact tidal banks, which must be carefully distinguished. Robinson shows that the offshore fluvioglacial features can be related to stages of deglaciation on land.

Another classic area where the outwash fans are associated with terminal moraines is in the northern Alpine foothills. This is the area where the original glacial sequence of A. Penck and E. Bruckner (1909) was worked out by studying the moraines and associated outwash deposits. The outwash of the Alpine valleys is called 'Schotter', which means gravel. The so-called 'Deckenschotter', which belong to the two older glaciations (Günz and Mindel), were true outwash plains, spreading extensively over the country in front of the moraines. In the later glaciations (Riss and Würm), however, rejuvenation during the Mindel-Riss interglacial restricted the outwash to the valleys. Thus valley trains were formed in the later glaciations. The Munich plain is covered by outwash from the Alps, brought down by the Inn and Isar rivers. The gradients of these deposits, varying from 1 to 1·2 per cent, are steeper where the outwash is confined within valleys than where it can spread more widely as in the North European plain. Outwash deposits are not confined to the northern fringe of the Alps. They also occur in the Po valley where they reach a thickness of 237 m at Cremona. The outwash is now buried by deposits owing to subsidence in the area.

A sandur-delta moraine is associated with the second Salpausselka moraine of Finland (P. Fogelberg, 1970). The sandur-delta lies in front of proximal moraine ridges which formed at the active ice margin. The ice then retreated 1·5 km and dead ice remained. A new sandur-delta was in turn deposited proglacially at the new marginal position and the same water level. The ice then gradually stagnated without further oscillation. The end-moraine complex consists in places of three parallel moraine ridges with plateau-like fluvioglacial deposits between them. They are composed of till with both rounded and angular stones and boulders, and their distal slopes are steeper. These moraine ridges were formed during active phases; transverse to them, eskers were deposited in marginal crevasses. Outwash accumulated during halt phases while the moraine ridges, some of which are push-moraines composed of proglacial deposits, formed during slight advances.

The outwash plains of North Germany are very extensive. They form wide sandy heaths, such as the Lüneburger Heide. The material gradually changes from gravel near the moraines (Fig. 12.9) to sand and then to finer clay deposits southward. Other areas where there are sandar of large dimensions and considerable thickness are found to the east of the Baltic. In Russia, they occur in the Dnieper valley. There are also very extensive outwash deposits, or modified drift, in North America. The effects of meltwater on the drift are apparent in the New England states westward to Minnesota, the Dakotas and Manitoba. The Mississippi River at one time carried a large load of glacially derived material. Its valley drained a 3000 km stretch of the Laurentian ice margin at a time when the precipitation was higher, during the maximum of the Wisconsin glaciation. This coarser material can be traced to the delta of the river 1120 km from the nearest ice edge.

3 References

ARNBORG, L. (1955), 'Hydrology of the glacial river Austerfljot', *Geogr. Annlr* **37**, 185–201

(1955), 'Ice-marginal lakes at Hoffellsjökull', *Geogr. Annlr* **37**, 202–28

CHURCH, M. (1972), 'Baffin Island sandurs: a study of Arctic fluvial processes', *Bull. geol. Surv. Can.* **216**, 208 pp.

FOGELBERG, P. (1970), 'Geomorphology and deglaciation at the second Salpausselka between Vääksy and Vierumäki, southern Finland', *Soc. Sci. Fennica Phys. Math.* **39**, 7–90

GAGE, M. (1958), 'Late Pleistocene glaciation of the Waimakariri valley, Canterbury, New Zealand', *N.Z. Jl Geol. Geophys.* **1**, 123–55

HJULSTRÖM, F. (1955), 'The ground water', *Geogr. Annlr* **37**, 234–45

JONSSON, J. (1955), 'On the formation of frontal glacial lakes', *Geogr. Annlr* **37**, 229–233

PENCK, A. and BRUCKNER, E. (1909), *Die Alpen im Eiszeitalter* (Leipzig)

PRICE, R. J. (1969), 'Moraines, sandar, kames and eskers near Breiðamerkurjökull, Iceland', *Trans. Inst. Br. Geogr.* **46**, 17–43

(1971), 'The development and destruction of a sandur, Breiðamerkurjökull, Iceland', *Arct. alp. Res.* **3**, 225–37

RIST, S. (1955), 'Skeiðarárhlaup 1954', *Jökull* **5**, 30–6

ROBINSON, A. H. W. (1968), 'The submerged glacial landscape off the Lincolnshire coast', *Trans. Inst. Br. Geogr.* **44**, 119–32

SPEIGHT, J. G. (1963), 'Late Glacial historical geomorphology of the Lake Pukaki area, New Zealand', *N.Z. Jl Geol. Geophys.* **6**, 160–88

SUGGATE, R. P. (1958), 'Late Quaternary deposits of the Christchurch metropolitan area, New Zealand', *N.Z. Jl Geol. Geophys.* **1**, 103–22

(1963), 'The fan surfaces of the central Canterbury Plains', *N.Z. Jl Geol. Geophys.* **6**, 281–7

19

Glacial lakes and lacustrine deposits

The great glacier of Aletsch gives rise ... to a small lake called the Märjelen See, which ... is periodically drained by changes which take place in the internal structure of the glacier. Rents or 'crevasses' in the ice open and give passage to the waters, which escape in a few hours, producing destructive inundations in the country below. The Märjelen See was about two miles in circumference when I visited it in August 1865, and about forty feet below its normal level.... Such a state of things gave me an opportunity of examining a point of great geological interest, namely, the ... large terrace or line of beach which encircles the lake basin all round its margin, and which constitutes its shore when full, and when its surplus waters flow over to the Viesch valley. I satisfied myself that this terrace is a counterpart of one of those ancient shelves or parallel roads, as they are called, of Glen Roy in Scotland.... (C. LYELL, 1866)

Glacial lakes are not today so common, nor do they attain such dimensions as in the Pleistocene period. Nevertheless, enough survive and in a sufficient variety of circumstances to throw much light on the nature and effects of the Pleistocene lakes. Present and former glacial lakes are responsible for distinctive landforms and deposits, including shoreline features such as strandlines, terracettes, deltas, and bottom deposits such as laminated clays. The term 'glacial lake' really embraces a whole family of lake forms. There are both lakes actually dammed by glacier ice, and lakes where glacier ice forms part of the lake margin though not the actual barrier. Relationships with the ice are as varied as in the case of meltwater channels. Some glacial lakes are supraglacial, some are marginal in various ways, and some exist subglacially or englacially although direct study of lakes in this latter category is rarely possible. The methods by which glacial lakes drain is equally diverse, but it is well established that the majority periodically empty, often catastrophically, by subglacial or englacial routes.

It is logical to begin by looking at present-day glacial lakes, and then to review the morphological and depositional features relating to such lakes and their Pleistocene equivalents.

1 Modern glacial lakes

1.1 *Lakes in lateral valleys dammed by ice in the main valley*

One of the biggest concentrations of glacial lakes in the world today is to be found in southern Alaska, especially in the coastal strip (K. H. Stone, 1963). Nearly all lie in ice-free tributary valleys that have been blocked by glacier ice in the trunk valley, the trunk glacier often sending a short distributary arm up into the side valley to form the actual dam. The largest lake, covering 124 km², is Lake George; in 1963, there were 52 other lakes in an area 1200×150 km. All drain under or into the ice—no instance of the lake waters overtopping the ice dam has been recorded. Until 1963, and since 1918 at least, Lake George emptied annually. As in all such regions, the number of lakes is slowly diminishing, the result largely of the current glacier recession, but the former lake sites are easy to recognize. R. P. Sharp in the St. Elias Range of Yukon Territory, Canada, describes the site of North Fork Lake, in a tributary valley dammed by Wolf Creek Glacier, where the wave-cut shoreline terrace and the delta built by the North Fork River indicate a maximum water depth of 50 m. Depth data for existing Alaskan lakes are few; the deepest measurement is 73 m for Lake Tulsequah in 1958, but this lake is known to have reached a depth of 195 m in 1910.

Plate XXXVI
Ice-dammed lake in lateral valley, Breiðamerkurjökull, Iceland. The ice dam is to the right, off the picture; numerous icebergs cover the lake, and those stranded on the hillslopes reveal a recent fall in lake level. *(C.E.)*

Another area where glacial lakes are still relatively numerous is that of Vatnajökull, Iceland. Here again the majority lie in tributary valleys blocked at their lower ends by glaciers. Such is the case with Graenalón, the largest, covering 18 km² in 1935 and 150 m deep, dammed by Skeiðarárjökull on the east. Until recently, Graenalón emptied subglacially every four years, but later every second year. Its maximum possible level has been determined by the col at 635 m leading south to the Nupsvötn River, the lake overflowing through this col in May 1939, for example. Farther east along the margin of Vatnajökull, another outlet glacier, Heinabergsjökull, has also been responsible for blocking some small tributary valleys (Fig. 19.1). Vatnsdalur, one of the latter,

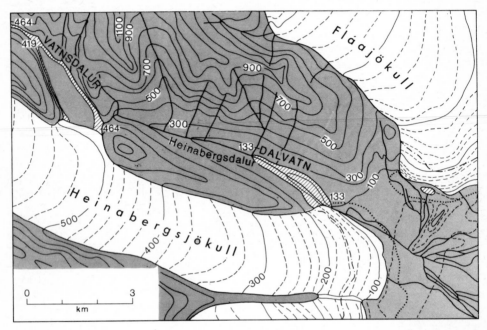

Fig. 19.1
Map of the area around Heinabergsjökull, Iceland, showing ice-dammed lake in Vatnsdalur and the site of formerly ice-dammed Dalvatn (S. Thorarinsson, 1956)

contained a lake up to a level of 464 m in the 1870s, which drained over a col at that height into a lower ice-dammed lake, Dalvatn. In 1898, Vatnsdalur lake emptied subglacially in a sudden and violent manner, and the overflow col was abandoned. Until the 1940s, it drained each year thereafter in a similar manner though the violence of the bursts decreased with the gradual shrinkage of Heinabergsjökull which became less and less capable of impounding any sizeable lake. A further result of the diminishing efficiency of the ice dam was that the date of drainage each year became earlier— from late September in 1899, late August in 1910, to July in the 1930s. In the 1950s, Vatnsdalur was draining two or three times each year, its level not exceeding 400 m.

In Europe, the most celebrated ice-dammed lake, also occupying a tributary valley, has undoubtedly been the Märjelensee, held up by the Aletsch glacier (Fig. 3.10). It attained its greatest dimensions in July 1878, when the water rose to 2366 m and the lake was 1600 m long. Its depth may have been as much as 80 m. It has long been known as a peculiarly variable lake, changing greatly in size and shape from one year to the next. When full, it has been on rare occasions observed to spill over the ice, but usually the ice is too crevassed to permit this, and its normal mode of emptying is into crevasses in the impounding ice wall, or beneath the ice. The last method empties the lake particularly rapidly. From 1813 to 1900, the lake emptied subglacially nineteen times. In 1873, 10 million m³ in 8 hours were said to have been released, causing a flood-wave on the Rhône. As the lake continued to attain unprecedented levels in the 1870s, and its bursting created flood problems in the valley below the Aletsch snout, construction of a tunnel from the lake into the Seebach was put in hand. After completion in 1896, it soon became superfluous, for slow shrinkage of the Aletsch glacier has since that date gradually lowered the maximum lake level. The 1957 map (Switzerland, 1:10,000) of the Aletschgletscher showed that the lake had all but disappeared, a small pond at 2300 m encased by moraine and draining into the ice at a level of 2288 m being the only survivor. As the edge of the Aletsch glacier does not exceed 2314 m at any point south of this, the possibilities of lake formation are now very limited. N. E. Odell (1966) recorded the Märjelensee as 'empty' in July 1965.

So far, the discussion has dealt only with lakes impounded against temperate glacier ice. There are also many examples of lakes dammed by sub-polar glaciers which usually drain in a completely different manner. Subglacial or deep englacial escape routes for the water are no longer possible because of sub-zero temperatures in the ice and the fact that the ice at the margins is frozen to bedrock; therefore, drainage takes place by overflow of the ice barrier or of a rock barrier, or else it takes a marginal or sub-marginal route. Axel Heiberg Island in the Canadian Arctic illustrates these points (H. Maag, 1969). As Table 19.1 shows, the area studied by Maag contains a high concentration of ice-dammed lakes—it should be noted that the area covers only one-eightieth of that examined by Stone (1963) in Alaska. Most lakes here drain either supraglacially or marginally and sub-marginally. There are a few possible instances of englacial escape, but these may have been either at shallow depth or by concealed sub-marginal routes (Fig. 12.6). In the Thule area of north-west Greenland, two large lakes dammed by cold ice normally drain supraglacially (V. Schytt, 1956).

Table 19.1 Comparison of ice-dammed lakes in three contrasting environments

	Axel Heiberg Island Expedition Area, Canada (Maag, 1969)	Southern Alaska (Stone, 1963)	Aletsch glacier, Switzerland
Types of glacier:	Sub-polar	Temperate	Temperate
Total number of lakes			
>400,000 m² in size:	4	53	0
>500 m² in size:	120	no data	8
Average number of lakes per 100 km² of ice/firn:	24	no data	6

1.2 *Lakes in main valleys dammed by ice from lateral valleys*

A second category of ice marginal lake concerns situations, often causing highly danger-ous lake bursts, where a side glacier advances into and blocks a main valley. Examples of Pleistocene age in western North America were described in Chapter 12. Large-scale present-day examples are known from the Karakoram (K. Mason, 1935), especi-ally among the upper Shyok glaciers. In 1924–5, for instance, the Chong Kumdun glacier suddenly advanced to form a complete block across the Shyok river, the lake bursting in October 1926 with catastrophic effects. The process was repeated several times subsequently, until the glacier ceased to advance in the 1930s. Another glacier, the Biafo, has been responsible for similar temporary lakes at times when it has advanced out of its side valley to block the main valley of the Braldu (K. Hewitt, 1964). The lake bursts cause severe floods and damage in the upper Indus valley. In Switzer-land, the Mattmark See was, until the 1920s, an example on a smaller scale (R. Keller, 1950). The steeply sloping tongue of the Allalin glacier extending across the Saaser Visp valley impounded the Mattmark lake in the latter to a depth of 30 m or more at times; in 1834 the lake was 2·5 km long. It has had a history of violent outbursts in the past—26 since 1859—but by 1909, the Allalin glacier had retreated sufficiently to unblock the main valley. Its lateral moraine was cut through by the lake overflow while the lake itself rapidly filled with sediment as it became shallower.

A slightly different example is provided by the Moreno glacier in Patagonia. This is one of the rare glaciers in the world that has been steadily advancing this century (R. L. Nichols and M. M. Miller, 1952). In 1917 it entered the main valley in which lies Lago Argentino and divided the latter. The part of the lake behind the Moreno ice barrier is known as Lago Rico; the water level of the latter was raised 17 m above Lago Argentino in 1942. The ice dam normally does not survive more than a year before rupturing, and after the burst, which commences subglacially and later follows an open ice-walled channel when the subglacial tunnel has collapsed, the water level settles down again to that of Lago Argentino.

1.3 *Glacier confluence lakes*

Small lakes frequently develop at the angle between confluent glaciers, fed by converg-ing marginal drainage or by local runoff trapped at this point. J. D. Forbes (1843) showed how, at times when melting is adequate, a small lake formed by the Tacul promontory between the Géant and Leschaux glaciers, north-east of Mont Blanc, and noted that marked variations in level suggested a subglacial outlet. Another example, for which records extend back to the seventeenth century, is the Gornersee, located at the junction of the Grenz and Gorner glaciers, south-east of Zermatt. A third glacier (the Mont-Rose) contributed to the impounding of the lake until 1955, but has since retreated. The lake drains into or beneath the ice when it is full or nearly so, usually between June and September each year (A. Bezinge *et al.*, 1973). Since 1948, the Gornersee and its emptying have been studied in some detail in connection with hydro-electric power development. During a burst, the lake water travels about 7 km subglaci-ally or englacially to the ice margin. Amounts of discharge measured range between

2 and 6 million m³, of which only a quarter to a third is present or visible in the Gorner-see even when the lake is full. The rest of the flood must therefore consist of water from a subglacial or englacial reservoir.

1.4 *Frontal lakes*

The term was employed by L. Arnborg (1955) to describe lakes formed between the ice terminus (or glacier snout) and a barrier such as moraine downstream. Glacial ice thus contacts the water but the lake is not actually ice-dammed. The glacial Great Lakes of North America (up to the Algonquin stage) were examples of frontal lakes on a huge scale. Well-known minor examples today occur at the foot of Brixdalsbreen (Norway) and the Steingletscher (Switzerland). Arnborg has investigated two frontal lakes at Hoffellsjökull (Iceland): Svinafellslón and Hoffellslón. The former is held by moraines dating from 1890 and 1903, and lies across minor terminal ridges dating from the 1930s. Such lakes are deepest next to the ice front, but soundings here are hazardous because of calving ice. J. Jonsson (1955) has suggested that some frontal lakes may occupy basins or a series of furrows eroded by subglacial streams when ice covered the site of the lake (Chapter 18). Ice collapsing into the furrows prevented them from being infilled with sediment, and when the ice melted back, meltwater filled the hollows. This may explain how the floors of some frontal lakes come to lie at a lower altitude than that of the outwash plain surface downstream, and how some frontal lakes form without morainic barriers.

P. J. Howarth and R. J. Price (1969) have described the proglacial lakes of Breiða-merkurjökull and Fjallsjökull in Iceland, the floor of one of which (Breiðárlón, 84 m deep) descends to 56 m below sea level. These lakes have come into existence since the retreat of the ice from the 1890 moraine; the oldest of them dates back to about 1932. The basins may have been formed by glacial erosion or, following Jonsson, by subglacial stream erosion (see p. 524).

1.5 *Minor marginal lakes*

Small lakes may develop in marginal channels blocked by slides or alluvial cones but these are of a very temporary nature.

A more important group comprises lakes forming next to nunataks or valley sides because of heat reflected from the darker-coloured rock. An alternative origin for such lakes has been put forward by M. Seppälä (1973), who has examined the sites and forms of small marginal lakes on the Juneau Icefield, Alaska. Some are situated against the foot of north-facing slopes, which would exclude a hypothesis of origin by reflected heat. The forms of some of the small basins suggest that they may have originated by wind eddying at the foot of steep slopes preventing snow accumulation. Any melt-water would collect in such hollows and enlarge them by melting. Wind-blown or other sediment trapped in the water would be partially exposed with changes of lake level, and the decreased albedo of the lake shores would further enhance melting here and widening of the lake.

1.6 *Lakes on glaciers and ice sheets*

Factors involved in the formation of supraglacial lakes include: 1. Local reversed slopes of the ice surface; 2. collapse of subglacial or englacial cavities; 3. different ablation; 4. blocking of crevasses or moulins. The Martin River glacier, Alaska, possesses in its marginal portions numerous roughly circular sink-holes with standing water, up to 90 m deep and 300 m in diameter (J. R. Reid and L. Clayton, 1963). The lakes are apparently connected with subglacial tunnel systems, and their periodic draining and refilling results from ice movement opening or closing the tunnels. R. P. Sharp (1947) noted similar features on the Wolf Creek glaciers; he also described 'dust basins' on the ice surface formed by localized melting under thin layers of dark silt, some as much as 5 m deep. Once such surface ponds are formed, the relative warmth of the water quickly enlarges them.

1.7 *Subglacial and englacial lakes*

The existence of such water bodies at depths within temperate glacier ice is well known and some aspects of such water-filled cavities have been discussed in Chapters 4 and 12. Sudden increases in the flow of subglacial streams emerging from glacier tongues (as preceded the Allalin glacier catastrophe of 1965: see Chapter 4) are likely to result from the tapping of englacial water bodies; and local subsidence of the ice surface over presumed subglacial or englacial cavities accompanied again by sudden increases in meltwater discharge from the glacier also points to the existence of internal lakes. A most impressive example is that of Grimsvötn in the heart of Vatnajökull, Iceland. This is an ice-filled depression, 300 km^2 in area and normally 400 m deep; beneath a thickness of ice, a lake accumulates periodically in this depression owing to ablation and to slow subglacial melting related to solfatara activity. When approximately 7 km^3 of water have accumulated subglacially, the lake bursts, escaping by tunnels 55 km long to the edge of Skeiðarárjökull. The bursts occurred at almost regular 10-year intervals until 1934 (10 years' annual snow accumulation at Grimsvötn totals about 7·5 km^3) and each was accompanied by a volcanic eruption at Grimsvötn. S. Thorarinsson (1953) argues that the eruptions are triggered off by the sudden release of water pressure when the lake bursts, for the maximum discharge of water just precedes the earth tremors accompanying the eruption. Since 1934, the bursts have tended to occur at intervals of 4 to 6 years. During and after the burst, the ice surface of Grimsvötn sinks more than 200 m into the drained cavity beneath. The most recent burst occurred in 1972 (see *Ice* **42** (1973), 3).

2 The problem of glacial lake drainage

In the case of lakes dammed by temperate glacier ice, the usual mode of glacial lake drainage is subglacial or englacial. T. G. Bonney emphasized this as long ago as 1896. A few lakes, as already noted, drain occasionally by overflow cols: this certainly happened in the case of many Pleistocene ice-dammed lakes (see Chapter 12) and no further discussion is required on this point. Drainage of a glacial lake over the ice barrier is rare except in the case of sub-polar glaciers where sub-zero ice temperatures preclude subglacial or englacial escape routes. Table 19.2 presents some data on the dimensions of ice-dammed lakes and the magnitude of their outflow during sudden draining.

There is considerable controversy over the mechanism by which glacial lakes drain subglacially, especially in regard to the sudden bursts or 'jökulhlaups'. The lake water undoubtedly finds its way into crevasses, into gaps between the ice and bedrock, or

Table 19.2 Comparison of glacier lake bursts

Lake	Region	Year of burst	Maximum volume of lake, 10^6 m³	Depth at ice dam, m	Maximum instantaneous discharge m³/s
Strupvatnet	N	1969	4.6	29	150
Ekalugad valley	B	1967	4.8	120	200
Demmevatn	N	1937	11.6	79	1,000
Gjánúpsvatn	I	1951	20	20	370
Vatnsdalur	I	1898	120	188	3,000
Tulsequah Lake	BC	1958	229	73	1,556
Summit Lake	BC	1965/7	251	200	3,260
Graenalón	I	1939	1,500	230	5,000
Lake George	A	1958	1,730	40	10,100
Grimsvötn	I	1934	7,000	c. 200*	50,000
Lake Missoula	M	Pleistocene	2,000,000	610	1,870,000

Source: mainly J. J. Clague and W. H. Mathews (1973). It should be stressed that the data are of varying reliability, and that in some cases conversion to metric units gives a false impression of accuracy. The lakes are ranked in order of size.

A Alaska, B Baffin Island, BC British Columbia,
I Iceland, M Montana, N Norway

* Subglacial

perhaps soaks through subglacial gravels, and may travel for great distances beneath the ice—note the 55 km length of the Grimsvötn escape route. The problems involved are several: 1. Why do glacial lakes often drain regularly and periodically? 2. How are subglacial outlets first opened and how are they kept open until the lake has emptied? 3. How are such long escape routes opened in such a short time? 4. What factors control the actual time when the burst commences?

S. Thorarinsson (1939) contended that when the depth of water in the lake reached a critical figure (about nine-tenths the height of the ice barrier), the ice barrier would begin to float. The height (h) of the ice barrier would thus control the maximum possible depth of the lake, and can be computed from the formula:

$$h = \frac{x - m}{1 - D}$$

where x = the height of the barrier above the lake surface, m = an allowance for crevassing and unevenness of the barrier surface, and D = ice density (about 0·9). Applied to the Vatnsdalur lake levels of 1903 and 1938, this gives very reasonable results; and Thorarinsson claims that during the Grimsvötn jökulhlaups, a large part of the edge of Skeiðarárjökull floats for a brief period. But it has also been pointed out that, on this hypothesis, one would expect the outflow of water to be intermittent and not a sudden rush, for once lake drainage had begun, lowering of lake level would reduce the pressure on the ice barrier which would then cease to float, until the lake once more filled to the critical level. Moreover, it is very difficult to conceive of several kilometres, let alone 55 km, of ice floating simultaneously. J. W. Glen (1954) has put forward a different hypothesis. Since the density of water is greater than that of ice, water of more than about 200 m depth resting against an ice dam or filling a hole in a glacier will exert a stress on the ice sufficient to cause deformation of the ice at the base of the water mass. This hypothesis, however, also encounters problems. The depths of glacial lakes are often nowhere near 200 m when bursts occur—the head of water preceding the 1958 burst of Lake Tulsequah, British Columbia, was no more than 73 m (M. G. Marcus, 1960). In this case, too, it appears that tunnels in or beneath the ice and 7 km long were opened up in the space of a few days, which is difficult to equate with the slow rate at which ice deforms (for instance, compare known rates of contraction of man-made tunnels at depths of 50 m or more in glaciers: Chapter 4). Furthermore, there is the same problem encountered by Thorarinsson's hypothesis that as soon as lake depth was reduced, pressure on the ice tending to enlarge an escape route would also diminish.

O. Liestøl (1955) thinks that enlargement of the escape route by melting is a most important factor. Initial opening of a subglacial passage is achieved by ice flotation, following Thorarinsson; the passage is then kept open and enlarged by melting until the lake has emptied. Calculations show that water at 1°C and flowing at only 1 m³/s can theoretically melt over 270 m³ of ice in 24 hours. This idea has found much support among other workers. R. Gilbert (1971) has applied it to the periodic draining of Summit Lake, British Columbia. After the 1965 burst of this lake, observations were kept on the rate of filling, which seem to indicate that water was beginning to leak through or under the Salmon Glacier at least three months before the next draining in 1967. Measurements of lake water temperatures supported Liestøl's view that melting could enlarge and maintain an escape route. For Summit Lake there are no possibilities of it draining over the ice dam or by flotation of the ice dam. A. K. Higgins (1970) discusses possible drainage mechanisms for some lakes in south-west Greenland near Frederikshåb. Leistøl's hypothesis is supported, but for one of the lakes, Tordensø, variations in level of 160 m suggest a maximum depth in excess of this, approaching the point at which Glen's mechanism may come into play. In this case, plastic deformation may hasten the point at which water begins to seep into the potential escape route, after which enlargement by melting takes over. It will be noted in Table 19.2 that there are a few other known examples where water depths approach or exceed Glen's critical figure of 200 m.

Another possibility for maintenance of an open escape route during emptying is suggested by N. Aitkenhead (1960). After the ice barrier has been lifted by flotation and lake level has dropped, the ice will never fall back exactly in its former position on a bedrock floor which is likely to be uneven. Moreover, icebergs wedged in the tunnels will help to prevent collapse. Marcus (1960) makes similar suggestions.

W. B. Whalley (1971) outlines some further possibilities for lake drainage in connection with Strupvatnet, Norway, where there is no evidence of lifting or flotation of the ice dam (though this does not, of course, rule out such a mechanism elsewhere). He proposes that it is the internal drainage system of the glacier itself that initiates the burst. T. Stenborg (1969) has shown that the capacity of such systems increases in summer (p. 327). Enlargement of the passages, or possibly changing flow rates and stress patterns in the ice, may cause a lake to be tapped, followed by rapid drainage. Another possible means of establishing an interconnection is if a crevasse becomes filled with water. J. Weertman (1973) has shown that an isolated water-filled crevasse could theoretically deepen until it reached the bed of a glacier (p. 148).

There are thus many possible suggestions to account for the manner in which ice-dammed lakes drain subglacially or englacially. The two most plausible hypotheses for initiating a lake burst are those of Thorarinsson (a flotation mechanism) and Whalley (tapping by the enlargement of the englacial or subglacial drainage system). Glen's hypothesis seems to demand greater depths of water than are usually found in the lakes. Once the burst has started, the hypotheses of Liestøl and Aitkenhead can adequately explain the maintenance and enlargement of tunnels to allow the lake to drain rapidly and completely.

2.1 The nature of jökulhlaups

The following account of the Grimsvötn burst of 1934 by S. Thorarinsson (1953) is typical of the large Icelandic jökulhlaups:

> The river Skeiðará, which in March and April normally has its smallest discharge, started to rise on March 22nd. The rise was slow at first, but on the 24th it had reached approximately the normal summer high-water level. On the 28th the water started forcing its way out from under the glacier at several places, breaking up its border. On the morning of March 31st, the glacier burst reached its climax. Forty to fifty thousand cubic metres of muddy grey water plunged forth every second from under the glacier border bringing with it icebergs as big as three-storeyed houses. Almost the whole of the *sandur*, some 1000 sq. km in area, was flooded. At 17.30 hr the same day the burst suddenly started to abate, and by the following morning the discharge of the Skeiðará was normal.

It is impressive to recall that the source of this water escaping from the edge of Skeiðarárjökull is Grimsvötn, 55 km to the north. Jökulhlaups are associated with other glaciers in Iceland. The most intense and dangerous bursts are the 'Katlahlaups' from Myrdalsjökull (Thorarinsson, 1957) where the peak discharge may exceed 100,000 m³/s (Fig. 19.2). The sudden release of such quantities of ice-impounded water has far-reaching influences on the morphology of the valley or floodplain over which the water escapes; vast quantities of sediment are carried away or redistributed, and in the course of a few days the whole form of the outwash plain may be radically altered (Chapter 18). Although present-day jökulhlaups have been studied in Iceland, Norway, North America and elsewhere, recognition of possible Pleistocene jökulhlaups and their effects has advanced little. The bursts of Pleistocene Lake Missoula (Chapter 12) are an exception, and also the bursting of glacial Lake Tahoe in Nevada, postulated by P. Birkeland (1964). Birkeland describes the evidence for such floods in the Truckee valley below Reno, where 3 m granitic boulders lie stranded many kilometres beyond

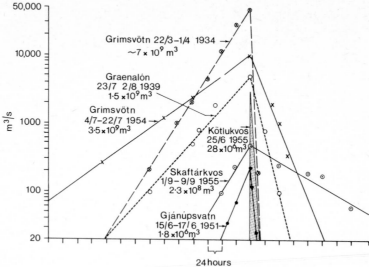

Fig. 19.2
Estimated discharge graphs of some Icelandic jökulhlaups. The vertical scale is logarithmic. The date
and estimated total volume of water discharged is shown for each event (S. Thorarinsson *et al.,*
On the geology and geomorphology of Iceland, Int. Geogr. Congr. Norden 1960, Excursion E.I.1)

the limits of any glaciation. In Britain, T. F. Jamieson (1892) tentatively suggested
that the glacial lakes of Glen Roy, Scotland, suffered a sudden burst when the lake
fell from the higher to the middle level (see p. 543).

The morphological effects of smaller lake bursts are by no means negligible. Maag
(1969) documents some of the activity associated with the resulting floods in Axel Heiberg Island, Canada, pointing to the massive amount of work accomplished by the
outflowing water in a few days, helped in this arctic area by thermo-erosion of the
frozen and shattered bedrock (Table 19.3).

Table 19.3   Estimates of material eroded and moved by Between
River during outflow of Between Lake, July, 1961 and
1963

	1961	1963
Number of days until two-thirds of the lake had drained:	4	$2\frac{1}{4}$
Volume of discharge in this period, million m³:	7	6.5
Peak discharge, m³/s	40	140
Bed load eroded/transported, t	60,000*	110,000*
Suspended load, t	17,500*	6,500*
Dissolved load, t	1,400	450

Source: H. Maag (1969), Table 11, p. 91.

* Most of this was picked up from 600 m of the channel bed, and relatively little came from the lake.

3 Glacial lakes as indicators of glacier oscillations

As Thorarinsson (1939) demonstrated in the case of Icelandic ice-dammed lakes, they can be of the highest value in recording changes in ice thickness. If the flotation hypothesis is accepted, the lake level prior to draining provides a means of accurately determining the ice thickness in the barrier. The lake shorelines provide the date of lake emptying. On the evidence from lake fluctuations alone, Thorarinsson reconstructs the recent glacial history of Iceland: 1700–50 general glacier advance; 1750–1800 stagnation or slight recession; readvance, probably to maximum extent in historical times between 1800 and 1850; and beginning about 1890, a general recession, which has gradually accelerated and threatens extinction of many glacial lakes.

4 Features associated with present and former glacial lakes

4.1   Strandlines and beach deposits

Studies of modern glacial lakes show that within surprisingly short periods of time, lake waters standing at one particular level are able to create a strandline provided that there is sufficient unconsolidated sediment available. From a distance, strandlines appear as horizontal markings (except where later isostatic movements have caused them to be tilted or warped), often at successive positions on the hill slopes to record changes in the lake levels. Closer examination reveals that the strandlines are features caused both by erosion by lake wave attack and by deposition from the lake waters. Lake ice may also play a part. Remnants still frozen to the shore after a lake empties in summer will disturb the beach material along the former water level. In the case of larger lakes, ice-push features on the shores will further demarcate the level (p. 545).

It is common to find that strandlines, particularly those of Pleistocene age, become less and less distinct as one approaches them; it may be exceedingly difficult to recognize and locate them on the ground over which one is walking, especially in wooded country. Sometimes they are represented only by slight changes in vegetation, related to differences of soil and drainage at the actual shoreline, and such vegetational changes, though easily discernible over a large area viewed from a distance, may be almost undetectable and inseparable from other variations in ground cover at close quarters. Tracing of former shorelines is further complicated by their discontinuity, by isostatic tilting, and by the fact that their sequence on the ground is not necessarily a chronological one.

The general morphological and depositional features that together form a strandline are those normally associated with the margins of any water body, and will vary in scale with the dimensions of the lake.

The Pleistocene Great Lakes of North America offer the most impressive lake-shore beaches to be found anywhere in the world. Recognized as old lake beaches as long ago as the 1830s, J. W. Spencer in 1890 described them thus: 'In ascending from the modern lakes to the highlands, several old shores must be crossed. The country may be described as a series of terraces or steps, whose frontal margins are moulded into hills, and whose surfaces are plains, most commonly of clay, although sometimes of gravel or sand, at the back of which there may be found the beach in some form. These gently rising terrace plains may each be several miles in width …' (p. 76). In places, the beaches consist of boulder pavements up to 1 km wide; the larger boulders were probably deposited by lake ice. The strandlines and beaches have been traced continuously for hundreds of kilometres in some cases (see Chapter 5, p. 174 and Chapter 12, p. 334); as expected, they terminate abruptly at the former ice margins, though successively lower strandlines extend farther and farther north and east, and the lowest completely enclose the present lakes showing how the ice gradually wasted away.

Even in the case of such huge lakes, wave-cut notches and cliffs are usually only present where the lake washes against unconsolidated materials such as glacial drift, or other soft formations. R. P. Sharp (1947) states that the wave-cut terrace of the former North Fork Lake in the Yukon Territory is 23 m wide in till and up to 5 m in bedrock, but the nature of the latter is unspecified (p. 51), nor is the duration of the lake known. Much more important in the definition of strandlines are the associated deposits, built into lake shore terraces. The deposits range from very fine silts, such as composed the small terraces recorded by G. de Boer (1949) for a former glacial lake beside Leirbreen, Norway, to coarse material including cobbles and boulders in the case of some of the larger Pleistocene glacial lakes. Much depends on the nature of the material locally available, on the size of the lake, determining strength of wave action, and its duration. If morainic debris is available, wave action may wash out the finer fraction from one place, leaving a lag deposit of boulders, and redeposit it elsewhere. Changes in lake level will also have important effects, as R. F. Flint (1971, p. 194) suggests. The more massive accumulations of cobbles and boulders may be related to a rising lake level and advancing shoreline, concentrating and reworking the beach material, whereas the smaller pebble ridges are perhaps to be associated with stable or even receding shorelines.

Glacial lakes which have recently drained often reveal on the surrounding slopes a series of minor shoreline features (sometimes described as terracettes). Lake George, Alaska, in 1951 just after emptying showed more than twenty micro-shorelines below the high-water mark of that year. All were evenly spaced and horizontal; some were no more than 0·5 m apart vertically (K. H. Stone, 1963). Gjanupsvatn beside Hoffells-jökull in Iceland drained suddenly in June 1951 and its sides presented terracettes at 193, 184, 175, 170, 153, 136, and 132 m (L. Arnborg, 1955). De Boer (1949) noted traces of about twenty terracettes below the two main shorelines of the Leirbreen marginal lake. Such micro-shorelines are clearly formed by the lake water and are mainly depositional features. Some have been recorded which are only 5 cm high and 5 cm wide. A possible explanation for the smallest is that they reflect a daily rhythm of lake lowering during drainage. Diurnal variations of air temperature and insolation may cause a decrease in the amount of water flowing *into* the lake each night, whereas the outflow, being subglacial or englacial, will remain constant. The micro-shorelines would then be linked to daily periods when the water level was stationary or only falling slightly, whereas each night there would be a more definite fall. If this is correct, micro-shorelines would be related to episodes of slow lake drainage in contrast to sudden bursts (when no micro-shorelines could form), or to episodes of slowly accelerating drainage preceding a burst. Further field studies are needed on these points.

Other features associated with glacial lake strandlines include spits and bars, lake ramparts and dune systems. These are to be found on the margins of the large Pleistocene lakes only. The finest examples of spits and bars ever described for Pleistocene lakes were along the shores of Lake Bonneville, Utah (G. K. Gilbert, 1890), which although not usually regarded as a glacial lake, did in fact receive glacial ice at certain points from the Wasatch Mountains. Gilbert's monograph depicts on numerous close-contoured maps magnificent examples of spits, tombolos, bay bars and shore terraces.

Lake ramparts require rather more consideration. Such features were recorded by F. Leverett and F. B. Taylor (1915) in their classic monograph on the Pleistocene history of the Great Lakes. Glacial Lake Maumee, for instance, is margined in places by peculiarly crooked shore ridges of coarse material 3–5 m high, which are almost impossible to interpret as normal beach ridges. Leverett and Taylor suggest their formation by lake ice which, after beginning to break up in a thaw period, has been driven on-shore by strong winds. If the shore gradient is gentle, the ice will slide up out of the water, pushing beach material in front. The suggestion has been amply confirmed by many observers on Arctic coasts today (for instance, J. D. Hume and M. Schalk, 1964). In the case of Arctic lake shores, A. L. Washburn (1947) described at first hand the way in which floating ice, driven by winds or currents, can push up ridges of angular and some rounded boulders to heights of 1·5 m. One such ridge was 20 m long and several metres wide. The beach material may be displaced several hundred metres inland, and, most important, up to 7–10 m vertically above water level. Hence, Pleistocene lake boulder ridges should be used with caution in estimating former lake levels.

Dune systems beside glacial lake shores need little comment. They are well known from the margins of such large Pleistocene lakes as glacial Lake Chicago, where they attain heights of up to 20 m above the Glenwood strandline, for instance.

In Britain, few examples of glacial lake strandlines and beach deposits exist. Those of the Glen Roy area in Scotland are unique as regards their clarity and extent and deserve some detailed consideration (Fig. 19.3). L. Agassiz (1842) correctly identified the horizontal markings or 'Parallel Roads' as old glacial lake shorelines; he claimed that the series of lakes was impounded by a 'lateral glacier projecting across the glen near Bridge Roy and another across the valley of Glen Speane' (p. 332). T. F. Jamieson (1863, 1892) was responsible for the first detailed field examination and interpretation. The Roads lie at three distinct principal levels. In Glen Roy itself, the three may be traced along a 10-km section of the Glen, while elsewhere in the Glen and in neighbouring Glens Gloy and Spean, only one or two of the strandlines may be observed. Commenting on the nature of the strandlines, Jamieson remarks that 'in the narrower landlocked parts, the little terraces jut sharply out from the hill, almost perfectly flat...; where however the valley is wide, ... the terraces are broader, ruder, and more shelving' (1863, p. 240). The strandlines tend to die out where they pass over resistant rock outcrops, but elsewhere their beaches may be 12–20 m broad. They usually slope valley-wards at 5 to 30 degrees (1892, p. 17). Rounded beach pebbles are found on them in places. Levelling by the Ordnance Survey, extended by Jamieson himself, gave the mean heights of the strandlines as 376·6, 350·1, and 280·5 m, the actual range of measurements being, respectively, 375·3–378·9, 348·5–353·3, and 278·9–282·8 m. The cols through which the lakes drained (circled on Fig. 19.3) and which determined the lake levels differed by 1–2 m from the mean heights of the respective strandlines. The variations have no significance, however, for as Jamieson pointed out (1892, p. 17),

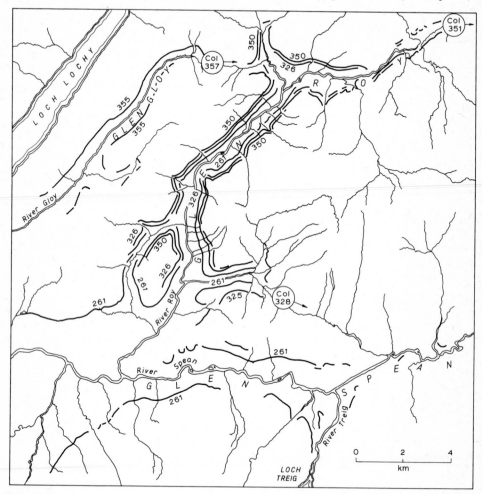

Fig. 19.3
The parallel roads of Glen Roy and adjacent valleys, Scotland (*Mem. geol. Surv. Scotland*, 1935, Crown Copyright reserved, reprinted by permission of the controller, HMSO)

the Ordnance Survey attempted to level along the median lines of the beaches, a procedure which would be expected to give slightly varying results for their levels, considering the ill-defined nature of the beaches in places and their transverse slopes. Allowance must also be made for accumulation of peat in the cols since lake waters drained through them. Lower lake levels of 138 m and 98 m were recognized later in Glen Spean by G. Wilson (1899).

There are many possible reasons why so few examples of glacial lake strandlines are known in Britain or in several other parts of the world. As already noted, modern glacial lakes are characterized by sudden and frequent changes of level, precluding the development of strongly marked strandlines. Hard rock will resist wave action, which will itself be limited in power if the lake was of small dimensions. The possibility of the lake being frozen for considerable periods must not be forgotten, nor the ease with which strandlines consisting mostly of soft unconsolidated deposits may be effaced by superficial movements or mass wasting, especially if subjected to periglacial climatic conditions. The length of time needed to form effective strandlines is a quantity variable with lake size, constancy of level, and shore materials, and few data are available. Vatnsdalur in Iceland (Fig. 19.1) provides one example, though, for in 1898 it first established a water level of 419 m. When visited by S. Thorarinsson (1939) forty years later, the lake had drained and a well-marked shoreline at 419 m was apparent. The lake area was no more than 1 km².

The most important factor determining the clarity of glacial lake strandlines is undoubtedly their age. In this respect, the Glen Roy strandlines are particularly significant, for it is now known that they date from the Late-glacial (Zone III) episode, terminating less than 10,000 years ago. The strandlines of glacial lakes formed in areas outside the limited range of Zone III glaciation in Britain are considerably older and this is the prime reason for their indistinctness or, in some cases, complete effacement. At the same time, great caution must be exercised in postulating the existence of such glacial lakes in the absence of any evidence in the form of strandlines.

4.2 Deltas

The sudden checking of velocity where streams enter glacial lakes results in the formation of deltas. Deltas will be particularly large where they are deposited by meltwater flowing either directly (for instance, subglacially) from ice into the lake, or indirectly by a normal river channel, for the rate of discharge and the load carried will be high. The typical form of a delta not laid down in contact with the ice bordering the lake will consist of a relatively steep and lobate frontal slope, and a top surface gently inclined lakewards. The junction of these two surfaces will approximate closely to lake level, though sub-aqueous extensions are known, and the lake level is liable to fluctuate. Positive changes of lake level will result in submergence and burial of the former delta by new deposits; negative changes will cause emergence of the whole or part of the delta, the incision of the stream feeding it (resulting in varying degrees of dissection), and the construction of a new delta beyond. Repeated changes of water level will thus produce complex forms.

The internal structure of a simple delta comprises the well-known sequence of bottom, foreset, and topset beds, but such a simple arrangement is by no means common. Changes of lake level already mentioned will produce complex structures, slumping and settling of the sediments will result in contorted bedding, and further complications will arise in the case of ice-contact deltas (see below).

North America holds the largest glacial lake deltas. Warren Upham's monograph (1896) on glacial Lake Agassiz records the evidence for the existence of that vast body of water (see Chapter 12) into which immense deltas were built out by meltwater. The Assiniboine delta, for instance, covers no less than 5000 km² (one-fifth as large as the Nile delta) and has an estimated volume of 80 km³. Its front rises 100 m, while wind-blown dunes up to 25 m high diversify its top surface. The foreset beds include waterworn cobbles up to 20 cm in diameter, testifying to the velocity of the currents which swept the material into the lake. From what we know of the duration of Lake Agassiz, the average annual increment of sediment for this delta must have been about one-fifth that of the present Mississippi delta.

Deltaic accumulations support the interpretation of the Parallel Roads of Glen Roy as glacial lake shorelines. Jamieson described one such delta on the eastern side of Glen Roy, 'bulging out like an artificial mound, compact and clear in outline as when first formed, save that the stream ... has now cut a great gash through its midst' (1863, p. 241). The delta surface in this case is related to the level of the lowest principal strandline. The largest of the Glen Roy deltas lies at the western end of Loch Laggan,

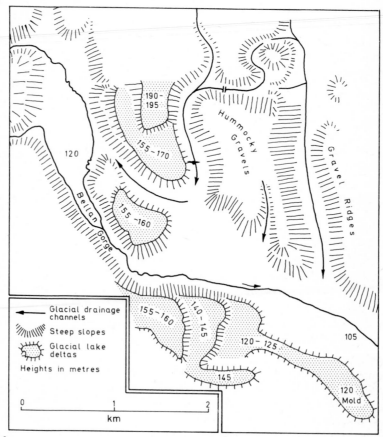

Fig. 19.4
Glacial lake deltas in the Alyn valley, north Wales. The lakes were impounded between high ground on the west and south-west, and decaying ice on the east and in the Alyn valley

a clearly defined mass of silt and gravel whose front rises 6 m above the present loch, and whose top surface slopes gently upwards to a height of about 15 m. Jamieson draws attention to the fact that deltas are mostly associated with the lower of the three principal shorelines; he suggests that at the earlier and higher levels, more of the surrounding land was covered with ice or snow and hence less debris might have been available to enter the lakes.

Fig. 19.4 shows a fine suite of glacial lake deltas near Mold in north-east Wales. Their surfaces rise steplike at levels of 120–5, 140–5 and 155–60 m; sections reveal delta-bedded sand and gravel. The meltwater stream carrying these materials entered through and partly excavated the Bellan gorge upstream. The lakes may not have been extensive, for there are no strandlines evident and the deltas prove only the existence of standing water at the places where they accumulated (C. Embleton, 1964).

Of somewhat different form are ice-contact deltas, sometimes termed esker deltas or delta moraines and already mentioned in Chapter 16. Fig. 16.10 provides an example. These features possess an ice-contact slope at the back, although the frontal portion building out into the lake is normal. Owing to the proximity of the ice, kettle-holes often appear on the top surface and along the rear portions; slump structures characterize the bedding as the ice withdrew its support; and eskers may be present at the rear, representing the courses of the subglacial streams which injected the material into the lake. Fine examples from Britain have been mapped by F. M. Trotter (1929) in the Vale of Eden (he, however, uses the ambiguous term 'outwash delta'), enabling reconstructions to be made of ice margin positions during deglaciation. Ice-contact slopes of the deltas may attain heights of 45 m and are considerably steeper than the frontal slopes. One series of deltas extends for 8 km along an ice margin; each individual delta of the series seems to be associated with a slight embayment in the ice margin where presumably the englacial or subglacial stream issued. Such streams appear to have sapped back the ice edge, enabling the deltas to be extended rearwards and attain widths approaching 1·5 km. In some cases, eskers run back from the ice-contact slope.

4.3 Lake-floor deposits

There is no sharp line dividing lake-floor and lake-margin deposits, nor is it possible to draw a distinction on the basis of grain size for all lakes, for conditions vary so much from one lake to another. However, although both coarse and fine sediments may be found in beach deposits and deltas, lake-floor deposits beyond the limit of delta growth are predominantly fine, and become increasingly so as the central or deepest parts of the lake are approached. The finest silts and clays represent the water-borne rock flour of glacial origin which is seen to discolour the water of lakes and streams and which requires extremely long periods and quiet water conditions to settle. E. M. Kindle (1930) found that even after five months, samples of water containing rock flour and obtained from glacier-fed Lake Cavell in the Rocky Mountains still remained discoloured except for the top 25 cm in a tube 1·6 m long. The action of turbidity currents (P. H. Kuenen, 1951) is now thought to be of considerable importance in spreading sediment over the whole lake floor, for the cold silt-laden meltwater is slightly denser than lake waters.

Lake depth is an important factor affecting the nature of the deposits. Deep lakes provide the best opportunities for the trapping and deposition of the finest material. In the case of shallow lakes fed by meltwater, the existence of more powerful currents will prevent the settling of fine material and often cause it to be washed towards and through the lake outlet resulting in its loss from the lake. This is the reason why the infill of former lakes frequently shows a gradation towards coarse material at the top, apart from the fact that the last stages of infilling will be carried out by a braided and rapidly aggrading stream.

Occasionally, in the fine lake bottom deposits, isolated pebbles or boulders appear. R. S. Tarr (1897, p. 151) described the exposed bed of a drained lake beside the Cornell glacier, Greenland, which showed fine clays and scattered blocks of rock. The boulders are clearly ice-rafted or dropped into the lake from the ice margin. Their weight and impact are responsible for folds in the lake floor sediments (R. P. Sharp, 1947).

A very common and highly significant characteristic of glacial lake-floor deposits is their laminated nature. The laminations are caused by sudden changes of grain size between successive thin layers, from finest mud to slightly coarser silt; there is also an accompanying colour change, the finest laminae being darker (and also thinner) than the coarser. These regularly banded deposits have been termed 'rhythmites'; an individual pair consisting of one fine and one slightly coarser lamina is known as a 'couplet'. The thin dark layer of a couplet which consists of very fine and partly colloidal material represents a phase of slow deposition under very quiet water conditions, such as would be expected in winter when the lake was frozen at its surface and little or no meltwater was entering. Kindle's work (1930) already quoted showed that rock flour will remain in suspension in still water long enough to allow deposition to continue for many months even if no further sediment enters the lake during this time. The light-coloured silt comprising the coarser member of a couplet represents more rapid accumulation of lake-floor sediment under more disturbed conditions, as when melt-water was entering the lake and lake currents, especially turbidity currents, were strong enough to spread the silt over the whole lake floor. B. K. Emerson in 1898 was the first to attribute the deposition of the coarser laminae to turbid underflows of sediment-laden water, and more recent detailed work by Kuenen (1951) has confirmed this general picture. Fig. 19.5 shows a typical pattern of density currents entering a

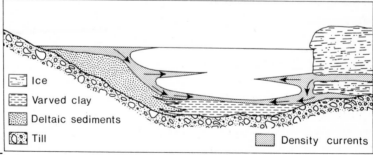

Fig. 19.5
Idealized pattern of density currents entering a proglacial lake as underflows and interflows from overland, englacial and subglacial streams (T. C. Gustavson, *Technical Rep. Coastal Res. Centre, Massachusetts*, 1972)

proglacial lake with underflows and interflows. It seems that rhythmites will readily develop in cold freshwater lakes into which meltwater streams intermittently flow; they are rarely found in salt or even brackish water where the salt aids flocculation of the fine particles and an unbanded deposit results. Most rhythmites of Pleistocene age indicate the nearby presence of glacial ice at the time they were laid down. Non-glacial rhythmites are known in some Pleistocene sequences and in many pre-Pleistocene geological formations, but the following discussion will be limited to glacial lake rhythmites.

Couplets may vary in thickness from less than 1 mm to as much as several dm, but 1 to 5 cm is the normal range. Many local factors will affect the thickness—the load and volume of the meltwater stream entering the lake, the depth of the lake and the relief of its floor, the strength of the currents in the lake and the distance from the point of entry of the meltwater stream. It is therefore impossible to correlate rhythmites in one lake with those in another on the basis of their absolute thickness.

Since the first half of the nineteenth century, a number of workers had suspected that the couplets in rhythmites were seasonal. A. Smith (1832) hinted at this possibility in a very early study of laminated clays in the Connecticut valley, the couplets here being 0·8–1·2 cm thick. G. de Geer from 1882 onwards developed the theory of the annual nature of couplets, and introduced the term 'varves' to describe such annual couplets in 1912. There is indeed strong evidence that many couplets are annual, though it is also now recognized that not all are, and great care must be taken to prove the point one way or the other if rhythmites are to be employed as a means of dating, as de Geer did. De Geer regarded the fine lamina of a couplet as the result of deposition in winter when the lake was frozen and meltwater entry limited or non-existent. The abrupt break at the top of the fine lamina represents the spring thaw when new coarser silt once again re-entered the lake. Confirmation of this notion has come from study of the pollen found in rhythmites, which provides evidence of seasonal changes in vegetation surrounding the lake during the time interval represented by one couplet. Further confirmation has been provided by studies of modern glacial lake-floor deposits, such as that made by W. A. Johnston (1922) in the case of Lake Louise in Alberta. This lake, about 0·8 km square and up to 70 m deep, receives meltwater from the Victoria glacier which then terminated about 1·6 km from the lake. The lake freezes for about seven months of the year, but some meltwater continues to flow into the lake. The meltwater has a temperature close to 0°C, and is appreciably colder than the lake in summer when water temperatures vary from a minimum of 2°C at the delta to as much as 4°C higher at the far end of the lake. Johnston calculated the approximate average thickness of the annual sediment layer to be expected on the lake floor from the known discharge and load of the meltwater stream as 4 mm. A core from the deepest part of the lake showed faint banding, the spacing being about 4–5 mm, therefore strongly suggesting a seasonal rhythm.

Cores from the bottom deposits of proglacial Lake Malaspina, Alaska, also show that varved sediments are presently being deposited. Although the lake remains frozen until mid-June on average, meltwater was observed to flow into it all year round in 1970–71 (T. C. Gustavson, 1972), the point of entry of marginal glacier streams remaining ice-free. Two large surface streams, one supraglacial, feed the lake, and there is

evidence of numerous subglacial and englacial streams also entering. The water from
the supraglacial stream was visibly more turbid than the lake water. Gustavson shows
that turbid underflows are responsible for spreading out the sand and silt layers of
the summer, while the finer particles in suspension may settle out year round, whenever
and wherever the water movement is slow enough. Some underflows cease in winter.
In areas not reached by underflows, deposition of suspended fine particles is continuous
and leads to the formation of massive clays.

Non-annual rhythmites (which should not be termed varves) are known to occur.
They may arise from sudden fluctuations of discharge and load on the part of the
meltwater stream, sometimes caused by the bursting of glacial water-bodies upstream;
from warm and cold spells of a non-annual nature; and from the action of periodic
storms stirring up the lake waters at an otherwise quiet period. But if, as seems likely
in much of southern Sweden, non-annual rhythmites are rare, then the laminations
provide a valuable time-scale for the late Pleistocene.

De Geer was the first to attempt to make regional correlations of varve sequences
(see his accounts published in 1912 and 1940), by measuring varve thicknesses on different sections, plotting these thicknesses on a time-scale assuming each varve to be
annual, and visually matching the curves so obtained. By this means, localities up to
1 km apart were correlated, a varve sequence established over a distance of almost
1000 km, and a chronology established over nearly 17,000 years. The method of correlating different sections is unreliable if the varves are thin or too regular in thickness—
there is a limit to the possible variations in the form of the graphs—but de Geer's work
is now supported by radiocarbon dates as far back as about 12,000 BP (N.-A. Mörner,
1969). The idea behind the correlation based on relative varve thicknesses is that such
variations reflect fluctuations of climate (and hence rates of summer ablation) which
should similarly affect all localities in a region of limited size. M. Sauramo's (1923)
work on varves in southern Finland used not only variations in varve thickness for
correlation but also variations in grain-size composition. He showed that relying simply
on varve thickness could lead to errors in correlation. However, recent checks on de
Geer's sequence by B. Järnefors and E. Fromm (1960) in Sweden have shown that
only minor corrections were needed.

De Geer's method was applied in eastern Canada and the United States by E. Antevs
(1925). A useful summary of the varve chronology gradually built up over many years
was given by Antevs in 1957. It is based on discontinuous varve series measured between
New York City and Cochrane (240 km south of James Bay) and on estimates of the
gaps in the varve series, the total chronology extending back over the last 19,000 years.
With the advent of radiocarbon dating, doubt has been cast on the validity of this
record, suggesting that some of the varves may not be annual or that false correlations
between different varve sections have been made. Antevs (1957, 1962), however, maintains that the radiocarbon dates are more likely to be in error than those based on
varve counts. An example of the discrepancy is given by the varve dating of the Two
Creeks Forest Bed, a critical horizon in the late-glacial chronology of North America,
as 19,000 BP, compared with a radiocarbon age of 11,000–12,000 BP for this formation.
Most workers now prefer to rely on the radiocarbon data.

One of the principal weaknesses of the traditional method of varve correlation has always lain in the process of visually matching varve sequences from different areas. Application of statistical techniques here holds considerable promise. R. Y. Anderson and D. W. Kirkland (1966) have recently applied statistical correlation techniques to some non-glacial varved sediments, finding that thickness variations of varve laminae have correlation coefficients higher than $+0.8$ over the area of varve formation at any given time. D. J. Schove (1971) has examined the best statistical correlation between the Swedish and Finnish varve sequences, and even applied similar methods in an attempt to bridge the trans-Atlantic gap between the eastern North American and Baltic sequences.

Extensive deposits of rhythmites are not known in Britain, but the sites of some late Pleistocene glacial lakes of limited extent are occupied by laminated clays. Trotter (1929) noted 2·4 m of such clays in the Carlisle district. K. M. Clayton and J. C. Brown (1958) recorded up to 6 m of laminated silts and clays near Ware which were laid down in Lake Hertford impounded by the last ice to reach the London Basin. C. A. Baker (personal communication, 1974) has investigated an unusually thick sequence of fine-grained sediments, including up to 25·6 m of laminated beds, laid down in a proglacial lake dammed by pre-Anglian (or possibly intra-Anglian) ice against the Chalk escarpment near Newport in the upper Cam valley, Essex. The sediments were examined in 28 deep boreholes penetrating altogether 54 m of lake deposits. They out-crop over 1 km² but there are no natural surface exposures. Analysis of undisturbed samples from a pit in the laminated beds showed the following results:

> *Clay laminae*: clay 40 per cent, silt 57 per cent, sand 3 per cent.
> Carbonates 46·1 per cent, Fe 2·2, pH 8·7.
> Munsell colour 10YR/4/3 (dark brown)
> *Silt laminae*: clay 20 per cent, silt 70 per cent, sand 10 per cent.
> Carbonates 39·5 per cent, Fe 1·8, pH 8·7.
> Munsell colour 10YR/6/3 (brownish yellow)

The differentiation of the layers was not biological, nor primarily chemical, but pre-dominantly mechanical in origin, in accordance with a hypothesis of origin as varves. The laminations showed varied thickness and structure; there were semi-regular coup-lets 5–10 mm thick (clay 2–5 mm, silt 0·5–5 mm); disrupted laminations showing high-angle bedding, brecciation and micro-faulting and folding, possibly caused by later slumping into subglacial channels cut after the lake had disappeared; and micro-lami-nations from less than 1 mm to 4 mm in thickness. Overall, the clay laminae tended to be thicker than the silt laminae. Contacts were sharp above the clay and below the silt laminae; the silt–clay transition in an upward direction was more gradual. Plate XXXVII shows part of one core. It has been claimed by R. G. Carruthers (1939, 1940) that many of the laminated clays said to have been laid down in extensive ice-dammed lakes, particularly in northern and eastern England, are not lacustrine or water-laid deposits at all, but the banded dirts of englacial detritus, released by rising bottom melt (Chapter 13). There has been much opposition to Carruthers' interpretation of the laminated clays as 'pressed clays' or 'pressed melts', though he was undoubtedly correct in maintaining that Pleistocene ice-dammed lakes in Britain were not so com-mon nor so large as has been suggested. Examining the specimens of laminated clay

Plate XXXVII
Part of a core in varved deposits, Pleistocene glacial lake Newport, Essex (courtesy of C. A. Baker).
Centimetre scale on the left.

obtained by Carruthers, E. B. Bailey and S. E. Hollingworth (1940) were equally convinced that they were true lacustrine deposits; and that the contortions found in the laminae recorded flow distortion of bedded mobile mud that had accumulated on the floor of a glacial lake, or that they resulted from the settling of boulders released from floating ice.

5 Conclusion

Glacial lakes include lakes dammed by ice and lakes in which glacier ice forms part of the lake margin though not the actual barrier. They occur today in lateral valleys dammed by trunk glaciers, in main valleys blocked by ice emerging from a tributary valley, and in other marginal, supraglacial, subglacial or englacial positions. Most lakes dammed by temperate glacier ice drain subglacially or englacially, though a few sometimes overflow through cols cut in rock or drift if the lake level is high enough, whereas lakes dammed by sub-polar glacier ice normally drain over the ice barrier or escape by marginal or sub-marginal routes. The method by which lakes drain subglacially is not fully understood, especially in connection with the occurrence of sudden lake bursts or jökulhlaups. Thorarinsson's hypothesis (1939) which involves temporary flotation of the ice barrier seems more likely to apply, with certain modifications, than Glen's hypothesis (1954) which demands water of sufficient depth (more than 200 m) to cause ice deformation. Once drainage of a lake has commenced, it usually proceeds rapidly and completely as the relative warmth of the escaping water enlarges the tunnels through or under the ice.

Features associated with present and former glacial lakes include strandlines, beach formations, lake ramparts (formed by floating ice impinging on the shore), deltas and lake-floor deposits. Strandlines of former lakes may be absent or indistinct if the lake level was variable, if the lake was too small to permit effective wave action, or if sufficient time has subsequently elapsed to allow their effacement by mass wasting. Deltas may be preserved for much longer periods and are particularly useful in determining former lake levels. Ice-contact deltas or delta moraines are valuable indicators of ice margin positions.

Glacial lake-floor deposits frequently include rhythmites (varves), comprising a succession of couplets which are in most cases of an annual nature. On this basis late-glacial chronologies in Scandinavia and eastern North America have been established, though in the latter area, radiocarbon dating has shown some major disagreements. Rhythmites must be carefully distinguished from laminated deposits resulting from the melting-out of englacial or subglacial silt layers.

6 References

AGASSIZ, L. (1840–41), 'On glaciers, and the evidence of their having once existed in Scotland, Ireland, and England', *Proc. geol. Soc. Lond.* **3**, 327–32

AITKENHEAD, N. (1960), 'Observations on the drainage of a glacier-dammed lake in Norway', *J. Glaciol.* **3**, 607–9

ANDERSON, R. Y. and KIRKLAND, D. W. (1966), 'Intrabasin varve correlation', *Bull. geol. Soc. Am.* **77**, 241–55

ANTEVS, E. (1925), 'Retreat of the last ice-sheet in eastern Canada', *Mem. geol. Surv. Can.* **146**

(1957), 'Geological tests of the varve and radiocarbon chronologies', *J. Geol.* **65**, 129–48

(1962), 'Transatlantic climatic agreement versus C14 dates', *J. Geol.* **70**, 194–205

ARNBORG, L. (1955), 'Ice-marginal lakes at Hoffellsjökull', *Geogr. Annlr* **37**, 202–28

BAILEY, E. B. and HOLLINGWORTH, S. E. (1940) in discussion on 'Northern glacial drifts', *Q. J. geol. Soc. Lond.* **96**, 249–69

BEZINGE, A., PERRETEN, J. P. and SCHAFER, F. (1973), 'Phénomènes du lac glaciaire du Gorner', *Symposium on the hydrology of glaciers* (Cambridge, 1969), *Int. Ass. scient. Hydrol.*, Publ. **95**, 65–78

BIRKELAND, P. (1964), 'Pleistocene glaciation of the northern Sierra Nevada, north of Lake Tahoe, California', *J. Geol.* **72**, 810–25

BOER, G. DE (1949), 'Ice margin features, Leirbreen, Norway', *J. Glaciol.* **1**, 332–6

BONNEY, T. G. (1896), *Ice work present and past* (London)

CARRUTHERS, R. G. (1939), 'On northern glacial drifts: some peculiarities and their significance', *Q. J. geol. Soc. Lond.* **95**, 299–333

CLAGUE, J. J. and MATHEWS, W. H. (1973), 'The magnitude of jökulhlaups', *J. Glaciol.* **12**, 501–4

CLAYTON, K. M. and BROWN, J. C. (1958), 'The glacial deposits around Hertford', *Proc. Geol. Ass.* **69**, 103–19

COLLET, L. W. (1926), 'The lakes of Scotland and Switzerland', *Geogrl J.* **67**, 193–213

EMBLETON, C. (1964), 'Sub-glacial drainage and supposed ice-dammed lakes in north-east Wales', *Proc. Geol. Ass.* **75**, 31–8

EMERSON, B. K. (1898), 'Geology of old Hampshire County, Massachusetts', *U.S. geol. Surv. Monogr.* **29**, 790 pp.

FLINT, R. F. (1971), *Glacial and Quaternary geology*

FORBES, J. D. (1843), *Travels through the Alps of Savoy* (Edinburgh)

FULTON, R. J. (1965), 'Silt deposition in late-glacial lakes of southern British Columbia', *Am. J. Sci.* **263**, 553–70

GEER, G. DE (1912), 'A geochronology of the last 12,000 years', *C. r. 11th int. geol. Congr. (Stockholm, 1910)*, **1**, 241–58

(1940), 'Geochronologia suecica principles', *K. svenska Vetensk-Akad. Handl.* Ser. 3, **18**, 6. Text and Atlas

GILBERT, G. K. (1890), 'Lake Bonneville', *U.S. geol. Surv. Monogr.* **1**, 1–438

GILBERT, R. (1971), 'Observations on ice-dammed Summit Lake, British Columbia, Canada', *J. Glaciol.* **10**, 351–6

GLEN, J. W. (1954), 'The stability of ice-dammed lakes and other water-filled holes in glaciers', *J. Glaciol.* **2**, 316–18

GUSTAVSON, T. C. (1972), 'Sedimentation and physical limnology in proglacial Malaspina Lake, Alaska', *Coastal Res. Centre, Massachusetts, Tech. Rep.* **5**-CRC, 47 pp.

HEWITT, K. (1964), 'The great ice dam', *Indus* **5**, 18–30

HIGGINS, A. K. (1970), 'On some ice-dammed lakes in the Frederikshåb district, south-west Greenland', *Meddr dansk geol. Foren.* **19**, 378–97

HOWARTH, P. J. and PRICE, R. J. (1969), 'The proglacial lakes of Breiðamerkurjökull and Fjallsjökull, Iceland', *Geogrl J.* **135**, 573–81

HUME, J. D. and SCHALK, M. (1964), 'The effects of ice-push on Arctic beaches', *Am. J. Sci.* **262**, 267–73

JAMIESON, T. F. (1863), 'On the parallel roads of Glen Roy, and their place in the history of the glacial period', *Q. J. geol. Soc. Lond.* **19**, 235–59

(1892), 'Supplementary remarks on Glen Roy', *Q. J. geol. Soc. Lond.* **48,** 5–28

JÄRNEFORS, B. and FROMM, E. (1960), 'Chronology of the ice recession through middle Sweden', *Rep. 21st int. geol. Congr.* (Norden, 1960), **4**, 93–7

JOHNSTON, W. A. (1922), 'Sedimentation in Lake Louise, Alberta, Canada', *Am. J. Sci.* **204**, 376–86

JONSSON, J. (1955), 'On the formation of frontal glacial lakes', *Geogr. Annlr* **37**, 229–233

KELLER, R. (1950), 'Niederschlag, Abfluss und Verdunstung in Schweizer Hochgebirge', *Erkunde* **4**, 54–67

KERR, F. A. (1934), 'The ice dam and floods of the Talsekwe, British Columbia', *Geogrl Rev.* **24**, 643–5

KINDLE, E. M. (1930), 'Sedimentation in a glacial lake', *J. Geol.* **38**, 81–7

KUENEN, P. H. (1951), 'Mechanics of varve formation and the action of turbidity currents', *Geol. För. Stockh. Förh.* **6**, 149–62

LEVERETT, F. and TAYLOR, F. B. (1915), 'The Pleistocene of Indiana and Michigan, and the history of the Great Lakes', *U.S. geol. Surv. Monogr.* **53**

LIESTØL, O. (1955), 'Glacier-dammed lakes in Norway', *Norsk Geogr. Tidsskr.* **15**, 122–49

MAAG, H. (1969), 'Ice-dammed lakes and marginal glacial drainage on Axel Heiberg Island', *Axel Heiberg Island Res. Rep., McGill Univ. Montreal*, 147 pp.

MANNERFELT, C. M. (1945), 'Några glacialmorfologiska formelement', *Geogr. Annlr* **27**, 1–239 (with English summary)

MARCUS, M. G. (1960), 'Periodic drainage of glacier-dammed Tulsequah Lake, British Columbia', *Geogrl Rev.* **50**, 89–106

MASON, K. (1935), 'The study of threatening glaciers', *Geogrl J.* **85**, 24–41

MÖRNER, N.-A. (1969), 'The late Quaternary history of the Kattegatt sea and the Swedish west coast', *Sver. geol. Unders. Afh.* C, **640**, 487 pp.

NICHOLS, R. L. and MILLER, M. M. (1952), 'The Moreno glacier, Lago Argentino, Patagonia', *J. Glaciol.* **2**, 41–50

ODELL, N. E. (1966), 'The Märjelen See and its fluctuations', *Ice* **20**, 27

REID, J. R. and CLAYTON, L. (1963), 'Observations of rapid water-level fluctuations in ice sink-hole lakes, Martin River glacier, Alaska', *J. Glaciol.* **4**, 650–2

SAURAMO, M. (1923), 'Studies on the Quaternary varve sediments in southern Finland', *Bull. Commn géol. Finl.* **60**, 164 pp.

SCHOVE, D. J. (1971), 'Varve teleconnection across the Baltic', *Geogr. Annlr* **53A**, 214–34

SCHYTT, V. (1956), 'Lateral drainage channels along the northern side of the
    Moltke glacier, north-west Greenland', *Geogr. Annlr* **38**, 64–77

SEPPÄLÄ, M. (1973), 'On the formation of small marginal lakes on the Juneau
    Icefield, south-eastern Alaska, U.S.A.', *J. Glaciol.* **12**, 267–73

SHARP, R. P. (1947), 'The Wolf Creek glaciers, St Elias Range, Yukon Territory',
    Geogrl Rev. **37**, 26–52

SISSONS, J. B. (1958), 'Supposed ice-dammed lakes in Britain, with particular
    reference to the Eddleston valley, southern Scotland', *Geogr. Annlr* **40**, 150–87

SMITH, A. (1832), 'On the water courses and the alluvial and rock formations of the
    Connecticut River valley', *Am. J. Sci.* **22**, 205–31

SPENCER, J. W. (1890), 'Ancient shores, boulder pavements, and high-level gravel
    deposits in the region of the Great Lakes', *Bull. geol. Soc. Am.* **1**, 71–86

STENBORG, T. (1969), 'Studies of the internal drainage of glaciers', *Geogr. Annlr* **51**A,
    13–41

STONE, K. H. (1963), 'Alaskan ice-dammed lakes', *Ann. Ass. Am. Geogr.* **53**, 332–49

TARR, R. S. (1897), 'The margin of the Cornell glacier', *Am. Geol.* **20**, 139–56

THORARINSSON, S. (1939), 'The ice-dammed lakes of Iceland, with particular
    reference to their values as indicators of glacier oscillations', *Geogr. Annlr* **21**,
    216–42

    (1953), 'Some new aspects of the Grimsvötn problem', *J. Glaciol.* **2**, 267–74

    (1957), 'The jökulhlaups from the Katla area in 1955 compared with other
    jökulhlaups in Iceland', *Jökull* **7**, 21–5

TROTTER, F. M. (1929), 'The glaciation of eastern Edenside, the Alston Block, and
    the Carlisle Plain', *Q. J. geol. Soc. Lond.* **85**, 549–612

UPHAM, W. (1896), 'The glacial lake Agassiz', *U.S. geol. Surv. Monogr.* **25**

WASHBURN, A. L. (1947), 'Reconnaissance geology of portions of Victoria Island and
    adjacent regions, Arctic Canada', *Am. geol. Soc. Mem.* **22**

WEERTMAN, J. (1973), 'Can a water-filled crevasse reach the bottom surface of a
    glacier?', *Symposium on the hydrology of glaciers* (Cambridge, 1969), *Int. Ass.
    scient. Hydrol.*, Publ. **95**, 139–45

WHALLEY, W. B. (1971), 'Observations on the drainage of an ice-dammed
    lake—Strupvatnet, Troms, Norway', *Norsk geogr. Tidsskr.* **25**, 165–74

WILSON, G. (1899), work described in *Mem. geol. Surv. Summ. Prog.* (1899), 159–62

Index